BUSINESS/SCIENCE/TECHNOLOGY DIVISION
CHICAGO PUBLIC LIBRARY
400 SOUTH STATE STREET
CHICAGO, IL 60605

R00208 33994

CHICAGO PUBLIC LIBRARY
HAROLD WASHINGTON LIBRARY CENTER
R0020833994

Form 178 rev. 01-07

INSTRUMENT ENGINEERS' *Handbook*

Supplement One
of the
Instrument Engineers' Handbook
(Volumes I and II)

BÉLA G. LIPTÁK, Editor

CHILTON BOOK COMPANY
PHILADELPHIA NEW YORK LONDON

REF
TS
156.8
.I56
Suppl. 1
V. 1-2
cop. 1

Library of Congress Cataloging in Publication Data

Main entry under title:

Instrument engineers' handbook.

Includes bibliographical references.

CONTENTS: v. 1. Process measurement.—v. 2. Process control.

______Supplement one of the Instrument engineers' handbook, volumes 1 and 2.
Includes bibliographical references.

1. Measuring instruments—Handbooks, manuals, etc.
2. Process control—Handbooks, manuals, etc.
I. Lipták, Béla G., ed.

TS156.8.156Suppl. 620′.0028 73-80445
ISBN 0-8019-5658-7 (Suppl.)

Copyright © 1972 by Béla G. Lipták
First Edition *All rights reserved*
Published in Philadelphia by Chilton Book Company
and simultaneously in Ontario, Canada,
by Thomas Nelson & Sons, Ltd.
Designed by William E. Lickfield
Manufactured in the United States of America

THE CHICAGO PUBLIC LIBRARY
APP. JAN - 8 1973 B

R0020833994
BST REF
BUSINESS/SCIENCE/TECHNOLOGY DIVISION
CHICAGO PUBLIC LIBRARY
400 SOUTH STATE STREET
CHICAGO, IL 60605

to my wife, MÁRTHA

35.00

CONTRIBUTORS

Wendall M. Barrows
Senior Applications Coordinator,
Union Carbide Corporation
(Section 1.12)

Christopher P. Blakeley
BSChE
Market Manager
Honeywell, Inc.
(Section 5.1)

H. L. Cook, Jr.
BSEE, MBA
Vice President,
The Ohmart Corporation
(Section 1.1)

John R. Copeland
BSEE, MSEE, PhDEE
Partner,
Jackson Associates
(Sections 2.1 and 2.2)

Lewis V. Corsetti
Chief Instrument Designer,
Crawford & Russell, Inc.
(Sections 6.1 through 6.6)

Nicholas O. Cromwell
BSChE
Senior Development Analyst,
The Foxboro Company
(Section 3.5)

Paul M. Glattstein
BSEE
Senior Electrical Engineer,
Crawford & Russell, Inc.
(Sections 1.17 and 6.7)

John D. Goodrich, Jr.
BSME
Engineering Supervisor,
Bechtel Corporation
(Sections 1.1, 1.2 and 1.3)

JOSEPH A. GUMP
BSChE
Product Manager,
Scam Instrument Corporation
(*Section 3.4*)

H. N. HILL, JR.
BA Physics
Assistant Development Manager,
Union Carbide Corporation
(*Section 1.15*)

STUART P. JACKSON
BSEE, MSEE, PhDEE, PE
General Partner,
Jackson Associates
(*Sections 2.1 and 2.2*)

VAUGHN A. KAISER
BSME, MSE, PE
Member Technical Staff,
Profimatics, Inc.
(*Section 2.3*)

CHANG H. KIM
BSChE
Senior Group Leader,
UniRoyal Chemical
(*Section 1.6*)

DONALD W. LEPORE
BSME
Design Engineer,
The Foxboro Company
(*Sections 3.2, 3.3 and 3.6*)

TRUMAN S. LIGHT
BSCh, MSCh, PhDCh
Senior Research Chemist,
The Foxboro Company
(*Section 1.8*)

BÉLA G. LIPTÁK
ME, MME, PE
Consultant and Chief Instrument Engineer,
Crawford & Russell, Inc.
(*Introduction, Section 1.10 and Miscellaneous*)

HARRY E. LOCKERY
BSEE, MSEE, PE
Vice President, Engineering,
BLH Electronics, Inc.
(*Section 1.7*)

THOMAS J. MYRON, JR.
BSChE
Senior Systems Design Engineer,
The Foxboro Company
(*Section 4.2*)

RICHARD T. OLIVER
BSCh, MSCh, PhDCh
Senior Systems Design Engineer,
The Foxboro Company
(*Sections 1.13 and 1.14*)

GLENN A. PETTIT
Manager, Plastic Instrumentation,
Rosemount Engineering Company
(*Section 4.1*)

DIETER RALL
BSME, MSME, PE
General Manager,
Trans-Met Engineering, Inc.
(*Section 1.4*)

HOWARD C. ROBERTS
ABEE, PE
Professor Emeritus,
University of Illinois
(*Sections 5.4 and 5.5*)

DOUGLAS D. RUSSELL
BSEE, MSEE
Group Leader,
The Foxboro Company
(*Section 3.5*)

CHAKRA J. SANTHANAM
BSCh, BSChE, MSChE, ChE, PE
Senior Process Engineer,
Crawford & Russell, Inc.
(*Sections 5.2 and 5.3*)

ROBERT SIEV
BSChE, MBA, CE
Engineering Specialist,
Bechtel Corporation
(*Section 1.5*)

J. WILLIAM SUGAR	BSChem, MSEnvChem Environmental Project Scientist, Union Carbide Corp. (*Section 1.16*)
MAURO G. TOGNERI	BSEE Senior Consultant, Industrial Computer Services (*Section 3.1*)
WILLIAM H. WAGNER	BSChE, PE Instrument Engineer, Union Carbide Corp. (*Section 1.11*)
NORMAN S. WANER	BSME, MSME, ME, PE Chief Engineer and Director, Hallikainen Instruments (*Section 1.9*)
RICHARD E. WENDT, JR.	BSEE, MSEE, EE, PhD Consultant (*Section 4.3*)
ROBERT A. WILLIAMSON	BSME, BA, PE Supervisor, Electromechanical, Packaging Group The Foxboro Company (*Sections 3.2, 3.3 and 3.6*)

PREFACE

B. G. Lipták

Dr. Edward Teller in his preface to the first volume of this handbook noted that it is in the nature of reference books to require periodic expansion and updating. The Instrument Engineers' Handbook—reflecting the state of the art in a young and growing professional discipline—is certainly no exception.

If the nature of a handbook is such that with the passage of time its contents become outdated or erroneous, then it is necessary to publish revised editions at relatively short intervals. This method of updating places a substantial financial burden on the reader.

Fortunately, the character of the Instrument Engineers' Handbook allows us to follow another approach. Because passing time does not affect the value and reliability of the contents of the two volumes they need not be revised at frequent intervals. On the other hand, time does cause coverage to become incomplete. This is unavoidable in a field where new concepts, techniques and components are constantly being introduced. Therefore, it is important to publish a supplement volume every few years in order to keep the handbooks complete and up to date.

This first supplement adds the following new topics to subjects covered in Volumes I and II: pollution instrumentation, human engineering, physical properties analyzers, ion selective electrodes, and instrument installation materials. Furthermore, it reports on the new developments in the already discussed areas of process measurement, computers, displays and control systems.

Because this supplement will be used by the reader as part of his set of earlier published volumes, care has been taken to organize these three volumes into a convenient single reference source, in which emphasis is on quick access to specific information, and repetition is eliminated by the extensive use of cross referencing.

CONTENTS

CHAPTER II

CHAPTER III

CHAPTER IV

CHAPTER V

CHAPTER VI

List of Important Tables and Charts

INSTRUMENT ENGINEERS'
Handbook

Chapter I

NEW DEVELOPMENTS IN PROCESS SENSORS

W. M. Barrows, H. L. Cook, Jr.,
P. M. Glattstein, J. D. Goodrich, Jr.,
H. N. Hill, Jr., C. H. Kim,
T. S. Light, B. G. Lipták,
H. E. Lockery, R. T. Oliver,
D. Rall, R. Siev,
J. W. Sugar, W. H. Wagner, and
N. S. Waner

CONTENTS OF CHAPTER I

1.1 NEW DEVELOPMENTS IN LEVEL MEASUREMENT

Level Detector Types: (a) Continuous radiation gauges.
(b) Armored tubular gauges.
(c) Tuning fork level switches.
(d) Tilt switches.

Note: In the feature summary below, the letters (a) to (d) refer to the listed sensor types.

Design Pressure: Atmospheric (d), to 140 PSIG (c), from vacuum to 600 PSIG (b), unlimited (a).

Design Temperature: To 140°F (c), to 425°F (b), to 750°F (d), unlimited on the process side (a).

Range: Point sensor (c and d), to 25 ft. (b), to 50 ft. (a).

Cost: $40 to $60 per foot (b), $60 to $150 (d), $175 to $225 (c), $1,000 and up (a).

Partial List of Suppliers: Endress-Hauser, Inc. (c), Industrial Nucleonics Corp. (a), In-Val-Co., Div. of Combustion Engineering, Inc. (a), Jogler Inc. (b), Kay-Ray, Inc. (a), Machinery Electrification Inc. (c), Monitor Manufacturing Inc. (d), Ohmart Corp. (a), Robertshaw Controls Co. (a), Texas Nuclear Div. of Nuclear Chicago (a).

Other suppliers are listed in Volume I.

This section reviews developments in level detection technology that have occurred during the last few years and thereby supplements Chapter I of Volume I. With these two sources the reader will find coverage of all level sensors, including those most recently introduced. In addition, he will find a discussion of new features that only recently became available on level gauges, such as the radioactive and the ultrasonic.

Radiation Type Level Sensors

Early models of nuclear radiation level switches used a thyratron tube as the triggering element and corona discharge regulators in the high voltage power supply. As solid state components became less expensive they replaced the thyratron and corona regulator.

Sensitivity has remained about the same because of the basic limitations of the Geiger-Mueller tube which is used as the radiation detector. Reliability has improved significantly because of the use of solid state electronics.

Sensors

Ionization chambers and Geiger-Mueller tubes continue to be the most frequently used sensors in continuous nuclear radiation level gauges.

Early level gauges of the ion chamber type used short detectors which were "stacked" to provide the desired height of measurement. Recently introduced detectors are continuous through the range of desired level measurement, a feature that eliminates connection between units and reduces the required number of hermetic seals. Thus, ruggedness and reliability are improved.

Amplifiers

Early ion chamber type continuous level sensors used an electrometer tube to amplify the ion chamber output current to a usable level. In newer amplifiers the electrometer tube is replaced with a metal oxide semiconductor, field-effect transistor (MOS-FET). See Figure 1.1a. The most advanced designs use a dual MOS-FET on a single chip in a balanced amplifier circuit. For more on transistors, see P. 501 of Volume II. Use of the dual MOS-FET results in greater reliability, lower zero drift and freedom from microphonics.

Early amplifiers supplied only a millivolt output signal. The newer amplifiers have not only a millivolt output but also a 0 to 10 volt output and an optional milliampere output available as a plug-in circuit board.

Linearity Considerations

Figure 1.1b shows the radiation detector response for various configurations of radioactive sources and detectors.

A single point source and detector produce a smooth, concave curve. This is because the radiation field strength is greater at the top of the level range since the source is closer to the detector at the top. Thus, sensitivity is greater at the top.

Multiple point sources produce multiple loops in the response curve. A

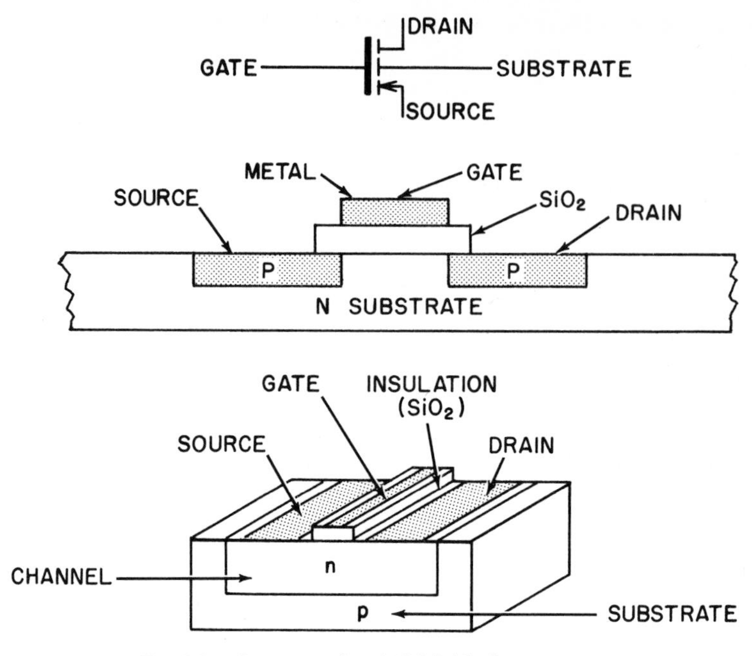

Fig. 1.1a Structure of an MOS field-effect transistor

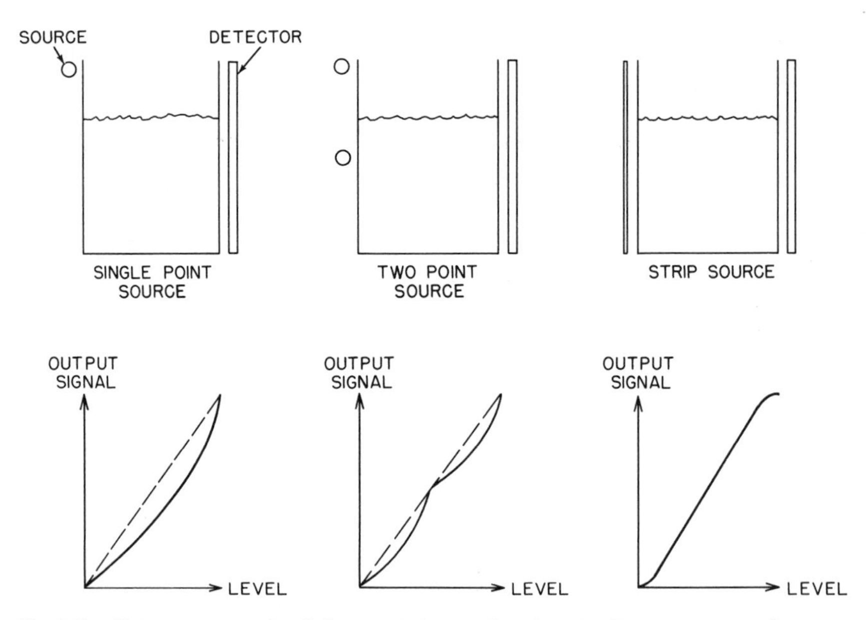

Fig. 1.1b Detector output signal characteristics as a function of radiation source configuration

strip source, of course, results in a linear response, but is usually more expensive than the point sources.

The preferred method of correcting the non-linear response of the single point source gauge is by a function generator whose response is the mirror image of the detector response as shown in Figure 1.1c. Usually, this function generator is available as an optional plug-in circuit board on the amplifier chassis.

Another method of linearization uses a tapered lead absorber in front of the detector. The absorber is thicker at the top than at the bottom so that the radiation falling on the detector is uniform throughout its length when the vessel is empty. Ordinarily, this technique requires field fabrication of the absorber. Continuous level gauges using Geiger-Mueller tubes are usually "stacked" to achieve the desired height of measurement, just as the early ion chamber level detectors were stacked. To improve linearity for this type of detector, three or four tubes are used at the bottom of the range where the radiation field intensity is low; two or three tubes are near the center of the range; and one tube is at the top.

Interaction Between Multiple Installations

When multiple, adjacent vessels are equipped with nuclear level sensors, orientation of the radiation sources must be considered to prevent interaction between the measurements. For two adjacent vessels, the sources should be adjacent (Figure 1.1d). This arrangement will preclude the radiation field from one measurement affecting the other measurement. When three or more

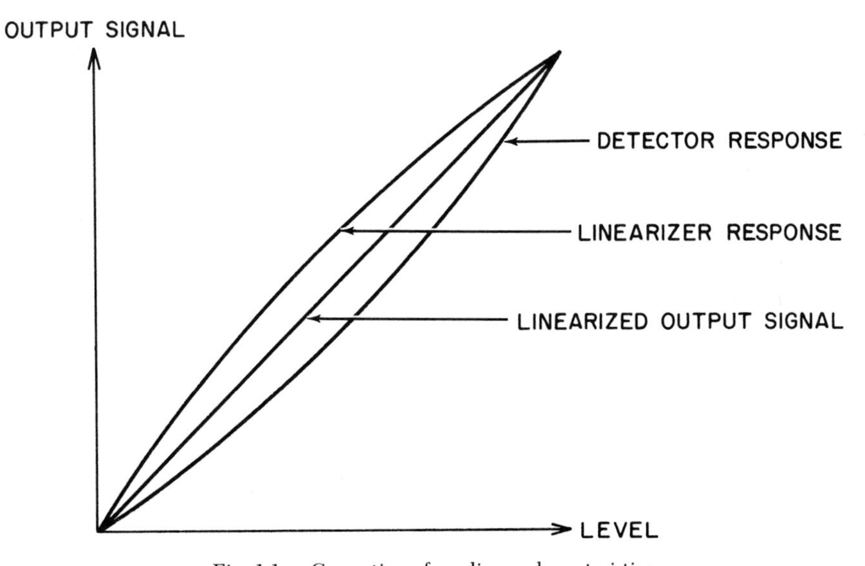

Fig. 1.1c Correction of nonlinear characteristics

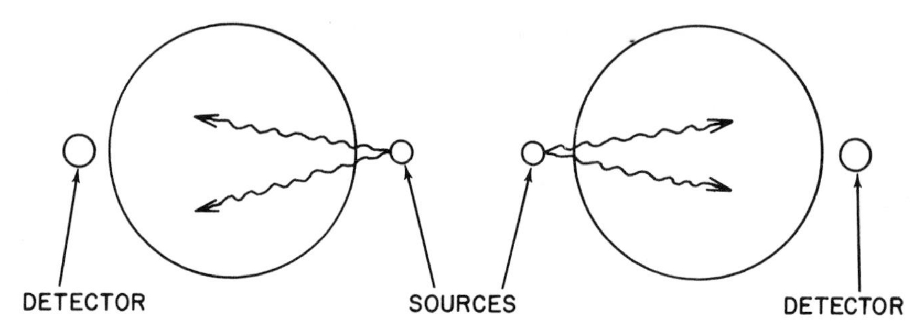

Fig. 1.1d Source location for adjacent vessels

adjacent vessels are equipped with nuclear level gauges, careful consideration must be given to interaction.

Sizing of Radiation Sources

Choosing the millicurie value of a radiation source for a given application can be complicated. Factors involved are radiation field intensity; linearity of radiation over the desired height of measurement; location of the source(s) and detector(s); consideration of the geometry, whether broad beam, narrow beam or somewhere in between; and the build-up factor, which depends on geometry.

With all these considerations it is wise to look for past successful installations and use them as a guide in selecting a source for a new application. Most manufacturers maintain data on past installations and utilize them to solve new-user problems. Some manufacturers have developed proprietary computer programs which take all of these factors into account when sizing sources.

Radiation Safety Considerations

Radiation level sensors are safe if properly installed and maintained. However, the most frequently overlooked aspect of radiation safety is the possible entry of personnel into large vessels for routine maintenance or repair. For these installations all means of ingress into the vessel should be marked with signs warning that nuclear radiation level devices are being used, or the portals should be locked and the key retained by a supervisor.

The safety procedure is very simple in that the radiation field intensity can be reduced to a negligible value merely by turning the source holder shutter to the CLOSED or OFF position. For units that use a source in a source well, it is relatively easy to remove the source and place it in a storage container.

Radiation Source Licensing

The licensing procedure with the Atomic Energy Commission (AEC) has not changed appreciably. However, in some states a state agency licenses radioactive material. These states have entered into an agreement with the AEC to regulate radioactive materials, and are termed Agreement States. The state licensing procedure is about the same as that of the AEC. The major difference is that the Agreement States require a license for the naturally radioactive materials, such as radium, as well as for the man-made materials such as Cs^{137} and Co^{60}.

As of June 1971, the following states (instead of the AEC) license radioactive materials:

Alabama, Arizona, Arkansas, California, Colorado, Florida, Georgia, Idaho, Kansas, Kentucky, Louisiana, Maryland, Mississippi, Nebraska, New Hampshire, New York, North Carolina, North Dakota, Oregon, South Carolina, Tennessee, Texas and Washington.

Armored Tubular Level Gauges

A recently marketed level gauge (Figure 1.1e) is now offered, incorporating some of the design features of the tubular gauge and of the armored metal-body gauge. The primary objection to (and limitation of) the classic tubular gauge is its fragility; the glass tubes are long and the protecting rods afford inadequate protection for most industrial applications. Also, any relative movement of the connections tends to stress the glass.

The design objective of the armored tubular gauge is to overcome these shortcomings and to gain the advantages of utilizing a tube. As in armored

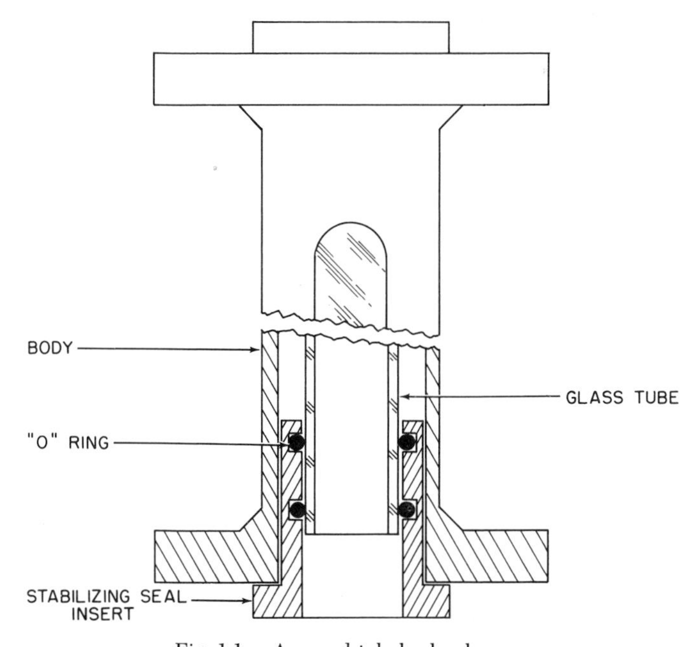

Fig. 1.1e Armored tubular level gauge

reflex and transparent gauges, this design has a heavy, rigid body (normally metal) which completely surrounds the glass except for the viewing slits. The body is a single piece from one connection to the other; therefore, relative movements of the connections primarily stress the body, not the glass.

Other advantages claimed for the design are that 1.) longer uninterrupted visible lengths, up to 12 ft, are possible; 2.) longer bodies are possible and where necessary, fewer vessel connections are required, resulting in a simpler and more economical installation; 3.) body is not in contact with the process fluid; therefore, economies can be realized in corrosive fluid applications because only atmospheric corrosiveness need be considered; 4.) glass sealing areas are greatly reduced and sealing is generally simplified; and 5.) components of the gauge are few and can be easily removed or replaced.

Tuning Fork Type Level Switches

In this instrument the sensing element consists of a membrane and two tines which are vibrated at their resonant frequency, thereby creating a piezoelectric effect. When material covers the tines, the oscillations are damped and the piezoelectric effect is either reduced or eliminated. Solid state electronic circuitry amplifies the piezoelectric effect and a relay is either energized or de-energized, depending on whether or not the tines are covered.

The instrument is suitable for either solid level sensing when the material size ranges from powder to $^3/_8$-inch diameter or for liquid levels. The instrument can detect low density solid materials and is available as either a high or low level switch. It has no adjustment, the trip level being established by the elevation at which the device is installed.

Tilt Switches

This is a simple and inexpensive device used as a high level switch for solids. It is suspended by a cable or rigid hanger and is connected by an eye bolt or swivel. As material rises and presses against the bottom of the switch it becomes unstable and tilts, thereby tripping a switch. The device is rugged and can be used on services where large particle size materials such as coal, ore and rock are present.

In some designs the lower end may be fitted with vanes or paddles to detect the presence or flow of solid materials.

Innovations in Ultrasonic Level Detectors

This type of detector (see Section 1.5 of Volume I) is probably the most promising developed to date for continuous measurement of solids levels. Nevertheless, in some applications, problems have been encountered precisely because of the measuring principle employed. In applications involving dry, dusty solids, the clouds of dust and suspended particles are sometimes sufficiently dense to reflect the transmitted sound wave, thereby "deceiving" the receiver and causing a false transmitted signal.

To combat this problem, circuitry has been developed to check the "validity" of the reflected sound wave received. This is accomplished by 1) the frequency being checked against the transmitted wave frequency to assure that the received wave does not come from some other source; 2) the pulse duration being compared against transmitted signal pulse duration (a transmitted pulse will be completely reflected by the material surface, but not necessarily by a dust cloud); and 3) the amplitude being compared against transmitted signal amplitude (dust clouds tend to absorb more signal energy than the material does and the wave amplitude reflected by the former is weaker).

The instrument will reject invalid signals and retain the output at the value corresponding to the last valid signal received. Problems have also been encountered in the orientation and location of the sensors. Pivotable sensors provide flexibility of orientation. Problems caused by mislocating instruments cannot be overcome by hardware improvements. The instrument engineer must be more careful in locating solids level sensors than in locating ones for the detection of liquid levels. Occasionally, a solids surface is not flat, as is a liquid surface. Solids level profiles depend on fill and withdraw point locations, material angle of repose, tendency to bridge and "rat-hole" and other conditions.

Additional innovations in equipment design include a liquid interface sensor which may either be mounted at a fixed elevation in the vessel or attached to a float so that the sensor is a definite distance below the free liquid level. The latter mounting permits monitoring the thickness of the layer of the lighter liquid. A low cost sump level transmitter with alarm contacts has also been marketed recently.

1.2 NEW DEVELOPMENTS IN PRESSURE MEASUREMENT

Pressure Detector Types:	(a) Piezoresistive. (b) Torque tube. (c) Piston. (d) High-precision manometer. (e) Multiple-pressure scanner. (f) Deadweight testers. (g) Seals. (h) Protectors. *Note:* In the feature summary below, the letters (a) to (h) refer to the listed designs.
Range:	2″ H_2O to 100″ Hg (d), 20 to 1,000″ H_2O (b), 5 to 100 PSID (c), 10 to 1,500 PSIG (a), to 5,000 PSIG (c), to 10,000 PSIG (e), vacuum to 20,000 PSIG (b).
Design Temperature:	All under 200°F.
Materials of Construction:	Aluminum (c), brass (b) (c), bronze (b), hastelloy C (a), Monel (a), Ni Span C (b), plastics (c), stainless steel (a,b,c,d), steel (b), teflon (e).
Accuracy:	±0.004″ Hg (d), ±0.5% FS (a,b,e), ±5% of setting (c).
Cost:	Switch only: $60 (c), pneumatic transmitter from $250 to $320 (b), electronic transmitter from $350 to $400 (a), from $500 to $580 (b), digital indicator from $800 to $3,000 (d), scanner for 48 points $2,500 (e).
Partial List of Suppliers:	American Instrument Co. (f), Ametek/U.S. Gage (f), Amthor Testing Co., Inc. (f), Applied Hydro-Pneumatics, Inc. (c), Bailey Meter Co. (a), Chandler-Engineering (f),

Dwyer Instruments, Inc. (d), Fisher Controls Co. (b), Ideal-Aerosmith, Inc. (d), Mid-West Instrument (h), Parks-Cramer Co. (g), Pressure Products Industries (f), Rutherford Co., Inc. (g), Scanivalve, Inc. (e), Schwien Engineering, Inc. (d), Stathan Instruments Inc. (d), Volumetrics, Inc. (d,f), Wallace & Tiernan, Inc. (d).

Other suppliers are listed in Volume I.

This section reviews the recent developments in pressure detection technology and thereby supplements Chapter II of Volume I.

Recent innovations made in this area are not so much in the discovery and application of new principles of measurement, as they are in improvement of accuracy, development of smaller, faster-responding transducers, increasing the variety of readout devices, perfecting of scanner-transducers, development of auxiliary devices and sophistication of primary standards.

Piezoresistive Transmitter

The piezoelectric effect is the change in the electrical characteristics of a material caused by pressurizing or deforming the material. An electronic pressure transmitter is now available which employs a piezoresistive sensor; this is the first widely marketed industrial instrument utilizing a piezoelectric effect. Pressure is applied to one side of a silicon wafer in which a piezoresistor has been diffused. Pressure changes cause resistance changes which are converted and amplified to produce a DC milliampere signal proportional to the applied pressure. Feedback circuitry minimizes errors which are caused by component, load and supply voltage variations. The two connecting wires conduct the output signal and also supply power to the transmitter. All electronic components are solid state. The design meets requirements for intrinsic safety and for Class I, Group D, Division I explosion-proof areas.

The instrument is quite small and light, weighing only about 4.5 pounds. It is simple and has no moving parts. Its principle of operation constitutes one of the most direct means for transducing variations in the process variable to variations in an electrical value. The wafer is coated so that the material in contact with the process fluid is 316 stainless steel, Monel or hastelloy C. Five different pressure elements are available for pressure ranges from 10 to 1,500 PSIG. For a given element, span is adjustable over a five to one range; suppressed range and elevated zero are available. Higher pressure elements are under development.

Torque Tube Type Sensor

These instruments incorporate small torque tubes that function like the torque tubes in displacement type level instruments. The tube is subjected to torsion which is proportional to the force (or movement) developed by the sensing element. For pressure instruments, a bourdon or bellows element produces a linear movement proportional to pressure. For differential pressure a diaphragm element produces a linear movement similar to that illustrated in Figure 1.2a. This movement is applied to the torque tube lever arm, which causes an angular displacement of the output shaft. This shaft rotation becomes the input of a motion transmitter which provides a milliampere or pneumatic output signal.

These are motion transmitting devices and are not supplied as direct-indicating types. All elements produce the same angular movement; consequently, all transmitters of a given type (i.e., pneumatic or electronic) are identical. Zero and span adjustments are made in the transmitter. Electronic transmitters use two wires which conduct both the signal and transmitter power. These transmitters meet Class I, Group D, Division I explosion-proof and intrinsic safety requirements.

Piston Type Sensor

The piston type sensor (Figure 1.2b) is currently available as a low-cost, moderately accurate indicator, switch or indicating switch. A common application for it is to monitor differential pressure across filters.

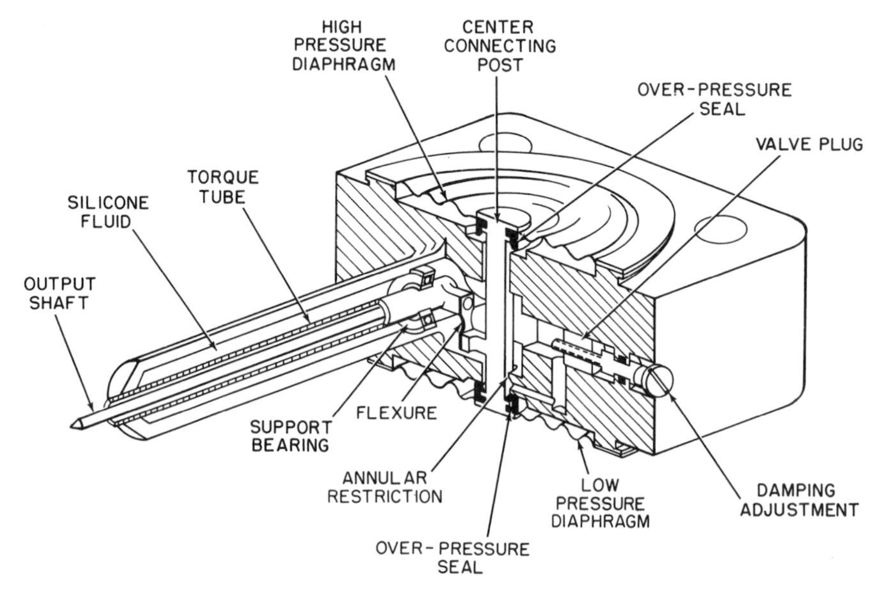

Fig. 1.2a Differential pressure element and torque tube

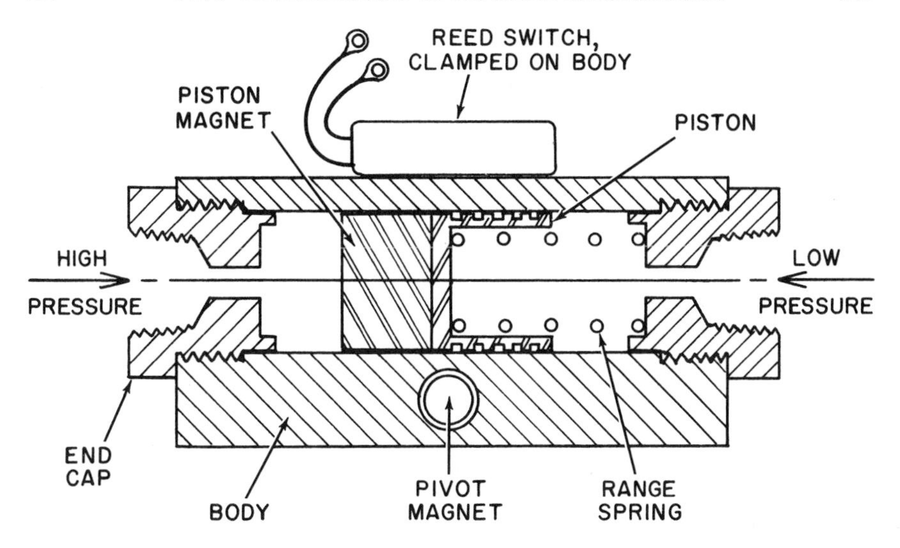

Fig. 1.2b Piston type sensor

The high pressure is applied to the top of a piston and the low pressure to the underside of the piston against which a spring presses. A given differential pressure produces a corresponding spring movement. A magnet is attached to the piston which, through a non-magnetic body, positions a pivot magnet, thereby driving the pointer or tripping as many as four switches, or both. Since piston movement is not transmitted mechanically, there are no associated moving seals. The only seals required are the piston seals and the static seals for the end caps.

Precision Manometers

The manometer is the simplest and probably the oldest means of measuring vacuum, pressure and differential pressure of limited range, and because of its simplicity, accuracy and directness of measurement, has long been an excellent primary standard. Manometers are available with errors of ± 0.0005 inch of Hg for absolute pressures, gauge pressures and differential pressures to about 100 inches of Hg. Automatic temperature and gravity compensation as well as digital read-out are both available and necessary for high-precision measurements. Read-out is obtained by ultrasonic or capacitance type devices which measure the distance to the mercury meniscus of each leg, and an electronic circuitry computes the difference. Generally, these devices can generate pressures for calibrating purposes and measure unknown pressures. The most sensitive instrument utilizes a hook gauge manometer and measures differential pressures within ± 0.00025 inch of water over a 2-inch water column range.

Multiple Pressure Scanner

This device, illustrated in Figure 1.2c, drives a scanning valve that sequentially connects a number of process pressure signals to an electronic or pneumatic transducer. The transducer then provides, for a portion of the scanning time interval, a signal proportional to the connected pressure.

The cost advantage derived from this device is that only one transducer is required for a large number of pressures, thus saving as much as $500 per transducer for each pressure. This, of course, means that as the number of pressures to be scanned increases so does the cost saving.

Several factors must be taken into account in determining the applicability of the scanner-transducer. As the number of scanned pressures becomes greater, the delay in "updating" a given pressure reading also increases. Inherently, the device provides only an intermittent signal for each pressure and is therefore not applicable when continuous signals are required. Installation costs are influenced by the fact that all pressures must be connected to the scanner-transducer. The scanning valve can only retain the required tightness when handling a clean gas.

The scanner-transducer was used initially in the aircraft and aerospace industries, primarily in testing facilities. More recently it has been used in the process industries, and the classic application has been in an existing central control room with pneumatic instruments when the decision has been made to install a computer-based data acquisition or control system. In this

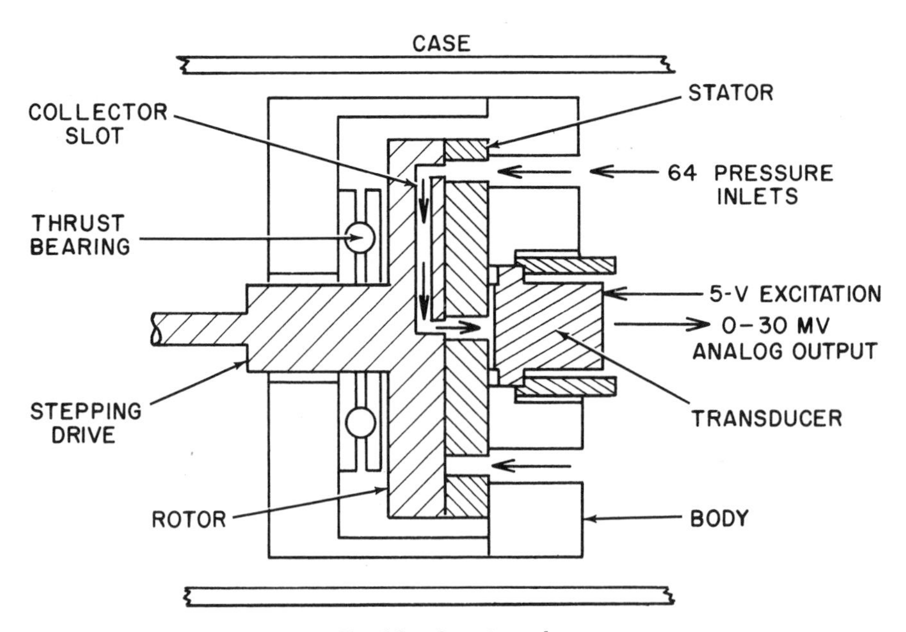

Fig. 1.2c Scanning valve

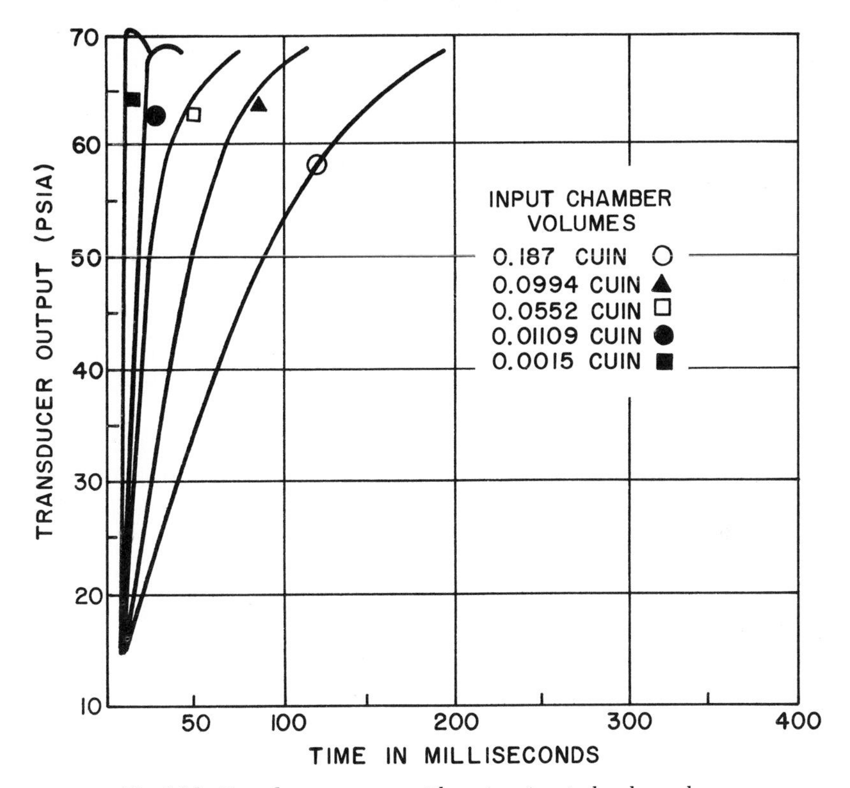

Fig. 1.2d Transducer response with various input chamber volumes

case there are many conditions which make the scanner/transducer applicable:

1.) the pressures have already been run to a central location;
2.) pneumatic flow, level and temperature transmitters already exist so that pressure signals are available for these process variables;
3.) because of its inherent capabilities, the computer can program the scanner-transducer and conveniently use the intermittent signals from it.

One primary limit on scanning rate is that in order to achieve satisfactory signal accuracy, maximum equalization must be achieved between the actual pressure to be sensed and the pressure on the transducer. Equalization time is made as brief as possible by having a small volume between the scanning value port and the transducer sensor, by filling this volume with oil and by utilizing sensors having minimal displacement. Figure 1.2d illustrates transducer response.

Leakage in the system is a source of error; therefore, valve leakage is kept below the value that would introduce an error significant in transducer accuracy. The scanning valve sealing surfaces are ground optically flat; materials used are stainless steel and teflon. The valve is capable of enduring 2,600,000 scans and still retain its guaranteed tightness.

Accessories

The problem of making valid and useful pressure measurements in industrial installations involves more than a good sensor. The instrument must be reliably calibrated. Readout must be selected for operator convenience. The instrument, as installed, must be protected against adverse effects caused by the process fluid, against freezing and plugging of impulse lines and against rapidly pulsating pressures; in sum, instruments must be installed so that they function properly. The following developments in equipment and devices are intended to achieve these ends.

Improvements in deadweight testers have reduced their error to ±0.02%, and other innovations have made them easier to use. Pneumatic handling devices for loading the weights and torque devices for spinning them are now available. Some testers can measure pressure continuously, by utilizing a calibrating weight suspended in a fluid. Balance is achieved automatically by varying fluid level and thereby varying buoyant force. Digital readout is provided.

Chemical seals are frequently used to protect pressure sensors. One disadvantage in using them, particularly on low pressures, is that they tend to desensitize the instrument due to their spring constant. A very flexible bellows-type seal fabricated from teflon is now available, which minimizes desensitization. For diaphragm-type seals an adapter is available in line sizes up to 8 inches which inserts between process line flanges. The diaphragm is placed as directly as possible in contact with the flowing fluid so as to avoid dead volumes which would tend to plug. The jacket permits a heating medium to be used.

Several pressure gauge manufacturers have recently offered alternate means of protecting gauges against pressure pulsations, vibration and corrosion: the cases are filled with oil so that the pressure element, linkages and gears are immersed. In some designs, element movement is transmitted to the pointer by two magnets, one on either side of the case front, which can therefore be solid. Another gauge protector is available which serves the combined functions of variable-restriction snubber, high pressure shut-off and isolating valve.

About forty suppliers offer electronic systems with digital, analog and cathode ray tube readouts. See Chapter III of this volume for additional details. The latter can display simultaneously a number of measurements in bar-graph form, thus furnishing a pressure profile. System configurations vary considerably. Capabilities offered are direct tie-in with computers or loggers, scanning of multiple inputs, control and multiple ranging.

1.3 NEW DEVELOPMENTS IN DENSITY MEASUREMENT

Density Detector Types:	(a) Sound velocity meters. (b) Vibrating spool sensors. *Note:* In the feature summary below, the letters (a) and (b) refer to the listed sensor types.
Design Pressure:	Up to 10,000 PSIG (a,b).
Design Temperature:	Up to 300°F (b), −20°F to +390°F (a).
Materials of Construction:	316 stainless steel (a,b), epoxy (a), hastelloy B & C (a), titanium (a).
Accuracy:	±0.1% of density span (a), ±0.1 SG units (b).
Cost:	Electronic transmitter $4,100 (a), with counter readout only $2,300 (a).
Partial List of Suppliers:	J. Agar and Co., Ltd. (b), NUSonics, Inc. (a).

This section covers recent developments in density detection technology, thereby supplementing Chapter III of Volume I.

Newly developed density sensors are more sophisticated than the traditional ones, and their principles of measurement tend to be more inferential than those of some of the originally developed sensors that directly measure the mass-per-unit volume. As a result, calibration is more empirical and a calibrating liquid of known composition and at a controlled temperature is required. Calibration by other standards, such as a known weight or differential pressure, is not possible. Also, because of the principles of measurement involved, the sensors are essentially "blind" and do not directly provide a movement or displacement that is an index of density. Presently, they are rather special instruments, and their application, calibration and maintenance require more qualified personnel than many of the traditional types.

Sound Velocity Meters

Sound velocimeters are used to measure the density of certain liquids and they are also used as analyzers. The equation for the speed of sound in a fluid or any medium is $C = \sqrt{E/\rho}$, where E is the bulk modulus and ρ is mass density. Since the velocity of sound is a function both of bulk modulus and of density, the velocimeter can only be used as a densitometer for a process liquid the bulk modulus of which does not vary significantly due to the changes in composition causing the density variations to be measured. The magnitude of the effect of variations in bulk modulus can be inferred from the equation cited. Compensation can be provided for applications in which variation in pressure and temperature have significant effects on sonic velocity.

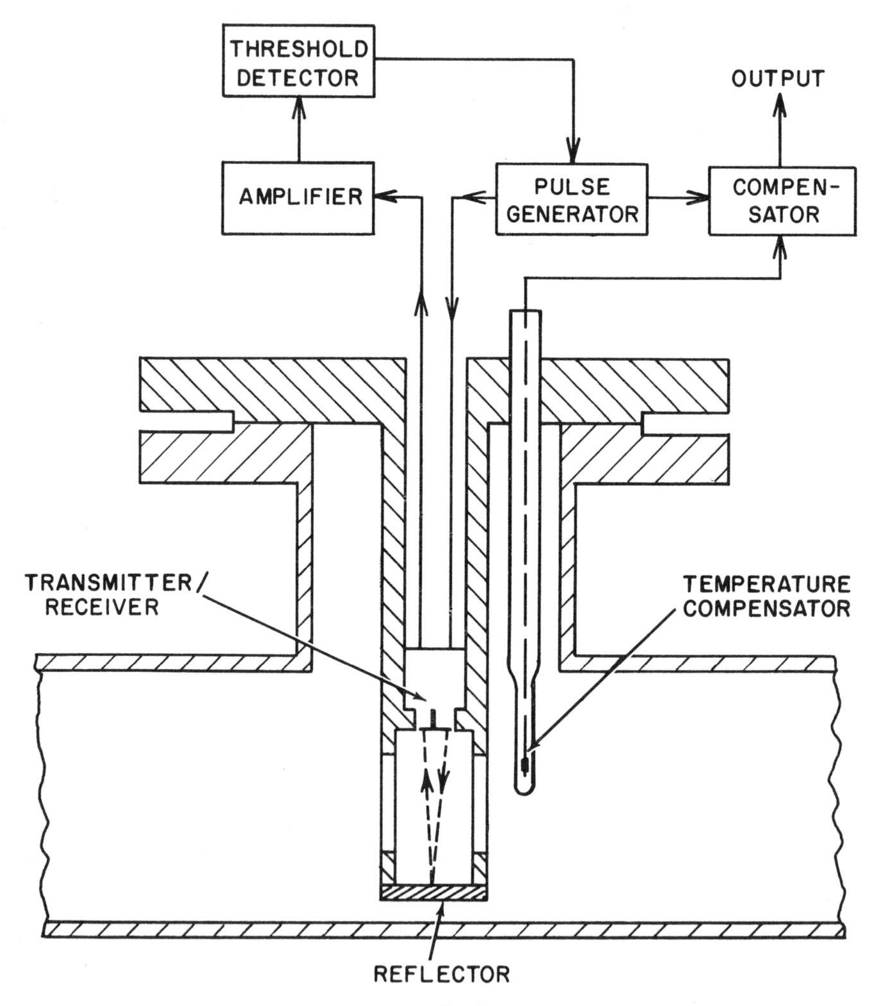

Fig. 1.3a Sound velocity meter

In a process velocimeter probe assembly (Figure 1.3a); the transmitter emits a 3,600 Hz sound wave toward the reflector, which directs the wave toward the receiver. The received signal is amplified and passes to a detector the output of which activates a pulse generator, causing another wave to be emitted by the transmitter. Thus, a "sing-around" cycle is continuously repeated. Since for a given sensor the length of the path traveled is fixed, the cycle repetition frequency is a function of sound velocity and consequently of density, if the bulk modulus does not vary significantly. A pulse converter provides a continuous analog signal that is a function of sound velocity.

The equation relating velocity and cycle repetition frequency is

$$C = \frac{Af\,(1 + \alpha t)}{1 - Bf \times 10^{-6}} \qquad 1.3(1)$$

where C = sound velocity.
A = length of sound path,
B = time intervals in microseconds between arrival of wave at receiver and emission or pulse by transmitter,
α = coefficient or thermal expansion of the velocimeter probe material,
t = temperature in °C,
f = cycle repetition frequency in Hz.

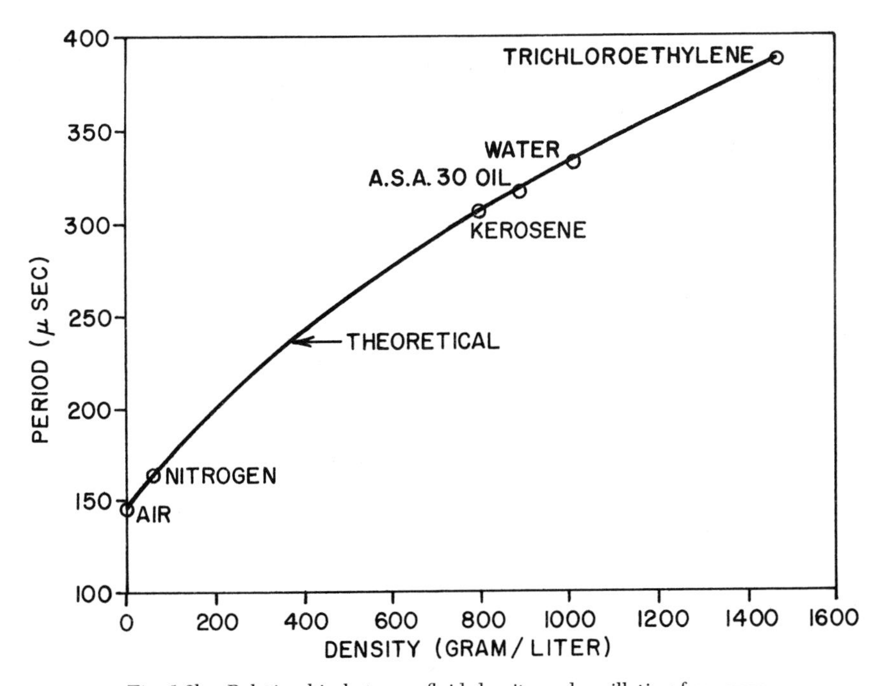

Fig. 1.3b Relationship between fluid density and oscillating frequency

Calibration is done by immersing the sensor in distilled water whose temperature is closely controlled at intervals between 0° and 75°C, and by measuring cycle repetition frequencies to an accuracy of 0.001 percent. Using published data for sound velocity in distilled water and the measured frequencies, the constants A and B in equation 1.3(1) are computed.

The sensor has several limitations that must be taken into account in determining its applicability. The liquid must be clear; emulsions, dispersions and slurries scatter the sound waves and cannot be monitored. Similarly, undissolved gases and bubbles (even those as small as several thousandths of an inch in diameter) cause errors and occasionally render the instrument inoperable. The limitation caused by variation in bulk modulus has already been mentioned.

The instruments have been successfully used to detect the interface between two different hydrocarbon products in pipelines.

Vibrating Spool Type Density Sensor

If a cylindrical spool is immersed in a fluid and if circumferential oscillation normal to the spool is induced and sustained, the spool will vibrate at a frequency that is a function of its stiffness and the oscillating mass. Since

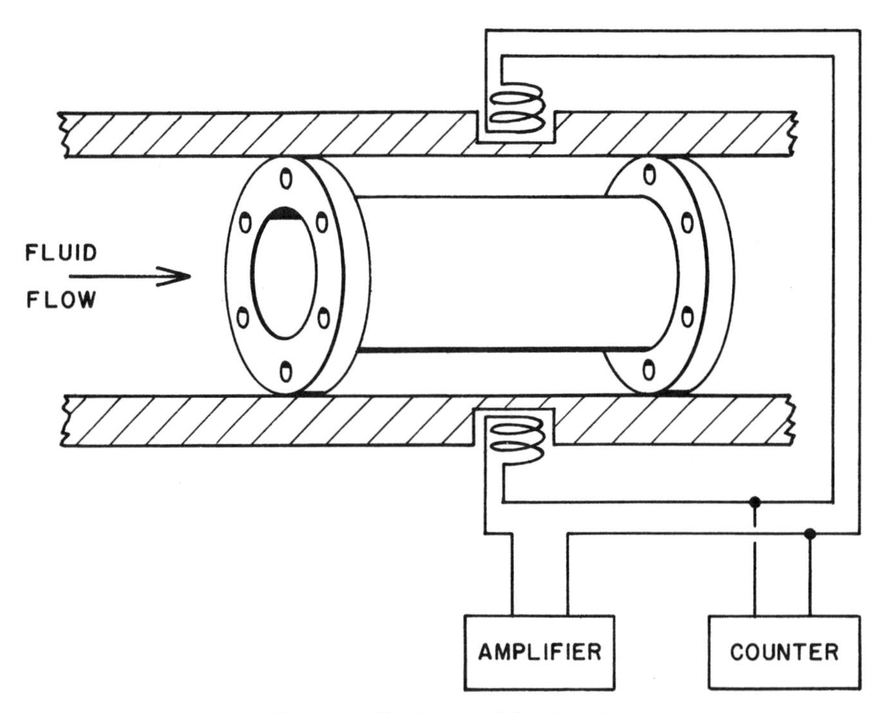

Fig. 1.3c Vibrating spool densitometer

the fluid surrounding the spool is caused to oscillate, the mass of the entire system in vibration is composed of the mass of the spool and of the fluid. The system can be treated as a lumped parameter damped harmonic oscillator over a small range of frequencies. If a loop closed around the oscillator exhibits a 90° phase shift, vibration is sustained at the natural frequency irrespective of fluid viscosity. Thus, for a spool of a fixed stiffness and mass, variations in oscillating frequencies are due solely to variations in fluid densities. This relationship is shown in Figure 1.3b.

In the instrument shown in Figure 1.3c, oscillations are induced and sustained by a feedback amplifier. A predetermined number of oscillations is counted and the elapsed time is measured by a high-frequency "clock"; the signal is then developed from these data. Developing the signal in this manner provides a faster response than does measurement of frequency.

Variations in spool stiffness and natural oscillating frequencies are minimized by several means. Temperature effects on spool dimensions and elastic modulus are kept as small as possible by alloys like Ni Span C. Fluid pressure is exerted on both surfaces of the spool, so that pressure variations have no influence on oscillation. Material is stressed considerably below fatigue levels because oscillation amplitude is very slight.

1.4 NEW DEVELOPMENTS IN TEMPERATURE MEASUREMENT

This section discusses the recent developments in temperature measurement technology, thereby supplementing Chapter IV in Volume I. The developments in non-contact temperature detection are discussed at length.

The International Practical Temperature Scale

The International Practical Temperature Scale (IPTS) is the basis of most present-day temperature measurements. The scale, which has been in effect for the past two decades, was established by an international commission in 1948 with a text revision in 1960.[1] A revision of the scale was formally adopted in 1968 and is reproduced in Table 1.4a. The scale is defined by reproducible temperature points established by physical constants of readily available materials. Interpolation between these fixed points is made by several standard measuring instruments summarized in Table 1.4a.

For a complete graphical presentation of the conversion factors to bring temperatures based on the (IPTS) of 1948 into agreement with the (IPTS) of 1968, the reader is referred to NBS Special Publication 300 in Volume II.[2]

Factors Affecting Temperature Measurements

A description of all types of temperature measuring instruments has previously been presented in Volume I from the view point of configuration, performance and accuracy.[3] Here the emphasis will be on how to choose an instrument so as to assure obtaining useful measurements under various applications.

All temperature measurements fall into two basic categories including measurements of fluid temperature (liquid, gas, vapors and slurries) and measurement of temperature on or within solid bodies. The discussion immediately following has been limited to sensing devices employed by attaching to or immersing in the substance of interest.

Temperature Range

Specification of temperature range can identify the type of sensing device which must be used. This is apparent from Table 1.4b. For example, if the temperature range for a measurement is 1,000°F to 3,000°F, the choice of

Table 1.4a
PRIMARY TEMPERATURE POINTS AND VALUES OF IPTS OF 1968

Temperature °C	*Defining Fixed Point*	*Interpolating Instrument*	*Accuracy of Realizing (IPTS) °C*
–183.09	Oxygen, liquid-vapor equilibrium	Platinum resistance thermometer	0.02
0.00	Water, solid-liquid equilibrium	Platinum resistance thermometer	0.0002
0.01	Water, triple point	Platinum resistance thermometer	0.0002
100.00	Water, liquid-vapor equilibrium	Platinum resistance thermometer	0.0005
419.58	Zinc, solid-liquid equilibrium	Platinum resistance thermometer	0.002
444.67	*Sulfur, liquid-vapor equilibrium	Platinum resistance thermometer	0.002
961.92	Silver, solid-liquid equilibrium	Platinum-platinum 10% rhodium thermocouple	0.2
1,064.43	Gold, solid-liquid equilibrium	Platinum-platinum 10% rhodium thermocouple	0.2

* Zinc point preferred

temperature sensors is narrowed to several types of commercially available thermocouples. If the temperature range is from 0°F to 200°F, the choice includes several types of thermocouples, liquid filled bulbs such as mercury thermometers, resistance thermometers, thermistors and bimetallic thermometers.

Accuracy

Accuracy requirements impose an additional restriction on the choice of sensing device. In considering accuracy two factors must be reviewed. First is the inherent accuracy of the measuring device, which is summarized in Table 1.4b. This generally represents the highest accuracy that can be achieved with the device under ideal conditions. The second consideration which in general is a greater limitation than the inherent accuracy of the device, is the inaccuracy of actual measurement due to environmental effects, method of installation and use.

Table 1.4b
TEMPERATURE DETECTOR ACCURACY AND RANGE

Description of Sensors	Temperature Range	Inaccuracy (Error)	
		ISA Standard Limits	ISA Special Limits
I Thermocouples			
a. Chromel/-Alumel	−300°F to +2500°F	±4°F, 0°F to +530°F ±3/4%, +530°F to +2500°F	—
b. Iron/Constantan	−310°F to +1400°F	±4°F, 0°F to 530°F ±3/4%, +530°F to +1400°F	±2°F, 0°F to +530°F ±3/8%, +530°F to +1400°F
c. Copper/Constantan	−310°F to +750°F	±2%, −150°F to −75°F ±1 1/2°F, −75°F to +200°F ±3/4%, +200°F to +700°F	±1%, −300°F to −75°F ±3/4°F, −75°F to +200°F ±3/8%, +200°F to +700°F
d. Platinum 10% Rhodium/Platinum	+32°F to +3200°F	±5°F, +32°F to +1000°F ±1/2%, +1000°F to +2700°F	—
e. Tungsten/Tungsten 26% Rhenium	+32°F to +5200°F	±8°F, +32°F to +800°F	—
f. Tungsten 5% Rhenium/Tungsten 26% Rhenium	+32°F to +5200°F	±1%, +800°F to +4200°F	—
II Resistance Thermometers			
a. Platinum (high purity strain free element)	−435°F to +900°F	±0.01°F limited Range, ±0.05% full Range	
b. Platinum (commercial purity element)	−320°F to +2000°F	±0.2% full Range	
c. Nickel	−320°F to +800°F	±0.4% full Range	
d. Balco	−320°F to +500°F	±0.4% full Range	
e. Semiconductor (germanium)	−458°F to −400°F	±0.01°F	
f. Semiconductor (thermistors)	−40°F to +300°F	±0.05°F limited Range, ±1°F full Range	
III Bimetallic Thermometers	0°F to +1000°F	±2°F limited Range, ±10°F full Range	
IV Liquid Filled Thermometer (glass stem)	−328°F to +800°F	±0.05°F limited Range, ±1°full Range	
V Liquid Filled Thermometer (metal bulb)	−150°F to +700°F	±2°F limited Range, ±5°F full Range	

Environmental Influences

In considering the validity of a measurement, it should be remembered that temperature sensing devices respond only to the temperature which they experience. It may be considerably different from the temperature one is attempting to measure if the sensor is of improper size or configuration, or is installed in a manner such that it interferes with the temperature that would exist if the sensor was not present, or if the sensor is not adequately coupled thermally to the media whose temperature is being measured.

Measuring the Temperature of Solids

The next consideration is the allowable size and configuration of the sensor so as to assure a useful measurement. This requires some knowledge of the heating or cooling conditions together with an estimate of the magnitude of the temperature gradients that are likely to exist in the region in which the measurement is to be made. A simple, rule-of-thumb indicator to determine if significant gradients are likely to be present is the magnitude of the Biot modulus (hL/K), where h is the surface heat transfer coefficient, L is the smallest dimension of the solid and K is the thermal conductivity of the solid. If this modulus is $hL/K > 0.2$, significant temperature gradients are likely to exist in the solid, and care should be exercised in choosing the size, location and orientation of the sensor within the solid. If the Biot modulus is less than 0.2, no significant gradient is expected and a measurement anywhere on or within the solid should give identical results regardless of size or configuration of the sensor. If significant gradients are likely to exist, the maximum rate of heat transfer to the surface of the solid must be known or estimated, and the maximum gradient at the point of measurement must be determined. The following relationship allows the maximum gradient at the surface of a solid to be calculated.

$$\frac{\Delta T}{\Delta X} = \frac{q}{K} \qquad 1.4(1)$$

where $\frac{\Delta T}{\Delta X}$ = temperature gradient at the surface

q = heat transfer rate per unit area at the surface
K = thermal conductivity of solid

Under certain conditions of heating or cooling, if measurements at points other than the surface are important, it may be necessary to evaluate anticipated heat transfer conditions and resulting temperature gradients.[4,5]

On the basis of this gradient, it is possible to establish limits on the size of the sensing device. For example, the length of any one of the three dimensions of the sensor (lead wires excluded) should not be greater than the distance between two points of the process that are different in temperature by more than the acceptable measurement error. It is assumed that the sensors are in satisfactory thermal coupling with the process material, which is not always the case. If the thermal coupling is poor, the sensor will not reflect the true temperature history experienced by the solid, a condition which can produce dynamic errors.[6] The best thermal coupling is achieved by direct bonding of the sensor, such as welding a thermocouple to the solid surface or into a cavity within the solid. The bond line between the sensor and the solid should be kept as thin as possible and should not fracture or fail during thermal cycling. Various epoxy and ceramic cements, with fillers to improve their conductivity, have been successfully used for such bonding. For example, a flat resistance thermometer bonded to a surface with an epoxy bond line 0.005 inch thick will produce a lag time of about 1 second, which will produce a dynamic error equal to the rate of temperature rise of the surface, i.e., a dynamic error of 25°F for a rate of surface temperature change of 25°F per second.[6]

Methods of installing temperature sensors on or within solid bodies are shown in Figure 1.4c.

Measuring the Temperature of Fluids

The fundamental problem of measuring the temperature of a fluid is one of assuring strong thermal coupling. For a fluid temperature measurement to have meaning, the sensor must come to equilibrium with the temperature of the fluid. The difference between the equilibrium temperature of the sensor and the fluid temperature is a direct error. Consequently, with rapidly changing temperatures the rate of heat transfer between the sensor and process fluid must be sufficient to overcome the thermal capacity of the sensor in order that it can follow the fluctuations in fluid temperature. Under such conditions one can write the following equation which expresses the temperature, T, experienced by a sensor which was initially at temperature T_1,[4]

$$\frac{T_1 - T}{T_1 - T_0} = e^{-(hA/WC_p)\theta} \qquad 1.4(2)$$

where T = temperature
e = base of natural logarithm
h = heat transfer coefficient at surface of sensor
A = surface area of sensor
W = weight of sensor
C_p = specific heat of sensor
θ = time after immersing sensor in fluid

A sensor, to respond rapidly to changes in fluid temperature, should have a large surface area-to-mass ratio. Furthermore, the heat transfer coefficient, which is a direct function of the fluid mass flow rate over the sensor, should also be as large as possible. Forced convection or rapid flow of the fluid over the sensor is therefore desirable.

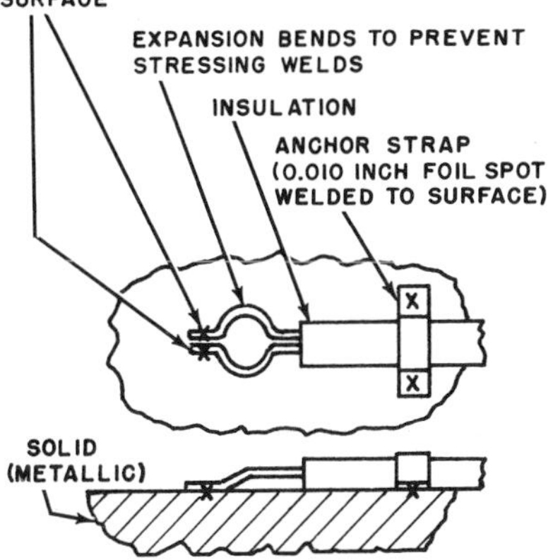

A-TYPICAL THERMOCOUPLE INSTALLATION FOR JUNCTION FORMED THROUGH SOLID BY DIRECT SPOT WELDING TO SURFACE

CHARACTERISTIC:

1. FAST RESPONSE
2. SIMPLE TO INSTALL
3. MAY CAUSE DISTURBANCE TO FLOW PATTERN UNDER HIGH SPEED FORCED CONVECTIVE, COMPRESSIBLE FLOW

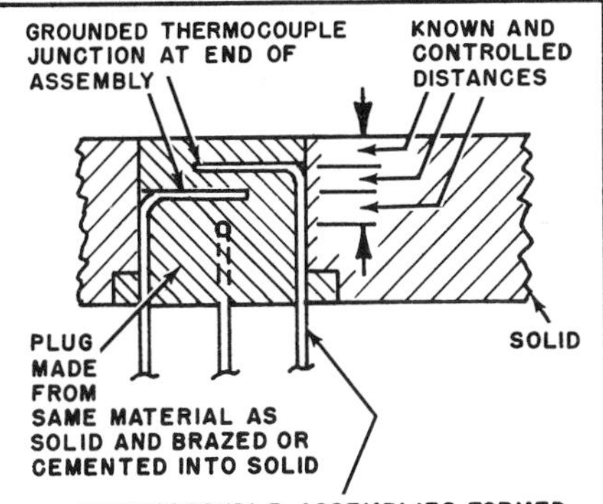

THERMOCOUPLE ASSEMBLIES FORMED FROM 0.025 INCH OD METALLIC CLAD THERMOCOUPLE WIRES WITH SWAGED CERAMIC OXIDE INSULATION. ASSEMBLIES EMBEDDED IN PLUG BY BRAZING, CEMENTING OR ELECTROFORMING TO ASSURE INTIMATE THERMAL CONTACT WITH PLUG.

C-TYPICAL EMBEDDED THERMOCOUPLE INSTALLATION IN SOLID

CHARACTERISTICS:

1. ALLOWS MEASURING GRADIENTS WITHIN SOLID
2. NO DISTURBANCE TO FLUID FLOW PATTERN AT HEATED SURFACE

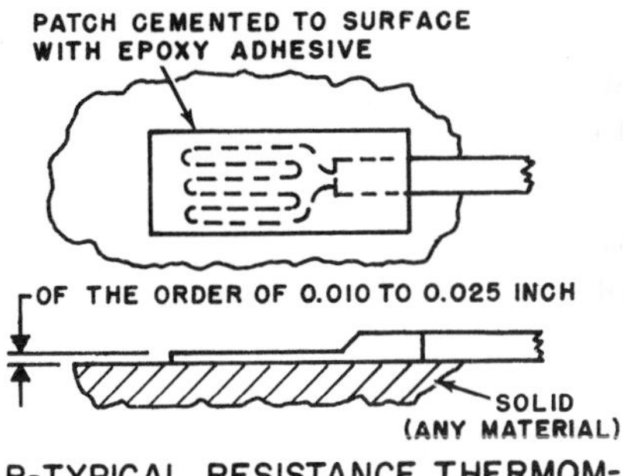

B-TYPICAL RESISTANCE THERMOMETER PATCH INSTALLATION ON SURFACE OF SOLID

CHARACTERISTICS:

1. SAME AS "A"

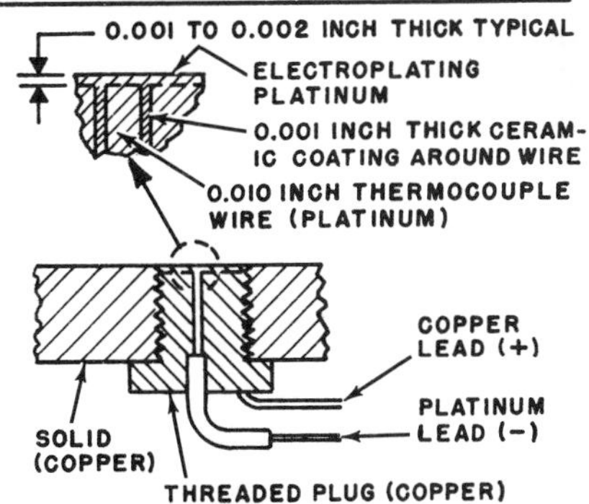

D-SURFACE THERMCOUPLE FOR HIGH SURFACE HEATING RATES

CHARACTERISTICS:

1. RESPONDS AT SAME RATE AS SOLID AT SAME DEPTH
2. NO DISTURBANCE TO FLUID FLOW PATTERN AT HEATED SURFACE
3. MINIMUM DISTURBANCE TO NATURAL HEAT FLOW IN SOLID
4. APPLICABLE UNDER EXTREMELY HIGH PRESSURE
5. USEFUL UP TO MELTING POINT OF SOLID

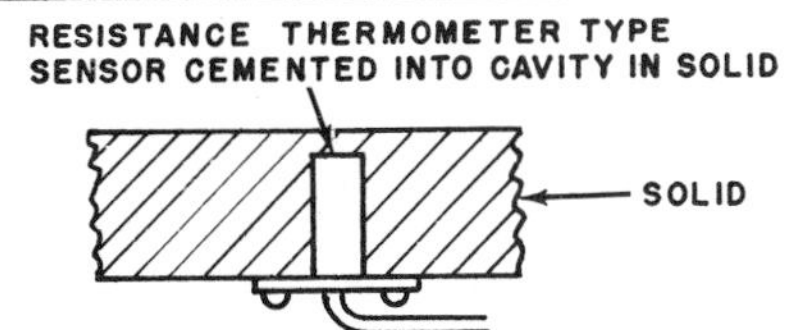

E-TYPICAL RESISTANCES THERMOMETER EMBEDDED IN SOLID

CHARACTERISTICS:

1. SLOW RESPONSE
2. INACCURATE KNOWLEDGE OF LOCATION OF TEMPERATURE MEASURED

Fig. 1.4c Some selected methods of installing temperature sensors on or within solids

A practical measure of sensor sensitivity is its time constant. Since the sensor approaches change in fluid temperature asymptotically, it is difficult to determine when it has reached that temperature. The time constant of a sensor is the time required to accomplish 63.2% of the step change. This point is reached when the exponent of equation 1.4(2) is 1. Thus, the time constant of a fluid stream temperature sensor (τ) is defined as,

$$\tau = \frac{1}{hA/WC_p} \qquad 1.4(3)$$

When measuring the temperature of a gas stream in a heated duct or furnace where temperature differences between the sensor and its surroundings exceed 1,000°F, significant errors can occur due to radiation exchange between the sensor and its surroundings. Under such conditions the sensor must be shielded against thermal radiation exchange. This can cause disturbances in the flow of the fluid around the sensor and hence affect the directional response characteristics.

Another thermal effect in measuring the temperature of gas streams at high velocities is the recovery factor, which results from an increase in the temperature of the gas at the sensor due to compression heating as the gas is brought to stagnation against the sensor.[7,8] Fluid stream sensor configurations are shown in Figure 1.4d together with some installation and application oriented guidance.

Atmospheric Effects

Adverse atmospheric conditions can cause problems with temperature sensors. For example, in a highly humid or very moist environment it is essential that the element of a resistance thermometer or the bead of a thermistor be well insulated electrically. If moisture contacts the resistance element or thermistor bead, it may cause a short. Thermocouples in general are less sensitive to moisture than are resistance thermometers. If null balance potentiometric or high input impedance readout devices are used, insulation resistance between legs of the couple as low as ten thousand ohms can be tolerated without serious error in the indicated temperature. If low input impedance current measuring readout devices are used, a high insulation resistance between legs of the couple becomes as important as with resistance thermometers and should be in the megohm range.

Corrosive, reducing or oxidizing environments also create problems. If a sensing device is exposed to such environments, it must be protected by some form of envelope or coating. It should be remembered that protection applied to a sensor will increase its mass and thus, in general, adversely affect its response characteristics.

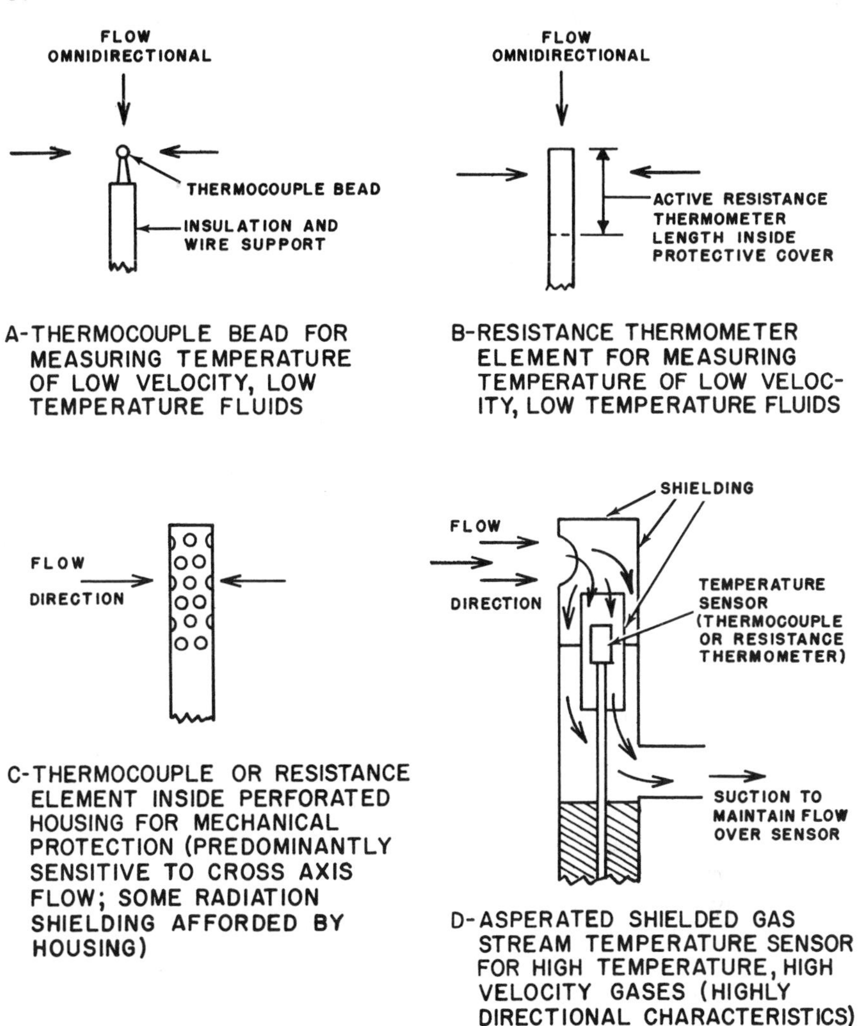

Fig. 1.4d Fluid stream sensor configurations

Non-Contact Temperature Measurement

Non-contact temperature measurement of a moving surface is needed when measuring webs of paper, plastic, textile and metal or rotating cylinders such as calender rolls or drier cans. Very small thermocouples or resistance elements placed close to the surface have had some success but, in general, are too vulnerable to physical damage and ambient influences. Various types of radiation devices such as infrared detectors have also been employed with some success, depending on the environmental conditions under which the measurements are made. For example, changes in emissivity of the surface and contamination of the required optical elements directly affect the accuracy and repeatability of such measurements.

A new principle, the convective-null-heat-balance concept,[9,10] involves the null-balance of convective heat transfer between a sensing head of known temperature and a surface of unknown temperature, in which temperatures are compared rather than directly measured. If, for example, two surfaces are placed in close proximity to one another, their temperatures can be compared by measuring the rate and direction of heat transfer between them. The direction of heat transfer is a definite measure of which surface is at the higher temperature and if the heat transfer between them is zero, their temperatures must be equal.

If one of these surfaces contains a heat transfer sensor and its temperature is known (T), one can readily determine if the temperature of the other surface is either below, equal to or above T by simply observing whether the heat transfer is zero or not and what its direction is. It is not necessary to measure the magnitude of the heat transfer. The technique, of course, works only if the point at which heat transfer is being sensed is isolated from surrounding influences. On a flat surface, isolation is accomplished by simply locating this point at the center of a relatively large isothermal plate (typically 6 inches in diameter) spaced approximately $\frac{1}{16}$ to $\frac{1}{8}$ inch from the surface being measured.

The basic concept of null-heat-balance can be used either to monitor or to control surface temperature. In the monitoring mode, the temperature of a sensing head is automatically varied to keep it at null-heat-balance (at the same temperature) as the moving surface. The head temperature, which is equal to the surface temperature, can then be measured.

In the control mode, the head is held at the desired set point temperature at which the surface is to be controlled and the signal from the heat flow sensor is employed to manipulate the surface temperature.

In another arrangement, the signal from the heat flow detector is calibrated for temperature difference between the sensor, which is at a constant known temperature, and the surface being measured.

REFERENCES

1. Stimson, H. F., "International Practical Temperature Scale of 1948, Text Revision of 1960." NBS Monograph 37, 1961.
2. NBS Special Publication 300, Volume II, January 1969.
3. Lipták, B. G., (ed.) "Instrument Engineers' Handbook" (Vol. I), Ch. IV. Philadelphia: Chilton, 1969.
4. Giedt, W. H., "Principles Of Engineering Heat Transfer." New York: Van Nostrand, 1957.
5. Carslow, H. S. and Jaeger, J. C., "Conduction Of Heat In Solids." London: Oxford, 1947.

6. Rall, D. L. and Hornbaker, D. R., "A Rational Approach To The Definition Of A Meaningful Response Time For Surface Temperature Transducers." New York: 21st Annual ISA Conference, October 22–27, 1966.
7. Glawe, G. E., Simmons, F. S. and Stickney, T. M., "Radiation and recovery corrections and time constants of several chromel-alumel thermocouple probes in high-temperature, high-velocity gas streams." NACA TN-3766, October, 1956.
8. Scadron, M. D. and Warshawsky, I., "Experimental determinations of time constants and Nusselt numbers for bare-wire thermocouples in high-velocity air streams and analytic approximations of conduction and radiation errors." NACA TN-2599k, January, 1952.
9. U.S. and foreign patents have been issued and others are pending, Trans-Met Engineering, Inc., La Habra, California.
10. Hornbaker, D. R. and Rall, D., "The Convective Null-Heat-Balance Concept For Non-Contact Temperature Measurements Of Sheets, Rolls, Fibers And Wire." Fifth Symposium On Temperature—It's Measurement And Control In Science And Industry, June 21–29, 1971, Washington, D.C.

1.5 NEW DEVELOPMENTS IN FLOW MEASUREMENT

Flow Detector Types:

FLOWMETERS:
(a) Vortex shedding.
(b) Noise detector.
(c) Annubar.
(d) Solids transmitter.
(e) Insert type turbine.
(f) Vibrating mass flow sensor.
(g) Pump with zero ΔP.
(h) Differential conductivity.

FLOW SWITCHES:
(i) Paddle.
(j) Valve body.
(k) By-pass.

Note: In the feature summary below, the letters (a) to (k) refer to the listed sensor types.

Line Size: ½ inch to 30 inch (i), ¾ inch to 6 inch (j), ½ inch (k), 1 inch and up (a).

Capacity: 0.05 GPH to 100 GPM (g), 0.2 to 15 GPM (k), 0.6 to 112 GPM (j), 1.2 to 1,300 GPM (i), 20 to 200,000 GPM (c), 0.2 to 500 ft per sec (e), 50 to 50,000 lbs per min (d), unlimited (b,f,h).

Cost: $40 and up (i), $60 and up (k), $120 and up (j), $200 to $400 (b), $50 to $500 (c), $500 to $2,000 (e), $1,000 to $2,500 (a,g), $2,000 (h), $2,500 (d,f).

Partial List of Suppliers: Astro Dynamics Inc. (g), Beckman Inc. (h), DeLaval Turbine Co. (i,k), Dieterick Standard Corp. (c), Eastech Inc. (a), Flow Technology Inc. (e),

R. B. Jacobs Assoc. (f), Magnetrol Inc. (i,j), Power Engineering and Equipment Co. (i), Scarpa Laboratories Inc. (b).

Other suppliers are listed in Volume I.

Section 5.20 in Volume I discusses the characteristics of an "ideal" flowmeter. Such a flowmeter has not yet been developed but in recent years several noteworthy and useful innovations in flow measuring methods and devices have appeared that should be brought to the attention of instrument engineers.

Some of these meters employ unique new measuring principles, whereas others incorporate significant improvements on principles known for some time. Still other meters measure flow-rate indirectly. Lastly, flow switches have gained general acceptance during the past few years, of which various types will be described. In Chapter V of Volume I a typographical error should be corrected by also including ρ (the density) under the square root sign in equation 5.4(1).

Vortex Shedding Flowmeter

This flowmeter uses a well-known but little understood hydraulic phenomenon to obtain information that is proportional to flow-rate. Namely, when a non-streamlined (bluff) obstruction is placed in a pipe, fluid does not flow past it smoothly. Rather, eddies (vortices) are formed at the surface of the obstruction, which grow larger and are eventually detached from the obstruction. The process of detachment is called shedding and occurs alternately at each side of the obstruction. The vortices form trails downstream of the obstruction as shown in Figure 1.5a called "Kármán vortex trails" after Dr. Theodor von Kármán who first described the phenomenon in the literature.

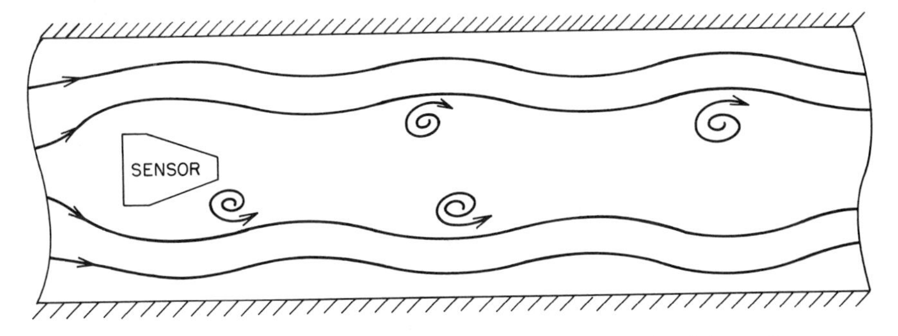

Fig. 1.5a Idealized Kármán vortex trail behind a non-streamlined obstruction

The shedding frequency of the vortices is directly proportional to flow-rate, and the instrument that employs this phenomenon to measure flow is called a "vortex shedding flowmeter."

In the version currently on the market the obstruction is "delta" shaped. This obstruction causes vortex shedding so intensive that the flow pattern upstream of the delta is affected. That is, the upstream face of the obstruction is struck by flow streams alternately from two directions. This cyclical pattern is a function of the vortex shedding frequency, and is thus proportional to flow-rate. In the commercially available flowmeter two self-heating thermistors on the upstream face of the obstruction are alternately cooled by the cyclically changing direction of the flow path (Figure 1.5b).

As the thermistors are cooled their electrical resistance changes, and this variation is detected by the electronics of the flowmeter. The frequency of changes in resistance is proportional to flow-rate. The electronics are very similar to those in resistance thermometry and in pulse counting techniques. An available AC or DC power source supplies power to the constant current source (which heats the thermistors) and to the amplifiers. A filtering network reduces extraneous noises, and a trigger converts the cyclical signal into an easily detected and recorded square wave pulse train.

It is interesting to note that the vortex shedding flowmeter incorporates features from two conventional flow instruments. The obstruction is analogous to an orifice plate, and the output signal is similar to that of a turbine-type flow transducer. The vortex shedding flowmeter senses flow velocity and is, therefore, a volumetric measuring device. The density of the fluid must be known to obtain mass flow. The vortex shedding flowmeter cannot be used in laminar flow; the Reynolds number must be at least 10,000. Also, it does not function properly above a fluid velocity of 20 ft per second in liquid and 100 ft per second in gaseous service. It has a wide range, universal calibration and no moving parts.

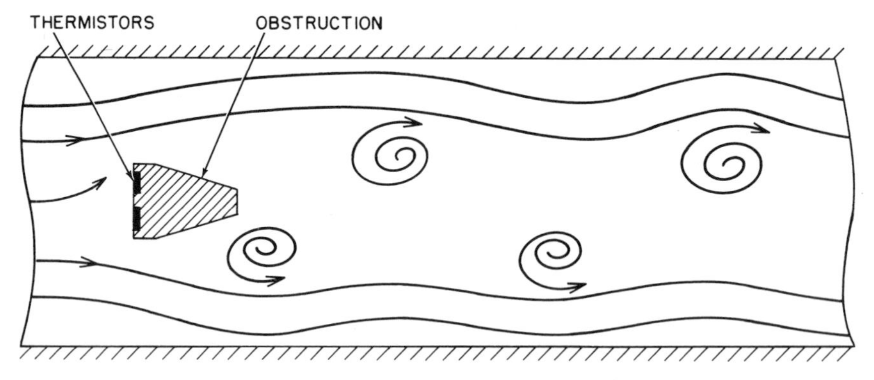

Fig. 1.5b Strong vortex shedding affecting direction of upstream flow

Noise Detection Meter

Another flow measuring device based on a never-before-used physical principle is the noise detection flowmeter. This instrument measures the ultrasonic noise developed by the turbulence and shear present in a fluid stream as it passes through a pipe. The instrument consists of two basic components: a sensor (or transducer) and a read-out device. The sensor is clamped to the outside of the pipe and connected via shielded coaxial cable to the read-out device. The cable length between the two components cannot be longer than 200 ft, and the read-out device is a switch or an indicator, or both. The noise detection meter is used most frequently as a flow switch. When employed as a flow measuring device, the user must perform his own calibration. The noise detection meter can be used on gaseous, liquid or solids flow service. The sensor can furnish either a current or an emf output signal.

Annubar Flow Element

According to Bernoulli's theorem the relationship that exists for any fluid flowing in a pipe is

$$(\text{fluid velocity})^2 = \text{velocity head}$$

A tube inserted into the pipe so that its entrance faces upstream will measure the fluid's pressure head plus velocity head. A second tube attached at the wall of the pipe and at a right angle to the fluid stream will measure only the fluid's pressure head. If these two tubes are connected, the differential pressure measured will be the fluid's velocity head. From this the fluid's velocity is determined, and by knowing the cross-sectional area of the pipe the fluid's flow-rate is calculated. This is the basic principle of the Pitot tube flowmeter, a flow measuring instrument that has been used in the laboratory and in industry for many years. It is a variable head meter, having the classic square-root relationship between flow and differential pressure. Its permanent pressure drop is very small.

Its principal disadvantage is that it is difficult to determine the location along the pipe's diameter where the tube should be inserted to measure both the fluid's pressure head and velocity head. This difficulty arises because the *average* velocity head must be measured, and only by lengthy experimentation and calculations can this representative average location be accurately obtained.

The annubar flow element operates on the Pitot tube principle just described. However, it incorporates a significant improvement. Namely, the tube measuring the fluid's pressure and velocity heads extends throughout the pipe and has four holes, located so as to measure the average velocity head. Figure 1.5c shows the construction of this improved Pitot tube type of flow element. Note that the tube measuring the fluid's pressure head (static pressure) is incorporated into the flow element structure and faces downstream.

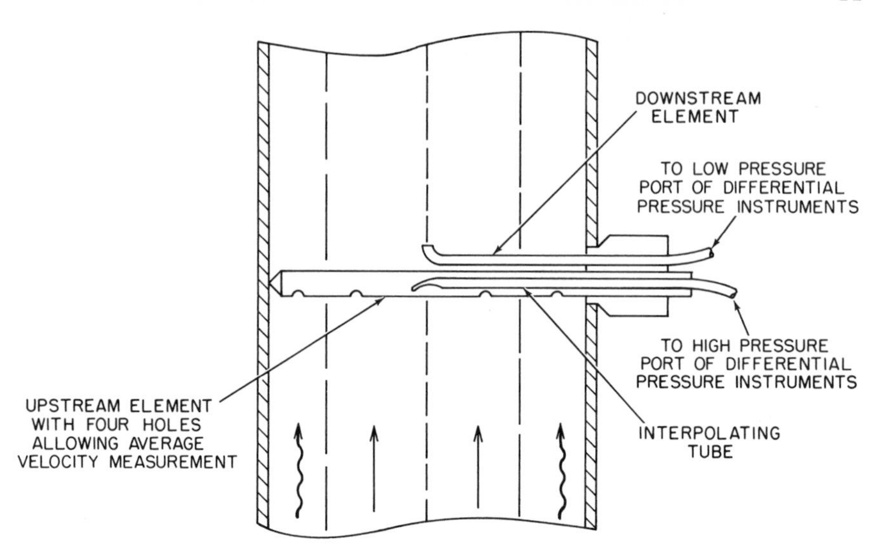

Fig. 1.5c Annubar flow element inserted in pipe

Any conventional differential pressure measuring instrument such as a manometer or differential pressure indicator or transmitter, or both, can be used with the annubar flow element.

Solids Flow Transmitter

The principles of impulse and momentum have long been used in liquid flowmeters such as the target or drag body flowmeters and the angular momentum flowmeter. They are based on Newton's second law of motion and the concept of conservation of momentum. These principles have now been successfully applied to solids flow measurement. The solid particles are allowed to fall by gravity on a calibrated spring loaded resistance, the displacement of which caused by the force of the falling particles is a function of the mass flow-rate of the solids. It is measured with a position transducer or transmitter. Figure 1.5d shows this type of instrument schematically.

These solids flow transmitters have a variety of applications such as continuous weighing, flow control or recording, or both, batching and alarming. Almost all conceivable types of solids can be measured, such as sugar, salts, cement, ores and the like.

Insert Type Turbine Meter

The turbine flowmeter is a very accurate and precise flow measuring device. One of its disadvantages, however, is that it is basically a low-flow (or small-pipe-diameter) instrument. As pipe size increases, the turbine flowmeter's weight increases drastically and its price becomes prohibitive. A recent

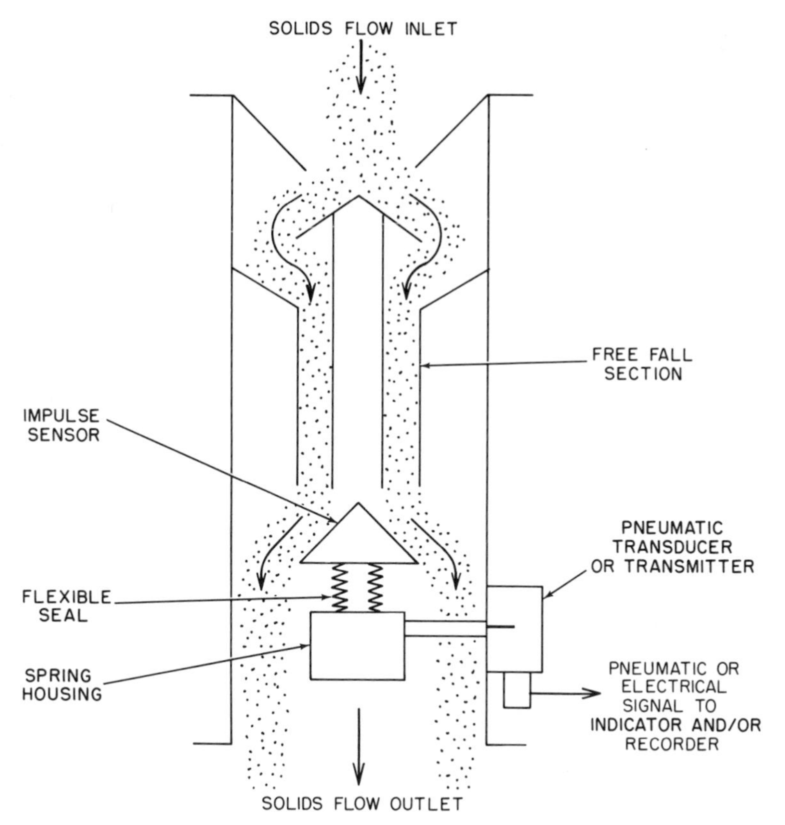

Fig. 1.5d Solids flow transmitter

innovation has reduced the problem considerably. Namely, a small turbine-type flow transducer is installed in the center of the large diameter pipe in which the fluid flow-rate is to be measured. A typical installation is shown in Figure 1.5e. The turbine flowmeter used in this manner measures fluid velocity, not volumetric flow-rate. As with the old-style Pitot tube, it is difficult to determine where along the pipe diameter the average or representative fluid velocity occurs. If installed at the center of the pipe's cross-sectional area the flowmeter will probably measure the fluid's maximum velocity. Owing to this difficulty, this type of flow measurement cannot be considered to have absolute accuracy. However, it has several desirable features such as repeatability, linearity and comparatively low costs for large-pipe-diameter service. It can of course be made very accurate by performing an in-place calibration. It is also possible to withdraw the unit from the pipe for maintenance, without exposing the process to the atmosphere.

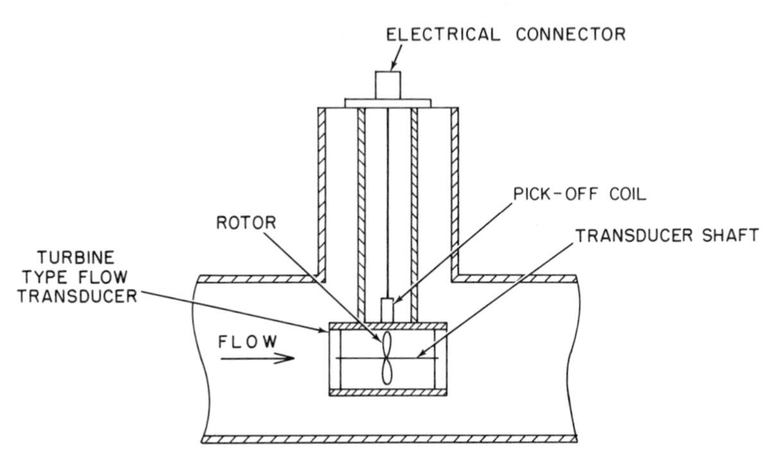

Fig. 1.5e Insert turbine type flow transducer

Indirect Measurements

Several flowmeters have recently become available that measure flow-rate by temporarily changing a physical characteristic of the fluid, or by injecting another fluid into the stream whose flow-rate is to be measured. For lack of a better term these devices or methods are here classified as indirect flow measurements.

Vibrating Mass Flowmeter

This recently developed flowmeter measures the stagnation pressure of a vibrating fluid. It can be shown mathematically that the value of a component of this pressure is directly proportional to mass flow-rate. A small pressure transducer is submerged into the pipeline, and a vibrator oscillates the transducer parallel to the direction of flow at a constant frequency and amplitude. The transducer thus measures the stagnation pressure of the vibrating fluid and supplies a signal which is a function of mass flow-rate.

Metering Pumps with Zero Pressure Drop

Positive displacement pumps are fairly accurate flow measuring devices. The volumetric displacement per stroke can be accurately determined and the number of strokes per unit time can easily be counted with any conventional counter. The source of inaccuracy in a positive displacement pump is its slippage which is not constant for any given pump, but is a function of fluid density, viscosity and the speed at which the stroking or displacement occurs. The slippage is reduced to a minimum if the pump can be operated so as to develop no pressure head. For any given combination of fluid density and viscosity a pump speed exists at which the pressure drop across the pump (or developed head) is practically zero. Therefore, if the positive displacement pump can be operated at zero pressure drop, its RPM will be directly proportional to volumetric flow-rate because slippage has been eliminated.

In a recently developed flow measuring system this is accomplished by mounting an accurate differential pressure transducer across the inlet and outlet ports of a positive displacement pump which has a variable speed motor, the speed of which is controlled by an amplifier and control circuit. The latter receives an error signal from the differential pressure transducer whenever the pump starts to develop a differential pressure. Thus, the pump always operates at a speed at which it has no pressure drop and thus, essentially no slippage. The pump speed is measured as a function of volumetric flow-rate.

This instrument is useful in test-stands and laboratories. It is economical only at comparatively low flow-rates because of the need to use a pump. However, it can measure flow-rates accurately in a pulsating piping system where other flowmeters might be inaccurate and unreliable.

Differential Conductivity Flow Measurement

The recent development of very accurate conductivity measuring devices has brought about the "differential conductivity flow measurement technique." In this method the conductivity of the fluid whose flow-rate is to be established is measured simultaneously at two points in the piping system. At a location between these two points a fluid of high conductivity is injected into the pipe. This fluid, usually a sodium chloride solution (brine), is injected at a known concentration and flow-rate. The conductivity cell downstream of the injection point measures a conductivity that is considerably higher than that measured upstream of the injection point. This difference in conductivities is a function of flow-rate. The higher the flow-rate the lower this difference, because brine is added at a fixed rate. It is not necessary to record the conductivity at each point. Instead, the two conductivity measuring instruments are connected in series and the difference of the "bucked" signal is a measure of the change in conductivity. This in turn is directly proportional to flow-rate (Figure 1.5f).

The differential conductivity flow measurement method is particularly applicable when large flow-rates at little or no pressure drop must be measured, such as in open channel flow measurements—common in water and sewage treatment plants.

Flow Switches

Frequently, it is important to generate a signal when flow has stopped or started, or when a certain flow-rate has been reached. Typical examples of such applications are in the lubricating and cooling systems of turbines, pumps and other expensive rotating machinery. If the lubricant or cooling fluid, or both, to such components ceases to flow or if its flow-rate is reduced beyond a safe point, the operator must be informed immediately by an alarm signal, or the rotating machinery must be stopped through an automatic

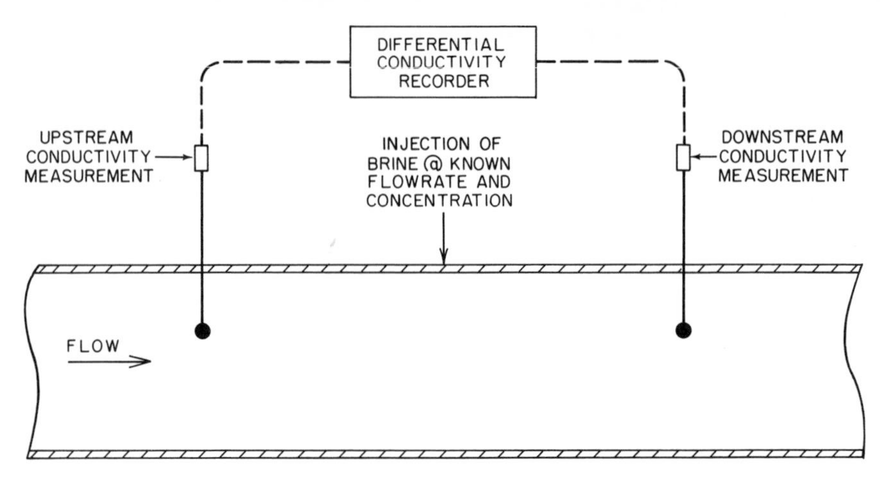

Fig. 1.5f Differential conductivity flow measurement

shutdown or tripping circuit. Signals of this type can be obtained from flow indicators, transmitters and recorders. For example, a differential pressure indicator installed across a variable head meter such as an orifice or venturi-tube can have a switch that opens or closes when its indicating pointer moves to show either no or less than minimum required flow. A recorder can have a similar arrangement. Another way of obtaining a switch closure is to install a pressure switch in the pneumatic output signal of a differential pressure transmitter. The output signal is proportional to flow-rate, and the pressure switch can be set to open or close a circuit at the appropriate output pressure.

Frequently, however, a signal as just described is required in a system in which a flow measuring device cannot be installed, either for economic reasons or because it is not necessary to know the actual flow-rate. For these applications a large variety of "flow switches" is available. These devices are installed directly into the piping system. They neither indicate flow-rate, nor provide a signal that is proportional to it. Instead, as the name implies, they are switches that open or close an electrical circuit when flow has become larger or smaller than a predetermined value.

Paddle-Type Switches

One of the most common flow switches is the paddle-type. Paddles are inserted into the pipe as shown in Figure 1.5g. At "no flow" the paddle will hang loosely, vertical to the pipe in which it is installed. As flow is initiated the paddle begins to swing upward in the direction of the flow stream. This deflection of the paddle is translated into mechanical motion by a variety of techniques including a pivoting cam, a flexure tube or a bellows assembly. The mechanical motion causes the switch to be opened or closed. If a mercury switch is used the mechanical motion drives a magnetic sleeve into the field of a permanent magnet which trips the switch. A hermetically sealed switch will be directly actuated by the permanent magnet as it moves up or down according to the paddle movement. If a micro-switch is used the translated motion will directly cause switch actuation.

The range and actuation point of paddle switches can be changed and adjusted by changing the length of the paddle. For any given pipe size the flow-rate at which switch actuation occurs decreases as the paddle length increases. Some paddle switches can also sense a reversal in flow direction.

Valve Body Type Switches

These flow switches are built into a pipe fitting that resembles the body of a single seated globe valve. A flow disk is allowed to move in a vertical direction within what is normally considered the valve seat. A magnetic sleeve is mounted above the flow disk as shown in Figure 1.5h. As the disk is lifted upward due to initiation of flow, a mercury switch is actuated by the movement of the magnetic sleeve into the field of the externally mounted permanent magnet.

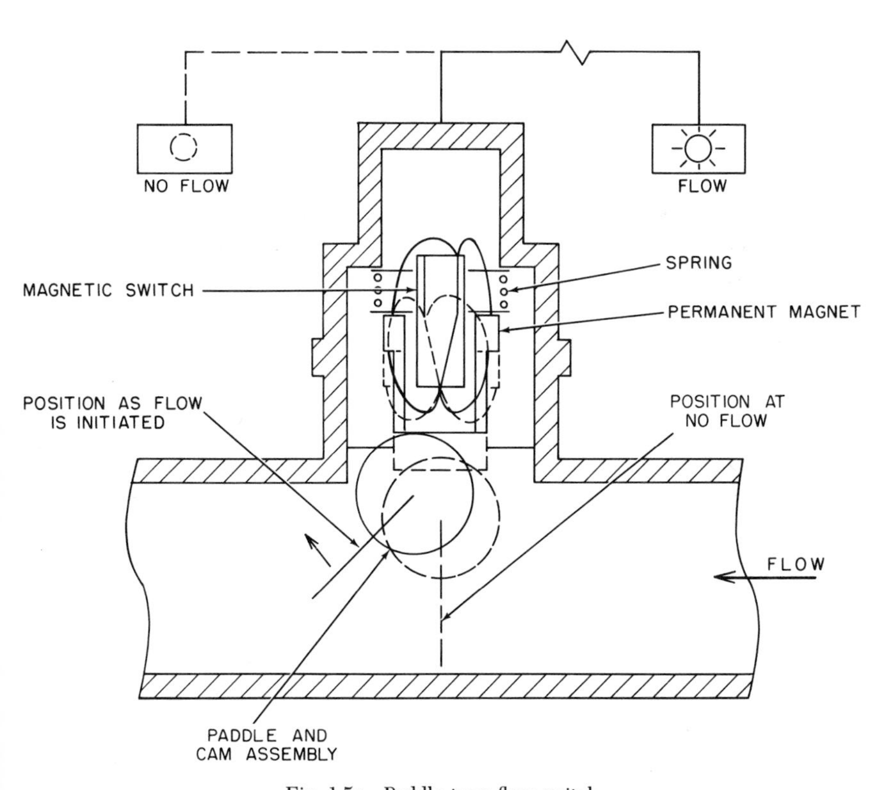

Fig. 1.5g Paddle type flow switch

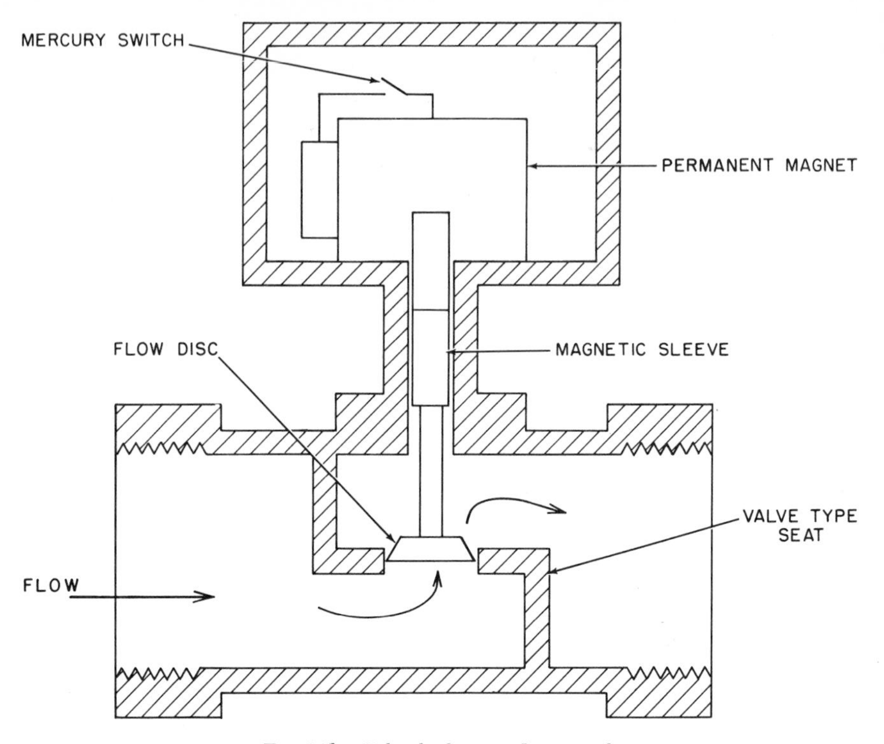

Fig. 1.5h Valve body type flow switch

By-pass Type Flow Switches

A by-pass type flow switch (Figure 1.5i) has an externally adjustable vane that creates a differential pressure in the flow stream. This differential pressure forces a proportional flow through the tubing that by-passes the vane. A piston retained by a spring is in this by-pass tubing and will move laterally as flow increases or decreases; this movement actuates a switch. This type of flow switch can be used for fairly low flow-rates and its ability to be externally adjusted is a very desirable feature.

Conclusions

One of the most difficult tasks for the instrument engineer is to decide what type of flowmeter should be used for any given application. Section 5.21 of Volume I offers some useful guidelines to help in the selection of the proper flowmeter. The recent developments in flow measurement technology have not changed the validity of these guidelines. The only change is that a more diverse selection of flowmeters is available.

Thus, for example, a Pitot tube would not have been considered for many applications because of its inaccuracy and fragility. The advent of the annubar

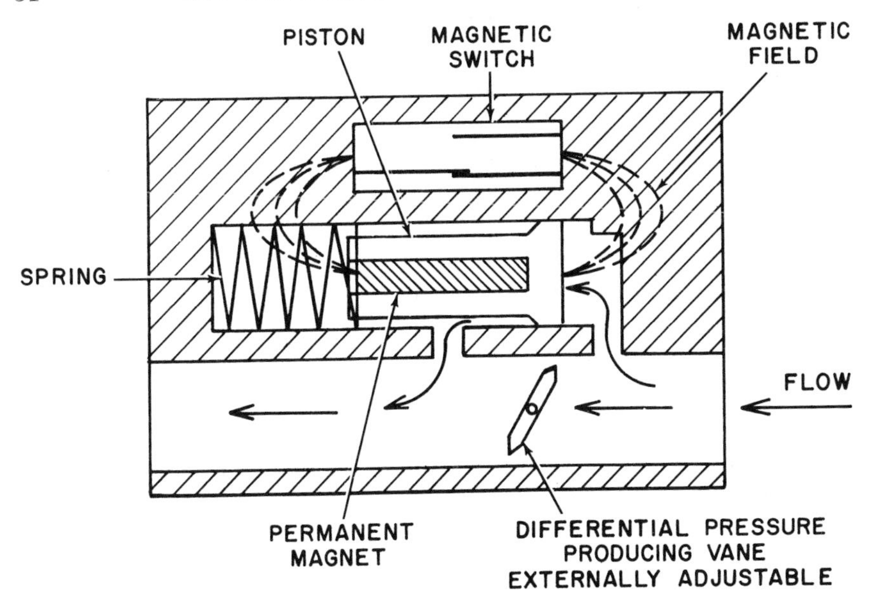

Fig. 1.5i By-pass type flow switch

flow element should change this attitude. A highly repeatable and digital output signal is now available from the vortex shedding meter without the moving part of a turbine type flow transducer. Also, a turbine type flowmeter is now economically available for large pipe sizes.

Finally, the instrument engineer should ask himself whether he really needs a "flow-rate" measuring device, or whether a flow switch will be more efficient and economical.

1.6 NEW DEVELOPMENTS IN VISCOSITY MEASUREMENT

Type of Viscometer:

(a) Torsional vibration.
(b) Automatic falling ball.
(c) Angular pulse-displacement.
(d) Magnetically coupled rotational.
(e) Falling slug.

Note: In the feature summary below, the letters (a) to (e) refer to the listed viscometers.

Operating Temperature and Pressure:

	Temperature, °F	*Pressure, PSIG*
(a)	−100 to 300	Atmospheric
(b)	Up to 350	Up to 15,000
(c)	Up to 600	Up to 100
(d)	Up to 570	Up to 1,600
(e)	Up to 300	Up to 200

Materials of Construction:

Wide selection of corrosion resistant materials and coatings.

Cost:

(a) \$1,500, (b) \$3,600, (c) \$1,500, (d) \$1,500 to \$3,600, (e) \$3,000 to \$5,000.

Precision:

	Accuracy	*Repeatability*
(a)	±1%	±0.1%
(b)	±0.1% to ±1%	±0.1%
(c)	±1%	±0.5%
(d)	±0.25%	±0.1%
(e)	±1%	±1%

Range:

(a) 0.1 to 1,000,000 centipoises.
(b) 0.1 to 200,000 centipoises.
(c) 0 to 50,000 centipoises.
(d) 1 to 4,000,000 centipoises.
(e) 10 to 1,000,000 centipoises.

Partial List of Suppliers:

Coleman Tulsa Co. (b), Datacon Corp. (c), Gam Rad, Inc. (e), Nametre Co.

(a), Olkon Corp. Contraves Viscometer Div. (d), Ruska Instrument Corp. (b)

The definition of viscosity, techniques available for measuring it and the application of viscosity measurements in various industries have been discussed in Chapter VI of Volume I.

This section will deal only with new developments in viscosity measurement and control. The two categories of new developments are 1.) automation to increase reliability, reduce labor involved in manual operation and convert laboratory types to industrial viscometers (especially falling ball type); and 2.) application of space age electronics and their spin-off technology to improve performance of in-line viscometers (angular pulse-displacement type.)

New Developments in Laboratory Viscometers

Torsional Vibration Viscometer

Torsional vibration viscometers are damped proportionally to the fluid friction and measure the viscosity of the fluid by the time of decay of the vibrations.

The amplitude of torsional vibration of a spherical, stainless steel tip immersed in a liquid is maintained constant, and the power required to do it is measured. The readings are converted to a viscosity-density product by chart. The chart reading is then divided by liquid density to obtain actual viscosity.

The spherical tip (Figure 1.6a) is firmly connected to the instrument housing by a cylindrical sheath. The center rod, surrounded by this sheath for the greater part of its length, is maintained in torsional vibration by an energized coil on a small magnet at one end of the cross-arm. The rod and sheath, joined at their lower ends by the spherical tip, are in torsional vibration about their concentric axis. Rod vibrations are controlled by the elasticity of the sheath. The power required to overcome fluid friction and to maintain a given amplitude of vibration is a measure of liquid viscosity. Amplitude is measured by a stationary coil at the other end of the cross-arm. The subsequent electrical signal is amplified and displaced on a microammeter. For each application there is a resonant frequency at which maximum vibrational amplitude is obtained for a given power input. Measurements are made at this resonant frequency, and readings to 1 part in 10,000 are possible.

One tip is used throughout the entire range of 0.1 to 1,000,000 centipoises. The viscometer is calibrated with standard liquids[1] of known viscosity and density and is easy to clean and to operate. The unit is suitable for both laboratory and on-line industrial use.

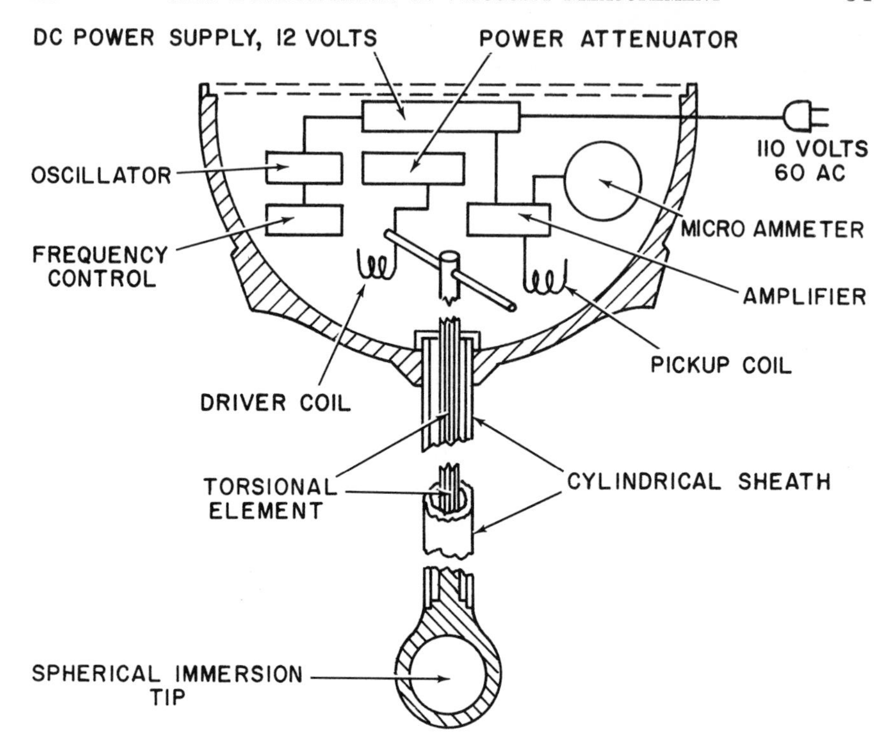

Fig. 1.6a Working components of torsional vibration viscometer

Automatic Falling Ball Viscometer

These viscometers are provided with automatic controls and timers to measure viscosities of 0.1 centipoises or higher, with a repeatability of 0.1%. Approximately 70cc of sample is required to charge the viscometer, which can operate at pressures as high as 15,000 PSIG and temperatures as high as 350°F.

The viscometer consists of two units: the viscosity test assembly (equipped with a precision bore tube, heating and temperature control components and the ball position detector), and the auxiliary control unit. The auxiliary control unit contains the temperature controller, electrical timer with reset and the electrical circuits. The viscosity test chamber, mounted on a trunnion, can be rotated to various angular positions. The electrical timer is activated by the ball release signal and when the ball closes the contact at the lower end of the tube, the timer stops. The timer is accurate to 0.001 minute. The fall time of the ball is proportional to the sample viscosity, and the readings are interpreted by calibration curves.

The automatic falling ball viscometer is simple and safe to operate and can measure fluid viscosity under simulated process conditions. High accuracy

(±0.1%) is obtained by judicious selection of the proper calibrated ball size and by adjusting the angular position of the precision bore tube.

New Developments in On-Line Viscometers

Angular Pulse-Displacement Viscometer

This unit is designed to operate continuously in the industrial environment and is equally applicable to Newtonian and non-Newtonian fluids. Viscosity is detected by rotating a drag member in the process fluid. The fluid shear force on the drag member is converted into a proportional output signal which is suitable for display or control, or both.

The unit can operate in any range from 0 to 10 centipoises to 0 to 50,000 centipoises with an accuracy of ±1% of full scale and with a repeatability of ±0.5% of full scale. Response time is less than 5 seconds for full scale change. The probe may be installed on pressurized or open vessels or in process pipe lines, and may be mounted at any angle up to 80° from the vertical. The probe is designed to operate at temperatures as high as 600°F (special transducers for up to 3,000°F). The electronic section should be purged with dry gases to maintain a slightly positive pressure on the transducer, which prevents liquid or vapor penetration. For in-line installation the process liquid flow should be constant, non-pulsating and laminar.

Due to its simple design, ease of cleaning and non-clogging feature this device has been installed in the drug, cosmetics, food, pulp and fiber, chemical, petroleum and paint industries.

Magnetically Coupled Rotational Viscometer

A recent development in the design of the rotating cone viscometer[1] is the introduction of a magnetic coupling between the electronic detector at atmospheric pressure and the rotating sensor, which is exposed to the process pressure (Figure 1.6b). With this separation between atmospheric and pressurized areas, no purging is required to keep the liquid away from the measuring instrument, and operating conditions at the levels of 2,850 PSIG per 20°C or 1,620 PSIG per 300°C can be tolerated. The magnetically coupled viscometer should not be used to measure fluids containing fiber, ferrite or abrasive materials because of interference with the operation of the magnetic coupling and of the stainless steel-sapphire bearings.

Falling Slug Viscometer

This instrument automatically measures the time required for a cylindrical slug of a specific density to fall a given distance in a vertical tube filled with the process liquid at a known, constant temperature.

The sample pump introduces the fresh sample and purges the system of the previous sample (Figure 1.6c). Two separate thermostats in the tempera-

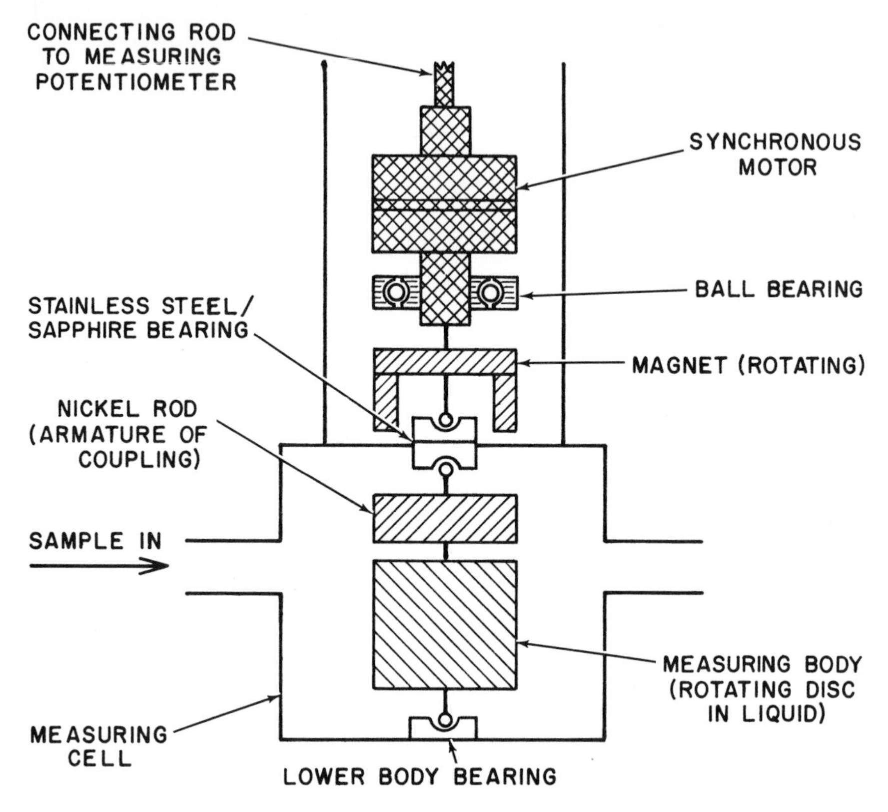

Fig. 1.6b Magnetically coupled rotational industrial viscometer

ture well control the purge and recirculation cycle by activating a three-way valve. The flow velocity raises the slug to the top of the fall tube, and when the sample temperature is reached, the pump and the sample stop, thereby permitting the slug to fall. As it does, it actuates a magnetic switch attached to the side of the fall tube, which in turn starts the recorder motor. The slug then actuates a second magnetic switch located at an adjustable distance (1 to 20 inches) below the first. This switch stops the recorder motor. The resultant time measurement is directly proportional to the viscosity of the sample. Actuation of the lower switch also initiates the system to the purging phase.

The total viscosity range is 10 to 1,000,000 centipoises. Specific ranges for a given process are field-selectable by adjusting the distance between the two magnetic switches on the fall tube. Also, recorder pen drive gears provide full scale indication from 10 to 250 seconds in 5 different steps. The full scale accuracy is about ±1%, and the reproducibility is about ±1% full scale,

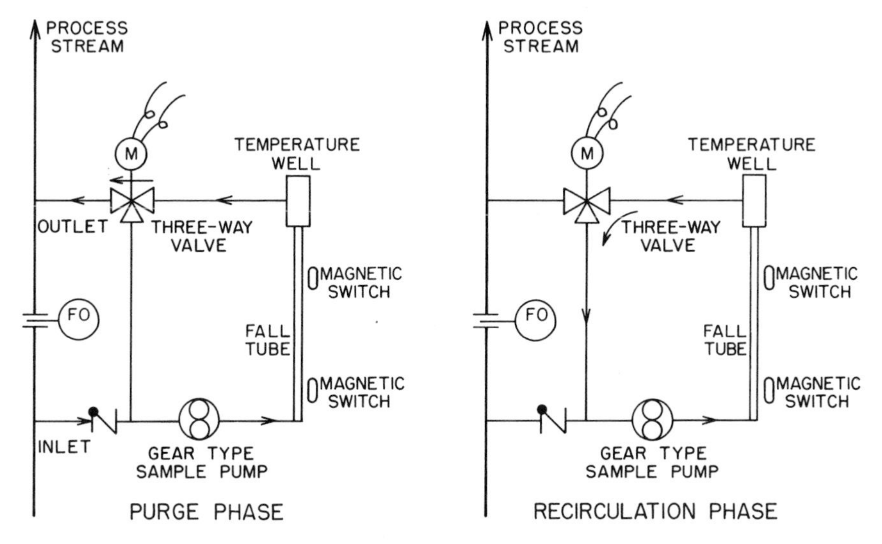

Fig. 1.6c Measuring cycle of falling-slug viscometer

depending on the accuracies of the thermostat and the recorder. The viscometer is designed to operate at temperatures up to 300°F and pressures up to 200 PSIG, and it is applicable to those continuous installations for which the three-minute maximum cycle frequency is sufficient. Depending on the fluid characteristics, the error from small variations in the process fluid temperature can be substantial. The falling slug viscometer is recommended only for clean streams that are not shear sensitive.

Conclusions

Viscosity measurements determine several fluid characteristics and properties. One of these is the behavior of plastic materials.[1] The Mooney viscosity value at a fixed shear rate is used in quality control, based on an assumption that the non-Newtonian behavior is the same for a given family of elastomers. In Figure 1.6d, curves 1 and 2 illustrate this assumption. Such, however, is not always the case and very different non-Newtonian characteristics were observed (for the same Mooney reading as elastomer 1) shown in curve 3. Therefore, it is necessary to distinguish between the various elastomer samples.

It is known that for non-Newtonian materials the difference in viscosity readings increases with decreasing shear rate and reaches its maximum in the Newtonian range. Thus, the Newtonian region viscosity or zero-shear viscosity, or both (extremely low rates of shear are attainable with the sliding-plate viscometer), characterizes elastomer processibility.[2]

The reader should keep in mind that this section contains only the updated

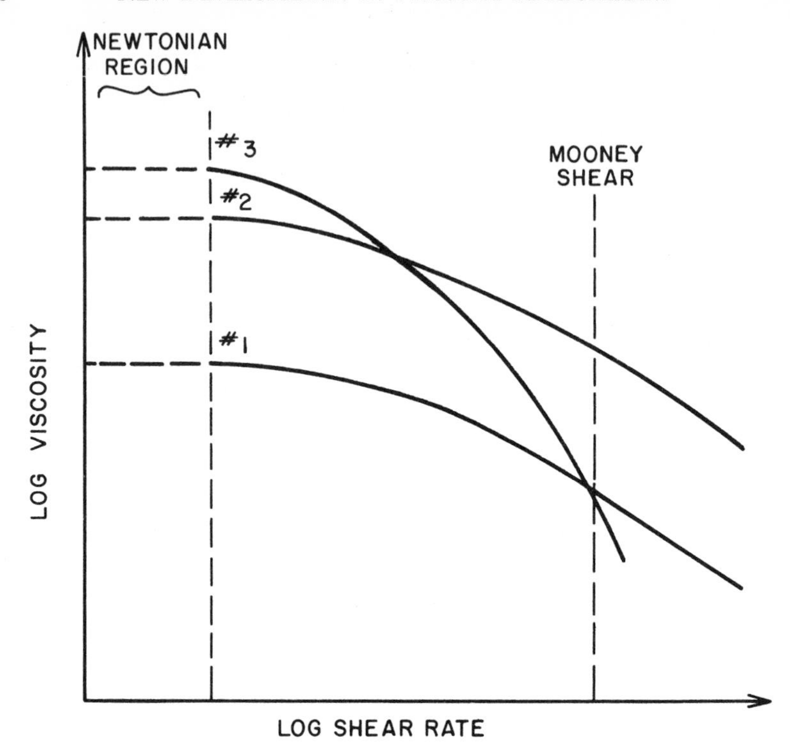

Fig. 1.6d Viscosity-shear rate relationship for three non-Newtonian elastomers

parts of topics covered in Volume I. The orientation table and selection guides provided in Volume I are also applicable to the viscometers discussed in this section.

REFERENCES

1. Lipták, Béla G., editor, *Instrument Engineers' Handbook, Volume I,* Philadelphia, Chilton Book Company, 1969.
2. Tokita, N., White, J., *J. App. Pol. Sci. 11* pp. 321, 1967.

1.7 NEW DEVELOPMENTS IN WEIGHT MEASUREMENT

For the first time in many years the weighing industry finds itself in a state of rapid and dynamic technological evolution. Influenced by the impressive advances made in computer technology and integrated circuitry complemented by the previous availability of electromechanical and electronic force sensing devices, the changes during the past several years have been dramatic. The purpose of this section is to review these advances and thereby update chapter VII in Volume I of the Instrument Engineers' Handbook.

New Sensing Techniques

For a more complete understanding and appreciation of the advantages and disadvantages of new weighing transducers or devices, it is advisable to be familiar with the basic sensing principles on which they are based.

Semiconductor Strain Gauge

The piezoresistive characteristics of germanium and silicon semiconductor materials were discovered by scientists at Bell Laboratories in the mid-1950's. It was discovered that the terminal resistance of these devices is highly sensitive to applied stress or strain. In fact, their gauge factors (unit change in resistance divided by unit strain) are more than fifty times that of their metallic wire or foil strain gauge counterparts.

While possessing very high strain sensitivity relative to that of metallic strain gauges, they also exhibit substantial non-linearity, and temperature effects on strain sensitivity and terminal resistances are also relatively high. The latter characteristics have limited their application. Nevertheless, semiconductor strain gauges are used in force measuring devices in which high output signal level and low system cost are the primary objectives.

Nuclear Radiation Sensor

This form of weight sensing is generally applied to in-motion weighing of bulk materials. It utilizes a radioactive source of gamma rays which are imposed on or directed through a certain section of the moving material. The material absorbs some of the gamma rays and allows others to pass through. The amount of radiation transmitted through the bulk material depends on the amount of material on the conveyor. A radiation sensor converts the transmitted radiation to an electronic signal which bears a known relationship to the amount of material on the weighing section of the conveyor.

The nuclear radiation form of weight sensing is applicable when the weight sensor should not contact the material or the conveying devices. Certain shortcomings of conventional belt scales can be avoided with this technique.

Inductive Sensing Technique

Inductive weight sensors use the change in inductance of a solenoid coil with changing position of an iron core. Two forms of the inductive sensing principle are illustrated in Figure 1.7a. In configuration A, motion of the iron core to the right increases the inductance of coil B and decreases the inductance of coil A. Arranging the two coils in a Wheatstone bridge with resistors completing the bridge network provides a means for developing a voltage signal proportional to the core position.

Configuration B utilizes three solenoid coils. Coils C and D are wound in opposite directions and surround an iron core, whereas coil E is placed between the two coils and is excited by an external AC voltage source. When the iron core is centrally located, voltages induced into the secondary coils (C and D) are equal and opposite, and no voltage appears across the output

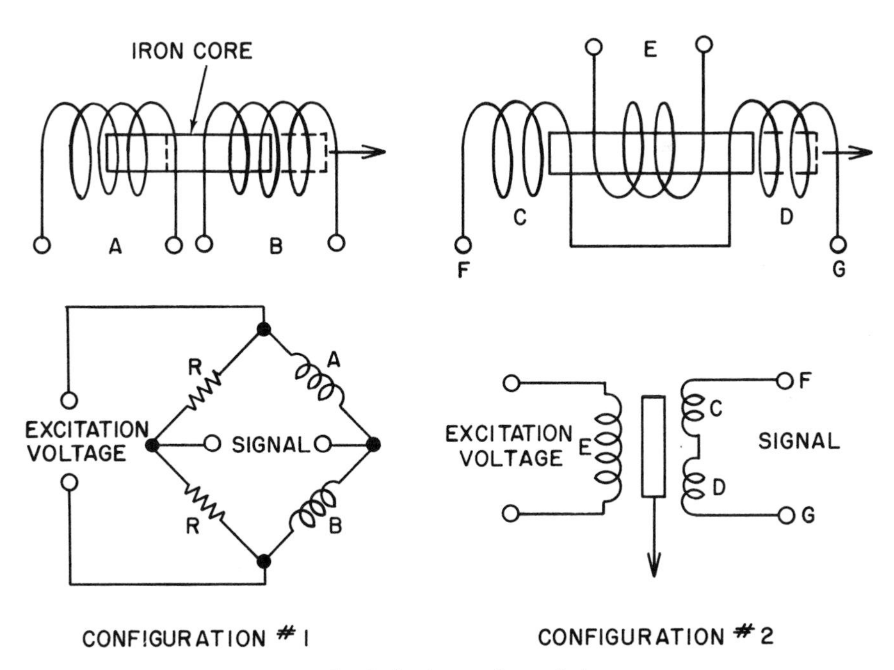

Fig. 1.7a Inductive sensing techniques

terminals (F and G). If the iron core is moved to the right, the voltage coupled into coil D is greater than that coupled into coil C and hence a voltage is developed at the output terminals. If the core were moved in the opposite direction by the same amount, a similar voltage of opposite phase would be developed. Other embodiments of inductive sensors are in current use. Those discussed here were for illustrative purposes only.

Inductive sensors furnish relatively high output signal levels and efficient null stability. Since their inertial masses are greater than strain gauge sensors, they are more subject to vibration.

Variable Reluctance Sensing Technique

Similar to the inductive sensing principle except that the inductance of one or more coils is usually changed by altering the reluctance of a very small air gap, this technique is illustrated in Figure 1.7b. Solenoid coils A and B are mounted on a structure of ferromagnetic material, and a U-shaped armature completes the magnetic circuit through air gaps 1, 2 and 3. Motion of the coil assembly to the right decreases air gap (2) while air gap (1) is increased. Air gap (3) remains constant during the translation of the coil assembly. As

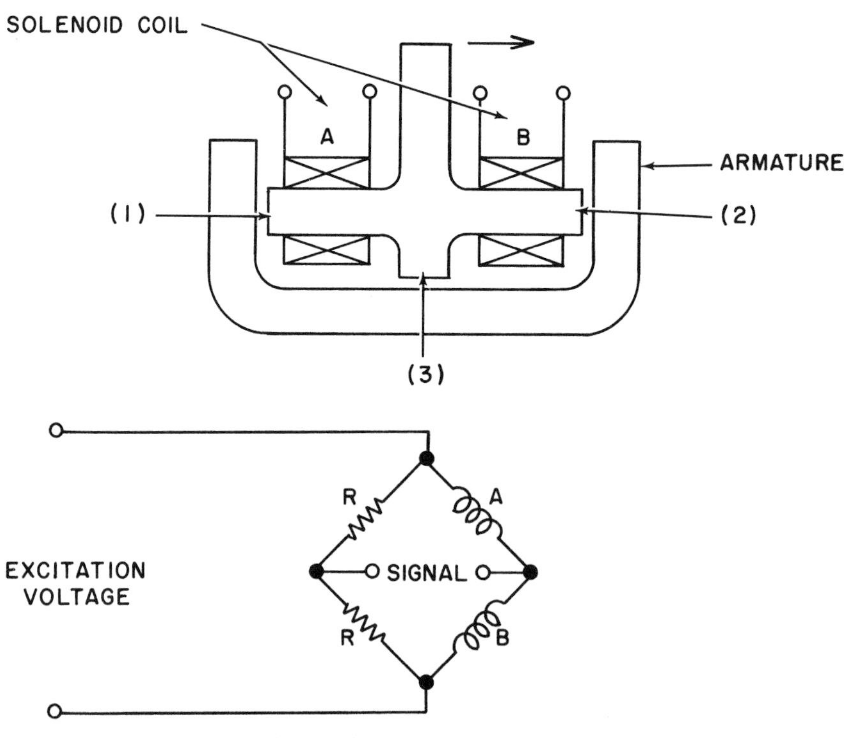

Fig. 1.7b Variable reluctance sensing technique

a result of horizontal translation, the inductance of coil B increases while that of coil A decreases. Incorporating the two coils in a Wheatstone bridge similar to that utilized in the inductive sensing principle permits development of a voltage proportional to the translation of the coil assembly.

The reluctive sensing principle also offers a relatively high output voltage and efficient null stability with the higher vibration sensitivity due to the relatively high inertial masses of the mechanical structure.

Magnetostrictive Sensing Technique

Based on the Villari effect, this technique utilizes the change in permeability of ferromagnetic materials with applied stress. A stack of laminations forms a load-bearing column (Figure 1.7c), and primary and secondary transformer windings are wound on the column through holes oriented as shown. Coil A is excited with an AC voltage and coil B provides the signal voltage. In the unstressed condition the permeability of the material is uniform throughout the structure and since the coils are oriented at 90° with respect to each other, little or no coupling exists between coil A and coil B. Hence, no output signal is developed. When the column is loaded, the induced stresses cause the permeability of the column to be non-uniform, resulting in corre-

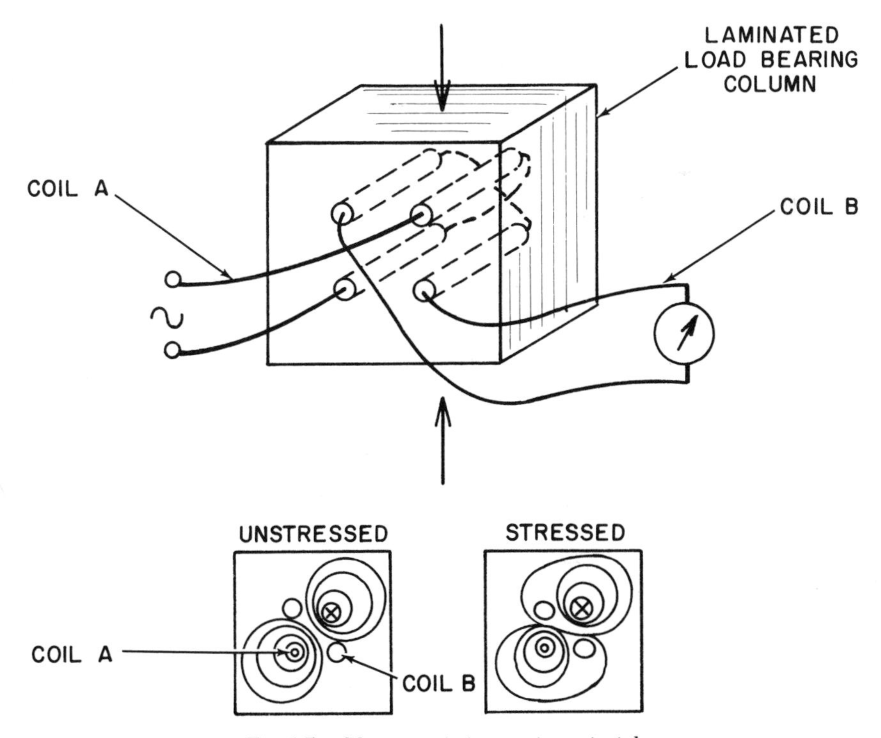

Fig. 1.7c Magnetostrictive sensing principle

sponding distortions in the flux pattern within the magnetic material. Magnetic coupling now exists between the two coils and a voltage is induced in the signal coil, providing an output signal proportional to the applied load.

The magnetostrictive principle produces relatively high output signal levels and offers extreme ruggedness in load cells incorporating this sensing principle.

New Developments in Weight Transducers

Although many of the electromechanical force sensing transducers in current weighing systems have been available for some time, improvements have been made over the past several years facilitating their application to new and improved weighing systems. Some of these transducers as well as certain force transducers which were not treated in Volume I will be discussed in the following sections.

Semiconductor Strain Gauge Load Cells

(Bonded strain gauge load cells were discussed in Section 7.4, Volume I.) Metallic wire and foil strain gauge units have long been used in weighing systems, furnishing accuracies of better than 0.1% of reading. Within the past five years semiconductor strain gauge load cells have been used in those systems requiring low cost and moderate accuracies.

Semiconductor strain gauges in load cell configurations provided units with rated output capabilities of 1.0 volt at 15 volts bridge excitation. As a result of the high signal level, semiconductor units were used in simple weighing systems with simple regulated power supplies and direct meter readouts. Sometimes an amplifier is interposed between the transducer and the meter display.

Typical performance characteristics of semiconductor load cells are listed in Table 1.7d together with those of their metallic strain gauge counterparts.

The moderately high cost of semiconductor load cells and the dramatic cost reductions in linear integrated circuitry have limited the use of the semiconductor load cell in low cost weighing systems. In other words, the cost of linear amplification required to raise metallic strain gauge load cell signals to the levels offered by their semiconductor counterparts is now less than the additional cost for semiconductor load cells.

Linearization of Column Type Load Cells

Column type strain gauge load cells in capacities above 10,000 pounds heretofore have suffered from a characteristic non-linearity of about 0.15% of rated capacity. The inherent non-linearity of these devices results from electrical bridge non-linearities caused by the fact that all strain gauges are not subjected to equal strain. Additional non-linearity also results from the column area change with increasing load. The characteristic column type load

Table 1.7d
PERFORMANCE CHARACTERISTICS OF SEMICONDUCTOR AND STRAIN GAUGE TYPE LOAD CELLS

Performance Characteristic	*Semiconductor Load Cell*	*Metallic Strain Gauge Load Cell*
Output (at 15 v)	1.0 v	30 mv
Terminal Linearity	0.25%	0.05%
Hysteresis	0.02%	0.02%
Temperature Effect on Zero Balance	±0.25%/100°F	±0.15%/100°F
Temperature Effect on Output	±0.5%/100°F	±0.08%/100°F
Selling Price Index	1.3	1.0

cell non-linearity is parabolic and lends itself to almost perfect compensation by utilizing a semiconductor strain gauge compensating element. Figure 1.7e shows a semiconductor strain gauge incorporated in series with the excitation terminals of the load cell bridge circuitry. From the curve of output voltage versus applied load, an uncompensated column type load cell exhibits a drooping concave downward characteristic when loaded in compression. The linearizing strain gauge senses column strain induced by the applied compressive load and due to its piezoresistive characteristics, its terminal resistance decreases with increasing load. The decreasing resistance with load causes the net excitation voltage applied to the bridge circuitry to increase with increasing load (dotted line), which compensates for the drooping characteristic of the uncompensated load cell and results in improved linearity (interrupted line). Adjusting the terminal resistance of the linearizing strain gauge almost exactly compensates for the inherent parabolic drooping characteristic, and terminal linearities of better than 0.02% of rated load can be provided.

Linearities of this magnitude not only eliminate external linearization within the instrumentation, but also reduce errors in multiple load cell weighing systems in which unequal distribution of total load between the individual load cells may be substantial. Unequal loading on non-linear load cells can cause serious system errors, even in systems in which load cell non-linearity compensation is included in the display instrumentation.

Beam Type Strain Gauge Transducers

A new generation of force measuring transducers that will find wide application in the weighing field consists of a slotted bending beam construction (Figure 1.7f). Strain gauges at the locations shown are arranged in a Wheatstone bridge configuration so that the output of the sensing element

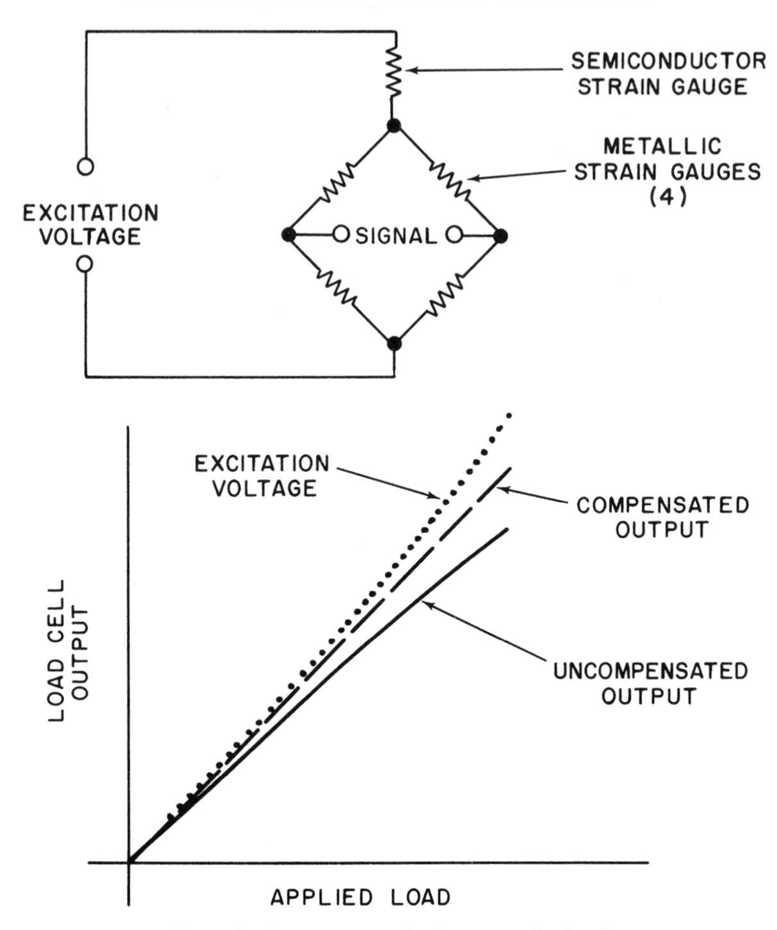

Fig. 1.7e Linearization of column type load cells

is independent of the position of the applied load. Electrical compensation of the bridge circuitry can reduce the load position sensitivity to virtually zero. As will be observed, this is a very important feature.

Protection from environmental effects is by a simple bellows arrangement (Figure 1.7f), thereby making unnecessary the conventional diaphragm and cylindrical casing. Strain gauge location and beam design provide inherent adverse loading sensitivity. This simplified load sensing configuration provides inherent linearity as well as very low creep and very high repeatability. Typical performance features are summarized in Table 1.7g.

Monorail Weighing Transducer

Another beam type weighing transducer is being used in monorail conveyor systems such as those in meat processing plants. Conventional monorail weighing systems support the "live" rail by load cell and flexure

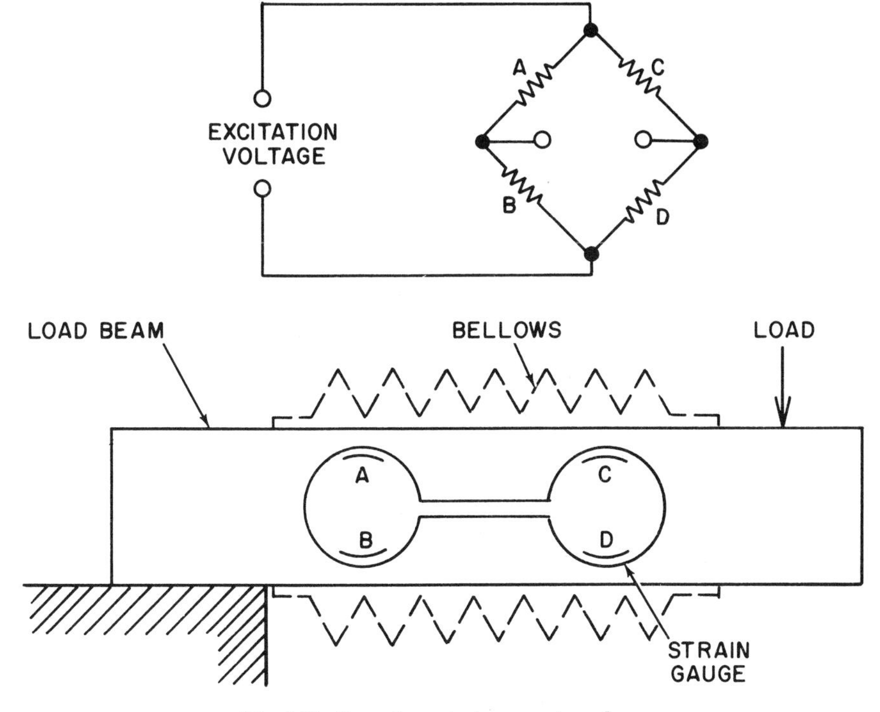

Fig. 1.7f Beam type strain gauge transducer

assemblies. The new monorail transducer replaces the live rail and is self-supporting.

A typical unit is shown in Figure 1.7h. Strain gauges at the designated location sense bending strains as the load traverses the transducer. The gauges interconnected in a bridge arrangement supply a constant output independent of load position on the transducer. The sloping arrangement on the upper edge of the transducer decouples the moving carrier from the "pusher" mechanism during the sensing period. That is, on the downward slope the carrier rolls free, eliminating contact with the pusher mechanism.

Table 1.7g
PERFORMANCE OF BEAM TYPE
STRAIN GAUGE TRANSDUCER

Rated Capacity Range:	10 lbs to 5,000 lbs
Output:	2 mv/v
Terminal Linearity:	±0.03%
Hysteresis:	0.015%
Creep (30 min):	0.02%
Temperature Effect on Zero:	0.15%/100°F
Temperature Effect on Output:	0.08%/100°F

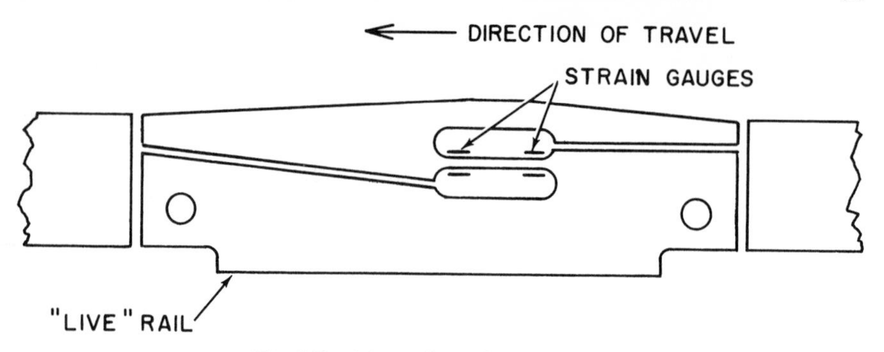

Fig. 1.7h Monorail weighing transducer

This simplified transducer not only provides greater measuring accuracy but also eliminates a second load cell and associated flexure and mounting arrangements, thereby reducing overall cost.

High Temperature Load Cells

As load cell weighing was applied to the metal processing industry, the need for devices to withstand high environmental temperatures became pressing. In recent years organic and inorganic bonded strain gauge backing and installation materials have become available and can withstand higher temperatures than conventional units. Bonded strain gauges with organic backings are now available for continuous operation at temperatures as high as 500°F.

On special applications high temperature strain sensing wire alloys have been installed with inorganic bonding materials such as ceramic cements and flame spray techniques, wherein molten aluminum oxide is sprayed on the sensing element and on the strain sensing grid to hold the latter firmly in place. These installations allow short-term operation at temperatures of 1,000°F but with some degradation in performance.

Direct Weighing of Tank Legs

Frequently, structures do not lend themselves to support by load cells of any type. They are already fabricated and erected, and their support by load cells would require extensive field modification.

One solution to the weighing of these structures is installing strain gauges directly on the supporting legs. The legs become the transducer sensing element to which the strain gauges are applied in full bridge configurations.

In a typical installation (Figure 1.7i), a pair of gauges is applied longitudinally, sensing the compressive stresses in the tank legs. Another pair is applied in the transverse direction, sensing the tensile strains due to the Poisson effect. The four gauges are connected in a Wheatstone bridge arrangement and leads are brought out from each leg to a summing box and from there

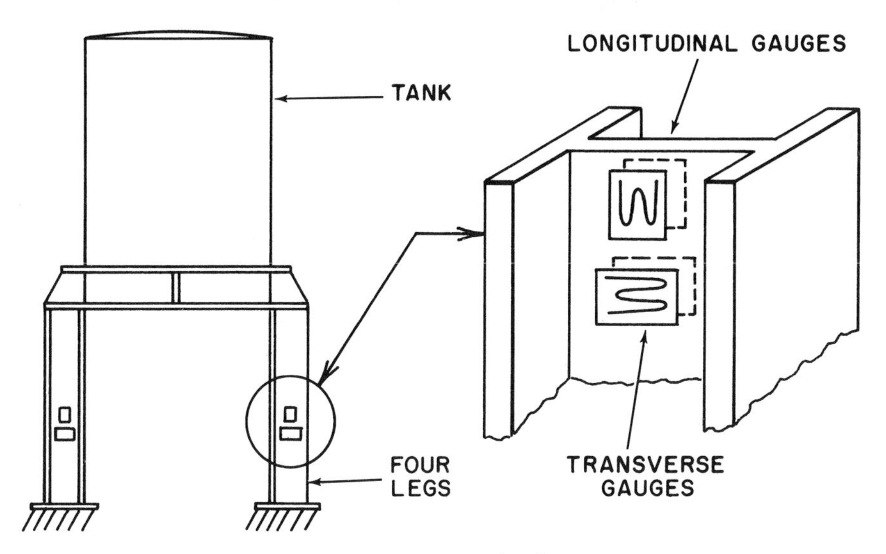

Fig. 1.7i Direct gauging of tank supports

to the read-out instrumentation. The installation is thoroughly protected with waterproofing materials.

Usually, the strains established in the supporting legs are very low and it is difficult to achieve perfect waterproofing permanently. As a result, the accuracy of such a weighing system tends to be relatively poor—3 to 5% of rated capacity.

Inductive and Reluctance Load Cells

Inductive and reluctance load cells incorporate the two basic sensing principles in the same way, i.e., the motion of a ferromagnetic core (inductive) or a coil assembly (reluctance) is converted to a voltage signal directly proportional to the displacement.

Various force sensing elements convert the applied force (weight) to a displacement to which the sensing element is coupled (Figure 1.7j).

These transducers furnish relatively high output signal levels and moderate to high accuracies, and cover a broad range of measuring capacities.

Magnetostrictive Load Cells

Magnetostrictive sensing load cells (pressductors) are finding use in industrial applications in which large output signals and ruggedness are desirable. Several typical configurations are shown in Figures 1.7 l and 1.7m.

The first configuration is for applications in which there are no bearing surfaces on the devices being weighed; in the presence of lateral loads the pressductor is very sensitive unless adequately protected. The vertical load

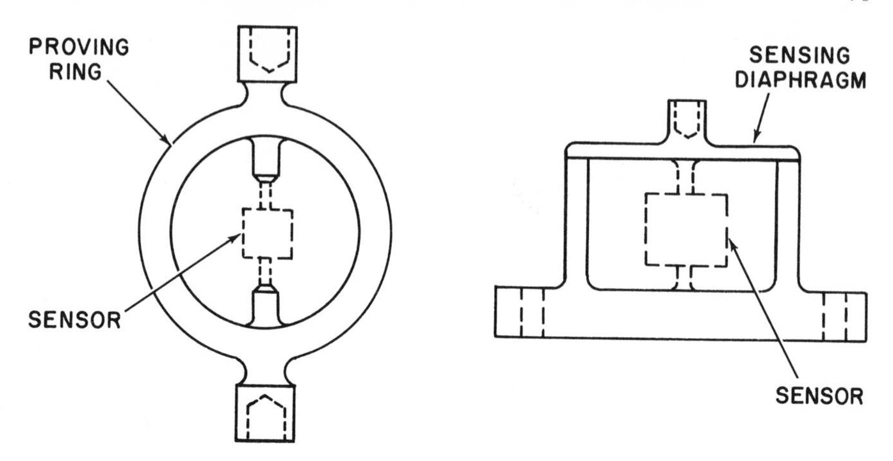

Fig. 1.7j Inductive and reluctive load cells

(Figure 1.7 l) is transmitted through the flexures, 1 and 2, to the sensing element, 3. The same flexures also transmit lateral forces to "ground" in a way so that the pressductor sensing unit is subjected to only a small portion of the adverse lateral loads.

The second embodiment (Figure 1.7m), designed for weighing during coiling operations, uses a similar construction with an additional overhanging member, 4, which supports the coiler shaft, and continuous weighing during coiling operations is provided. All units are adequately protected with watertight covers to accommodate applications in industrial environments.

New pressductor designs provide weighing accuracies of 0.1% of rated capacity. Output signal levels range from 1 to 20 VDC, with source impedances ranging from 0.5 to 25 ohms. Overload ratings as high as fifteen times the rated load are supplied. Although usable for weighing, the pressductor has greater applicability in the steel industry for the measurement of roll-forces in rolling mills and strip-tension in strip mills.

Table 1.7k
PERFORMANCE CHARACTERISTICS FOR INDUCTIVE OR RELUCTANCE LOAD CELLS

Capacity Range:	0.01 to 100,000 lbs
Rated Output Range:	5 to 200 mv/v
Linearity Range:	0.1 to 0.5%
Repeatability:	0.05%
Temperature Effect on Zero:	1%/100°F
Temperature Effect on Output:	1%/100°F

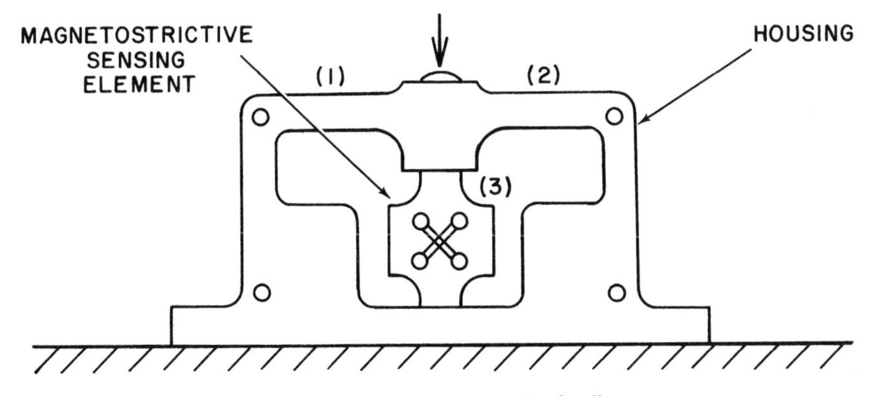

Fig. 1.71 Magnetostrictive load cell

Load Cell Weighing Accessories

Several load cell weighing accessories required for satisfactory application of load cells to weighing installations were discussed in Volume I. Others will now be discussed.

A New Load Cell Adaptor

As already noted, many load cell weighing installations involve large differential expansions which can impose severe horizontal forces on the installed load cells. Also, in vehicle scales large horizontal forces can be applied owing to deceleration and acceleration forces associated with bringing the

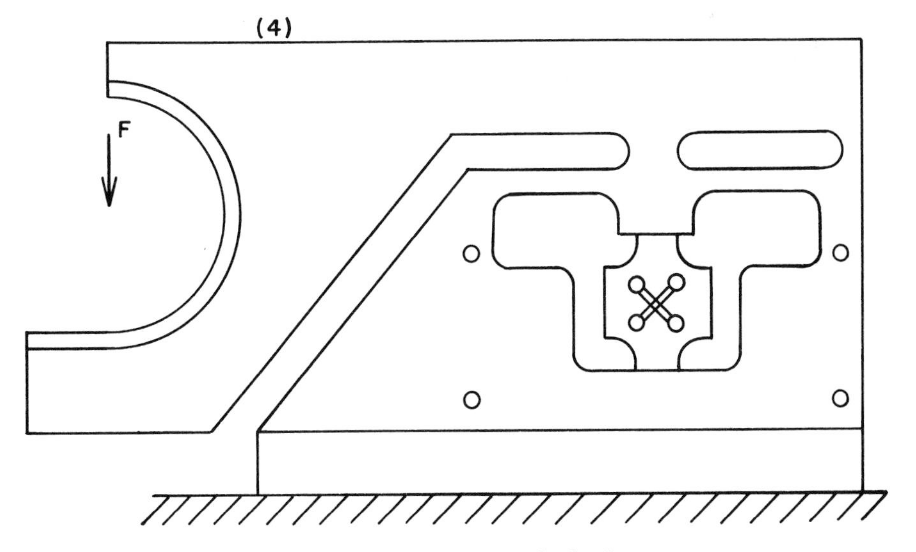

Fig. 1.7m Magnetostrictive load cell

vehicle on and off the scale. Development of a new load cell adaptor virtually eliminates such forces.

Primarily a mechanical arrangement, the active weighing platform is suspended from the top of the load cell by three suspension links (Figure 1.7n), and an upper plate and adaptor ring contact the load cell at the desired loading point. The upper plate carries the three links by link pins projecting radially from the upper plate. Hanging on the opposite end of the links is the lower plate which includes three additional link pins for engaging the lower end of the links.

The lower plate is connected to the active weighing platform thereby transmitting the weight through the links and upper plate to the top of the load cell. The load cell is supported by a base plate which rests on the foundation or ground structure. The base plate also serves to absorb heavy side loads when the horizontal deflection of the weigh-bridge exceeds the clearances provided between the base plate and the cutout portion of the

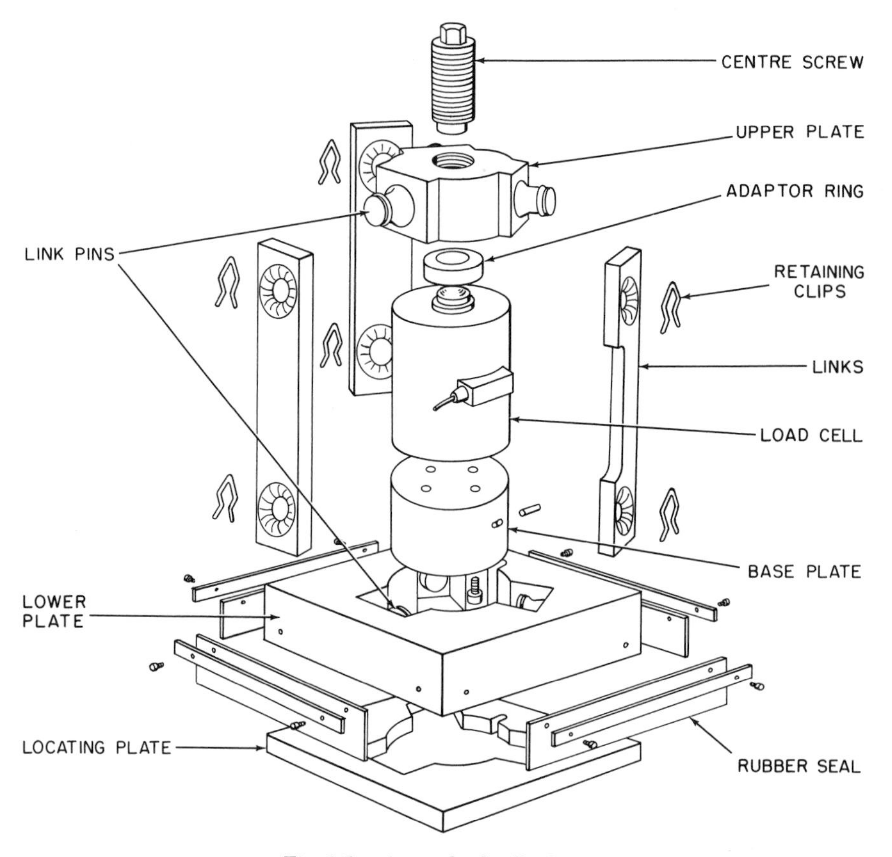

Fig. 1.7n A new load cell adaptor

lower plate. The height of the adaptor assembly can be adjusted by a center screw, enabling the equal distribution of total load among the several load cells in a given installation.

The structure provides a highly flexible load cell adaptor assembly which transmits virtually no side loads to the load cell caused by differential expansion of the weighing structure relative to the ground structure. The side loads that are transmitted to the load cell are from weigh-bridge deflections, imposing angular loads on the load cell. These are minimized by appropriate structural design of the weigh-bridge.

Rocker Assembly

Another load cell adaptor commonly used in weighing systems is the rocker assembly (Figure 1.7o). An adaptor is added to the bottom of the load cell which in effect provides a convex loading surface on the bottom as well as on the top of the unit. The load cell and adaptor are located in place by a stabilizer plate. Load is introduced to the load cell through the upper bearing block and transmitted through the load cell and the lower bearing block to the mounting plate. The stabilizer plate allows partial rotation of the load cell while at the same time restricting excessive lateral motion.

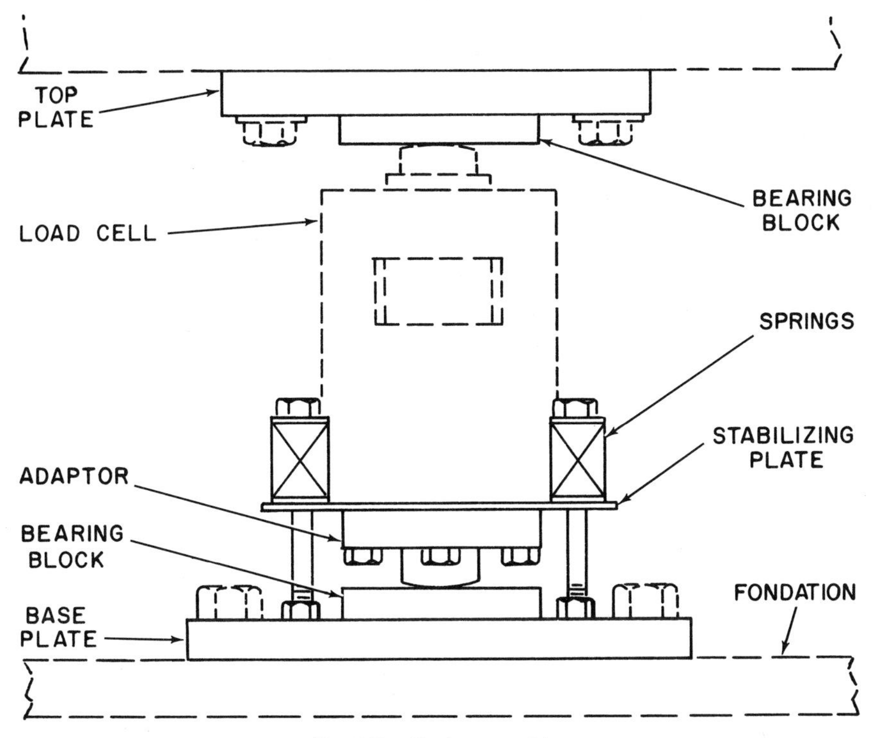

Fig. 1.7o Rocker assembly

Differential expansion between the structure being weighed and the foundation causes slight rotation of the load cell, greatly reducing the magnitude of the horizontal forces which would have been present in the absence of the rocker assembly. The load cell is thus protected from the adverse effects of large lateral forces caused by differential expansion in multiple cell weighing systems.

Various rocker assemblies for load cell capacities range from 20,000 to more than 300,000 pounds.

Integrated Weighing Devices

For several years a number of integrated weighing devices have become available combining force transducing elements and weigh-bridges or platforms in a single self-contained weighing structure.

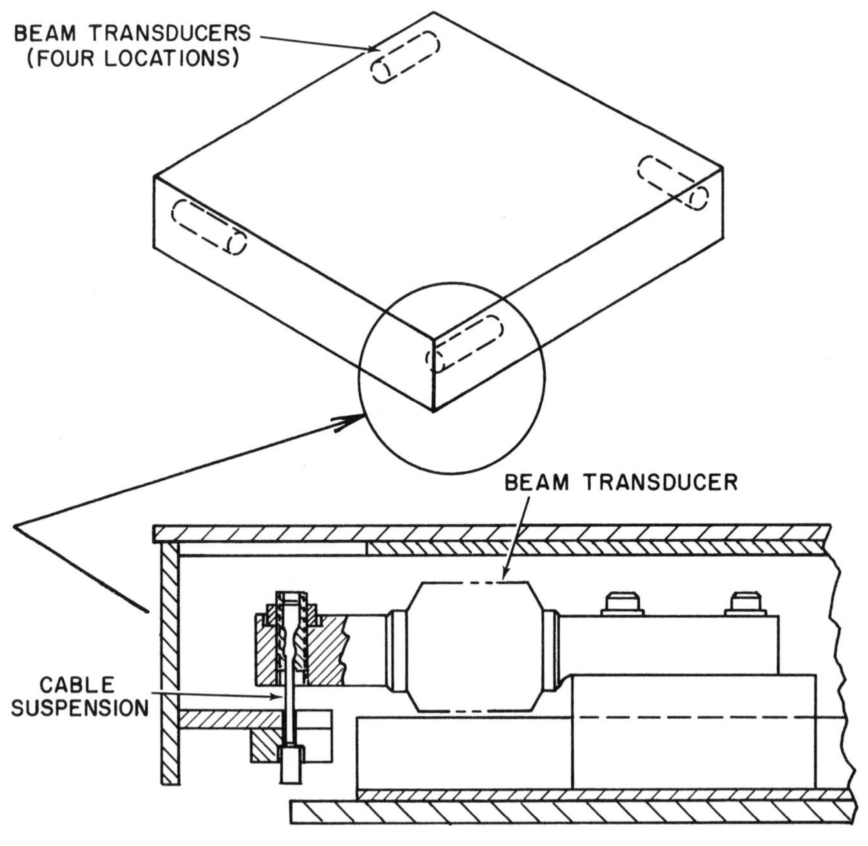

Fig. 1.7p Beam type weighing platform

Beam Type Weighing Platform

These transducers are now the basic sensing elements in medium capacity, low height, semiportable platform scales. Available in capacities from 500 to 14,000 pounds, it can be located almost anywhere within the process area without the need for pits and attendant preparation costs. Typical units, depending on capacity, are from 2 to 35 square ft in platform area and 1.5 to 6 inches in overall height (Table 1.7q).

Beam type transducers are in each corner of the weighing platform (Figure 1.7p). A steel cable or alternative flexible arrangements suspend the active weighing platform from the end of the transducer. The flexible connection allows the transducer to operate in its normal deflection mode, thereby capitalizing on its inherently high accuracy. The flexible connections also render the entire structure much less susceptible to uneven foundation conditions.

In addition to low height, high accuracy and relative insensitivity to foundation unevenness, the units are semiportable.

Smaller, lower capacity beam type platforms available in the range of

Table 1.7q
PERFORMANCE CHARACTERISTICS OF
BEAM TYPE WEIGHING PLATFORM

Rated Output:	1.5 mv/v
Terminal Linearity:	0.05%
Hysteresis:	0.03%
Repeatability:	0.02%
Corner Loading Sensitivity:	0.03%
Temperature Effect on Zero Balance:	0.15%/100°F
Temperature Effect on Output:	0.08%/100°F

10 to 250 pounds use two beam type sensing elements (Figure 1.7r). Features and characteristics are similar to those of platforms already described. One significant difference, however, relates to the use of only two beam type sensing elements. In this configuration the reduction of corner loading sensitivity or load placement errors must be approached differently. It is no longer a matter of simply dividing the load among four linear transducers but rather involves a reduction of the effect of changing load position on two transducers.

A changing load position results in a different bending moment applied to the beam type transducers (Figure 1.7s). However, as discussed in connection with beam type transducers, the strain gauges are applied and wired in a way such that the beam type element is insensitive to load position or bending moment.

This concept can be explained as follows. Strain gauges at location 1 sense

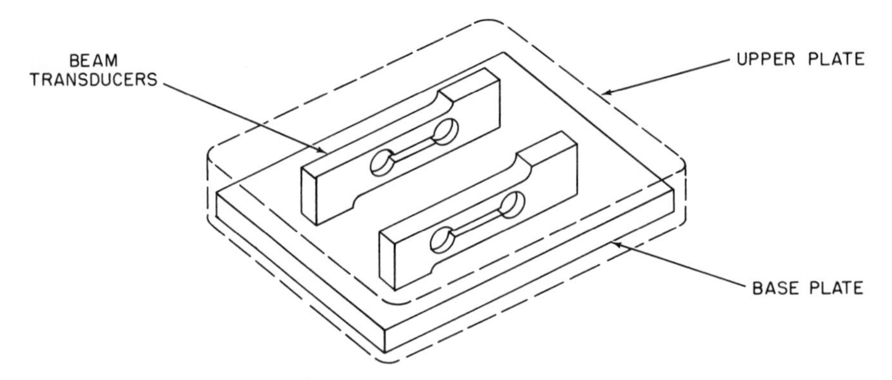

Fig. 1.7r Low capacity beam type platform

the bending moment there (M_1). Gauges at position 2 sense the bending moment at that position (M_2). The strain gauge bridge is arranged so that the output is proportional to the difference $M_1 - M_2$, and this may be expressed by the relationship

$$M_1 - M_2 = F(L + 1) - F(1) \qquad 1.7(1)$$

$$M_1 - M_2 = FL \qquad 1.7(2)$$

Hence, the output of the beam type transducer is proportional to the applied force and the distance between the strain gauge locations (L). It is insensitive to the placement of the load (1). This relationship is theoretically correct, assuming perfect strain gauge location and zero machining tolerances. Since this relationship seldom exists, electrical "tuning" methods tailor the sensitivities at locations 1 and 2 to give virtually perfect moment rejection.

The lower capacity platform configurations are available in capacities of 10 to 250 pounds with roughly 15 by 18 inches weighing area and height of about 3.5 inches.

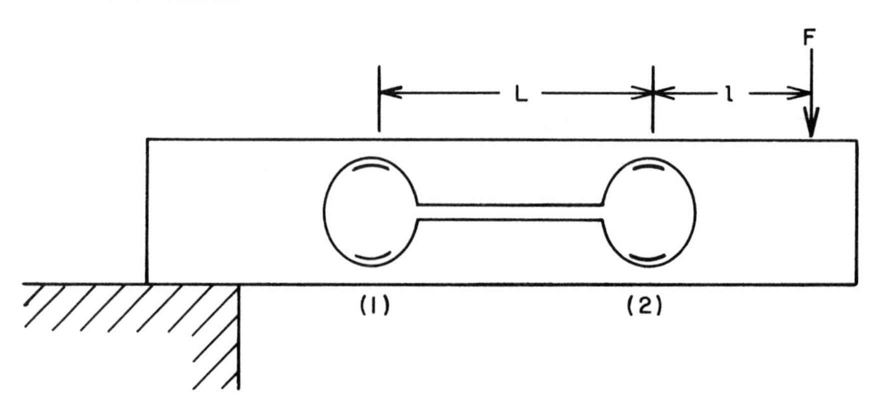

Fig. 1.7s Load position desensitization

Treadle Scales

Heretofore, scales for weighing vehicle axles on highways at toll booth locations utilized individual load cells, weigh-bridges and mechanical stabilization hardware, all of which required expensive pits and additional installation. An integrated treadle scale now eliminates some of these requirements and provides a self-contained unit which can be lowered into place in a shallow pit.

Conceptually, the treadle scale (Figure 1.7t) consists of two strain gauge sensing elements located at opposite ends of the bridging plate. The sensing elements measure shear forces by virtue of the sensing element configuration and gauge location. Strain gauges at each of the reduced sensing sections sense the principal tensile and compressive strains due to the applied shear forces.

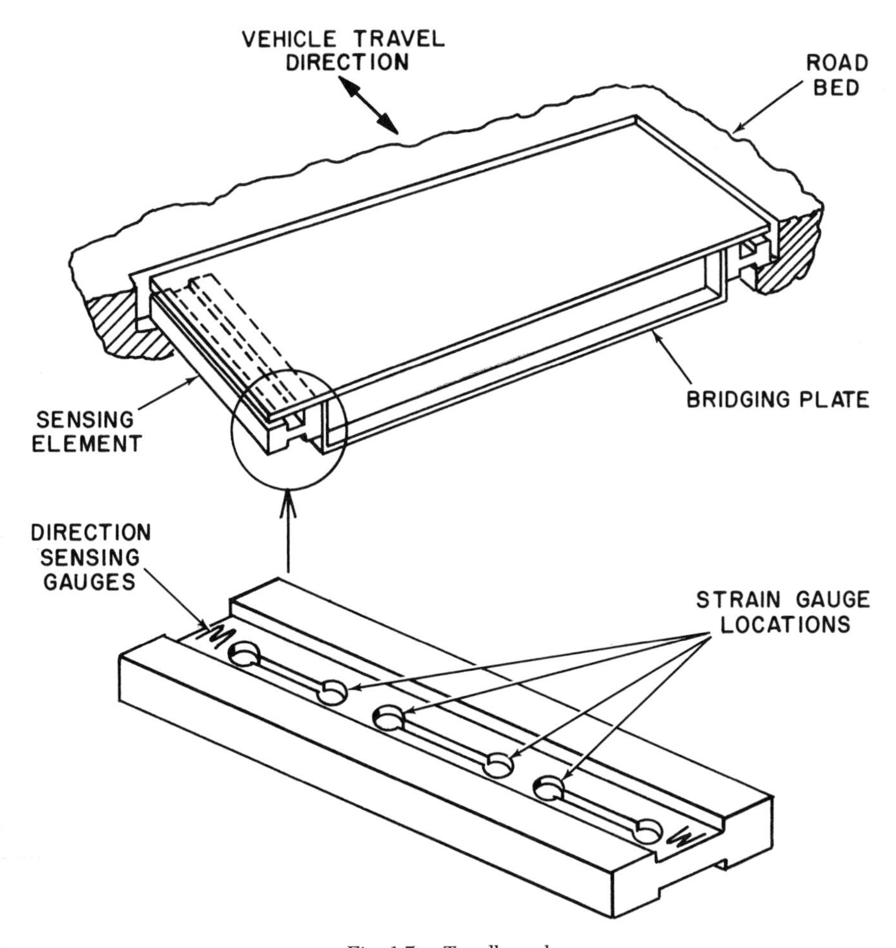

Fig. 1.7t Treadle scale

Additional strain gauges on the leading and trailing ends of the sensing elements are connected in a bridge configuration which provides direction sensing information for the associated instrumentation. Invalid vehicle entries are thereby detected.

Table 1.7u
TREADLE SCALE PERFORMANCE CHARACTERISTICS

Rated Capacity:	18,000 lbs
Rated Output (10 volts excitation):	0.25 v
Terminal Linearity:	0.75%
Hysteresis:	0.25%
Temperature Effect on Zero Balance:	1%/100°F
Temperature Effect on Output:	0.25%/100°F
Size:	15 in. × 60 in.
Weight:	200 lbs

Two units are generally installed end on end across a lane, spanning a distance of ten feet. Total capacity for the pair is 36,000 pounds. Currently, the units are used to sense overweight vehicles for subsequent weighing by law enforcement scales. Another use is axle weight measurement for toll assessment by weight classification.

Portable Platform Scales

Another portable platform scale (Figure 1.7v) is composed of a honeycomb weigh-bridge and four load cells, one at each corner. Adjustable leveling pads level the platform. Each pad includes a swivel joint for accommodating sloping surfaces. An entire structure weighs less than 140 pounds.

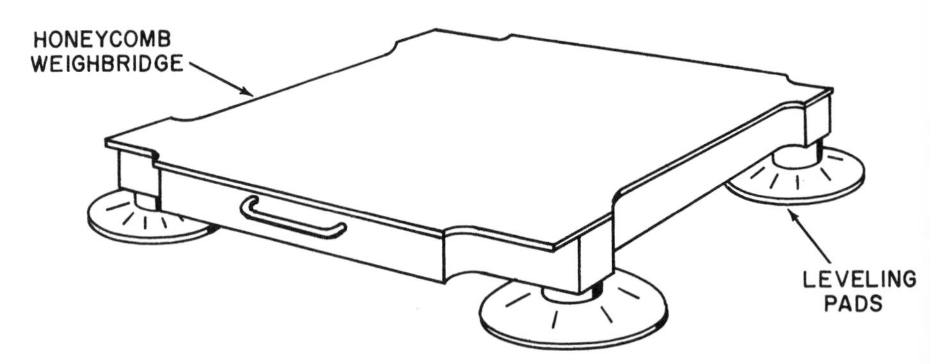

Fig. 1.7v Portable platform scale

Table 1.7w
PERFORMANCE CHARACTERISTICS OF PORTABLE PLATFORM SCALES

Rated Capacity:	50,000 lbs
Rated Output:	1.5 mv/v
Terminal Linearity:	0.05%
Hysteresis:	0.05%
Temperature Effect on Zero:	0.1%/100°F
Temperature Effect on Output:	0.15%/100°F

The units are available with ramps and spacers for use with various vehicle wheel configurations. They are being used in portable aircraft and vehicle weighing, to be described later.

Electronic Weighing Instrumentation

Electronic weighing systems became available shortly after the invention of the bonded strain gauge load cell by Dr. Arthur C. Ruge in the early 1940's. Early load cell weighing systems utilized servo-operated null-balance techniques. These incorporated rebalancing slidewires or variable capacitive elements to accomplish the automatic null-balance. Coupled to the balancing devices were disks and pointers for analog weight indication.

Over the years, servo-operated weighing instrumentation attained a high state of development with system accuracies of 0.1% of reading over 20% of full scale when care was exercised in system design, instrument adjustment and calibration. Although high accuracies were attainable, certain disadvantages remained.

The servo-operated instrument was slow, particularly when automatic zero suppression or range changing was used. This presented particular difficulty in batching systems in which moderately high operating speeds were required. Since the servo-instruments used slidewires, gear trains and other mechanical devices subject to wear and failure, instrument reliability was low and resulted in high maintenance costs. Analog instrument displays required interpolation by the operator, resulting in frequent reading error.

Evolution of Digital Instrumentation

During the later development of servo-operated weighing instrument, "odometer" type displays replaced the analog type, eliminating the disadvantages of analog displays; the other shortcomings, however, were still present.

In the early 1960's the first electronic digital indicators to use successive approximation, voltage comparison analog-to-digital conversion techniques were developed. Switching was by reed relays, which gave accurate digital read-out but were expensive, slow and erratic. Following this development, a host of fully solid state analog-to-digital conversion techniques became available.

Most of these encoding techniques advanced the state of the digital indicator art. The resulting instrumentation, however, was not yet applicable to the precision electronic weighing field because the cost for acceptable accuracy was too high. When linear and digital integrated circuits became available, costs decreased enough to allow serious consideration of solid state digital instrumentation for electronic weighing.

Analog-to-Digital Conversion Techniques

Many A to D conversion techniques are currently in use including:

DUAL SLOPE INTEGRATOR

An integrator operates on the amplified load cell signal and at the same time sends a train of pulses to a counter and storage unit. When the counter reaches a preset number of counts (usually 10,000), the upscale integration process is discontinued. The input to the integrator is then applied to a reference voltage source and integration proceeds in the descending direction until the integrator output reaches zero. During the downward integrating process, the same train of pulses is applied to the counter and storage unit which was reset to zero at the conclusion of the upward integration cycle. When the integrator output reaches zero, the gate is closed and the value in the counter and storage unit represents the measured variable. By using the reference voltage supply for load cell excitation as well as for the downward integration cycle, ratio operation is obtained.

DIGITAL SERVO-TECHNIQUE

Another conversion principle utilizes a solid state version of a servo-system in which a digital voltage reference source is continually adjusted to be equal and opposite to the voltage representing the measured variable. Any unbalance between the measured variable and reference source applies a signal to a voltage-to-frequency converter. The larger the unbalance the greater the frequency emanating from the V to F converter. The V to F converter drives an "up-down" counter which in turn is coupled to the digital reference source. Associated logic circuitry causes the counter to count upscale if the error signal is positive and downscale if it is negative. The counting ceases when the error signal returns to zero. The digital display coupled to the counter indicates the measured variable.

VOLTAGE-TO-FREQUENCY CONVERSION TECHNIQUE

A highly stable and linear voltage-to-frequency converter converts amplified DC load cell signals to an AC signal whose frequency is directly proportional and linearly related to the input load cell signal. The AC signal is gated to a counter and storage unit for a precisely fixed interval. The number of counts which are entered into the counter during the interval represents the average value of the measured variable during the counting.

SUBSTITUTION METHOD

The voltage representing the input signal is stored and connected to an error-sensing operational amplifier. Step voltages are applied to the same input with opposite polarity in order to bring the output of the error amplifier to zero. The step voltages are calibrated so as to be equal to decimal digits. The steps are entered, starting with the most significant digit first, and the most significant digit steps will continue to be entered until the output of the error amplifier reverses polarity. Reversal indicates that too many "most significant" steps have been entered. One step is removed and the steps next lower in order of magnitude are entered and added to the voltage produced by the highest order digit. These steps are smaller in amplitude by a factor of ten. The digitizing process continues until the least significant digits have been entered. The resulting digital information is stored in a counter and displayed in decimal form on the digital display. The complete digitizing process requires less than 0.1 second.

OPTICAL CONVERSION TECHNIQUES

An optical A to D conversion technique is finding considerable use on pendulum type mechanical scale indicators, particularly in Europe. Providing the A to D conversion is a transparent plate with digitally coded light and dark areas, coupled to the pendulum system of the mechanical indicator. Light is projected through the coded plate, falling on a bank of solid state photosensitive devices. In this manner the position of the mechanical indicator and hence the measured weight value is digitally coded for local or remote digital display of weight. Having accomplished the A to D conversion, many digital devices can be coupled to otherwise completely mechanical scale systems.

Performance Characteristics

Solid state digital instrumentation currently provides self-contained load cell excitation supplies (both DC and AC). Provisions are included for exciting more than one load cell in multiple load cell weighing systems. Binary coded decimal outputs are provided for operating auxiliary printers, tape punches, and the like. Currently available performance characteristics are listed in Table 1.7x.

Important Features

AUXILIARY OUTPUTS

Most digital indicators include a preamplifier prior to the analog to digital conversion. The analog output signal from this amplifier is usually made available at the instrument output terminals for activation of auxiliary analog

Table 1.7x
DIGITAL INDICATOR PERFORMANCE

Resolution:	1 part in 15,000
Linearity:	± 0.01% ± one digit
Short-term Stability:	1 count
Temperature Effect on Zero Balance:	0 to 0.25μr/°F
Temperature Effect on Span:	0.05%/100°F
Common Mode Rejection:	100 db
Maximum Sensitivity:	0.5 mv/count

devices. Current amplifiers convert the analog voltage to standard current signals of 1 to 5, 4 to 20 and 10 to 50 madc. Using this option, digital weighing systems can be interfaced with standard electronic process control systems for trend recording and continuous analog control.

ELECTRONIC AUTOMATIC ZERO

In many weighing applications using metallic strain gauge load cells, full scale signal levels as low as 5 mv must be accommodated. This level places difficult requirements on the preamplification hardware, and therefore some digital indicators include an automatic zero feature to improve zero stability.

In many currently available digital indicators during the A to D conversion cycle, the preamplifier is not used in the encoding process. During this period the input terminals of the preamplifier are shorted and the unit is electrically "rezeroed," resulting in zero voltage at its output terminals. Since there are usually several A to D conversions per second there are also a similar number of automatic zero operations per second. Hence, any zero offsets in the preamplifier due to temperature or aging are "zeroed out" several times per second.

Some instruments provide for rezeroing the entire system rather than the instrument alone. Gradual buildup of dirt, rain, snow or other foreign material is sensed periodically and an overall rezeroing operation brings the system back to zero.

REMOTE SENSING

In applications in which the load cell installation is a considerable distance from the readout instrumentation, the effect of interconnecting cable resistance must be considered. A significant amount of cable resistance in the signal leads is of little concern because of the high instrument input impedance. This is not the case, however, in the excitation leads because resistance in this location decreases the net excitation voltage applied to the load cells. Although this effect can be accommodated during initial system calibration, variations in cable temperature after calibration cause resistance changes which in turn change the load cell excitation voltage and hence system calibration. To eliminate this effect, some digital indicators incorporate a regulated bridge excitation supply that senses the voltage directly at the load cell terminals through an additional pair of leads. It regulates that voltage rather than the instrument terminal voltage. Hence, any change in load cell excitation voltage, as might be caused by temperature-induced cable resistance changes, is sensed by the regulator, and compensation is automatic.

ELECTRONIC LINEARIZATION

Previously linearization of load cells using semiconductor strain gauges was discussed. In another form of linearization, a small portion of the preamplifier output voltage is applied to the bridge excitation regulatory circuitry to vary the excitation voltage slightly. If the load cell signal (Figure 1.7y) tends to drop off parabolically with load (dotted line), the interaction between the small portion of preamplifier voltage and the regulator causes the bridge excitation voltage to vary as shown (interrupted line). The result is that the load cell output has practically no linearity error. The instruments can vary the amount of parabolic linearization introduced into the

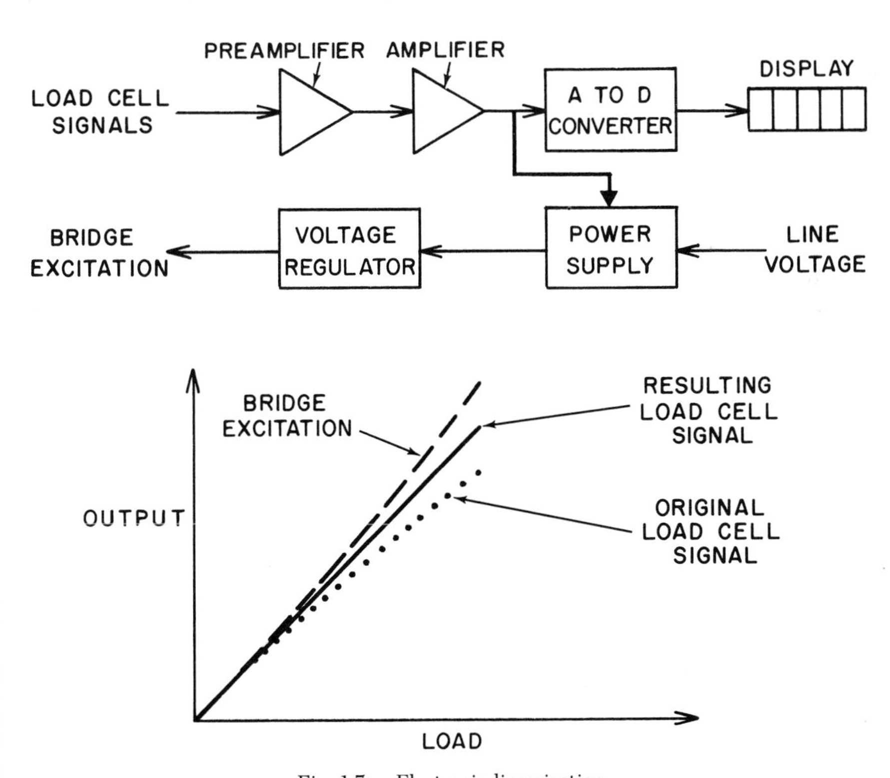

Fig. 1.7y Electronic linearization

system from −0.25% to +0.25%, supplying enough correction for most conventional load cells that require linearization.

RATIO OPERATION

One of the most important features of the older servo-instruments was that they provided for ratio operation in which supply voltage changes had little or no effect on system accuracy. Many currently available digital instruments provide for ratio operation by using the same regulating supply voltage for bridge excitation and analog to digital conversion. Hence, any supply voltage changes occurring at a rate slower than the digitizing period are automatically accommodated and the desirable ratio operation principle of the older servo-instruments is preserved.

DIGITAL MULTIPLICATION

One of the most serious drawbacks of conventional digital indicators in precision electronic weighing is limited resolution. A digital indicator with a full scale display of 10,000 counting by 1s when applied to a 10,000 pound system furnishes a system resolution of 0.01%. However, if the same indicator is applied to a 5,000 pound system, the resolution becomes 0.02% and the resolution decreases proportionally as the instrument is applied to lower capacities. Considerable accuracy consequently is lost when the full resolution of the indicator is not utilized.

Most indicators compensate for counting by 1s, 2s and 5s in an attempt to maintain full instrument resolution. One indicator incorporates a digital multiplication function in the logic circuitry which preserves the full 0.01% resolution for all rated capacities. The concept (Figure 1.7z) can best be explained by assuming a 3,000 pound rated capacity, an indicator with a 5 digit display over range and a basic resolution of 10,000 counts or 0.01%. In the conventional case, the system resolution would be 0.033% (0 to 3,000 counting by 1s). However, in this case a digital multiplier incorporated in the logic circuitry generates 3 pulses for each conventional pulse from the voltage to count converter, applies them to the count storage unit and the digital display reads 0 to 3,000.0 by 0.3s. Hence, the full instrument resolution is 1 part in 10,000 or 0.01%. By properly programming the multiplier, it is possible to count by any integer from 1 through 9. In this manner limited resolutions typical of conventional digital indicators are avoided for all system capacities.

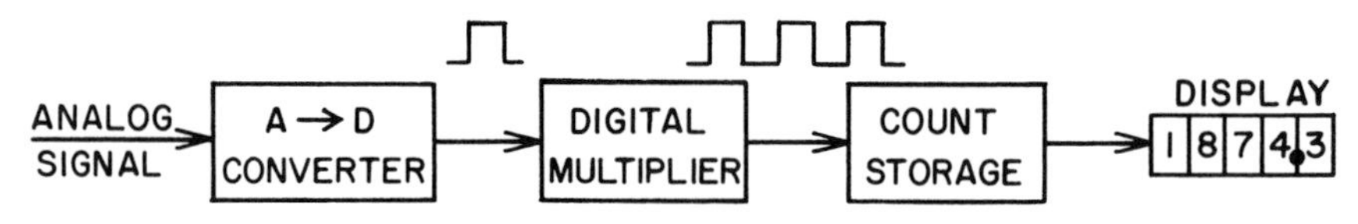

Fig. 1.7z Digital multiplication

REJECTION OF VIBRATION EFFECTS

In many weighing applications vibrations due to rotating machinery, agitators and vibrators introduce "noise" into the system, preventing accurate weight measurement. To reduce noise, most digital indicators have an optional plug-in low pass filter usually inserted between the preamplifier and the analog to digital converter. Various filter cutoff frequencies can accommodate the particular application.

PROGRAMMABLE SWITCHING UNITS

The flexibility of digital instrumentation affords the opportunity for using one indicator on several weighing channels. These can be mixed at will without regard to live capacity, tare weight or scaling. Modules for each channel, when switched into the system, function in several ways.

Each module, with a selected tare weight adjustment and fine zero control set for its particular channel, programs the indicator not only for the proper calibration setting (scale factor) for the live load in its channel, but also for the required transducer linearization in each channel, if that function is required.

Since rated capacity can vary widely over the different channels, the digital multiplication function is also programmed by each module to preserve the 0.01% resolution in all channels. Lastly, each module programs the indicator for decimal point location. Thus, multiple channels can be sensed by a single indicator completely programmed in each channel for tare weight, zero, scale factor, linearization, digital multiplication and decimal point location—an excellent example of the flexibility and versatility of solid state digital instrumentation.

DIGITAL TARE ADJUSTMENT

Tare can be adjusted more accurately on a digital basis than by analog devices, and can now be made to the least significant digit furnished by the indicator.

DIGITAL CONTROL POINTS

As with digital tare adjustment, so digital control points within the digital logic can also indicate when weight values have reached a predetermined level. Since the comparison is made on a digital basis, precision is enhanced.

Integrated Weighing Systems

In consequence of the advances made in weighing sensors and instrumentation, integrated systems are now being developed, generally to perform a specific repetitive function.

Mobile Electronic Weighing System

One integrated weighing system combines eighteen 50,000 pound load cell scales, associated ramps and spacers, roller conveyors and a multi-channel instrument for automatically determining the weight and lateral and longitudinal center of gravity of vehicles and bulk materials.

The platform scales may be in many configurations to accommodate aircraft landing gear, multi-wheel vehicles or bulk material transported over roller conveyors.

The instrument provides "dialing in" reference distance information necessary for center of gravity computations. This system has found considerable use in the weighing of materials prior to loading on aircraft and subsequent weighing of the loaded aircraft; it is also used for calibrating on-board aircraft weighing systems.

On-Board Aircraft Weighing Systems

During the past six years, considerable effort has been expended in developing integrally mounted aircraft weight and center of gravity detection systems, which eliminate the need for jacking the aircraft on load cells. One of these systems measures pressure on the landing gear strut which, except for friction effects, is directly proportional to the weight reaction at each landing gear. Other systems utilize transducers (Figure 1.7aa) mounted inside the hollow landing gear axles or (externally) on lugs projecting from the bogie

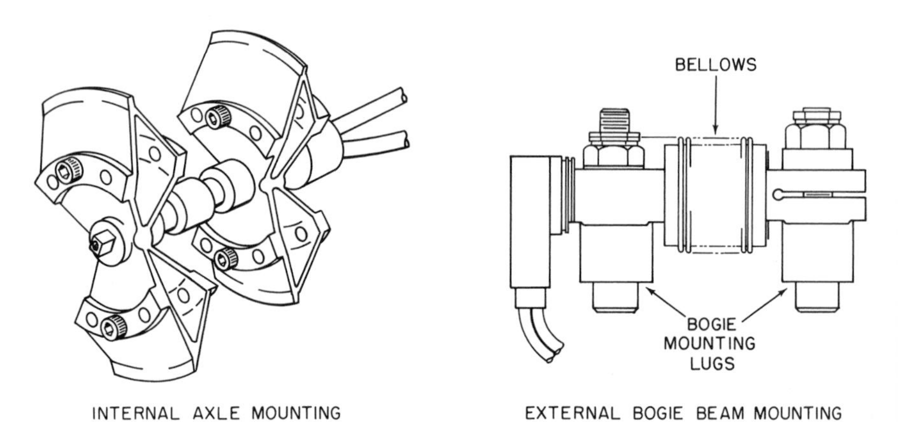

Fig. 1.7aa Transducers for aircraft weighing

beam. In the latter two examples, shear deflections in the axle or bogie beam are measured by the transducers which provide signals directly proportional to the reactions at each landing gear. System accuracy is 1% of gross weight and 1% mean aerodynamic chord under typical aircraft environments.

The development of integral aircraft weighing systems suggests the beginning of a new generation of integral weighing systems applicable to all types of vehicles.

Fork Lift Truck Weighing Systems

Several fork lift truck weighing systems afford the opportunity of weighing material carried by trucks and thereby preclude the bringing of the material to a locally mounted scale.

One system utilizes special tines on which strain gauges are installed (Figure 1.7bb) and interconnected in a Wheatstone bridge configuration, as illustrated in Figure 1.7f. The weight indication derived from the strain gauge bridge is therefore independent of location of the material on the tines.

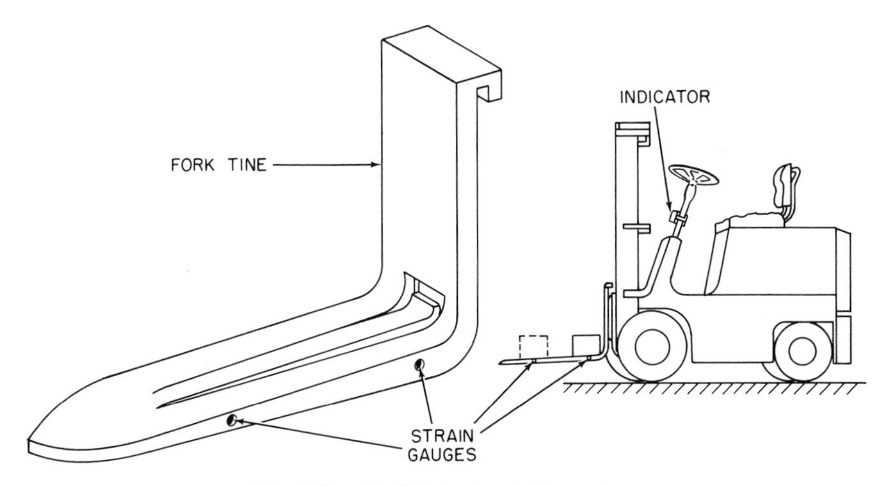

Fig. 1.7bb Forklift truck weighing system

An instrument mounted on the steering column or on the cowling of the truck indicates net weight and furnishes print out on a tape or ticket. Deadweight of pallets or tote boxes can be "tared out" with a calibrated digital tare adjustment, and the system accuracy is 0.5%.

Feed Lot Truck Weighing System

An integral weighing system is also available on feed lot trucks and similar vehicles. Four load cells are mounted to the truck frame and support the feed bin. Monoball check rods restrict lateral and longitudinal motion of the bin, and overload stops prevent transmission of shock loads to the load cells during normal travel.

A manually balanced digital servo-instrument indicates the weight. A wide range zero adjustment allows rezeroing the system before weighing out material at each feeding station. System capacity ranges from 10,000 to 50,000 pounds, with a resolution of one part in 1,000 and with a system accuracy of 0.25%. In the balanced position, the operator moves a lever to print the value of feed dispensed at each location.

Belt-Conveyor Scale

Belt feeder designs are described in Section 7.6 of Volume I. A new continuous weighing system with calibrating time of five minutes requires a single operator. The major components of the system are conventional. An inductive type load cell senses weight on the active portion of the belt and a belt travel pulser senses belt travel and speed. A controller operates the DC belt drive motor through a silicon controlled rectifier supply.

Calibration is as follows:

1. A standard weight is placed at a holder on the scale beneath the active portion of the belt.
2. With the conveyor running, a weight totalizer reading is recorded.
3. The function switch is turned to "calibrate" and the totalizer advances until the standard test footage of the belt is reached, at which time the belt is automatically stopped.
4. The integrated value is compared with a pre-established standard value.
5. The graduated span adjustment is linear and does not interact with the system zero.

Combined component accuracy including linearity, repeatability and sensitivity is within ±0.1% for scale and integrating system. Other factors affecting performance are belt tension variations, windage, idler alignment and scale location. With proper attention to these factors, system errors can be reduced ±0.25%.

Loss-in-weight Feeder

Loss-in-weight detectors are discussed in Section 7.7 of Volume I. The continuous feeder (5) shown in Figure 1.7cc can handle solids, slurries and liquids with "batch" accuracy because it puts weight sensing in a "control-before-delivery" instead of a "reacting-after delivery" class. The concept reduces transport lags and furnishes greater accuracy of delivery over short time increments. The equipment provides constant rate of delivery from high to low feed rates.

The system uses load cells (4) to sense the weight of material in the hopper at any instant. The hopper discharges the contained material to the process (7) at a controlled rate by the feeder (5). The control system (6) continually senses the decreasing weight in the hopper (3) and compares it with a pre-determined rate of emptying. The difference between the actual weight in the hopper and the pre-determined weight at any instant is sensed, and the speed of the feeder (5) is regulated to eliminate this difference. When the

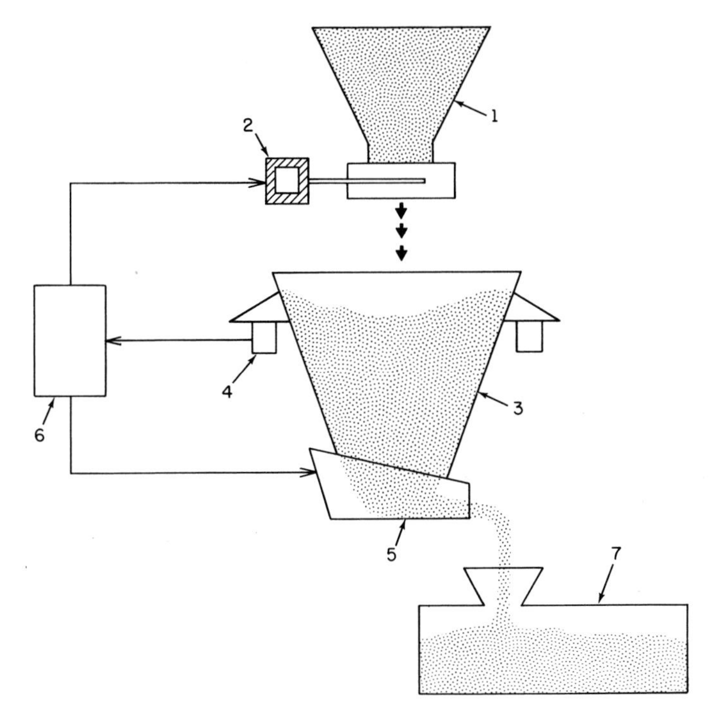

Fig. 1.7cc Loss-in-weight feeder

material in the hopper reaches a pre-set minimum weight level, it is replenished by temporarily opening a discharge valve (2) on the storage hopper (1).

The outflow rate of the feeder is not affected by the addition of new material because a "heel" always remains in the hopper (3) and because the rate is locked into the speed setting of the feeder that existed prior to refilling. Immediately after refilling, rate checking is resumed.

1.8 NEW DEVELOPMENTS IN pH MEASUREMENT AND REFERENCE ELECTRODES

Components:	Immersion and flow-through assemblies (see Section 8.8 in Volume I). pH electrodes. Reference electrodes. Amplifiers. Ultrasonic cleaners.
Design Pressures:	Electrode holders, 0 to 150 PSIG; liquid ion-exchange electrodes, 0 PSIG; solid state electrodes, 100(+) PSIG; glass electrodes, 150 PSIG; reference electrodes, 150 PSIG.
Design Temperature:	Electrode holders, 0 to 100°C; liquid ion-exchange electrodes, 0 to 50°C; solid state electrodes, −5° to 100°C (intermittent); glass electrodes, 140°C; reference electrodes, 140°C; amplifiers, 60°C; ultrasonic cleaners, 140°C.
System Accuracy:	±1 millivolt which corresponds to ±5% relative concentration for monovalent ions.
Range:	Low ppm to saturated solutions.
Cost:	Electrodes: $50 to $300. Systems: $1,000 to $5,000.
Partial List of Suppliers:	Beckman Instruments, Inc., Cambridge Instruments Co., Inc., The Foxboro Company, Great Lakes Instruments, Inc., Leeds & Northrup, Universal Interloc, Inc.

This section discusses the recent developments in pH measurement, supplementing not only Section 8.8 in Volume I, but also the discussion of ion selective electrodes in Section 1.13 in this volume, particularly reference electrodes. Although recent developments are examined, the emphasis is on reference electrodes and electrode cleaning devices.

pH Electrodes

The pH glass electrode, in measuring the activity of hydrogen ion in solution, is an almost perfect analytical tool. Its behavior corresponds to the classic Nernst equation better than the standard on which it is based (the standard hydrogen electrode) because it is much freer from interference. In earlier years, the glass electrode was believed to have both acid and basic solution errors and was therefore offered in separate models for acid and alkaline solutions, and for low and high temperatures. Current pH glass electrodes with suitable high-impedance amplifiers and reference electrodes can operate over the pH range of 0 to 14, temperatures −5°C to +110°C (23°F to 230°F) and pressures as high as 150 PSIG. Their perfection is marred only by an interference that occurs above pH 12 and is called the sodium ion error, although it is actually an alkali metal error. The magnitude of error is a function of the pH and temperature. Nomographs are furnished by the manufacturers for its calculation.

Special pH electrodes for temperatures as high as 140°C (284°F) may have thallium amalgam internal elements and therefore require matching reference electrodes. Another special form of the pH electrode is the so-called combination electrode (referred to as "dual electrodes" in Section 8.8 in Volume I). They are a concentric or closely joined combination of measuring and reference electrode in a common body and are attractive for many process applications because they require only one opening for insertion into a process stream or reactor. Models are available in which the reference electrode may be fitted to a reservoir.

Flow-through and Submersion Assemblies

No major improvement has occurred in this area of pH detection during the last few years. Flow-through systems are still available in stainless steel or inert plastic materials, with the former rated for pressures up to 150 PSIG and the latter for pressures up to 50 PSIG. Piped into a process line with by-pass valving, they may be disassembled for cleaning or calibration in standard buffer solutions.

Submersion assemblies are also made of stainless steel or inert plastic materials with ratings for maximum submersible depth of 100 feet for the stainless steel and 50 feet for the plastic units. Both the flow-through and submersion assemblies are still available with translucent pressurizable reservoirs for the electrolyte for the reference electrode.

Amplifiers

pH amplifiers have special requirements because they are associated with high electrical resistance measuring electrodes (thousands of megohms for glass electrodes at 0°C) and virtually zero measuring current (hundredths of picoamps) in order to limit measurement error to a few tenths of millivolts. Two significant developments include the location and type of amplifiers.

Location of the Amplifier

The inaccessibility of many process measuring points and their hostile environment (temperature, pressure and corrosive fumes and solutions) necessitate that amplifiers be mounted at a distance from the measuring electrodes. These distances are sometimes only 3 to 20 feet, although amplifiers with driven shields may be as much as several thousand feet from the electrodes. Transmission and amplification of high impedance signals are susceptible to stray pickup of spurious electrostatic noise or to loss by leakage to ground. High quality and expensive shielded cable connects the measuring electrodes to the amplifier. The "leadless" electrodes require special desiccated and sealed housings to prevent signal loss at the junction of cable and electrode. Electrodes with integral cables require a higher electrode replacement cost.

Advances in miniaturization of electronics and in strength of components have permitted the construction of amplifiers which are mounted in the head of the measuring electrode (Figure 1.8a). Depending on the design, the

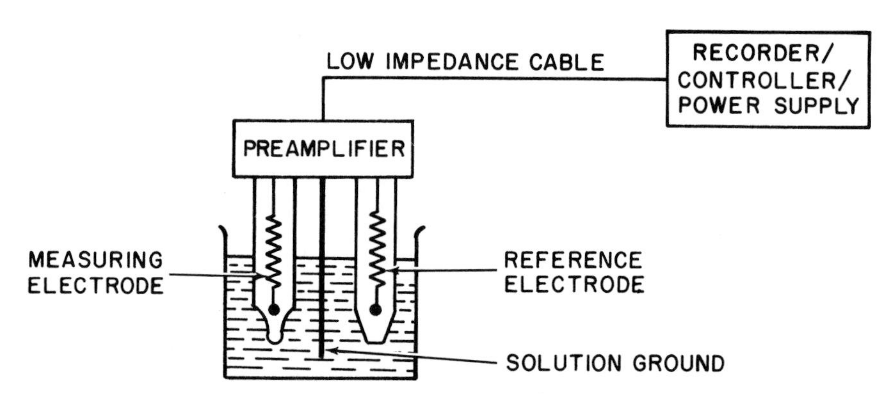

Fig. 1.8a Amplifier closely coupled or integrally mounted with electrodes to permit low impedance transmission lines

electrode may or may not be separated from the amplifier. The entire unit is sealed and in some designs may be submerged to 100 feet. The transmitted signal is of low impedance, thereby reducing the interference possibilities and shielding requirements. The cable contains the power supply wires in addition to the electrode wires. The upper temperature limit and the temperature stability of this unit are less than with remote amplifiers. Because of the encapsulated electronic components, servicing is limited to replacement of the sealed modular units.

Dual High-impedance Input Amplifiers

Amplifiers have progressed from the electrometer tube inputs of ten years ago to solid state components utilizing varactor diode bridges and field effect transistors (FETs). Poor voltage stability, long warm-up periods and susceptibility to microphonic noise are characteristics of the tube design. Varactor bridges are low drift devices. FETs have fair voltage stability and may require overvoltage protection on their gates. These varactor bridge or FET amplifiers can measure currents as low as 0.01 picoampere and are the basis for dual high-impedance input amplifiers commercially available.

In conventional amplifiers there is only one high-impedance input, and therefore the reference electrode operates at or near ground potential. Addition of a second high-impedance input for the reference electrode offers greater stability and accuracy. In addition, when these amplifiers are used with conventional reference electrodes, the system will operate properly even when the reference electrode resistance is increased from coating or plugging of the junction area.

Figure 1.8b shows a process amplifier with high-impedance input for both measuring and reference electrodes. Each of the two input amplifiers, 1 and 2, make a differential measurement relative to the solution ground, and the difference between the two measurements is the desired Nernst equation potential (amplifier 3). The solution ground electrode, which may be an inert metal or the pipeline containing the process, does not need to be at a well-

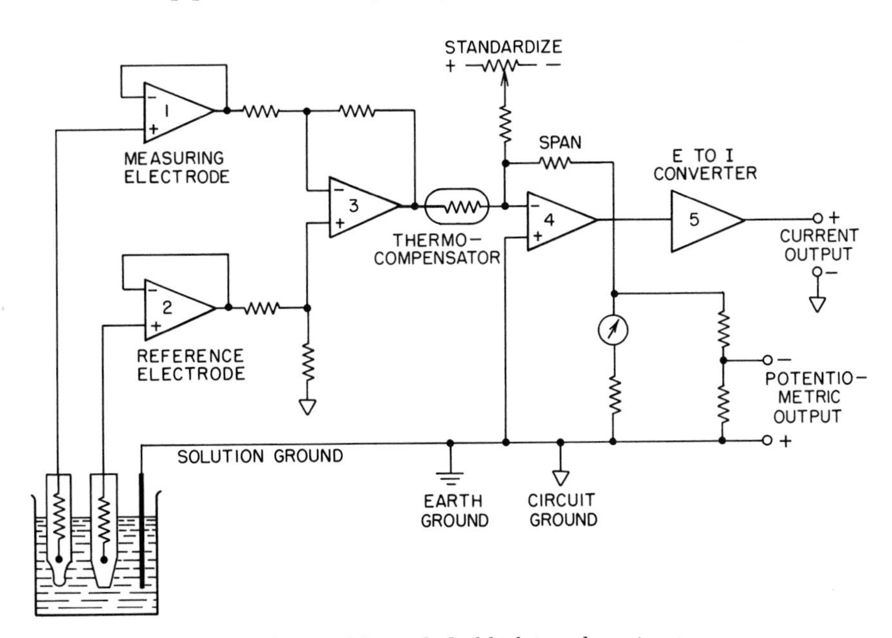

Fig. 1.8b Amplifier with dual high-impedance inputs

established potential because in the difference measurement its potential is cancelled. This solution ground is also connected to the earth ground and provides a low-impedance return path for any stray ground loop currents.

To illustrate the advantage of this design, consider the problem of maintaining a process solution, such as a one molar (1 M) strong acid at a constant concentration:

$$\text{Ag, AgCl, KCl (4 M)} \mid H^+ \text{(1 M)} \qquad 1.8(1)$$

The potential (E) generated by a glass electrode is given by the equation

$$E = E^\circ + (2.3\ RT/F) \log a_{H^+} + E_J \qquad 1.8(2)$$

where E_J is the liquid junction potential. However, it is possible to construct a new cell with a matched glass electrode as the internal element of the reference electrode.

The electrolyte surrounding this internal element is selected because it has the same 1 M acid value as that which is desired at the control point of the process:

$$H^+ \text{(1 M)} \mid H^+ \text{(1 M)} \qquad 1.8(3)$$

The emf of this cell would be given by

$$E = E^\circ_{asy} + (2.3\ RT/F)\ (\log a_{H^+} - \log (a_{H^+})_R + E_J \qquad 1.8(4)$$

In this equation, R represents the reference electrode and its electrolyte and E°_{asy} is the asymmetry potential observed when the two glass electrodes are immersed in the same solution (less than 5 millivolts for a matched pair of electrodes). Table 1.8c summarizes the characteristics of the two cells. Calculations for this table use liquid junction potential data from Table 1.8d and neglected activity coefficients.

Table 1.8c illustrates that when the process solution is at the desired concentration, the control point signal with the ion-selective internal cell is 0.0 millivolt. This "zero" number is more stable for control purposes and

Table 1.8c
COMPARISON BETWEEN CONVENTIONAL AND ION SELECTIVE INTERNAL CELL DESIGNS

	E, the Electrode Potential in Millivolts	*dE/dT, the Thermal Sensitivity in Millivolts per °C*	*Ej, the Junction Potential in Millivolts*
Conventional cell 1.8(1)	400	1.4	−14.1
Ion selective internal cell 1.8(3)	±0.0	±0.0	±0.0

Table 1.8d
LIQUID JUNCTION POTENTIALS[1]

Description of Solution X	*Junction Potential (Ej) in Millivolts at 25°C Between Solution X and the Noted Concentration of KCl*	
	Saturated KCl	*One Molar KCl*
HCl, 1 M (pH = 0)	−14.1	−26.8
HCl, 0.1 M (pH = 1)	−4.6	−9.1
HCl, 0.01 M (pH = 2)	−3.0	
HCl, 0.01 M, KCl, 0.09 M, (pH = 2)	−2.1	
KH phthalate, 0.05 M (pH = 4.01)	−2.6	
KCl, 0.1 M (pH = neut)	−1.8	
KCl, 1.0 M (pH = neut)	−0.7	0.0
KCl, 4.0 M (pH = neut)	−0.1	
KCl, sat'd (pH = neut)	±0.0	
KH_2PO_4, 0.025 M; Na_2HPO_4, 0.025 M (pH = 6.87)	−1.9	
$NaHCO_3$, 0.025 M; Na_2CO_3, 0.025 M (pH = 10)	−1.8	
NaOH, 0.01 M (pH = 12)	−2.3	
NaOH, 0.10 M (pH = 13)	0.4	4.4
NaOH, 1 M (pH = 14)	8.6	19.1

more sensitive to concentration changes than the 414 millivolt value of the conventional cell because its temperature coefficient and liquid junction potential are both at "zero" whereas the conventional cell has significant and possibly variable contributions from these.

Because the user has the opportunity to select a filling solution for his reference electrode to match his process, the ion selective internal reference electrode coupled with the dual high-impedance input amplifier offers great promise for controlling chemical composition.

Another example of utility of dual high-impedance input amplifiers involves conventional cells, in which, if the reference electrode becomes partially coated or clogged and its resistance increases, current leakage through the non-isolated reference electrode may occur. A moderate leakage-current of 0.1 microampere passing through a reference electrode which has increased its resistance from 3,000 ohms to 30,000 ohms would result in a false potential drop of 3 millivolts. This is equivalent to 0.05 pH unit. With a high-impedance input circuit on the reference electrode, the error would not exist.

Reference Electrodes

The observed emfs of the electrochemical cells used for the determination of ionic species in solution are given by the Nernst equation. See equation 8.8(1) in volume I for pH electrode systems and equation 1.13(8) in this volume for the ion selective electrode systems. Both the pH and the ion selective

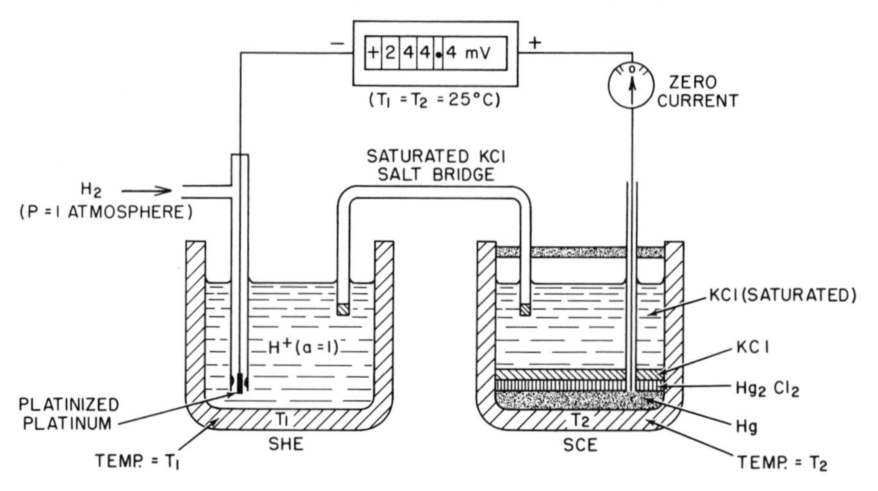

Fig. 1.8e Cell illustrating two standard reference electrodes, their mode of construction and measurement. Standard hydrogen electrode (SHE); saturated calomel electrode (SCE)

measurements require two electrodes: measuring and reference. The observed emf of the electrochemical cell E_{meter} (Figure 1.8e) can be represented by

$$E_{meter} = V_{meas} - V_{ref} + E_J \qquad 1.8(5)$$

where V_{meas} is the single electrode potential of the measuring electrode in the process solution, V_{ref} is the single electrode potential of the reference electrode, and E_J is the liquid junction potential. The single electrode potential is defined as the potential with reference to the standard hydrogen electrode and is denoted by the symbol V, whereas the emf of a cell composed of two working electrodes, as reflected by equation 1.8(5) is denoted by the symbol E. The function of the reference electrode is to maintain a fixed potential or a reference point for the measuring electrode. It should maintain that potential or point regardless of the species or concentrations of the ions present. The internal element of a reference electrode is in itself an ion selective electrode of potential V_{ref}, which operates in a controlled chemical environment. The electrical contact between its controlled internal solution and the variable process solution is the source of the electrical potential, E_J, the liquid junction potential at the tip of the reference electrode. This latter term is pertinent because the potential generated between the salt bridge solution and the process solution is neither zero nor constant, although in many cases both potentials can be made small.

Reference Electrode Principles

The potential of a reference electrode is in accordance with the Nernst equation. The principal difference between a reference electrode and a measuring electrode is the controlled nature of the chemical environment of the reference electrode. This condition is maintained by controlled concentrations of filling solutions. The reference electrode illustrations in this section show that the controlled internal filling solution is separated from the process. The point of contact is referred to as the junction potential zone. These junctions may be made of inert porous materials such as asbestos, nylon or cotton threads, porous ceramic, porous or cracked glass, wood, noble metals, porous plastic or other inert materials that permit controlled leaks.

For any reference electrode, the working potential, V'_{ref}, may be represented by adding the term for the liquid junction potential to the Nernst equation for the basic electrode system

$$V'_{ref} = V^{\circ}_{ref} + (2.3\,RT/nF)\ \log a_i + E_J \qquad 1.8(6)$$

where V°_{ref} is the standard potential of the reference electrode and a_i is the activity of the ion which determines its working potential.

LIQUID JUNCTION POTENTIALS

Liquid junction potentials arise when two solutions of different compositions are brought into contact. This is what happens when the tip of the reference electrode comes into contact with the process solution. The potential arises from the interdiffusion of the ions in the two solutions and is called the liquid junction or diffusion potential, denoted by E_J.

Since ions diffuse at different rates, the charge will be carried unequally across the solution boundaries, resulting in a potential difference between the two solutions. Liquid junction potential may be calculated from the mobility of ions across the boundary by using the Henderson equation.[1] Data of Table 1.8d reflect the magnitude of typical liquid junction potentials at the tips of reference electrodes in various "real" solutions. It is particularly large when the junction involves the extremely mobile hydrogen or hydroxyl ions. Saturated potassium chloride solutions, which are close to the ideal salt bridge solutions, have junction potentials that vary from −14 millivolts in a one molar (1 M) acid solution to 9 mv in 1 M alkaline solution but are nearly zero in neutral solutions. If, for example, a pH electrode system were standardized in a biphthalate buffer at pH 4.01, ($E_J = -2.6$ mv) and then used to measure a 1 M hydrochloric acid solution at pH = 0 ($E_J = -14.1$ mv), an error of −11.5 mv (−0.19 pH unit) would have been introduced even though the electrode system were functioning properly. For pH measurement these errors can be considered negligible but for ion selective measurements, in which 1 mv absolute measurement error is equivalent to 4% or 8% relative error in concentration (Figure 1.13 l), they are of greater significance. Table 1.8d also shows that changing the composition or concentration of the bridge electrolyte changes the junction potential. Smaller junction potentials are obtained with the 4 M KCl solution, which is considered to be the ideal salt bridge solution, than are obtained with the 1 M KCl solution.

Table 1.8d is based on a theoretical estimation. The actual magnitude of this junction potential is also affected by the construction of the junction. Extraneous and spurious potentials may be created by mechanical strain or chemical contamination of this junction. The flow at the junction tip may be fast or slow, the junction itself may be made of glass, ceramic, porcelain or plastic and precipitation or crystallization may occur on it. It is desirable that there be a small continuous flow of the reference electrode bridge filling solution into the sample solution.

The velocity of this flow should be sufficient to overcome the backflow of the process and thereby minimize backdiffusion effects. Otherwise, there would be a steady build-up of sample ions inside the junction, which might contaminate or plug it and then contaminate the internal electrolyte. The constant flow of bridge electrolyte may be maintained by a gravity head or by pressurizing the electrolyte reservoir.

The methods for eliminating or minimizing junction potentials include a reference electrode without a liquid junction. In practice, this means that the reference electrode is actually an ion selective electrode and that the process solution is constant in composition with reference to an ion other than the ion being measured. For example, if the pH is to be measured in a solution in which the fluoride ion activity is constant, then a fluoride ion reference electrode without liquid junction can be employed. In chemical processes, this condition seldom exists, but by using a reagent addition system to mix the process sample with a constant amount of standard reagent (fluoride), it may be attained. Another method to minimize the liquid junction potentials is a double junction reference electrode assembly (Figure 1.8 l). The conventional reference electrode is separated from the process by an auxiliary reference electrode tip and a bridge solution. The bridge solution is selected for high conductivity and compatibility with the process.

This remote junction assembly is very practical for process measurements and solves problems relating to temperature and solution interaction.

TEMPERATURE COEFFICIENTS

Isopotential data for electrode systems, which are necessary for manual or automatic temperature compensation, are derived from electrode temperature coefficients. In the usual operation, the two electrodes are at the same temperature (isothermal operation) and the temperature coefficient of the system is the difference between the coefficients of the individual electrodes. When a remote junction reference electrode is employed at fixed temperature (thermal operation) as in Figures 1.8e or 1.8 l, then only the temperature coefficient of the measuring electrode need be considered.

Both the measuring and reference electrodes are characterized by the Nernst equation:

$$V = V^\circ + (0.19841T/n)\log a_i. \qquad 1.8(7)$$

and the electrode temperature coefficient is then given by

$$(dV/dT) = dV^\circ/dT) + (0.19841/n)\log a_i$$
$$/ + (0.19841T/n)d(\log a_i)/dT \qquad 1.8(8)$$

This equation indicates that there are three characteristic terms influencing the temperature coefficient of any given electrode.

1.) A constant, (dV°/dT), is characteristic of the electrode.

2.) A term which varies with the activity of the ion in which the electrode is immersed is: $(0.19841/n)\log a_i$. For a reference electrode with its internal element in a captive solution of a strong electrolyte, it is a constant; for a measuring electrode, it depends on the activity of the test solution.

3.) A term which varies with the temperature coefficient of the solution in which the electrode is immersed is: $(0.19841T/n)d(\log a_i)/dT$. For the strong electrolyte of the reference electrode, this term is mainly due to the volume expansion of the solution with temperature and may be neglected; for the measuring electrode, it may be a complex function of the chemistry of the solution, especially if weak electrolytes are involved.

For reference electrodes, the temperature coefficient, $(dV/dT)_{ref}$, is considered to be constant within a few hundredths mv per °C over a moderate temperature range.

Electrodes may be characterized by two temperature coefficients—the isothermal temperature coefficient, $(dV/dT)_{isoth}$, and the thermal temperature coefficient, $(dV/dT)_{th}$. The mode of measurement and the significance of these coefficients are illustrated in Figure 1.8e. The isothermal temperature coefficient is computed from measurements made on an isothermal cell consisting of the electrode under test and the standard hydrogen electrode (SHE). The thermal temperature coefficient is computed from measurements made on a thermal cell in which the electrode is exposed to changing temperatures while the reference electrode is held at constant temperature. In the latter case, the constant temperature reference electrode may be the SHE or any other convenient reference electrode type including the electrode type under test. The two electrodes at different temperatures are connected by a thermal salt bridge and the thermal emf generated in this bridge, amounting to a few microvolts per °C, is usually neglected. The difference between the two temperature coefficients is fundamental. The isothermal temperature coefficient is a characteristic of both electrodes (the electrode under test and the SHE), whereas the thermal temperature coefficient is characteristic only of the electrode under test. The potential of the SHE is zero at all temperatures, by definition. This is an unfortunate choice because the SHE actually has its own thermal temperature coefficient of 0.871 mv per °C.

The thermal temperature coefficient is the more useful of the two because it is characteristic only of the electrode under discussion. Table 1.8f gives a list of temperature coefficients for commonly used reference electrodes.

The temperature coefficient of a cell utilizing a reference electrode and a pH, ORP or ion selective measuring electrode is given by

$$(dE/dT)_{cell} = (dV/dT)_{meas} - (dV/dT)_{ref}. \quad 1.8(9)$$

Although the temperature coefficients of conventionally constructed reference electrodes are available in Table 1.8f, temperature coefficient constants for measuring electrodes, which could be obtained from equation 1.8(8), are not generally available because of the proprietary nature of internal electrode

Table 1.8f
THERMAL TEMPERATURE COEFFICIENTS OF COMMON REFERENCE ELECTRODES

	Reference Electrode	*Thermal Temperature Coefficient (Millivolts/°C)*
SCE	Hg, Hg_2Cl_2 KCl (sat'd, 4.16 M)	+0.17
	Ag, AgCl, KCl (1 M)	+0.25
	Ag, AgCl, KCl (4 M)	+0.09
	Ag, AgCl, KCl (sat'd, 4.16 M)	+0.1*
SHE	Pt, H_2 (p = 1), H^+ (a = 1)/sat'd KCl	+0.87

* Estimated at about 25°C.
Note: A positive thermal temperature coefficient means that the warmer electrode is the (+) terminal in a thermal cell. Isothermal coefficients can be computed from the thermal coefficients by subtracting 0.871 mv per °C.

designs and filling solutions. Whenever possible, measuring electrode internals and reference electrodes should be matched to provide a minimum temperature coefficient. Most of the temperature compensators in the receiving instruments correct only for the temperature effect on the slope term of the Nernst equation (0.19841T/n) and not for mismatched measuring and reference electrodes or for solution temperature. However, units with adjustable calibration accommodate a wide range of isopotential points and specific solution effects.

Standard Reference Electrodes

STANDARD HYDROGEN ELECTRODE

The universally adopted primary standard reference electrode is the standard hydrogen electrode (SHE) (Figure 1.8e) represented by

$$Pt, H_2 (p = 1 \text{ atm.}), H^+ (a = 1)| \quad 1.8(10)$$

It consists of platinum wire or foil, coated with platinum black immersed in a hypothetical solution of unit hydrogen ion activity and in equilibrium with hydrogen gas at unit partial pressure. The potential of this standard electrode is defined as zero millivolts at all temperatures, even though its thermal temperature coefficient is 0.871 mv per °C, as was discussed earlier. The standard hydrogen electrode is not convenient to use in most process and laboratory solutions because it requires a hydrogen supply, it is easily fouled and many reducible substances interfere with its operation.

SATURATED CALOMEL ELECTRODE

One of the secondary reference electrodes which is widely used because of its stability in variable solutions and under differing conditions is the saturated calomel electrode (SCE) shown in Figure 1.8e and represented by

$$Hg, Hg_2Cl_2, KCl\,(sat'd.)| \qquad 1.8(11)$$

It consists of a layer of mercury covered with a paste of mercury, mercurous chloride and potassium chloride, all in contact with a solution saturated both in potassium chloride and in mercurous chloride. A laboratory standard SCE with readily flushable and reproducible liquid junction is shown in Figure 1.8g. The industrial version of this electrode is shown in Figure 8.8a in Volume I. The potential of the saturated calomel electrode is 244 mv when compared with that of the SHE at 25°C (Figure 1.8e).

The concentration of the potassium chloride solutions used as the electrolyte in the calomel electrode may vary. The saturated calomel electrode (SCE) and the normal calomel electrode (NCE) derive their names from the concentrations of the potassium chloride in each electrode (saturated and one normal [1 N] respectively). Calomel electrodes with potassium chloride concentrations of 3.8, 3.5, 3 and 2 M are also used from time to time as are mixtures termed equitransferent solutions. The user is warned against altering internal solution compositions because the absorbent construction of the internal element creates a "memory" for previous solutions, and weeks may be required for the electrode to assume a new equilibrium potential.

The SCE is not as widely used in the process industry as the silver-silver chloride reference electrode. The mercury and calomel internals are toxic and by regulation are forbidden for use in the food and pharmaceutical industries and by industrial requirements in the photographic industry. SCE at elevated and variable temperatures is less reproducible than are the silver-silver chloride electrodes. To achieve temperature stability and temperature compensation, the reference electrode should be matched to the internal reference of the measuring electrode (For more on temperature compensation, see Section 1.13 in this volume and Section 8.8 in Volume I). Many modern pH and ion selective measuring electrodes use silver-silver chloride internals as reference elements.

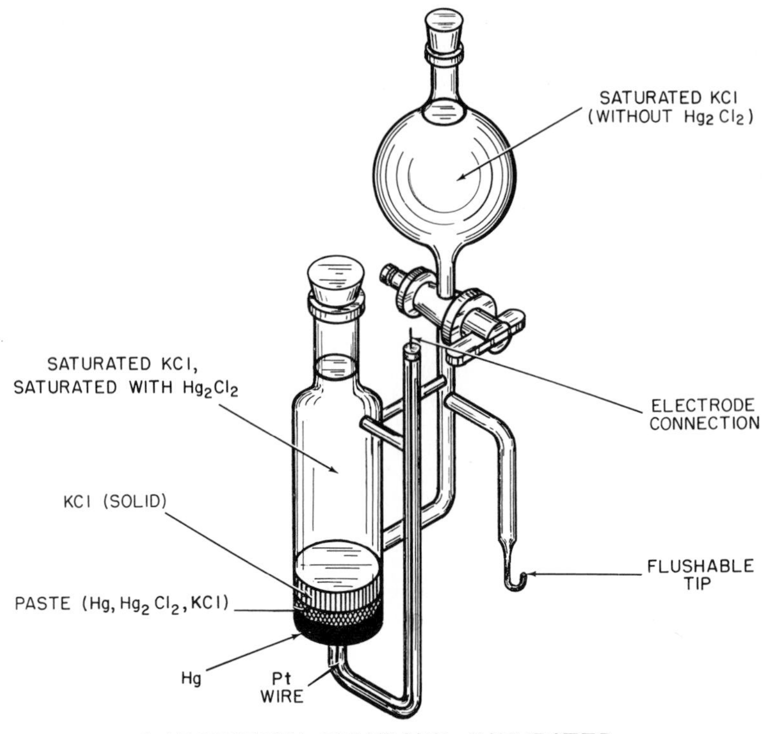

Fig. 1.8g Laboratory standard saturated calomel electrode (SCE)

SILVER-SILVER CHLORIDE, 4 M KCl ELECTRODE

This electrode (Figure 1.8h) consists of a silver-silver chloride (Ag, AgCl) solid internal in contact with an electrolyte solution of potassium chloride (KCl) saturated with silver chloride. The silver may be in the form of wire, rod or tube or may be electrolytically deposited on platinum. The silver chloride may be electrochemically or thermally deposited or mechanically packed around the silver.

The concentration of the potassium chloride electrolyte in both the silver-silver chloride and the mercury-calomel reference electrodes may vary, but the same warning against alteration of the filling solution applies. At ambient temperatures, and for practical purposes, the 4 M KCl solution may be regarded as saturated since it is actually saturated at 19.4°C, whereas a saturated KCl solution at 25°C is 4.16 M.

The saturated (or 4 M) potassium chloride solution is used with the most common reference electrodes because it is relatively easy to prepare and does not require accurate weighing or volume measurements. Also, at room temperature its concentration is relatively constant and its electrical conductivity is the largest possible for such a salt solution, thus contributing to an electrode with minimum electrical resistance. Potassium chloride solutions give minimum liquid junction potentials (Table 1.8d).

The main disadvantages include the change in concentration of saturated potassium chloride solutions with temperature and the high solubility of silver chloride in concentrated potassium chloride solutions. The concentrations of saturated solutions in the temperature range of 0° to 100°C are given in Table 1.8i. The change from 3.4 M to 5.8 M within this temperature range corresponds to an emf change of 28 mv (0.28 mv per °C). Excess solid KCl crystals must therefore be present; otherwise, as the temperature rises, the solution may become unsaturated. The internal electrolyte is saturated with respect not only to potassium chloride but also to silver chloride. Table 1.8i also gives

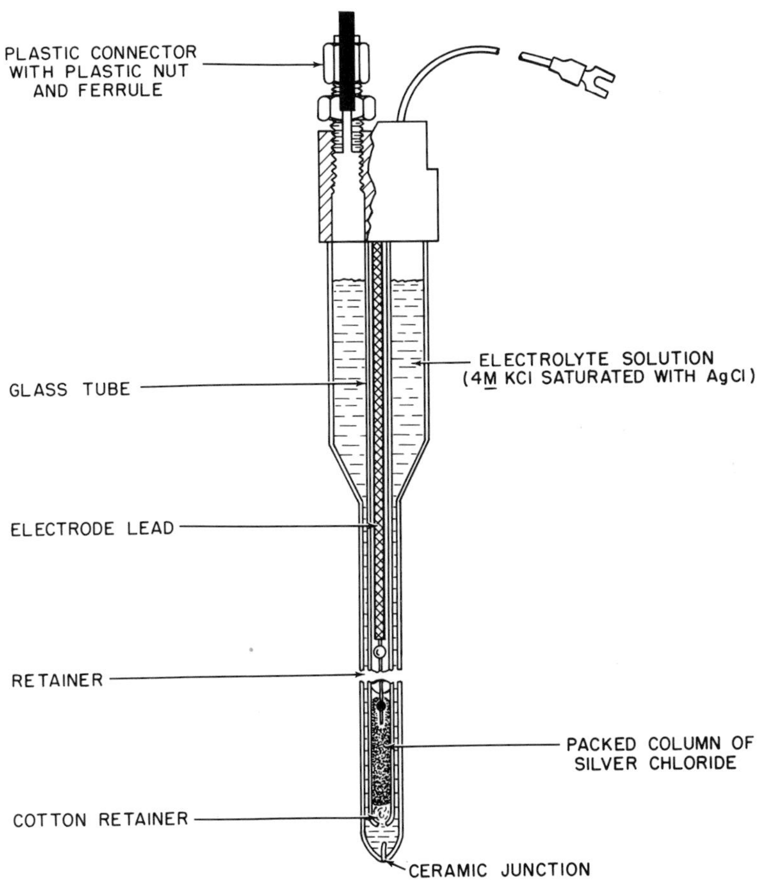

Fig. 1.8h Silver-silver chloride, 4 M potassium chloride process reference electrode

Table 1.8i
SOLUBILITIES AS A FUNCTION OF
TEMPERATURE AND CONCENTRATION

	Temperature (°C)					
	0	19.4	20	25	50	100
KCl Solubility in Moles/Liter	3.39	4.00	4.02	4.16	4.80	5.84
AgCl Solubility in Various 25°C KCl Solutions in Milligrams/Liter	0.10 M/L KCl Solution			4×10^{-4}		
	1.00 M/L KCl Solution			14.4		
	4.00 M/L KCl Solution			913		
	4.16 M/L KCl Solution			1,000		

the solubility of silver chloride in potassium chloride solution at 25°C (77°F). In water and in dilute potassium chloride solutions, silver chloride is considered to be an insoluble salt.

However, in concentrated potassium chloride solutions, a soluble complex silver chloride ion is formed. The data in Table 1.8i show that the solubility of silver chloride in 4 M potassium chloride solution at 25°C is 913 mg per liter. Solubility is 65 times greater than in a 1 M potassium chloride solution. When a reference electrode saturated with respect to both potassium chloride and silver chloride cools, both of these salts will crystallize in the body and in the junction region. If the process solution contains ions such as sulfide, carbonate or phosphate that form insoluble silver salts, the silver ion from the dissolved silver chloride may also form an insoluble precipitate at the electrode tip. Because of its greater solubility, much more silver ion or silver chloride is available for precipitation from a 4 M or saturated potassium chloride electrolyte than from more dilute solutions. Solid matter in the porous junction is obviously undesirable.

SILVER-SILVER CHLORIDE, 1 M KCl ELECTRODE

Low or varying temperatures and increased stability demands justify the selection of this design. Advantages of the 1 M KCl electrolyte include operation at temperatures as low as −5°C (22°F) without crystallization, substantial reduction in solubility of silver chloride and salt concentration still large enough to give low electrical resistance.

SILVER-SILVER CHLORIDE, NON-FLOWING ELECTRODE

Maintenance of a positive flow of potassium chloride across the junction of the reference electrode prevents contamination of its internal portions by the process fluid and ensures that nearly theoretical internal and liquid junction potentials are achieved under operating conditions. The quantity of flow needed to achieve this state varies depending on the operating conditions of the electrode tip. Rapid flow rate may be required if the process fluid is viscous or colloidal, tends to crystallize or leaves deposits on electrode surfaces. (Ultrasonic cleaning is discussed later in this section.) High-flow electrodes pass about 1 to 10 ml per day. In order to achieve a one- to six-month maintenance schedule, a corresponding reservoir size is needed. In processes at high or variable pressure, pressurizing devices are used. (See Figures 8.8d, e, and g in Volume I.) Low flow reference electrodes (0.1 to 1 ml per day) are utilized in many clean processes operating at atmospheric or low pressures. The newly introduced electrodes termed "no-flow" or "nonflowing" reference electrodes are sealed devices requiring no pressurization, no electrolyte reservoir, no maintenance for as long as one year and a very low electrolyte flow as shown in Table 1.8j.

Table 1.8j
CHARACTERIZATION OF FLOW-RATES FOR REFERENCE ELECTRODES

Classification	*Volume flow (ml/day)*	*Mass flow (mg KCl/day)*
High-flow	1 to 10	100 to 3,000
Low-flow	0.1 to 1	10 to 300
No-flow	0.01 to 0.1	1 to 30

The high-flow and low-flow reference electrodes (Figure 1.8h), combine porosity and pressure drop across the tip to control and regulate flow. The no-flow electrodes approach diffusion rates (Figure 1.8k). The electrode is made of plastic or composite material, usually a fluorocarbon. It is inert, resistant to coating and less fragile than glass. It is also opaque so as to eliminate photosensitivity of silver chloride internals, which has been a minor problem with transparent glass body electrodes. The junction may be composed of a thin porous wall of the same plastic material as the body or may be an inserted ceramic, wood or other inert porous material.

The electrode internal components are silver-silver chloride and the electrode is filled with a mixture of saturated potassium chloride, saturated silver chloride solution and a paste of these two chemicals. The sealed construction and thin wall of the electrode at the tip permit equalization of pressure inside and outside the electrode so that no net pressure difference exists to cause flow across the junction. Diffusion flow across this boundary (in both directions) is about 0.01 ml daily, or 4 ml per year. Several grams of solid potassium chloride in the sealed electrode is sufficient to saturate the small amounts of process solution which may diffuse in during the life of the

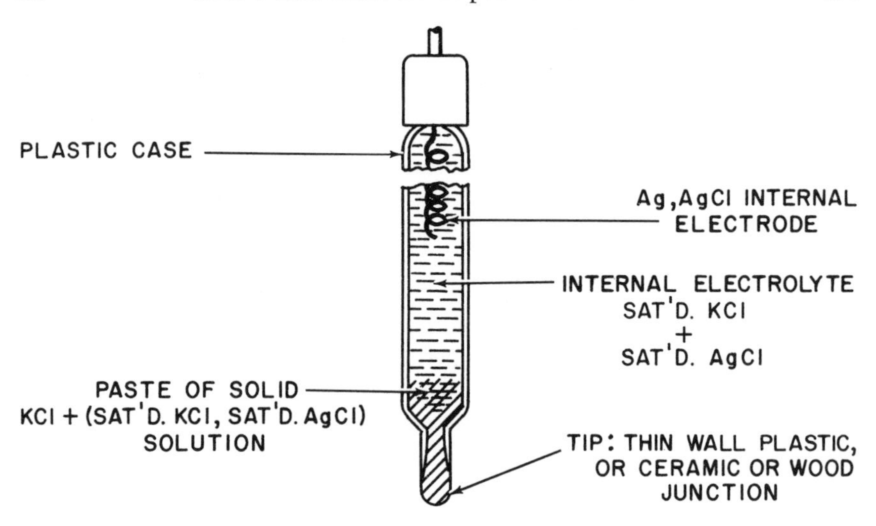

Fig. 1.8k Sealed "no-flow" reference electrode

electrode. Minor changes in other ionic constituents will affect the total ionic strength—and thus the activity of the chloride ion to a slight extent—but the potential of the internal electrode will remain reasonably constant.

Process solutions may contain ions which "poison" the internal electrode, i.e., react with silver to produce new electrode potential systems. Such ions include bromide, iodide, cyanide, sulfide, alkaline solutions of phosphate, carbonate and many others. The well designed no-flow electrodes will compensate for this interference because their internal elements are separated from the tip, and therefore from the process solution, by a protective layer containing 1,000 mg per liter of silver ions (Table 1.8i). These silver ions will precipitate ions that could change the potential of the internal silver-silver chloride electrode and permit it to maintain its proper potential for a long time.

The junction potential problems also apply to the no-flow sealed internal reference electrodes. It will be recalled that the junction potential has two portions. One is theoretically predictable and arises from the interface between the internal saturated potassium chloride solution and the process. This potential has been listed in Table 1.8d for a variety of solutions. Frequently, it is small and may be acceptable as long as it is constant. The second junction potential arises from the contaminants in the junction region—the "memory" of the preceding process solution. This would include crystallization of potassium chloride or silver chloride in the tip or the precipitation of materials that react with the relatively large concentrations of potassium chloride and silver chloride in the internal compartment of the electrode. This memory is not a serious factor with conventional flowing junctions which are self-cleaning and have indefinite life. The no-flow reference electrode may dry out if it is removed from the process solution, and as a consequence may develop very high resistance and require soaking in potassium chloride solution before being reusable. This difficulty is minimized if a dual-input, high-impedance amplifier is used.

The no-flow reference electrode should be very useful with pH electrodes which have similar silver-silver chloride, potassium chloride internals, and monitor relatively clean water at reasonably constant temperatures at atmospheric pressure wherein 0.1 to 0.2 pH units (6 to 12 mv) furnishes satisfactory accuracy. As the process solutions, temperatures and pressures become more inimical and as accuracy requirements increase, the reduced long-term stability of the no-flow reference electrodes must be evaluated against the advantages of simpler installation and minimum maintenance.

THALLIUM AMALGAM AND OTHER ELECTRODES

A variation on the construction of the saturated calomel electrode uses thallium amalgam and thallous chloride to replace the mercury-mercurous chloride internal system:

$$\mathrm{Tl(Hg),\ TlCl(solid),\ TlCl(sat'd.),\ KCl(sat'd.)}\,| \qquad 1.8(12)$$

It is known as the thalamid electrode, the advantages of which are its usability at higher temperatures and its greater freedom from temperature hysteresis. The potential of the 40% amalgam relative to the saturated calomel electrode at 20°C is −816 mv and relative to the silver-silver chloride, 4 M KCl electrode is −771 mv. The measuring electrode must have similar thallium amalgam internals to avoid a serious mismatching of the thallium amalgam reference electrode.

Other variations on the calomel electrode include the mercury-mercurous sulfate, saturated potassium sulfate electrode which is particularly useful when chloride is objectionable at the tip of the reference electrode. A remote junction reference electrode with saturated potassium sulfate bridge solution could probably solve the same problem (Figure 1.8 l). The potential of the mercury-mercurous sulfate (saturated) electrode at 25°C is 414 mv versus the SCE and 459 mv versus the Ag-AgCl, 4 M KCl reference electrodes.

Interest has also been revived recently in what is termed the mercury-mercurous carboxylate electrodes. Carboxylate, in this case, may mean acetate, oxalate, propionate or benzoate—sometimes in mixed solvents. Except for certain special cases, these electrodes are not likely to be considered.

A convenient method of comparing the reference electrodes with each other is to review the so-called potential tree of Figure 1.8m. This figure compares "working" electrodes with the primary standard electrode—the SHE—and permits ready comparison with the SCE which is generally accepted as the laboratory standard for reference electrodes. To determine

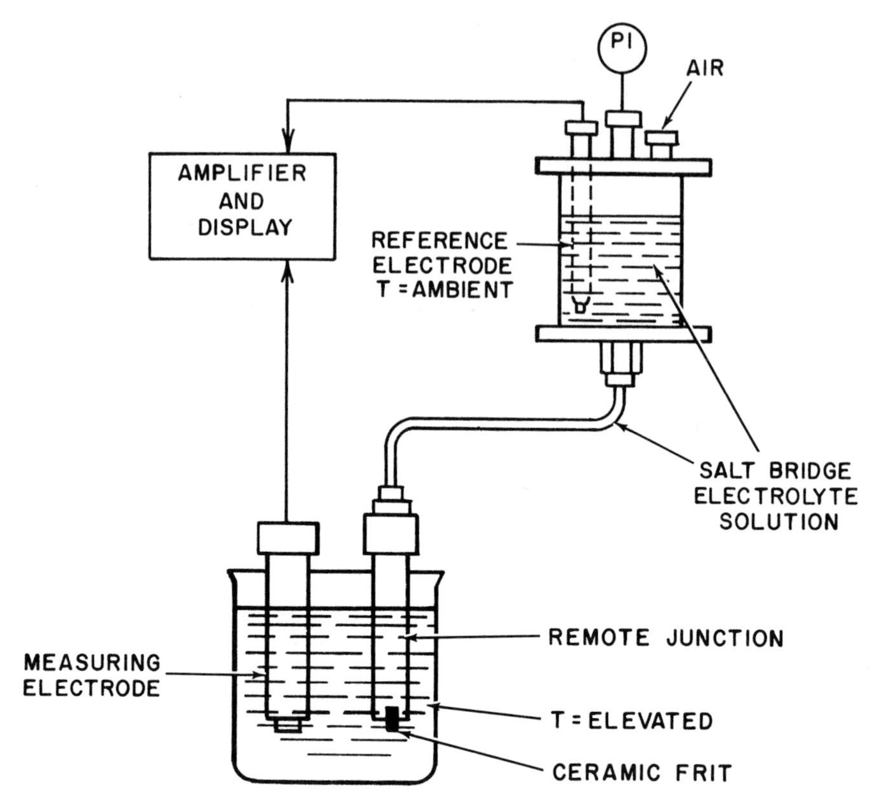

Fig. 1.8l Electrode measuring system using remote junction reference electrode

potentials at temperatures other than 25°C, Figure 1.8m should be used with the reference electrode temperature coefficients of Table 1.8f.

ION SELECTIVE INTERNAL REFERENCE ELECTRODES

The choice of the internal elements for the reference electrodes has been severely restricted by the requirements of potentiometric measurement systems until recent years. These requirements dictated that the reference electrode of the circuit be of low impedance, usually less than 10,000 ohms although the measuring electrode circuit could be 1,000 megohms or higher. Dual-channel, high-impedance amplifiers have removed this limitation. These amplifiers permit new reference electrodes to be designed by removing the internal element of a conventional reference electrode and inserting an ion selective electrode in its place. The internal filling solution is selected to be compatible with the internal electrode and with the process solution. The construction of the liquid junction potential region of the reference electrode remains the same. It is logical to apply this construction to the most common ion selective electrode—the glass pH measuring electrode. Figure 1.8n illustrates an ion selective internal reference electrode, utilizing a glass pH electrode. The internal electrolyte solution selected may be optimized to the control point of the process. Table 1.8c and the associated discussion illustrate the advantages of ion selective internal reference electrodes in combination with dual high-impedance input amplifiers.

Applications of the ion selective internal reference electrode appear to comprise a fertile field for imaginative users. At the time of this writing, more than twenty ion selective measuring electrodes are commercially available but the dual-channel, high-impedance amplifiers have not yet been marketed long enough to be reliably evaluated.

Selection Criteria for Reference Electrodes

The following guidelines are intended to aid in the optimum selection of a reference electrode for a particular application.

MATCHING ELECTRODE INTERNALS

The reference electrode internal element should be matched to the internals of the measuring electrode. For example, a pH-measuring electrode

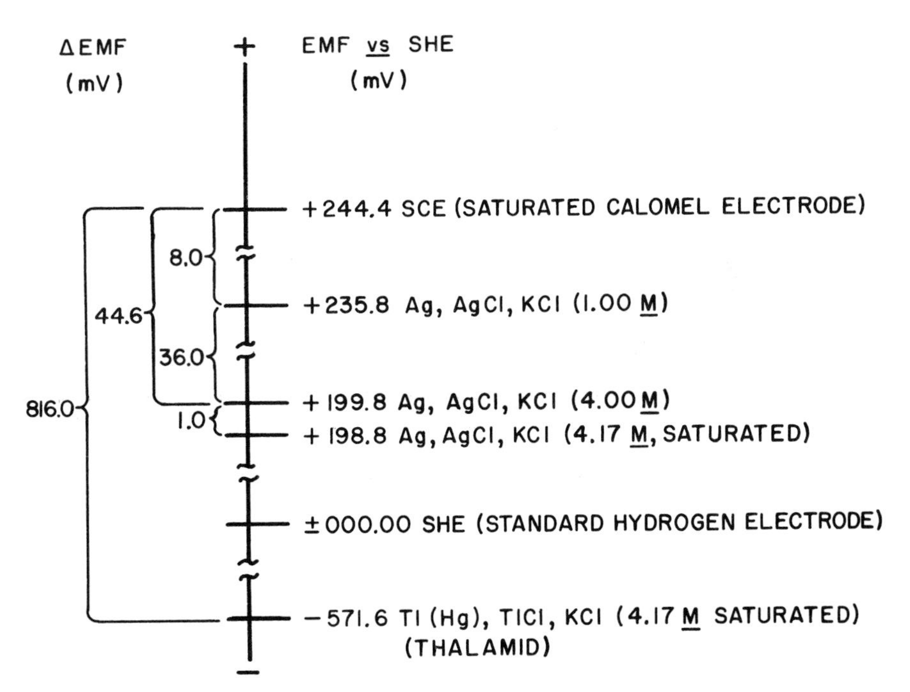

Fig. 1.8m Potential tree with potentials of commonly used reference electrodes (in millivolts at 25°C)

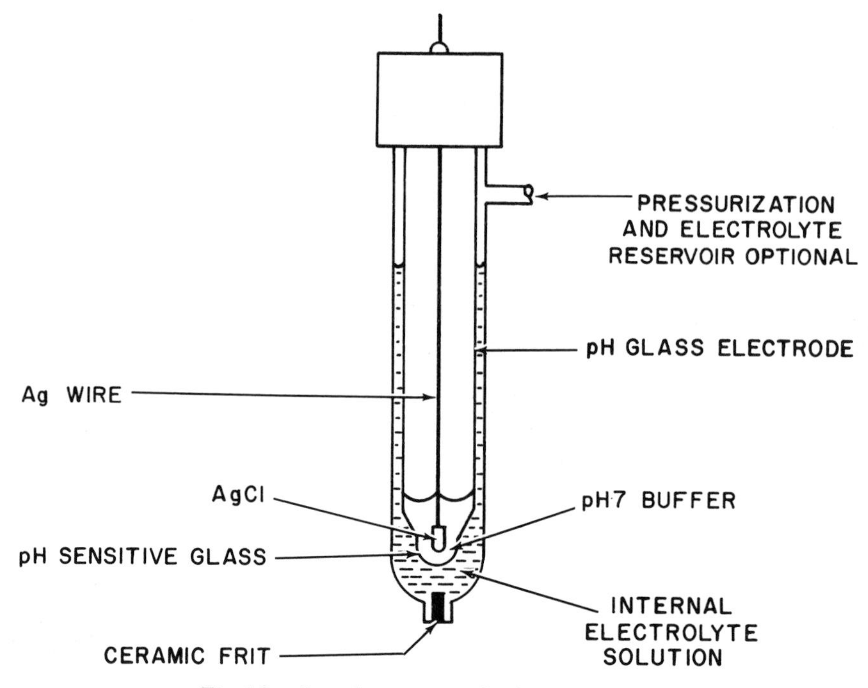

Fig. 1.8n Ion selective internal reference electrode

with silver-silver chloride internals contains a mixture of 4 M potassium chloride and a buffer at pH 7. It should be matched with a reference electrode with silver-silver chloride internals and a 4 M potassium chloride filling solution. With this combination the measuring system gives zero mv and zero mv per °C temperature coefficient in pH 7 solution. Extreme mismatching, as for example if thallium amalgam internals were used in only one of the electrodes, might result in emf differences sufficiently great (Figure 1.8m) as to lie outside the standardization capability of most instrumentation. Mismatching also causes increasingly larger errors because of the difference in temperature coefficients the measured emf departs from the isopotential point (Figure 8.8b in Volume I).

For ion selective electrodes, in which internals and filling solutions are sometimes proprietary, manufacturers' recommendations and data for zero potential and isopotential points need to be consulted.

Reference electrodes with filling solutions that are unsaturated over the range of operating temperatures will give optimum performance insofar as reproducibility, temperature hysteresis and junction potential deviations are concerned. The great solubility of silver chloride and mercurous chloride in strong potassium chloride solutions should be considered when the process solution contains halide, sulfide, phosphate or carbonate ions, with which they might react.

pH-measuring electrodes are available with a choice of silver chloride, calomel or thallium amalgam internals, and corresponding reference electrodes are available for each. Silver chloride internals are the first choice for process measurements, although in transparent electrodes they may exhibit variable photosensitivity. The calomel and thallium internals are both toxic substances, and the problems attendant on breakage in the process must be considered. Best performance at elevated or variable temperatures is attributed to the thallium electrodes, with the silver chloride and calomel electrodes following in that order.

FLOW RATE

The more rapid flow rates give more reproducible reference electrode potentials. The penalty for the faster flow rate is the increased likelihood of process solution contamination with reference electrode electrolyte and the need for more frequent refilling of the pressurized electrolyte reservoir.

The so-called no-flow or sealed reference electrodes may permit development of spurious or drifting junction potentials if there are rapid process changes or difficult conditions. Their saturated filling solutions and susceptibility to process contamination make them less reproducible (and accurate) than the conventional flowing and unsaturated electrodes. If low cost of maintenance is more important than high accuracy, this design is the proper choice.

JUNCTION MATERIALS

The junction region must satisfy the sometimes contradictory requirements of chemical and physical inertness, electrical neutrality, low resistance and, for most reference electrodes, flow restriction. Ceramic, glass and fluorocarbons are ordinarily inert to most environments, whereas wood, metal and fiber junctions must be carefully evaluated for the application. Noble metals can create oxidation potentials in solutions containing dissolved air. Miscellaneous filling materials, such as are used in some glasses and in many plastics, may also cause spurious junction potentials.

ACCURACY

The accuracy obtainable from potentiometric measurement systems may be related to the emf errors (Figure 1.13p). Overall accuracy of ±1 mv for prolonged intervals may be expected from well designed systems. This limit would correspond to ±4% relative error in univalent ion concentration measurements. To achieve this accuracy, a high-flow or low-flow reference electrode with provisions for keeping process liquids out of its internal compartment is required. Temperature control and ultrasonic cleaning may also need to be considered.

For many potentiometric measurements, accuracy as high as this is not required. For example, some pH monitoring applications are acceptable with accuracies at the ±0.1 to 0.2 pH unit level, corresponding to a ±6 to 12 mv system error. In such cases, saturated and non-flowing reference electrodes are satisfactory and offer the bonus of very low maintenance requirements.

Electrode Cleaning Devices

In Section 8.8 of Volume I the problems of both measuring and reference electrode fouling are discussed. The measuring electrode may have a deposit on its sensitive membrane that interferes with its ability to sense accurately the bulk ion activity in the solution and increases its electrical resistance to the point that it becomes sluggish or inaccurate or disappears altogether. The reference electrode may be subjected to similar interferences and, in addition, the coatings may create spurious and unstable liquid junction potentials that interfere with calibration and create erroneous interpretation, drifting and eventual failure of the system emf by open circuiting. In relatively pure water supply systems, the combination of hardness, alkalinity and phosphate or carbonate ions can cause slow deposition of calcium phosphates or carbonates on the surface of measuring and reference electrodes. The viscous syrup of the sugar processing industry, even at elevated temperature, crystallizes on the electrodes in a matter of hours. The use of lime to control pH in effluent treatment systems, the suspensions of pulp in the paper industry and the oils, colloids and emulsions in many waste streams have created pH measurement problems which necessitate the periodic cleaning of the electrodes.

Frequently, a simple wiping or immersion in a suitable chemical cleaning solution will restore them to optimum operating condition, but this requires measurement interruption and the attention of maintenance personnel. In critical pH monitoring and control, redundant pH sensors are installed to avoid interruption during regular maintenance cleaning which may be as frequent as once per eight-hour shift. The cost of this attention and the critical nature of the measurement have warranted a number of custom cleaning installations. They utilize periodic injection of suitable solvents (such as acid for cleaning calcium carbonate deposits) or periodic activating of mechanical devices (such as brushes in close proximity to the electrodes).

One approach that has been successfully employed in recent years is the use of ultrasonic energy. Units for ultrasonic cleaning of pH and ion selective measuring systems (both immersion and flow-through) are commercially available. Much reduction in electrode maintenance has been reported including a report of reduction in frequency of electrode removal from once every eight hours to once every three months.

Ultrasonic Principles

Sound energy with frequency higher than can be heard by human ears is referred to as ultrasonic. Ultrasonic frequencies may range from several kilohertz (kHz) to ten megahertz (MHz), although a common range of ultrasonic cleaning equipment is from 20 kHz to 80 kHz. Ultrasonic energy may be used in transmission of information or in transmission of energy. Examples of information transmission include thickness monitoring and flow or level measurements, detection of defects in metals, sonar devices for fish or submarine location and medical diagnoses. Applications of energy transmission include testing metals and other materials for fatigue, chemical processing, emulsification, bacterial destruction, gas and bubble removal, particle dispersal and dust collection.

Ultrasonic cleaning is an example of mechanical energy transmission. One kind is due to the "cavitation effect" produced by vibrations in which tiny cavities are first formed and then collapse within the solution. The implosions produce localized pressures that can be very high, the resulting shock waves tend to dislodge particles from the immersed object.

A second type of cleaning action is attributed to the acceleration of physical and chemical reactions at the interface between the dirt and the bulk liquid. The acceleration is explained by the increased surface available for reaction, agitation, bubble removal, emulsification and other factors caused by the generation of sonic energy.

Ultrasonic Electrode Cleaners

Figure 1.8o shows how ultrasonic transducers may be coupled to flow-through assemblies to produce pH or ion selective ultrasonic cleaning systems. Similar systems are available for immersion units. The ultrasonic transducer is generally a crystal vibrator and contains piezoelectric crystals (lead zirconate titanate or barium titanate) that oscillate at a characteristic frequency when a driving voltage is applied. The sound energy is transmitted to the solution by a stainless steel or titanium diaphragm. Electromechanical vibrators and induction converters may also be used as ultrasonic transducers. The ultrasonic energy, at a typical frequency of 40 kHz and at a power level of 10 to 100 watts, is directed near the tips of the probes. The ultrasonic transducer is coupled to the ultrasonic power generator by a cable for transmitting the excitation voltage. The power generation oscillator is fitted with a transistor and may be equipped with a timing device to permit intermittent operation, with typical times being 5 to 15 minutes per hour.

In the design of the ultrasonic pH cleaner the sonic energy must be large enough to produce a cleaning action on the sensitive electrode surfaces, but not so large as to "sandblast" the surface, shatter the electrodes or remove coatings of silver chloride from the internal reference electrodes. Some ultra-

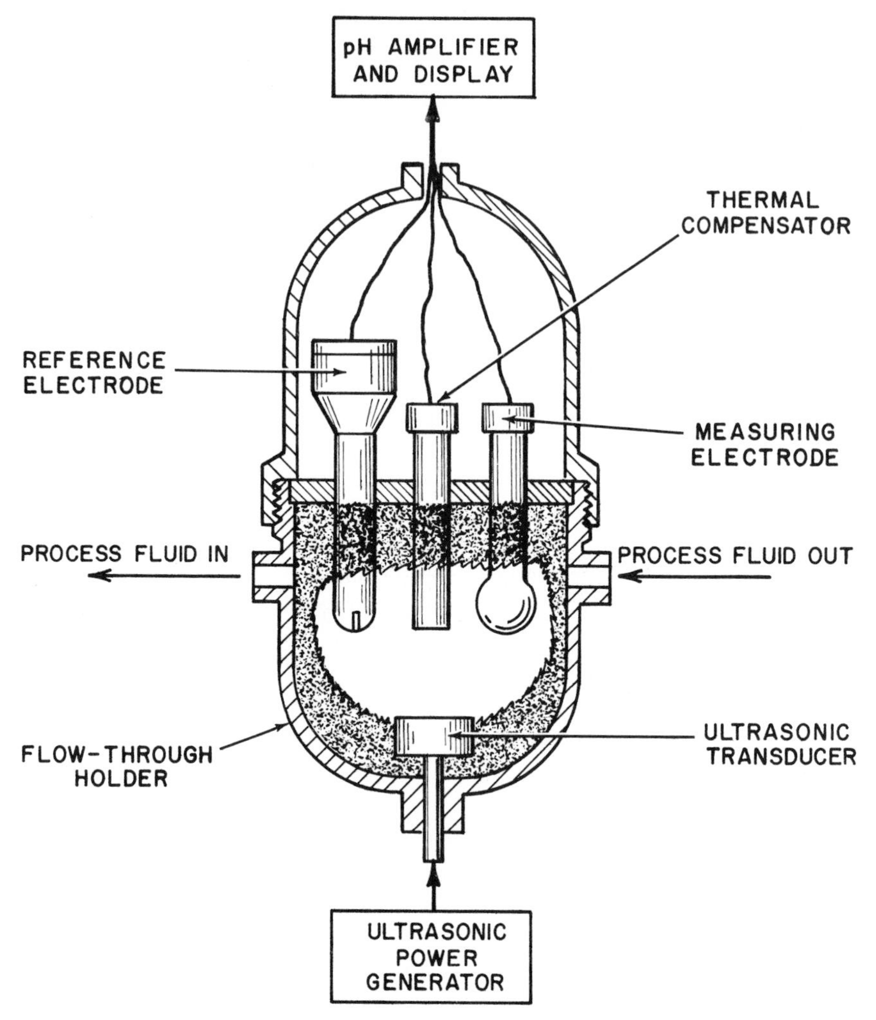

Fig. 1.8o Ultrasonic cleaner for flow-through pH assembly

sonic cleaner systems are designed to operate continuously at low power levels and thereby prevent film formation rather than remove the film after it has been deposited. Experimentation may be needed to establish the duty cycle—the minimum ratio of time on to time off to maintain satisfactory continuous operation of the electrode system.

Precautions

One set of precautions is concerned with the harmful effects of ultrasonic energy on the body, whereas the other has to do with the hazards associated with the high voltages used for the generation of ultrasonic energy. Ultrasonic vibration, even at the lowest energy levels, should not come into contact with the body. Ultrasonic vibrating devices or cavitating liquids can produce irreversible tissue destruction. The damage can be cumulative even though initial effects are not immediately recognizable. It is conceivable that an automated ultrasonic power generator could be activated while the pH system is being serviced. In such cases, the maintenance technician could be unknowingly exposed to ultrasonic energy unless suitable electrical interlocks are provided.

Equipment can be damaged if the unit is empty and dry when the ultrasonic generator is activated. Audible noise is usually at low levels and is not considered to be a hazard with ultrasonic cleaner systems.

At power levels up to 100 watts, voltage outputs of several hundred to a few thousand volts are not uncommon. Explosion-proof housings and conduits with proper electrical grounding are required. The wire and insulation should have proper ratings and the electronic circuitry should have the necessary isolation.

Electrode Cleaner Applications

Coatings that interfere with potentiometric electrode measurements include 1. coatings formed from insoluble salt crystals that adhere to electrode membranes and bodies; 2. coatings formed by the coagulation or settling of suspended material, solutions, gels or emulsions; and 3. coatings formed by oily or other nonaqueous films.

These coatings in turn may occur because of one of the following process operations:

1. Processes which involve saturated or nearly saturated solutions due to temperature variations or other causes.

2. Waste water disposal processes in which hydrates of many metals may be found. Frequently these hydrates, (iron and aluminum hydroxides) are gelantinous and "sticky".

3. Processes which produce precipitates as in the addition of lime to remove carbonates, phosphates or fluorides.

4. Processes with suspended solids in the flow streams such as in paper pulp and fiber manufacturing.

5. Emulsions of mixed aqueous and nonaqueous solutions, including oily contamination of waste streams.

Ultrasonic electrode cleaning systems have been satisfactory for many of these difficult process types.

Conclusions

Developments in pH measurements and in reference electrodes have been closely linked to ion selective electrode technology. The introduction of amplifiers which are mounted integrally with the pH electrodes and the development of dual high-impedance amplifiers have improved the performance of potentiometric measurement systems. The dual high-impedance amplifier has permitted the introduction of the ion selective internal reference electrode. This combination permits direct potentiometric measurements rivaling titrimetric procedures in accuracy, not only for pH determinations but also for the many analytical procedures that may now be applied to online monitoring and control by ion selective measuring electrodes.

Fast-flow, non-saturated reference electrodes permit the achievement of very high measuring accuracy, whereas non-flow saturated reference electrodes eliminate maintenance requirements for as long as one year. The introduction of continuous ultrasonic cleaners contributes to the reduction of maintenance by automatically removing the coatings and suspensions that interfere with electrode operation.

BIBLIOGRAPHY

Bates, R. G., "Determination of pH, Theory and Practice," New York: Wiley, 1964.

Bleak, T. M. and Sawa, K. B., "A new process pH analyzer design." In "Analysis Instrumentation," (Vol. 9). Proceedings of the 17th National ISA Analysis and Instrumentation Symposium. Houston, 1971.

deBethune, A. J., Light, T. S. and Swendeman, N. J., "The temperature coefficients of electrode potentials." *J. Electrochem. Soc., 106:* 616, 1959; also in Hampel, C. A. (ed.) Encyclopedia of Electrochemistry. Reinhold, N.Y., 1964.

Brand, M. J. D. and Rechnitz, G. A., "Differential potentiometry with ion-selective electrodes, a new instrumental approach." *Anal. Chem., 42:* 616, 1970; "Fast-response differential amplifier for use with ion-selective electrode." *Anal. Chem., 42:* 1659, 1970.

Covington, A. K., "Reference electrodes." In Durst, R. A. (ed.) "Ion-Selective Electrodes." Washington, D.C.: U.S. Government Printing Office, 1969.

Ives, D. J. G. and Janz, G. J., (eds.) "Reference Electrodes." New York: Academic, 1961.

Light, T. S., "An improved reference electrode for process measurement." In "Analysis Instrumentation," (Vol. 8). Proceedings of the 16th National ISA Analysis Instrumentation Symposium. Pittsburgh, 1970.

Light, T. S., "Industrial analysis and control with ion-selective electrodes." In Durst, R. A. (ed.) "Ion-Selective Electrodes." Washington, D.C., U.S. Government Printing Office, 1969.

Mattock, G., "pH Measurement and Titration." New York: Macmillan, 1961.

Neti, R. M. and Jones, R. H., "Performance and application of a new reference electrode for process potentiometric measurements." In Analysis Instrumentation (Vol. 8). Proceedings of the 16th National ISA Analysis Instrumentation Symposium. Pittsburgh, 1970.

1.9 PHYSICAL PROPERTIES ANALYZERS

Analyzer Types:

(a) DISTILLATION ANALYZERS,
(a1) packed column end point analyzer,
(a2) film evaporator type distillation analyzer,
(a3) packed column initial boiling point (IBP) analyzer,
(a4) atmospheric pressure IBP analyzer,
(a5) packed column vacuum distillation analyzer,
(a6) automatic process analyzer,
(b) VAPOR PRESSURE ANALYZERS,
(b1) air saturated type,
(b2) dynamic type,
(c) VAPOR/LIQUID (V/L) RATIO ANALYZERS,
(c1) batch type,
(c2) continuous type,
(d) POUR POINT ANALYZERS,
(d1) tilting plate type,
(d2) moving ball type,
(d3) dropping plummet type,
(e) CLOUD POINT ANALYZERS,
(e1) optical type,
(e2) convective heat transfer type,
(e3) crystal formation type,
(f) FREEZING POINT ANALYZER,
(f1) aviation fuels,
(f2) hydrocarbon purity sensor,
(g) FLASH POINT ANALYZERS,
(g1) low range,
(g2) integrated,
(g3) high range,
(h) OCTANE ANALYZERS,
(h1) standard engine comparitor,

(h2) continuous reactor tube type.

Note: In the feature summary below, the letters (a1) to (h2) refer to the listed analyzer types.

Sample Flow Rate (GPH): 0.1 to 0.2 (a4), 0.4 (a2,a6), 1.0 (g1,g2,g3), 2.0 (b1,h2), 3 to 4 (a5,c2,e1,e2,f1), 4 to 8 (a1,a3), 12 to 20 (b2), 2 PSID analyzer drop (f2).

Sample Pressure (PSIG): 10 to 200 mm Hg (a5), 0 to 140 (a2), 0.5 to 5 (g1), over 1.5 (d2), 10 to 100 (b1), 10 to 500 (g2), 15 to 50 (c1), 20 to 90 (e1,g3), 30 to 150 (a4), up to 35 (d1), over 40 (b2), 50 to 250 (e1), 60 to 300 (a1,a3,a5), up to 70 (h2), up to 200 (d3,e3,f1).

Sample Temperature (°F): 20° below bubble point (h2), 30° below flash point (g1), 50° below flash point (g3), 250° above flash point (g2), 40 to 90 (b1), 50° to 150° (a2,b2,d3,e1,e2,e3), up to 180° (a1,a3,a5,d2).

Wetted Parts: Copper (f2) stainless steel (a1, a2,a3,a4,a5,c1,e2,g1,g2,g3).

Area Classification for Field Components: Class I, Group D, Division 1 explosion-proof (a1,a2,a4,a5,a6,b1,b2,c1, d1,d2,d3,e1,e2,f1,f2,g1,g2,g3,h2).

Analysis Range: up to 20% evap. (a4), up to 50% evap. (a3), over 30% evap. (a1), 5 to 97% evap. (a2,a5), 98 to 100% purity (f2), IBM to end point (a6), 2 to 19 pounds RVP (b1), 0 to 20 pounds RVP for gasolines (b2), 0 to 150 PSIG for LPG (b2), 10 to 30 V/L (c1), 0 to 40 V/L (c2), to −90°F (f1), −50 to +5°F (d1), −40 to +60°F (d3,e3), −30 to +60°F (d2, e2), −12 to +50° F (el), 50 to 250°F (g1,g2), 250 to 600°F (g3), span of 2 octane numbers (h2).

Repeatability: per ASTM (a6,d3,g3), ±0.01°F (f2), ±0.5°F (f1), ±1°F (a2,e1,e2), ±1.5°F (d2), ±2°F (d1,g1,g2), ±1% of boiling range (a1,a3,a5), ±0.1

	pounds RVP (b1,b2), ±0.2 V/L (c1), ±0.5 V/L (c2), ±0.1 RON (h2).
Response or Cycle Time in Minutes:	0.5 (b2), 0.5 to 1 (c2), 1 (b1), 2 (e2,h2), 2 to 5 (a2,a4,g1,g2), 5 to 17 (d1), 6 (c1), 10 (e2,f2), 10 to 20 (a3), 16 (a1,a5), 20 to 30 (d2,d3,e3), 25 (a6).
Approximate Cost:	\$3,500 to \$4,000 (g1), \$4,000 (d1,f2), \$5,250 (a4), \$5,500 (e1), \$6,000 (a2), \$6,500 (a1,a3,e2), \$5,800 to \$6,900 (b2), \$6,900 (b1), \$7,500 (d2,g3), \$8,500 (g2), \$10,000 to \$12,000 (c1), \$13,000 (a5), \$15,000 (a6,h1), \$23,000 (c2), \$24,000 (h2).
Partial List of Suppliers:	E. I. du Pont de Nemours (h1), Ethyl Corp. (c2,h1), Hallikainen Instruments (a2,a4,b2,c1,d2,d3,e2,e3,f1, f2), Precision Scientific Development Co. (b1,d1,g1,g2,g3), Technical Oil Tool Corp. (a1,a3,a5,a6,e1), Universal Oil Products Co. (h2).

This section deals with onstream analyzers which measure a physical property of a process stream.

Prior to the introduction of onstream analyzers, laboratory analyses were made on grab samples at intervals, and the results were reported to the process unit operator, permitting set point adjustments of parameters such as flow, temperature, pressure and level. Continuous onstream plant analyzers offer many advantages including

1.) continuous measurement of stream, eliminating high time lags;
2.) reduction of errors caused by unrepresentative samples, or by changes in sample composition due to sample handling;
3.) elimination of human errors characteristic of non-automated laboratory procedures;
4.) ability to recognize process trends, permitting the direct control of a given process variable by closed loop control;
5.) cost reductions resulting from the minimization of laboratory analyses;
6.) closer control with narrower tolerances of final product specifications, and reduction in quality "give-away";
7.) feasibility of establishing inline blending systems, resulting in

economies from elimination of tankage and from increasing system flexibility and quality control;

8.) ability to provide continuous inputs to computerized process control systems for plant optimization; and

9.) direct measurement of process variables rather than detection of properties by inference.

Distillation Analyzers

Laboratory Measurements

Distillation analyzers were introduced to provide data on the volatility characteristics of process streams and separation efficiency of distillation units. ASTM methods D86—IP-123 and D216—IP-191 are the currently accepted laboratory standards for determining the boiling characteristics of petroleum products distilled at atmospheric pressure. Both methods employ batch techniques and approach a single plate distillation process without reflux. The petroleum products analyzed are complex mixtures of many components, and a low level of fractionation is achieved. True boiling point distillations, in columns with 15 to 100 theoretical plates, and at reflux ratios of 5 to 1 or more, produce greater separation of components. The apparatus and procedure for true boiling point determinations are not standard, are complex, take longer to perform and are not as widely used.

Distillation curves for a few hydrocarbons are shown in Figure 1.9a together with a comparison of curves generated by ASTM method D86—IP-123 and by true boiling point determinations for kerosene.

ASTM METHOD D86—IP-123

A 100 ml sample is heated in an Engler flask at a prescribed rate. Packing is not used and some refluxing occurs due to condensation. (Figure 1.9b.) The vapors produced flow through a condenser immersed in an ice-water bath, and the distillate is collected in a graduate cylinder. The initial boiling point temperature is defined as the reading of the thermometer at the instant when the first drop of condensate falls from the lip of the condenser tube. As the higher boiling fractions vaporize, condense and collect in the graduate, corresponding temperature readings are recorded to permit the plot of a curve of temperature versus percent of sample recovered. The end point or final boiling point is defined as the maximum thermometer reading observed during the test; it usually occurs when all the liquid has been boiled off from the bottom of the flask.

Usually, the percent recovered does not equal the 100 ml sample charge, partly due to the inability of the apparatus to condense the lightest fractions. A curve of temperature against percent evaporation is determined by adding the percent of light ends lost to each of the recorded percentages recovered.

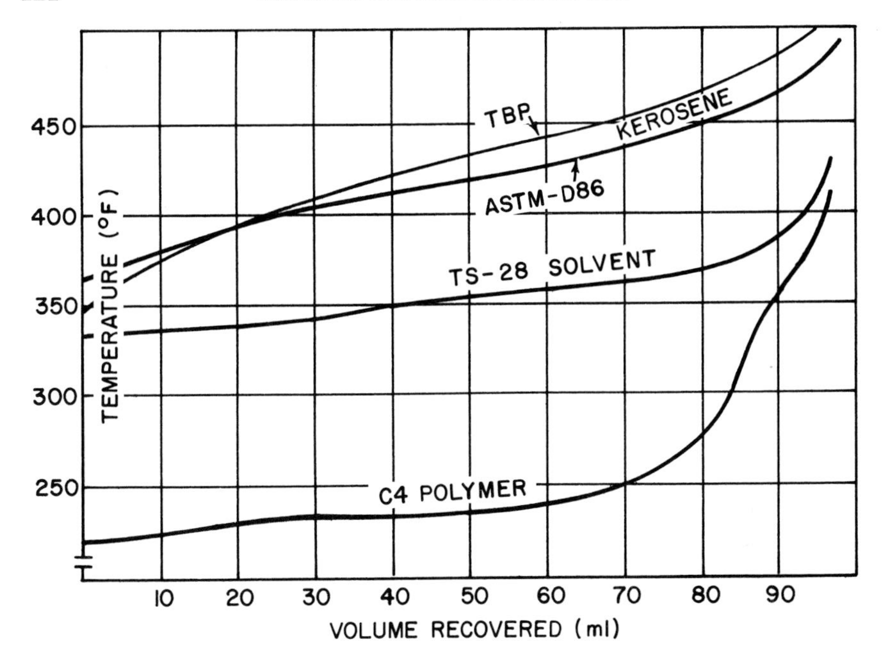

Fig. 1.9a Distillation curves

The precision of this method is a function of the rate of temperature change. Repeatability ranges from 1° to 5°F, and the reproducibility from 4° to 13°F.

ASTM METHOD D-1160

This method provides for measurements under vacuums, ranging from 1 mm Hg absolute to atmospheric, to a maximum liquid temperature of 750°F. Results are not comparable with other ASTM distillation tests although they may be converted to corresponding vapor temperatures at 760 mm by reference to Maxwell and Bonnell vapor pressure charts.

The sample must be moisture-free, and is equivalent to a volume of 200 ml at the temperature of the receiver in which the condensed overheads are collected. The sample charge is boiled at a rate which produces a recovered distillate of approximately 4 to 8 ml per minute. The overhead vapors are condensed in a jacketed condenser and receiver in which the circulating coolant is maintained within ±5°F in the range of 90° to 170°F. Measurement of the vapor temperature is by a special Kovar-tipped thermocouple located at the sidearm of the boiling flask leading to the condenser. Temperatures for the 5, 10, 20, 30, 40, 50, 60, 70, 80, 90 and 95 percent recovered volumes are reported unless the liquid temperature, measured by a mercury in glass thermometer positioned in the boiling flask, reaches a value of 750°F. In these cases the test is terminated and the percent recovered is also reported.

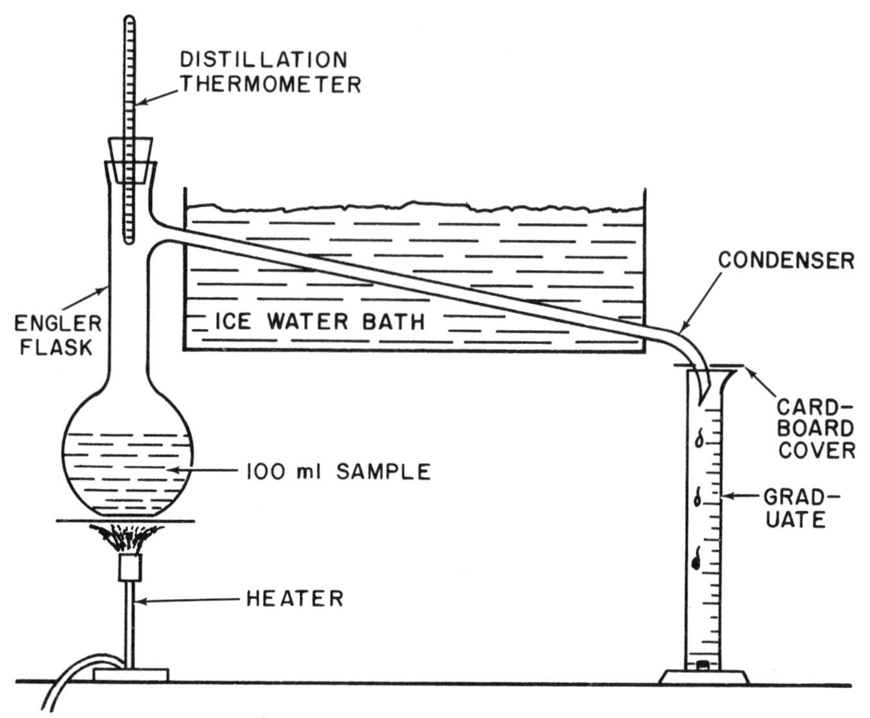

Fig. 1.9b Apparatus for ASTM D-86 distillation test

Depending on the percent recovered and the operating pressure, repeatability varies from 8° to 10°F and reproducibility from 15° to 30°F.

This method requires considerably greater skill to obtain optimum results and is more complicated than the more popular ASTM method D86—IP-123.

Plant Distillation Analyzers

Continuous plant distillation analyzers may be correlated with the results obtained by the ASTM laboratory methods but are not an exact duplication. In the laboratory (batch technique) the temperature (of a given percent evaporated value) is read off a rapidly rising vapor temperature as the sample is evaporated, whereas in the plant the analyzer measures the temperature of a continuous process, in which an equilibrium has been established for the given percent evaporated to be monitored.

PACKED COLUMN END POINT ANALYZER (a1)

As shown in Figure 1.9c, the analyzer is a miniature process unit. A conditioned and pressure regulated sample is delivered at a rate of approximately 4 gallons per hour to the top of a packed column through an inlet valve. The inlet sample flow rate is governed by a float in a boiling pot below

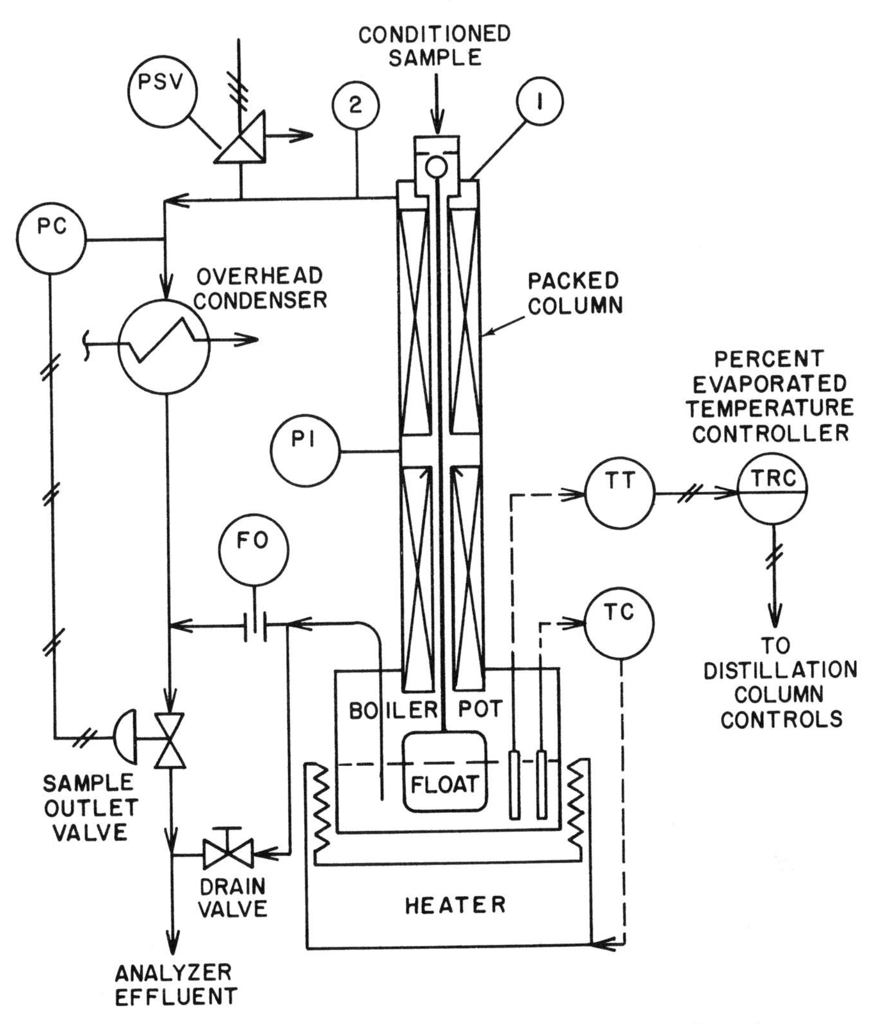

Fig. 1.9c Packed column end point analyzer. In an initial boiling point analyzer, the temperature is detected in location 1 and a restriction orifice is added in location 2.

the column which maintains an essentially constant level in the pot. A radiant heater boils the sample in the pot under an elevated pressure which is controlled. The overheads are condensed at this pressure. The bottoms flow from the boiler pot is metered by means of a restriction orifice upstream of the outlet control valve.

Because the orifice and the packed column are subject to the same differential pressure, the ratio of overheads to bottoms is fixed. The orifice size and heater wattage are selected for the particular sample and for the percent evaporated point to be monitored. The bubble point temperature of the bottoms can be correlated with the percent evaporated point to be monitored. As the sample's percent evaporated temperature changes, the distillation process within the analyzer correspondingly adjusts the temperature in the boiler pot.

Sample effluent may be returned to a pressurized line by an eductor or receiver and pump. The analyzer's thermocouple read-out is calibrated by determining the temperature differential between distillation under the elevated pressure and the results of ASTM method D86—IP-123 measurements carried out at atmospheric pressure. The analyzer may be used for measurements from 50 percent evaporated temperatures to roughly the end point temperature.

FILM EVAPORATOR TYPE DISTILLATION ANALYZER (a2)

This analyzer measures any single percent temperature or series of percent evaporated temperatures between the 5 and the 97 percent points over a temperature range of 150° to 600°F.

Figure 1.9d shows a single point analyzer. A conditioned sample is metered into the boiler section of the analyzer at a fixed rate by one head of a duplex diaphragm pump. The incoming sample is preheated by the exiting sample and by the condensing vapors inside its boiler chamber before discharging into a distributor or flash cup. For low boiling point temperatures and for very volatile products, a water cooling jacket is substitued for the preheat exchanger, and the internal heat exchanger coil is eliminated. The sample is allowed to spill onto a $\frac{1}{16}$-inch-diameter tubular heater wound in a tight spiral around a thin walled, vertical stainless steel tube. As the sample, following the heater's spiral path, flows downward, lower fractions boil off at atmospheric pressure and the remainder is collected in a bottoms cup.

The bottoms cup is connected to head 2 of the duplex metering pump. The pumping rate of the second head of the pump can be set to withdraw bottoms at any desired percentage of the input sample flow. A float in the bottoms cup is coupled to a light shield which regulates the incident light reaching the photocell as a function of the level. The photocell and a simple solid state circuit modulate the power to the heater so as to maintain a constant level in the bottoms cup. The percentage of the input sample vaporized (or overheads) is thus fixed, and the temperature of the residue or bottoms sample dropping off the end of the coiled heater is at its bubble point temperature, which is measured by a thermocouple immersed in the bottoms cup and which can be correlated with the corresponding percent evaporated temperature as determined by ASTM method D83—IP-123.

The pumping rate of the bottoms pump establishes the percent evaporated to be measured and is unaffected by the sample composition. As a result, several different streams may be monitored without changing the analyzer operating control settings. To change the percent evaporated point to be

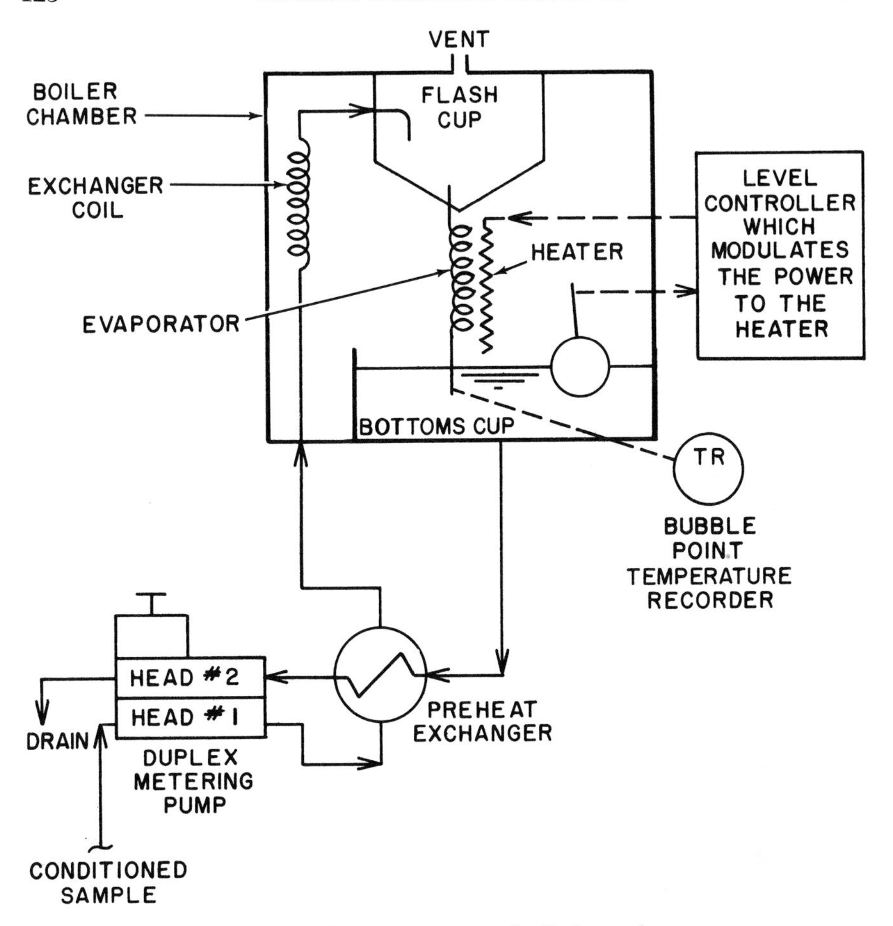

Fig. 1.9d Film evaporator type distillation analyzer

monitored the bottoms pumping rate may be changed. If the streams differ widely, the correlation factors with ASTM method D83—IP-123 may differ. For volatile samples such as gasolines, the heater may be partially coated with ceramic boiling chips.

A variation of this design permits programmed or remote selection of the percent evaporated temperature to be measured by pneumatically positioning the stroke adjusting mechanism of the bottoms pump. One programmer version cycles the analyzer to measure the 10, 30, 50, 90 and 95 percent evaporated temperatures at fixed periodic intervals.

PACKED COLUMN INITIAL BOILING POINT ANALYZER (a3)

The analyzer operates in much the same manner as the packed column end point boiling point analyzer described in Figure 1.9c. One difference is that the conditioned and pressure regulated sample first enters a preheater coil at the top of the boiler pot and is then fed to the top of the packed column. The inlet sample flow rate is controlled by a float-actuated valve in the boiler pot to maintain a constant level, and the pot is heated at a constant rate by a radiant heater. However, a restricting orifice is located in the overhead line at the top of the packed column. Where required, a restrictor may also be used in the bottoms line.

The sample is boiled at a controlled elevated pressure. The split of distilled sample into overhead and bottoms fractions is determined by the restrictions in these lines. Their ratio is constant since they are both subject to the same differential pressure. A thermocouple in the overhead line at the top of the column measures the overhead vapor temperature and serves as the analyzer read-out. By the suitable selection of restrictors and heater size, a given percent evaporated temperature between the initial boiling point to the 50 percent point can be measured.

ATMOSPHERIC PRESSURE INITIAL BOILING POINT ANALYZER (a4)

A conditioned sample is metered by a displacement pump and is preheated by the sample effluent in a tube-in-tube heat exchanger (Figure 1.9e). In the boiler the sample is partially vaporized at atmospheric pressure by a cartridge heater, the power output of which is fixed by a constant voltage transformer. The sample volume is also fixed by an overflow weir which is designed to serve also as a vapor trap. As sample overflows at the top of the boiler it is continuously replenished by fresh sample. The upper portion of the boiler consists of a water cooled condenser jacket and a coaxially positioned special thermocouple constructed so that its supporting sheath is heat sunk by the water jacket. The rising, light end vapors condense on the cooled walls of the upper water jacketed section and on the thermocouple tip.

Under steady state conditions a series of reflux rings are established with the lightest fraction condensing essentially at the tip of the thermocouple and at the lower portion of the condenser jacket. The thermocouple tip is positioned to the level at which the lightest vapors condense, thus providing a measure of the initial boiling point temperature. As with the mercury in glass thermometer used in the ASTM method, a heat path is established along the stem of the thermocouple to simulate the same dynamic response characteristics. A small air stream at the top of the condenser purges this section and prevents non-condensable gases from accumulating.

For higher percent evaporated temperatures, an adjustable glass stem Kovar-tipped thermocouple which provides a reduced heat sink path is used. The tip of the thermocouple is positioned closer to the boiling sample surface where the reflux ring of particular interest is situated. As many as four thermocouples may be used simultaneously to read four different percent evaporated temperatures up to the 20 percent point.

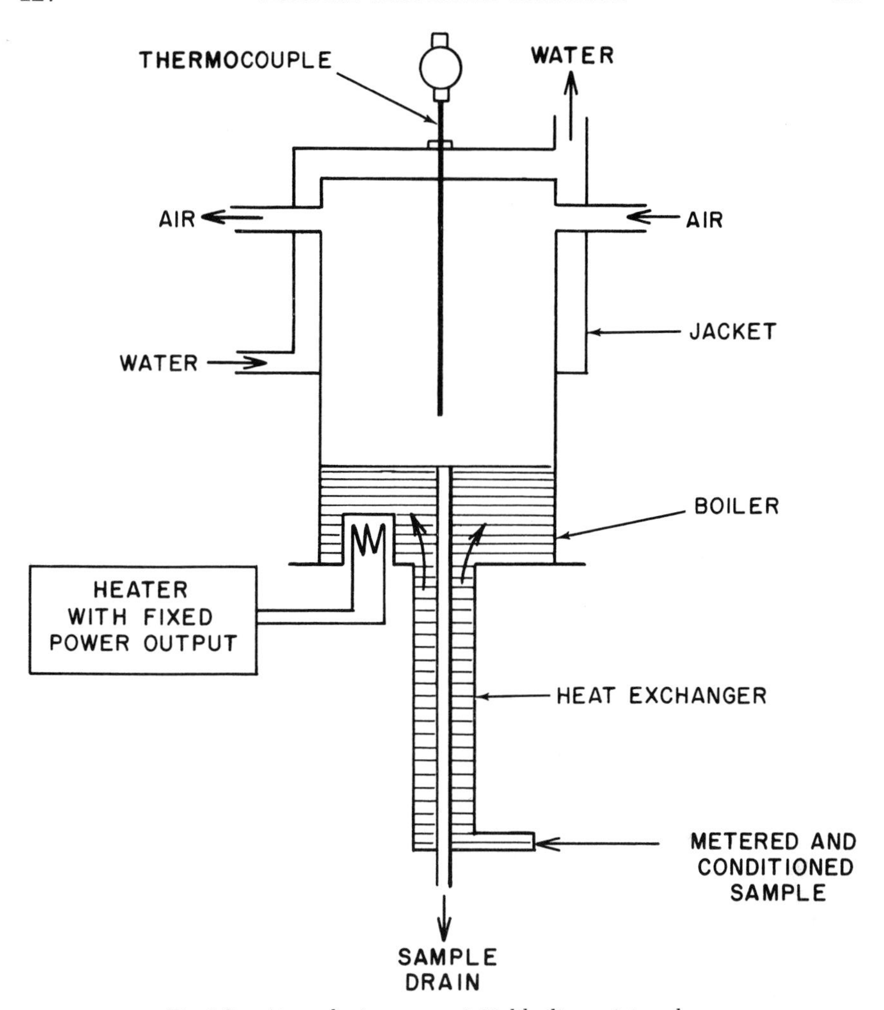

Fig. 1.9e Atmospheric pressure initial boiling point analyzer

For monitoring of low IBP temperatures, a mechanical refrigeration system chills the incoming sample and provides a low temperature coolant for the condenser jacket, which is maintained at a temperature of about 0°F.

The boiler may be operated at a reduced pressure for materials with very high IBP temperatures at atmospheric pressure. This version requires a modified boiler with pressure seals and a vacuum system and is usually steam traced to prevent plugging of sample lines.

PERCENT EVAPORATED ANALYZER

Some operators are more interested in the percent material evaporated at a given temperature (usually at 158°F) than in the inverse relationship discussed previously. Volatility control for inline gasoline blending systems is an example of such application.

This analyzer is an outgrowth of the film evaporator type distillation analyzer shown in Figure 1.9d. A constant flow of conditioned sample is metered to the spiral vertical heater and the residue is collected in the bottoms cup. The temperature of the bottoms is sensed by a thermistor and controlled by modulating the heater power. As the percent of evaporated material varies with sample composition, the level in the bottoms cup changes. The change is sensed by the float, and the photocell-light controller modulates the pumping rate of the bottoms pump (rather than the heater output as in Figure 1.9d).

The controller output is transduced to a 3 to 15 PSIG pneumatic signal which in turn operates a positioner fitted with a cam to vary the bottoms pump stroke so as to maintain a constant level in the bottoms cup (Figure 1.9f). The pneumatic signal is thus proportional to the bottoms flow rate and can be used as a direct analyzer read-out in addition to or in place of the current signals. Suppressed ranges, such as 20 to 40 percent evaporated, are attained by adjustments of the positioner actuated cam, wherein the slope of the cam determines the span, and its position on the positioner shaft determines the zero setting or range suppression.

A thermocouple adjacent to the thermistor temperature control probe displays the bottoms temperature on a meter relay which also provides alarm contacts for temperature control malfunctions or upsets.

PACKED COLUMN VACUUM DISTALLATION ANALYZER (a5)

Some hydrocarbon feed stocks have very high boiling points or they may decompose if boiled at atmospheric pressure. To avoid decomposition and to reduce the boiling point temperatures, the product may be distilled at a reduced pressure. Esssentially the same technique is used as shown in Figure 1.9c except that the column and boiler pot are operated under a controlled vacuum. A conditioned and pressure regulated sample enters the top of the column through a metering valve controlled by a float in the reboiler pot so as to maintain a constant sample level in the pot. The sample is preheated in a heat-exchanger above the reboiler pot before entering at the top of the packed column (Figure 1.9g). A fixed wattage radiant heater boils the sample. Overheads from the column flow through a water cooled condenser from which some of the condensate is withdrawn by a precision metering pump at a constant rate, and the remaining distillate is refluxed to the top of the column through an overflow tube.

A second metering pump withdraws the bottoms fractions at a constant rate so that the ratio of overheads to bottoms flow fixes the percent evaporated material to be measured. Both pumps are gear driven by the same motor and maintained at a constant temperature by immersion in individually heated oil baths. A thermocouple in the reboiler pot measures the bottoms bubble

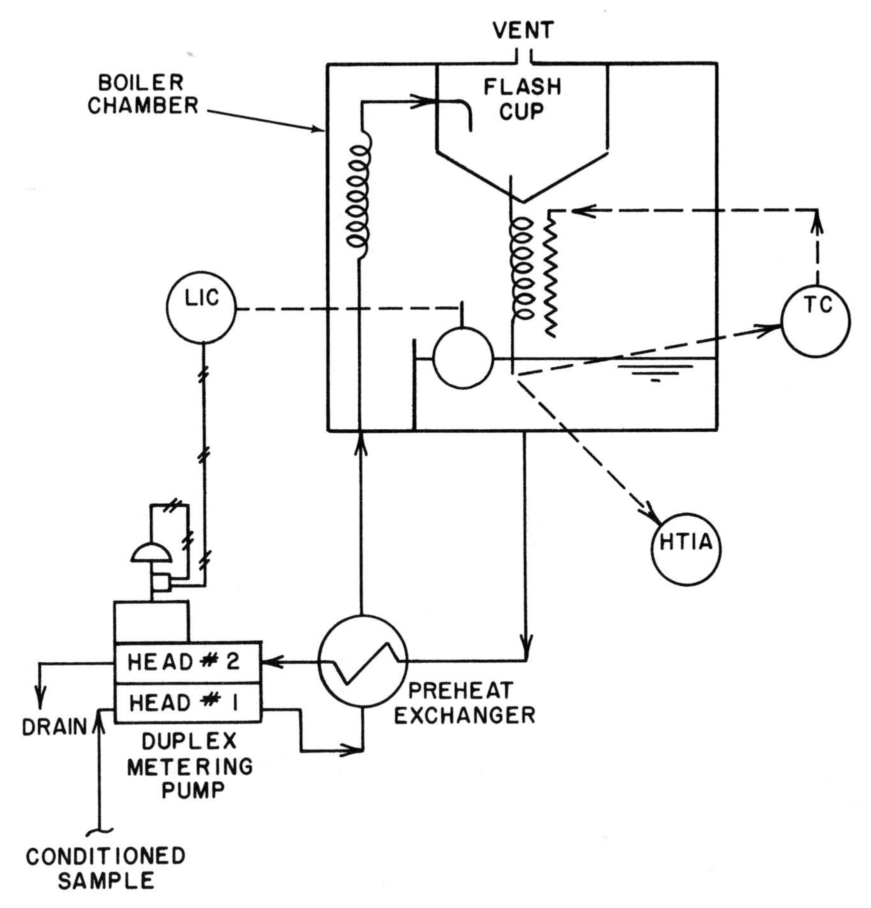

Fig. 1.9f Percent evaporated analyzer

point temperature, which may be correlated with ASTM method D-1160.

A vacuum pump removes the non-condensable vapors from the system, a pressure controller modulates vacuum pressure and a vacuum surge tank stabilizes the system to avoid excessive pressure fluctuations. If wide variations in product end point are expected, the heat input to the reboiler pot may be regulated by an autoformer in the heater circuit to avoid column flooding due to excessive refluxing. A sight glass in the reflux line permits observation of the reflux rate so that optimum column loading conditions may be established.

As the process boiling point temperature changes, the bottoms temperature increases or decreases accordingly.

Calibration is done by comparing analyzer readouts with ASTM D-1160 determinations for the same product drawn at the analyzer.

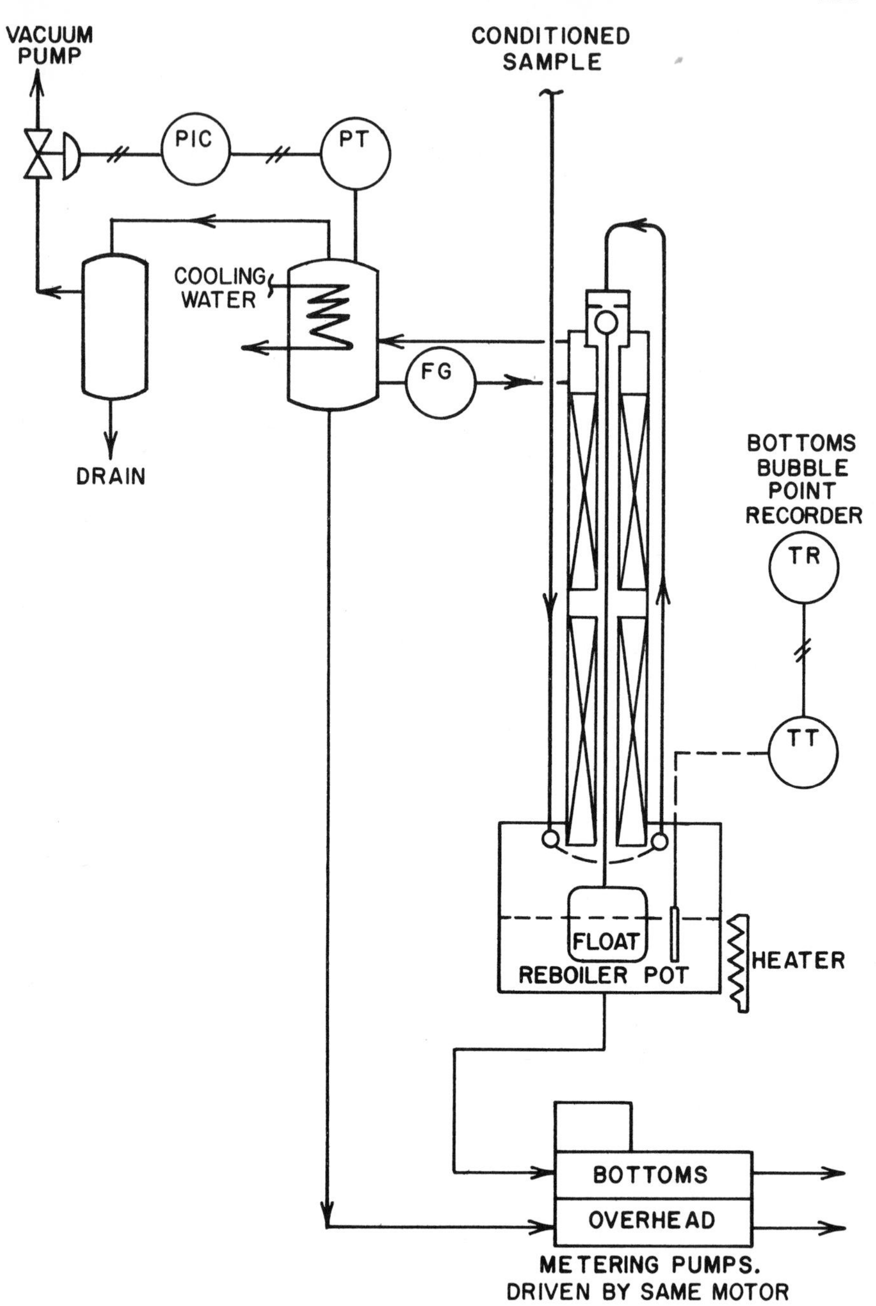

Fig. 1.9g Packed column vacuum distillation analyzer

AUTOMATIC ONSTREAM DISTILLATION ANALYZER (a6)

This analyzer is an automated version of the ASTM method D86—IP-123 designed for onstream plant use. It performs the same distillation test as the one illustrated in Figure 1.9b and produces a continuous temperature record with percent point lines and scribed at desired values (Figure 1.9a). Following a complete distillation, the cycle is repeated. Various safety and local indicators aid the operator, protect the analyzer and avoid false data recordings due to malfunctioning.

CALIBRATION

Techniques employed in the calibration of distillation analyzers are the standard sample method and the spot or grab sample method. Both are described in ASTM method D2891—70T. In the latter method, a sample is drawn off the process line and its properties determined (standardized) by replicate tests using ASTM method D86—IP-123. The standard sample is then introduced into the analyzer and the results are compared to the laboratory determinations. Sampling systems include a reservoir tank for holding a standard sample and the means for its introduction into the analyzer. In addition to establishing a correlation between the analyzer read-out and ASTM method D86—IP-123, it also serves to check analyzer performance in the event that its output is suspect. Care should be exercised to assure that the standard sample remains stable and unaltered for the length of use.

The spot or grab sample technique involves the collection of samples at the analyzer without disturbing its operation, and simultaneously notes the analyzer read-outs with due compensation for the system response time. Analyzer read-outs are then compared with laboratory determinations of the spot samples by ASTM method D86—IP-123. Again, care in collecting and handling the spot samples are of particular importance so as to assure accurate results.

APPLICATION

With the exception of the chromatograph, distillation analyzers have greater use in the control of petroleum refining processes than have any other onstream analyzer. In practically every refinery they have applications as varied as the production of fuel oils and the blending of gasolines; they are part of crude oil distillation units and alkylation units; and they are used for control of catalytic reforming feed stocks and for control of the reflux ratio in gasoline fractionating towers.[2]

Vapor Pressure Analyzers

Reid Method (ASTM D323-68)

The vapor pressure of petroleum products, except for liquified petroleum gases, is usually determined by the Reid method as given by ASTM method

D323-68. Sample handling is very critical with very volatile products, owing to the hazard of loss of light ends. For products having a vapor pressure less than 26 PSIA, a sample volume of at least 1 quart but not greater than 2 gallons is collected and as soon as possible is placed in a bath maintained at 32° to 40°F. For higher vapor pressure products, a special bomb approximately 2 inches in diameter and 10 inches high is used.

The Reid vapor pressure (RVP) bomb consists of a liquid sample lower chamber coupled to an air chamber having a volume approximately four times that of the liquid, and a Bourdon tube pressure gauge connected to the top of the air chamber.

Chilled sample is used to fill the liquid chamber and is quickly coupled to the air chamber with the attached pressure gauge. The assembly is then immersed in a water bath maintained at 100° ± 0.2°F. After a minimum of five minutes, the apparatus is removed and shaken vigorously and the pressure is noted. It is then quickly reimmersed in the bath, and at intervals of not less than two minutes the procedure is repeated at least five times until two consecutive pressure readings are noted. This reading is reported in pounds, RVP (PSIA). Although the pressure is approximately the true vapor pressure in PSIA for some products, it should be noted that RVP is lower than the true vapor pressure due to the loss of some light ends during sample handling. The ratio of true pressure to RVP varies from about 1.03 to 1.45 for different gasolines to as high as 9.75 for some crude cuts.[1]

Repeatability varies from 0.1 PSI for products of 0 to 5 pounds to 0.4 PSI for products having an RVP greater than 26 pounds.

Corresponding reproducibility varies from 0.35 PSI to 0.7 PSI.

Liquified Petroleum Gases (ASTM D1267-67)

ASTM method D1267-67 for the determination of the vapor pressure of liquified petroleum gases (LPG) is usually performed at 100° ± 0.2°F, but may be run at temperatures as high as 158°F and is limited to products the vapor pressure of which does not exceed 225 PSIG. Essentially the same apparatus is used as in the Reid method. The sample is not air saturated; however, 40 volume percent of the liquid sample is withdrawn from the completely filled bomb after a prescribed purging procedure to allow space for expansion of the product when it is immersed in the test bath.

The observed pressure is reported as LPG vapor pressure in PSIG, after corrections are made for gauge error and for standard barometric pressure at the test temperature.

Repeatability is given as 0.5 PSIG plus 0.5 percent of the mean value for a test temperature of 100°F, and 1.0 PSIG plus 1.0 percent of the mean value for a test temperature of 158°F.

Corresponding reproducibility is given as 1.0 PSIG plus 0.5 percent of mean value and 2.0 PSIG plus 2.0 percent of mean value.

Air Saturated Vapor Pressure Analyzer (b1)

Sample is metered by a positive displacement pump at a rate of 100 cc per minute through a heat exchanger immersed in a constant temperature bath at 100°F (Figure 1.9h). The sample is then sprayed into an air saturation chamber, also immersed in the bath, where it is saturated with air. A float controlled needle valve discharges the air saturated sample and vapors into the vaporizing chamber and also maintains a constant level of liquid in the chamber. The sample supply pump is double ended and designed so that the exhaust end withdraws a flow of 500 cc per minute of liquid and vapor mixture from the vaporizing chamber, thus establishing the 4 to 1 liquid air ratio prescribed by the RVP (ASTM D323-68) method. The pressure in the vapor chamber is sensed by a pressure transmitter with its 3 to 15 PSIG output calibrated to represent the RVP. The exhaust pump jacket is also maintained at the 100°F bath temperature.

Dynamic Vapor Pressure Analyzer (b2)

This analyzer may be calibrated to measure the vapor pressure of products covered by either ASTM method D323-68 or ASTM method D1267-67. The

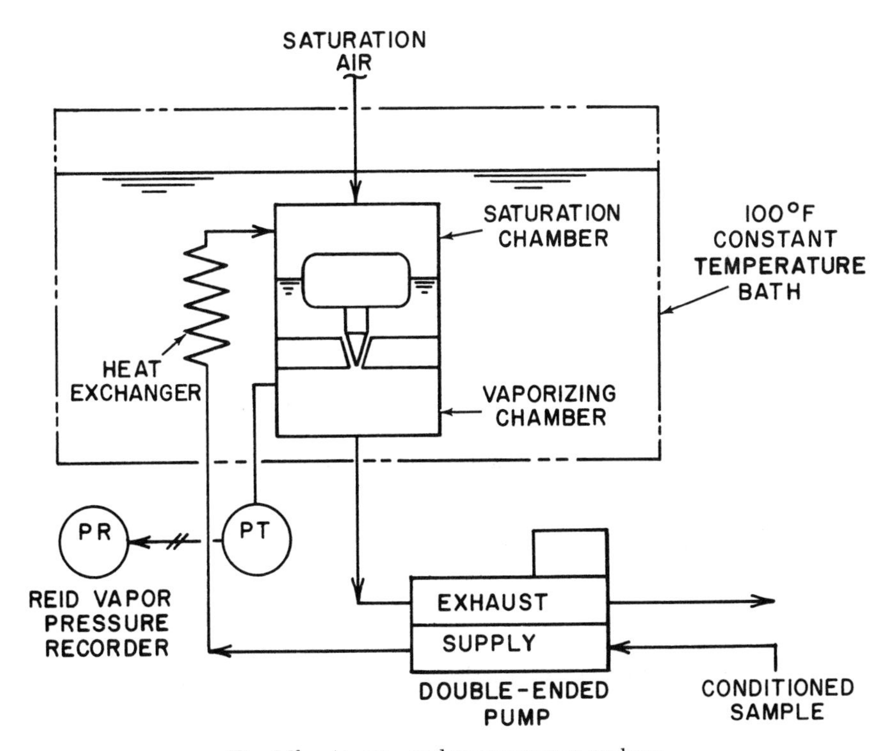

Fig. 1.9h Air saturated vapor pressure analyzer

analysis is continuous, and the sample effluent may be returned either to a pressurized line or to an atmospheric receiver tank. Incoming sample is filtered and maintained at a constant pressure (Figure 1.9i). The sample is brought to a temperature of 100° ± 0.1°F in the heat-exchanger, which is immersed in the constant temperature bath. The sensing device is a modified jet pump element with the suction side dead-ended into the cell of an absolute pressure transmitter. The velocity of the sample in the small diameter nozzle causes the pressure head to approach the vapor pressure of the least volatile component in the stream. This value is lower than the RVP as determined by ASTM method D323-68. By reducing the efficiency, the system simulates the selective vaporization which occurs in the 4 to 1 air to liquid volume (vapor-liquid ratio) in the RVP test apparatus.

One of the parameters affecting the efficiency of a jet pump, eductor or aspirator is the location of the tip of the nozzle with respect to the throat of the downstream venturi. By adjusting the location of the nozzle, the operating efficiency may be adjusted so that the analyzer read-out provides an essentially 1 to 1 correlation with the RVP. Figure 1.9j shows the relationship of the analyzer results against RVP for various hydrocarbons and blends.

When the analyzer has to redischarge into a pressurized line, a back pressure regulator is added to maintain a constant back pressure on the system and thereby eliminate the effects of varying return line pressure. An inlet

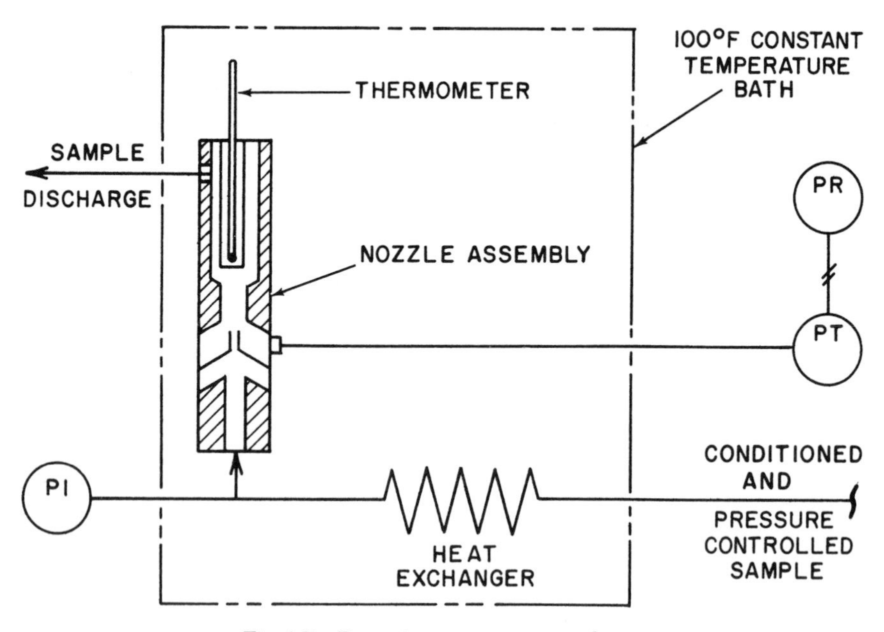

Fig. 1.9i Dynamic vapor pressure analyzer

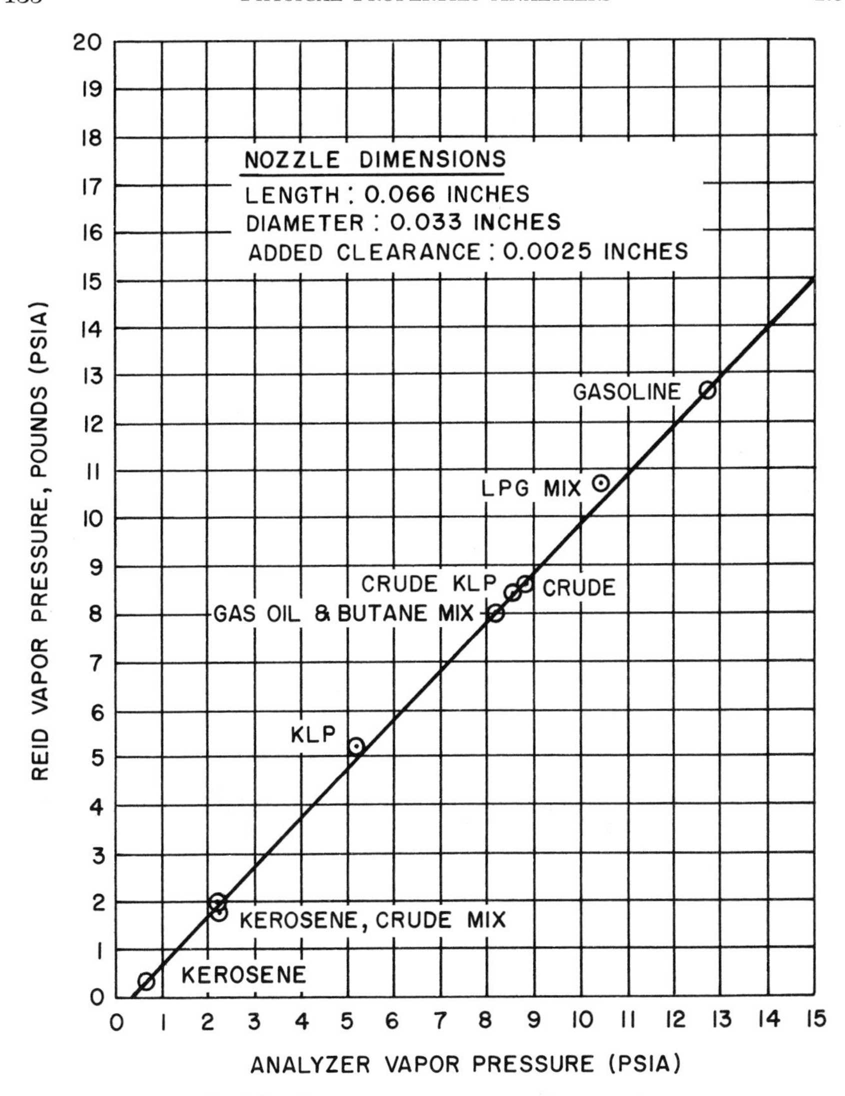

Fig. 1.9j Dynamic vapor pressure analyzer results

pressure of roughly 45 to 50 PSIG is required when the analyzer is discharging to atmosphere and the inlet pressure regulator is set at 40 PSIG. When the sample is returned to a pressurized line, the inlet pressure regulator is set at a value equal to 2.5 times the return line pressure plus 40.

Calibration

A standard is now under consideration by ASTM for a method of validating vapor pressure analyzers. The standard sample method and the spot or grab sample method are equally reliable if carefully performed. Sample collection and handling are critical in both cases as is the manner in which a stable and uniform standard sample is preserved during delivery to the analyzer.

Application

Increasing the quantity and quality of butanes and pentanes in the inline blending of finished gasolines represents one potential application of vapor pressure analyzers. They have been used with LPGs; more extensively they have been used to monitor the vapor pressure of pipeline transported products, thereby avoiding vapor locking the pumps and minimizing safety hazards during tanker loading operations. They may also be used to detect product interfaces at pipeline receiver stations.

Vapor-Liquid (V-L) Ratio Analyzers

Volatility Test (ASTM Method D2533-67)

Front-end volatility must be closely controlled in a gasoline blend to permit the greatest use of light blending components without incurring vapor lock during operation of a gasoline engine.

The vaporization of a fuel in the carburetor of an engine is predictable neither from distillation temperatures nor from pressure tests. The curves of Figure 1.9k show how four different fuels (with some similar properties) exhibit different volatility characteristics as determined by measuring their V-L ratios at various temperatures. In the past, front-end volatility was determined by utilizing indirect measurements such as the RVP, the 10, 20 and 50 percent evaporated temperatures and the temperature corresponding to a given V-L ratio (usually 20). This procedure and the required computations are time consuming, cumbersome and inaccurate.

The laboratory technique determines the V-L ratio by direct measurement for a given reference temperature. ASTM method D2533-67 utilizes a special buret, constructed with a stopcock at the top, 0.5 cc graduations, a short bottom arm fitted with a rubber septum and a long bottom arm connected to a 250 ml leveling bulb by rubber tubing. The buret is filled with pure dry glycerin, and a sample of 1 cc or less is injected into the buret through the rubber septum by a hypodermic syringe.

The buret is then placed in a controlled temperature bath. As the sample vaporizes, the 250 cc leveling bulb, open at the top to the atmosphere by a drying tube, is raised or lowered to maintain the generated vapor at a pressure of 1 atmosphere absolute, compensating for the prevailing barometric pressure. The volume of vapor is indicated on the buret's graduations, and since the liquid volume injected is known, the V-L ratio can be calculated.

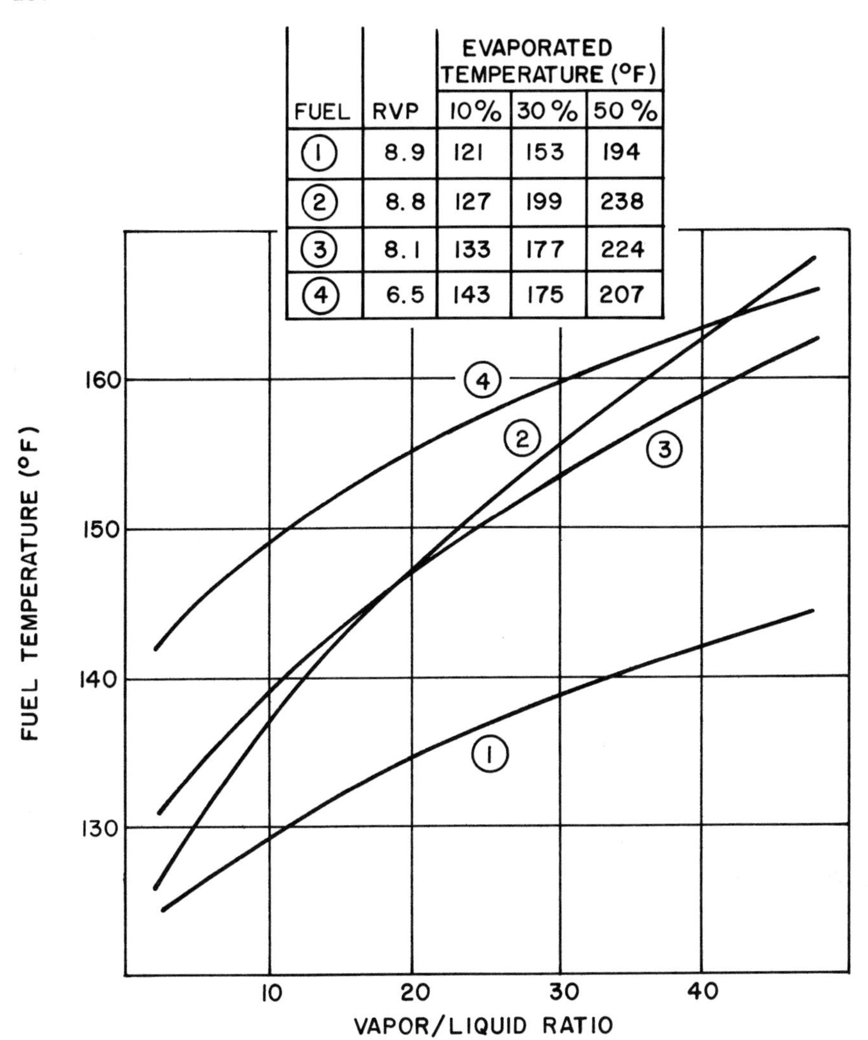

FUEL	RVP	EVAPORATED TEMPERATURE (°F) 10%	30 %	50 %
①	8.9	121	153	194
②	8.8	127	199	238
③	8.1	133	177	224
④	6.5	143	175	207

Fig. 1.9k Volatility effects on gasoline vaporization

Repeatability is given as 4.0 percent of the mean V-L ratio and reproducibility as 1.6 plus 8.0 percent of the mean V-L ratio.

Batch Vapor-Liquid Ratio Analyzer (*c1*)

A fixed volume of sample is injected into the head of a cylinder and allowed to vaporize at a constant temperature and at a pressure of 1 atmosphere absolute. The vapor volume generated is measured and used to compute the V-L ratio.

Most of the apparatus is immersed in a constant temperature bath (Figure 1.9 1). At the beginning of a cycle, the piston in an expansion cylinder is at top dead center. A continuously flowing sweepstream is diverted to flow through a heat-exchanger in the constant temperature bath and then through a fixed volume recess in the piston head, purging it of spent sample. The precise liquid volume of 1.8 cc is trapped by solenoid valves built into the cylinder head. (Valving is shown in this trapped position.) Simultaneously the sweepstream flow is redirected to the process line. The piston, actuated by a servosystem, withdraws as vaporization of the sample progresses. The servo is operated through a differential pressure switch with a sensitivity of ± 0.5 mm Hg.

One side of the switch is open to the vapor being generated in the cylinder, and the other is open to a reference pressure set at one atmosphere absolute. The pressure produced by the gasoline vapors closes the switch, activating a servomotor to drive the piston down. The piston movement is

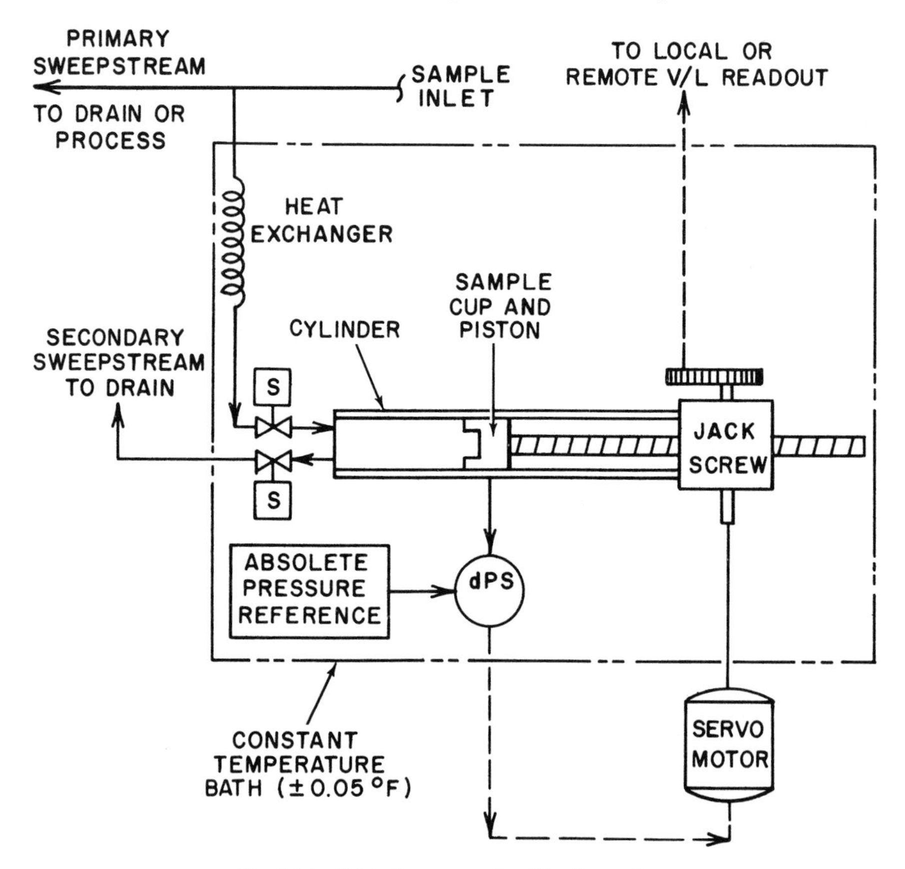

Fig. 1.9 1 Batch type vapor-liquid ratio analyzer

coupled to a mechanical counter for local indication, and to a potentiometer in an electronic circuit for remote display of the V-L ratio. Since the cylinder is in a horizontal position the liquid surface increases as the piston withdraws, promoting more rapid vaporization.

After a predetermined time, the cylinder is vented and the piston is driven up, exhausting the generated vapors. When the piston reaches top dead center, the vent line is closed and another fresh sample is diverted into the cylinder reference volume. Peak picking circuitry may be applied to permit the output signal to be fed directly to a closed-loop control system, computer or recorder.

***Continuous Vapor-Liquid Ratio Analyzer*[3] (*c2*)**

Figure 1.9m shows a block diagram of the analyzer. A slipstream from a sample loop is conditioned and cooled to approximately 35° to 40°F before entering a metering gear pump. The differential pressure across the pump is controlled to guarantee the constant delivery of liquid sample to a vapor-liquid separator at a rate of 25 cc per minute. The vapor-liquid separator is enclosed in a constant temperature bath controlled at the desired test temperature. Vaporization occurs at or near atmospheric pressure, with the residual liquid phase being discharged through a liquid seal, and the vapors are measured by a low pressure drop flow meter. The meter is maintained

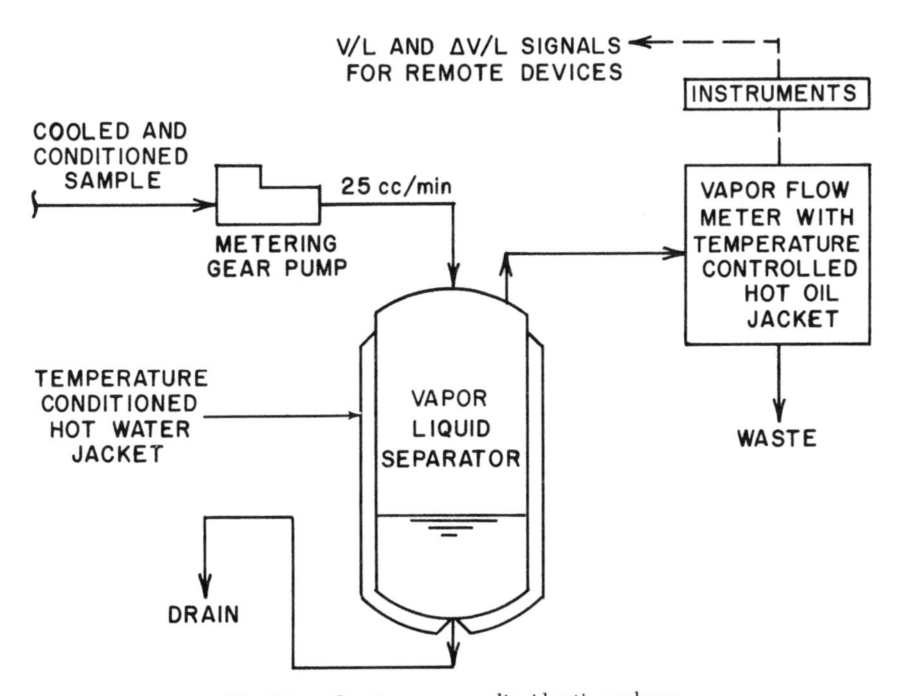

Fig. 1.9m Continuous vapor-liquid ratio analyzer

at 200°F to prevent condensation and to establish a temperature reference for vapor flow measurements. With the inlet liquid flow rate fixed, the V-L ratio may be computed after compensations are made for the vapor temperature and the barometric pressure. In addition to computing the V-L ratio, the deviation of V-L from a preselected V-L set point is also displayed and provided as an output signal.

Calibration and Application

A V-L analyzer may be calibrated either by the introduction of a sample of known volatility or by the grab sample technique.

These analyzers are primarily applicable to inline gasoline blending operations. An inline blending system offers decided economic advantages in addition to greater flexibility, speed and ease in switching blend formulas to meet production requirements. Although precise metering systems provide the means for implementing an inline blending system, the final blend properties are still only implied rather than precisely known. Vapor pressure alone or its combination with distillation characteristics has been shown to be insufficient to provide efficient control of front end volatility, because of inaccuracies and lags. The ability to maximize components like butanes and pentanes to meet seasonal and geographical requirements for volatility provides a very significant economic incentive. A V-L analyzer, in an inline gasoline blending system, may be used to reset the set point of metering pumps. This measurement may eventually supplant vapor pressure measurements for controlling additions of butanes and pentanes to gasolines.

Pour Point Analyzers

Pour Point Tests (ASTM Method D97-66, IP-15/67)

The standard laboratory procedure for measuring the flow characteristics of petroleum oils is given in ASTM method D97-66, IP-15/67. The sample must be heated without stirring to 115°F, or 15°F above the expected pour point temperature, before starting the test. A thermometer immersed in a jacketed sample test jar and a cooling bath are used. The cooling rate of the sample is fixed as it is examined, at 5°F (or 3°C) intervals to ascertain if it will flow when the test jar is tilted. When no sample movement is detected in the tilted jar after a five-second interval, the pour point is reported as 5°F (or 3°C) above the indicated temperature. This point corresponds approximately to a viscosity of about 500,000 centistokes. Repeatability is given as 5°F (or 3°C) and reproducibility as 10°F (or 6°C).

Pour point analyzers were developed in an attempt not only to automate a laboratory procedure for process control, but also to improve the accuracy of such measurements. These analyzers were reasonably successful; however, with materials containing pour point depressants, some types failed to correlate with laboratory determinations.

Tilting Plate Pour Point Analyzer (d1)

A conditioned sample is allowed to enter a shallow 2 cc test cell and overflow, thus flushing and replenishing it with fresh sample (Figure 1.9n). A cascade thermoelectric cooler chills the sample at a rate of 8°F per minute. A light beam focused on the surface of the sample is reflected to a photocell

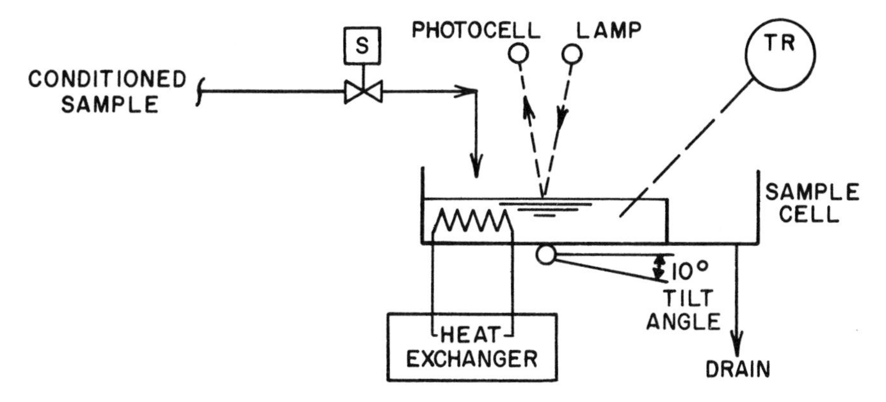

Fig. 1.9n Tilting plate pour point analyzer

which is one leg of a bridge circuit. The sample is tilted 4 times per minute through an angle of 10° with the horizontal. As long as the sample is fluid, the surface remains horizontal. When the sample has been cooled sufficiently, its surface follows the tilting platform, and the light beam is deflected from the photocell, unbalancing the bridge and signaling that the pour point temperature has been reached. This signal triggers the reversal of the thermoelectric cooler so that it warms the sample and opens sample solenoid valves to flush the sample cell and introduce fresh sample. A thermocouple in the cell assembly continuously measures the sample temperature which is reflected as a series of spikes on a recorder. The low points of these spikes represent the measured pour points of batch samples tested.

A peak picking accessory circuit may be used to signal the terminal temperatures of each cycle. If used with products containing pour point depressant additives or for residual fuels, the pour point determined by the analyzer will correspond to that of the base stock but will not agree with the ASTM test. Vibration may cause ripples in the oil surface of the sample cell. Similarly, if water is present in the sample it may accumulate in the cell and form a second reflecting surface. Both conditions are intolerable.

Moving Ball Pour Point Analyzer (d2)

This apparatus permits pour point measurements of products which do not contain wax, and is often referred to as the viscosity pour point detector.

The analyzer section is a slowly rotating (0.1 rpm) cup containing about 22 cc of sample which is cooled by four thermoelectric elements. A tungsten carbide ball with a thermocouple brazed to its surface is suspended in the sample at the end of a thin steel tube from low friction conical pivots. As the sample temperature decreases, the viscous drag on the ball increases.

When a precalibrated displacement has been reached, a photoelectric circuit is actuated by the movement of the ball suspension assembly and the following events are initiated: 1.) the temperature record is updated; 2.) the thermoelectric elements are reversed to heat the sample; 3.) the sample cup rotation is stopped.

The sample in the cup is drained and replaced by fresh preheated sample. The output from the measuring thermocouple is amplified for use as a transmitted signal and is also displayed on a local indicator.

The analyzer may be manually operated for individual tests or automatically programmed. It is mounted on a free standing rack which includes a sample conditioning system and sweepstream. It is unaffected by pour point depressants and correlates with the standard ASTM test.

Dropping Plummet Pour Point Analyzer (d3)

This analyzer consists of two sections: a solid state control unit in the control room; and an explosion-proof field unit containing the sample cell, refrigeration unit, sampling system, valving, relays and measuring thermistors (Figure 1.9o). Its performance is unaffected by pour point depressants.

After the pour point of a sample has been measured, a valve is opened to drain the tested sample. Hot and conditioned fresh sample from a sweepstream loop is then admitted for an adjustable period of time through a fill valve, purging the sample cell through the drain valve which remains open. The cell contains an overflow which fixes the volume of sample. The drain valve is then closed and fresh sample is introduced until it overflows. The sample inlet valve closes after an adjustable preset fill time. A sensing plummet weight containing thermistor A is then raised by an air operated bellows assembly and is held at its uppermost position in the cell. A mechanical, air cooled refrigeration unit cools the jacket.

Thermistor B in the jacket determines the differential temperature between the sample and the jacket so as to maintain the cooling rate at approximately $1\frac{1}{2}$°F per minute. The sample is cooled to an adjustable temperature of about 5°F below the previously recorded pour point temperature so that the sensing plummet is solidly frozen in position. The sensing plummet weight is then released and the sample temperature rises by the warm air in the

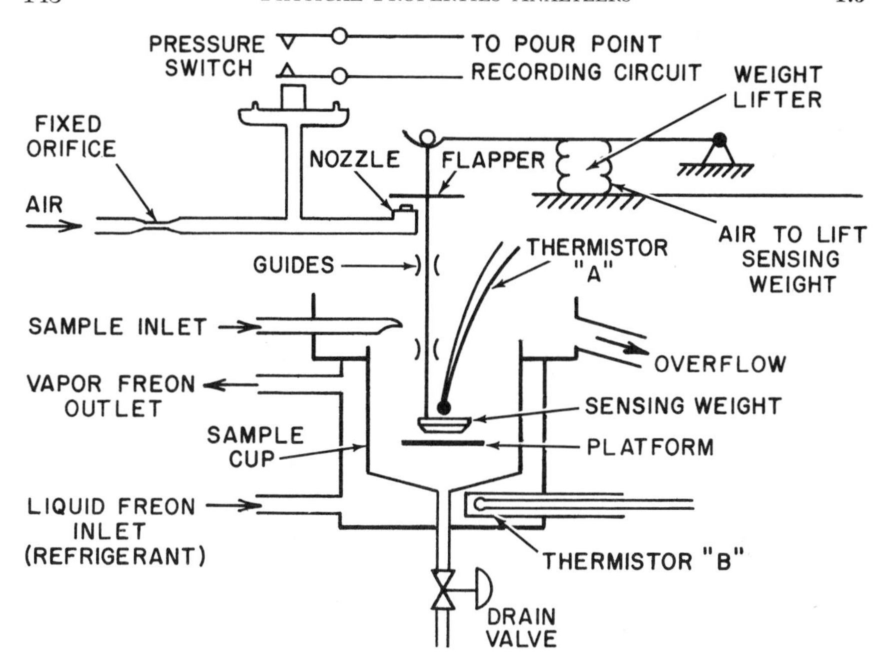

Fig. 1.9o Dropping plummet pour point analyzer

housing at a rate of $1\frac{1}{2}$°F per minute. As the sample becomes warm and melts, the sensing weight tends to drop. Since warming progresses from the walls of the sample cup to the center, a circular platform below the plummet supports the coaxial plug of solidified sample and prevents the weighted plummet from falling prematurely. When the central slug melts, the plummet drops. This transition occurs due to a change of 1°F and the movement of the plummet is sensed by a flapper that triggers a pressure switch, indicating that the pour point temperature has been reached.

The control unit energizes the temperature recorder drive for ten seconds, and the cycle is repeated for a new sample. During cooling and warming, the recorder is inoperative. Several local indicators and abort circuits are included to avoid false pour point measurements. A memory circuit holds the previously determined pour point temperature in order to establish the set-point for the next cooling cycle.

An improved version of this analyzer is under development. It is simpler and faster and permits the determination of the cloud or crystal formation point for the same sample. In this design, the weighted plummet is lowered in the sample cup by a synchronous motor-driven holder. As cooling progresses, the sample becomes sufficiently viscous to impede the fall of the plummet and separates from its holder. This in turn breaks an electrical contact, indicating attainment of the pour point temperature.

Cloud or crystal formation determination will be described in a later section.

Calibration and Application

The repeatability of ASTM method D97 for pour point is 5°F and its reproducibility is 10°F. A sufficient number of determinations to improve the accuracy of the ASTM results is advisable, because the process analyzer ordinarily exceeds the accuracy of the laboratory method. Either the standard or the grab sample method is suitable. The convenience of a locally available standard sample provides a rapid check on the analyzer's performance as well as serving as a means of calibration. The greater initial cost for this feature should be weighed against the delay incurred in obtaining a laboratory analysis of the spot sample.

Pour point temperature measurements are utilized more extensively in Europe than they are domestically, since the use there of furnace and fuel oils is less pronounced. When a product is sufficiently free of wax so that a cloud point determination becomes meaningless, the pour point may be used as an index of the temperature at which flow will be impeded due to semi-solidification rather than by the formation of wax crystals.

Cloud Point Analyzers

Cloud Point Tests (*ASTM Method D2500-66—IP-219/67*)

This method is applicable to products with a cloud point below 120°F, which are transparent in a layer 1½ inches thick.

A cylindrical, flat-bottomed, clear glass test jar 1¼ inches in diameter by 4¾ inches long, and having a scribed sample fill line 2⅛ inches above the inside bottom surface is used. A cork holds a thermometer coaxially in the test jar so that its bulb rests at the bottom.

A watertight jacket with a ¼-inch-thick cork or felt pad at the bottom holds the test jar when it is immersed in a cooling water bath. The sample must be dried at a temperature at least 25°F above the approximate cloud point so as to remove any moisture and so as to minimize trace water haze formation. The test jar is fitted with a cork or felt ring approximately 3/16-inch thick which is positioned one inch from the bottom to keep it centered in the jacket.

The test is begun by placing the jar vertically in a 30° to 35°F water bath so that the jacket projects no more than one inch from the bath liquid.

At intervals of 2°F, the jar is quickly (3 seconds at most) but gently removed and examined for formation of a wax haze. (A water haze is generally uniform throughout the sample, whereas a wax crystal haze always appears first at the bottom of the jar.)

If the cloud point is not detected when the sample reaches 50°F, the sample is transferred to a second bath also containing a test jacket and maintained at 0° to 5°F and the test is continued. Successive lower temperature baths are used as required for low temperature cloud point products. The temperature, expressed in increments of 2°F, at which a distinct wax haze is first observed is reported as the cloud point. For gas oils repeatability is 4°F and reproducibility is 8°F. For other oils, both repeatability and reproducibility are 10°F.

Optical Cloud Point Analyzer (e1)

The principle of the method is based on the observation that wax crystals have two widely disparate indices of refraction and can depolarize polarized light.

Light from an incandescent source is polarized and passes through a glass sample cell (Figure 1.9p). In the absence of wax crystals a second polarizer oriented at 90° to the initial polarizer blocks the light beam to a photocell. As the sample is cooled by two thermoelectric units and wax crystals are formed, they depolarize the incident beam and permit light to be transmitted through the second polarizer to the photocell, changing its resistance. This initiates a sequence of operations which records the crystal formation temperature by a thermocouple in the sample cell and switches a three-way solenoid valve so that the normally by-passed warm, conditioned slipstream passes through the cell, flushing and replenishing it with fresh sample.

Simultaneously, a peak picker circuit displays and holds this value on a recorder chart until the next cloud point temperature is measured. After suitable flushing the solenoid returns to the slipstream by-pass position, trapping the fresh sample in the cell, and the cycle is repeated. The thermoelectric coolers operate continuously.

Convective Heat Transfer Cloud Point Analyzer (e2)

When an oil sample is cooled gradually, the heat transfer mechanism undergoes a distinct change at the cloud point or at the instant when a wax crystal lattice is formed.

After a cylindrical sample cell has been purged by fresh warm sample, an inlet solenoid valve at the bottom of the cell closes, trapping a fixed volume of fresh sample (Figure 1.9q). A thermoelectric cooler surrounding the cell is then energized to initiate the cooling cycle. A convective thermal current is established, with chilled sample adjacent to the cell wall flowing downward. The coaxial upward flow of warmer sample is accelerated by extending a small section of the lower portion of the sample cell outside the thermoelectric cooling unit so that it may absorb heat from its surroundings. Two thermistors, one positioned at the top and the other at the bottom, sense the temperature

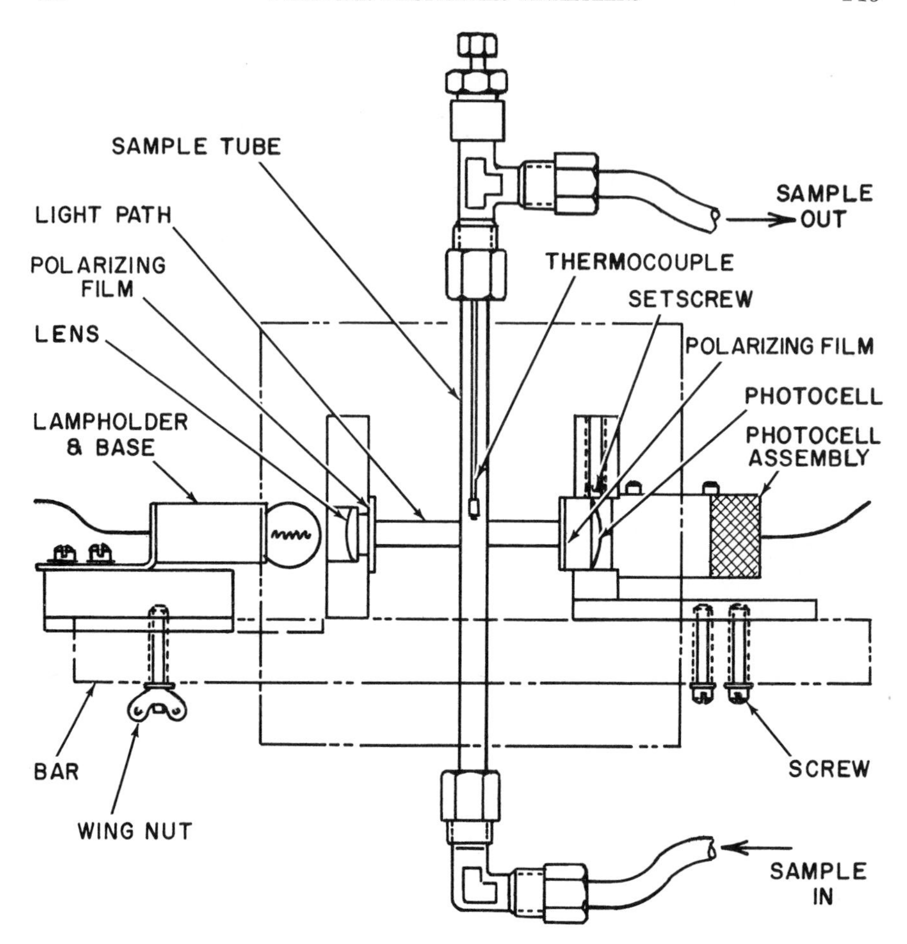

Fig. 1.9p Optical cloud point analyzer

differential produced by the convective thermal current. Upon the formation of a wax crystal lattice at the walls of the sample cell, heat transfer by convection is impeded but continues by conduction alone. Consequently, the two thermistors, cooled only by radial conduction from the walls of the sample cell, quickly reach the same equilibrium temperature.

The thermistors are connected into the opposite arms of a Wheatstone bridge which is unbalanced when a temperature differential exists. At the end of a cooling cycle reestablishing the bridge balance deenergizes the thermoelectric cooler and opens the sample inlet solenoid valve to introduce fresh warm sample. The cycle is then repeated to produce a new measurement.

A thermocouple in the center of the sample cell is used as the pour point temperature read-out and may be interpreted either from a sawtooth type

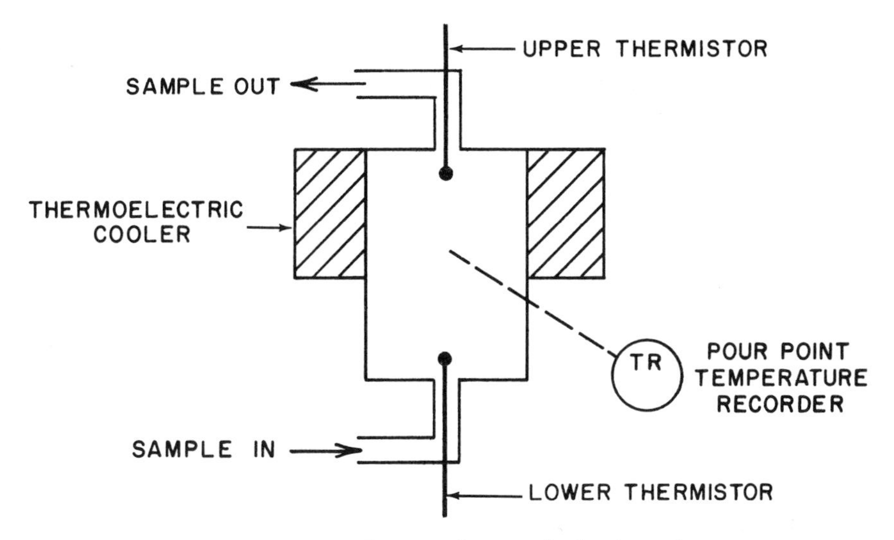

Fig. 1.9q Convective heat transfer type cloud point analyzer

record or from a record which prints out only the low peaks at the termination of each cycle, holding this value until the next analysis.

Temperature Rate of Change Cloud Point Analyzer

The dropping plummet pour point analyzer previously described (Figure 1.9o) may simultaneously be used to measure the cloud or crystal formation point temperature, provided that the product when cooled will first produce wax crystals prior to solidification. These points are often only a few degrees Fahrenheit apart.

During the sample cooling cycle, a distinct temperature plateau may be detected before the solid point temperature is reached. From a recorded curve of the cooling cycle, the crystal formation point temperature may be identified by the sharp change in slope or bend in the cooling curve. Automating this measurement entails the use of the first derivative of temperature which triggers a cloud point read-out when the sharp change in cooling rate occurs.

Calibration and Application

The techniques used to calibrate pour point analyzers are also used to calibrate cloud point analyzers. The measurement of the cloud point applies only to those petroleum oils that are transparent in layers $1\frac{1}{2}$ inches thick and contain paraffin waxes or other compounds capable of forming crystals prior to total solidification. Its major application has been to gas oils and cycle oils not only to meet specifications but also to facilitate product transport and to prevent filter clogging during cold weather.

Freezing Point Analyzer

***Aviation Fuel Tests* (*ASTM Method D2386-67 and IP-16/68*)**

The freezing point as defined by this method is the temperature at which the last hydrocarbon crystal (formed during cooling) melts after the sample temperature is allowed to rise. This temperature must be within 3°C of the temperature at which the appearance of hydrocarbon crystals is first observed.

A jacketed clear glass tube is filled with either dry air or nitrogen at atmospheric pressure. The outer tube OD is 30 mm and the inner tube ID is 18 mm, with a 2 mm space between tubes. The over-all length is 237 mm. The plug in the sample tube holds a total immersion thermometer and a stirring rod formed with three spiral loops at the bottom and positioned slightly below the thermometer bulb. The tube is filled with 25 cc of fuel and placed in a clear vacuum flask (70 mm ID and 280 mm long) containing a coolant such as alcohol or solid carbon dioxide.

The sample is stirred vigorously during cooling and the temperature at which hydrocarbon crystals first appear is noted, neglecting any haze which may form at about 14°F owing to dissolved water in the sample. The sample tube is then removed and allowed to warm up slowly while the sample is continuously stirred. The temperature at which the last crystal disappears is reported as the freezing point temperature, provided it is within 3°C of the crystal formation point temperature. Otherwise the test must be repeated.

Repeatability is 0.7°C and reproducibility is 2.7°C.

***High Purity Hydrocarbon Test* (*ASTM Method D1015-55*)**

This method is used in conjunction with ASTM method D1016-55 for determining the purity of hydrocarbons from freezing point measurements.

The freezing point is determined from a 50 cc sample and a precision platinum resistance thermometer, calibrated by the National Bureau of Standards.

The sample is placed in a freezing tube (25 mm ID and 50 mm OD) having a silvered inner wall. A brass cylindrical sheath with an asbestos pad at the bottom holds the freezing tube in a cooling bath. The plug at the top of the freezing tube supports the precision platinum resistance thermometer, a double spiral stirring rod and a tube with a spherical joint for admission of dry and carbon dioxide-free air. The space between inner and outer tubes is connected to a vacuum system. At the beginning of the test the bath is filled with a refrigerant suitable for the estimated freezing point temperature. Sample is introduced into the freezing tube by temporarily removing the top stopper and thermometer.

The purpose of the dry air flow into the freezing tube is to blanket the sample and prevent water vapor from entering. The sample is stirred and when the temperature approaches the freezing point (within 27°F), evacuation of the jacket space is begun. Time and temperature observations are made and the vacuum on the jacket is adjusted so as to achieve a cooling rate of 2°F in 1 to 3 minutes. The time (within 1 second) is recorded for a resistance thermometer change of 0.05 to 0.1 ohm. After subcooling, crystallization is induced by dipping a chilled special rod into the sample.

A cooling curve may be plotted which permits the determination of the freezing point temperature with a sensitivity of about 0.0001°C.

Repeatability is ±0.005°C and reproducibility is ±0.015°C.

Freezing Point Analyzer for Aviation Fuels (*f1*)

When a layer of frozen sample is formed around a temperature detector, such as a thermistor, and is then allowed to melt gradually, the last thin layer of frozen crust will be rapidly carried away by the convective thermal current, resulting in a sharp rise in temperature as the detector suddenly comes in contact with the liquid sample. The effect may be amplified if the first derivative of the temperature is measured. The analyzer consists of two sections: a panel mounted controller section, and a field unit.

Following the determination of the freezing point of a sample, the cell is flushed in two directions by fresh, warm and conditioned sample from a slipstream. As shown in Figure 1.9r, by programming the three-way sample solenoid valves, the fresh sample first flows down through the sample cell and

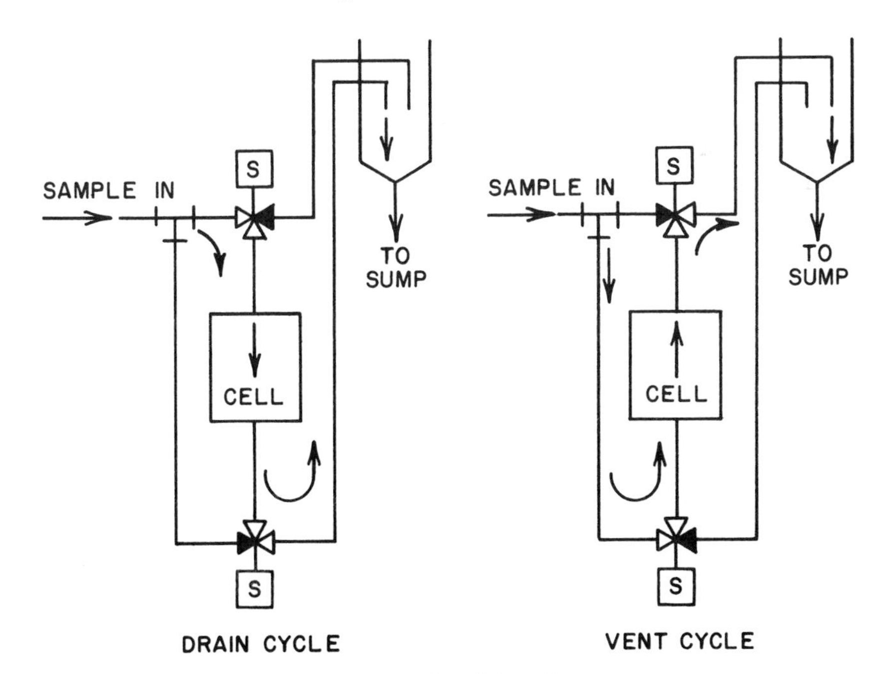

Fig. 1.9r Sample cell flow diagram

is then reversed to flow upward, thus insuring that the cell is full and free of any vapor bubbles or moisture.

The cell is about $^{13}/_{16}$ inch in diameter and $3^5/_8$ inches long and has a metallic wall and filled teflon end caps. The cell is jacketed to accept liquid refrigerant for sample cooling, and a heater is inserted in the jacket wall for subsequent sample warming. A thermistor and a miniature thermocouple are placed side by side about midway in the center of the cell. Following flushing and filling, the refrigeration system is started and the sample is cooled 2° to 8°F below the previously determined freezing point. The previously measured freezing point is stored in the controller's memory and is reset after each cycle.

At this subfreezing temperature the refrigeration system is shut off and the heater is energized. The thermal inertia of the system, however, continues to lower the sample temperature 16° to 22°F below the previously determined freezing point. As the sample becomes warmer, the temperature derivative circuit is triggered indicating that the last crystal layer has melted. An amplified output from the adjacent thermocouple provides a signal for recording or controlling. Simultaneously, the heater is deenergized and a new cycle is initiated.

Although stirring is not used (as in ASTM method D2386-67, IP16/68), the results are more precise than those obtained by the laboratory method. The analyzer package includes alarm and safety circuits, adjustable cycle sequences and local temperature indicators.

Freezing Point Analyzer for Hydrocarbon Purity (*f2*)

This analyzer was initially developed to measure the purity of p-xylene but has since been applied to other organic compounds such as benzene and phthalic anhydride.

The technique follows ASTM method D1015-55 closely except that stirring of the sample is omitted.

A two-pen temperature recorder with programmer, limit switches and thermistor bridge circuit is mounted in the control room. One pen (high span) triggers crystallization by jarring the sample when a predetermined supercooled state has been achieved. The second pen (narrow span) records the percent purity of the sample. The field unit contains the sample cell and related hardware.

Fresh sample is admitted at the bottom of a vertical cylindrical sample cell by a programmed sample inlet solenoid valve, purging the system for 30 seconds. Sample cooling is initiated upon closure of the inlet valve and trapping a 15 cc sample in the cell. Cooling continues until 9° to 15°F subfreezing is achieved. The high span pen will at this point actuate a limit switch and energize a plunger which jars the sample cell to initiate crystallization—an exothermic reaction, raising the sample temperature to its freezing point. Since the total latent heat of fusion was not extracted from the sample during supercooling, a small volume in the center of the cell remains in the liquid phase at a temperature in equilibrium with the solidified portion. A thermistor extending coaxially down from the top of the cell displays the sample's liquid phase temperature on the narrow span pen of the recorder. In a few minutes the heat absorbed from the surrounding air completely liquifies the sample, and the recorder chart displays the freezing point temperature on the calibrated chart. At this point a new cycle is initiated.

When the freezing point is above ambient temperature, the same technique is used except that the sample is heated and allowed to supercool by heat loss to ambient temperature.

Calibration and Application

Calibration of the aviation fuel analyzer should be based on the comparison of analyzer read-outs with multiple test results. Its principal application is in the processing of aviation fuels such as JP-4, kerosines and similar products.

Determination of purity is needed in the production of benzene, toluene, ethylbenzene, o-xylene, p-xylene and phthalic anhydride. These materials must be able to be supercooled and have a specified freezing point temperature.

Calibration of hydrocarbon purity analyzers is best achieved by using a sample of known purity. A certified standard thermistor may also be used if a sample of known purity is unavailable. Comparison with the ASTM methods can also establish the purity of a sample which can then be used for analyzer calibration.

Flash Point Analyzer

Flash Point Tests (ASTM Method D56-70, ASTM Method D93-66, IP-34/67)

The plant analyzers are intended to correlate with ASTM method D56-70 (Tag closed tester) and ASTM method D93-66, IP-34/67 (Pensky-Martens closed tester).

ASTM method D56-70 is for materials with a viscosity less than 45 SSU at 100°F and a flash point below 200°F. A sample of 50 cc is used at a temperature of at least 20°F below the expected flash point. The sample cup is immersed in a bath the temperature of which may be raised at a prescribed rate. Thermometers measure the bath and sample temperature. The sample lid prevents loss of sample vapors and directs a small flame into the cup periodically. The flash point is defined as the lowest sample temperature to cause ignition of the vapor above the sample at one atmosphere absolute.

Repeatability is 2°F for flash points below 140°F and 3°F for flash points between 140° and 199°F. Reproducibility is 6°F for flash points below 55°F, 4°F for flash points between 55° and 139°F and 6°F for flash points from 140° to 199°F.

The Pensky-Martens closed tester is for materials with an indicated flash point temperature as high as 700°F. Approximately 4.2 cubic inches of sample are used and the sample cup is heated directly by either a gas or electric heater at a prescribed rate. The sample cup lid is designed to support a sample stirrer, a mercury-in-glass thermometer and an apparatus for periodically exposing the vapor above the sample to a test flame. The repeatability is 10°F for materials with a flash point above 220°F, and the reproducibility is 15°F.

***Low Range Flash Point Analyzer** (gl)*

When the sample is heated to its flash point and its vapor ignited, the temperature in the vapor space increases. Sample is fed to the analyzer's heating chamber at a constant rate, and air is added at a rate of 600 cc per minute (Figure 1.9s). The air-sample mixture is heated at a controlled rate before it enters the flash cup and overflows to maintain a constant level. The vapor rises into the vapor space and is periodically exposed to a high voltage spark. Thermocouple 1 in the flash cup measures the temperature of the air-sample mixture, and thermocouple 2 in the vapor space responds to the temperature rise when vapor ignition occurs. This shuts the heater off and causes recorder

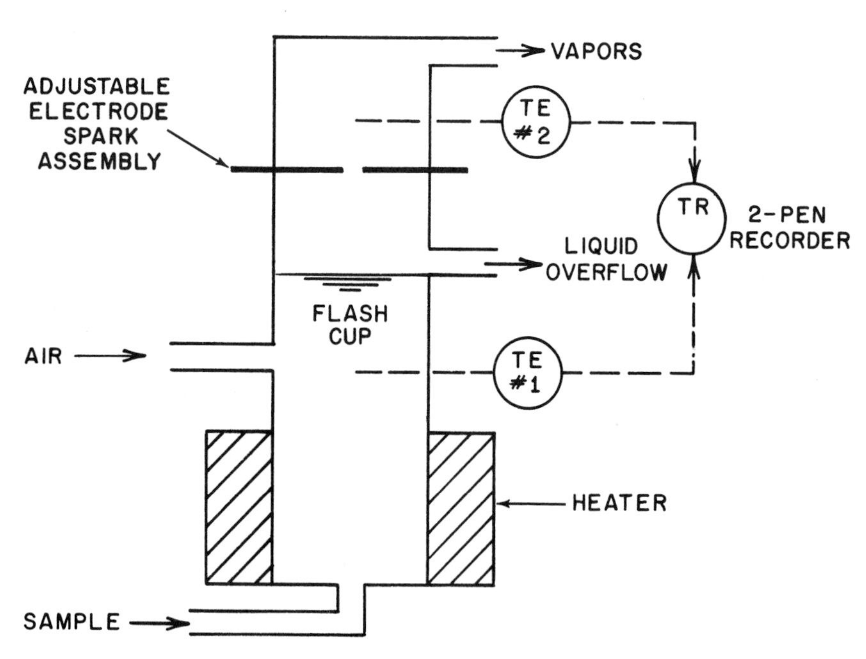

Fig. 1.9s Low range flash point analyzer

pen 1 (to which thermocouple 1 is connected) to be driven downscale. The peaks of the resulting sawtoothed record indicate the flash point temperature for each analysis cycle.

Integrated Flash Point Analyzer (g2)

This design is identical to the unit described in Figure 1.9s except that it is mounted on a frame with all the necessary accessory components piped and wired for field installation. It is particularly applicable for pipeline interface detection.

The accessories include

1.) a sample conditioning system to filter and coalesce free water from the sample, regulate sample pressure and indicate coalescer by-pass flow; 2.) a mechanical refrigeration system and temperature controller to cool the sample below its flash point; 3.) an air compressor, filter and flowmeter to supply combustible air; 4.) a duplex positive displacement pump to provide a constant rate of sample flow to the analyzer and return analyzer effluent and coalescer by-pass to the pressurized process line, and 5.) block, check, relief and backpressure valves for isolation and ability to withdraw sample for calibration.

High Range Flash Point Analyzer (g3)

Sample from a sweepstream is metered to the system at a constant rate by one head of a duplex positive displacement pump and is preheated to a fixed temperature below the flash point as determined from the preceding analysis. It is mixed with air at a rate of 1,500 cc per minute. A final heater provides the additional heat required to bring the air-sample mixture to the flash point temperature. The liquid entering the flash chamber is returned to process by the second head of the duplex pump, and the rising vapors are exposed to a high voltage spark every 10 seconds.

Ignition is detected by the deflection of a diaphragm caused by the combustion pressure pulse. The control circuit increases or decreases the final heater output, depending on whether or not ignition has occurred. At the same time the preheater controls are also adjusted to maintain the desired temperature differential. Flash point temperature is sensed by a thermocouple in the flash chamber liquid and displayed on a recorder chart.

Calibration and Application

Either the spot sample or standard sample method may be used. In either case, care must be exercised when a low flash point sample is used so as to prevent loss of light ends.

The analyzer can be applied to the control of vacuum distillation, dewaxing, solvent extraction and stripping, deasphalting, blending, residual fuel oil processing and pipeline interface detection.

Octane Analyzers

Laboratory Tests

Treatment will be brief owing to the vast complexity of this subject. Basically two methods are employed in which in a standard engine an unknown fuel is compared with a standard or reference fuel. One method yields a motor octane number (MON) in which the engine is run at 900 rpm, and the second method provides a research octane number (RON) in which the engine is run at 600 rpm. The difference in octane number by these methods (or "spread") is indicative of city driving at low speeds as compared to highway engine performance. Standard fuels are based on normal heptane (zero rating) blended with iso-octane (100 rating), with the octane number equal to the percent of iso-octane in the blend. The range has been extended by the addition of tetraethyl lead (TEL) to the iso-octane for ratings above 100. When the unknown fuel produces the same knocking as a standard fuel blend it is rated equal to the octane rating of the standard blend. The RON is higher than the MON, with the spread increasing with increasing octane numbers.

Standard Engine Octane Comparitor Analyzer (h1)

These analyzers serve to automate the ASTM procedure, using a standard engine with a modified carburetor fuel delivery system and standard detonation pickup and knockmeter. Comparison is made between the process stream and a prototype fuel, which serves as the standard or octane number reference point, from which an octane number difference is determined as the analyzer read-out. Figure 1.9t illustrates the equipment diagram for such an analyzer.

The accuracy of the systems depends on the performance of the standard engine, which must be properly maintained for optimum system operation. Also, the prototype fuel octane number should be determined to within ±0.1 or better since it serves as the reference for stream comparison measurements.

These analyzers can trim octane "give-away" during blending to approximately 0.05 octane above specification requirements.

Reactor Tube Continuous Octane Analyzer[4] (h2)

This analyzer monitors the reactions that precede engine knocking, the parameters of which may be controlled and correlated with octane number. A fuel and controlled air volume mixture is delivered at a rate of 1 cc per minute to a reactor tube maintained at an elevated temperature. Partial oxidation reactions in the tube produce a peak temperature the location of which is related to octane number. Higher octane fuels cause the peak temperature to move away from the tube inlet, whereas increasing the

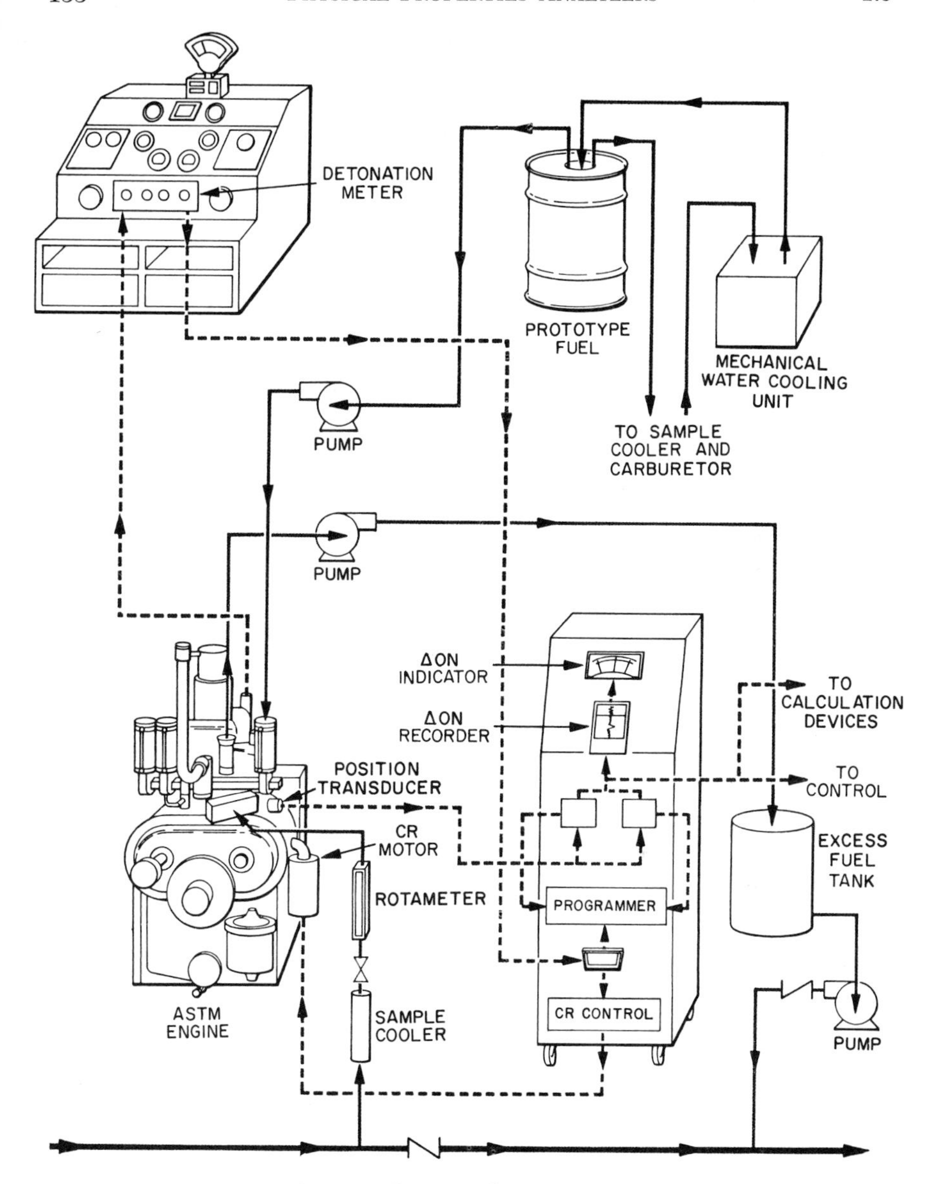

Fig. 1.9t Octane analyzer equipment

reactor tube pressure moves the peak closer to the tube inlet. Consequently, if the temperature peak location is fixed by varying the reactor tube pressure as fuel octane number varies, the pressure may be correlated with octane number and used as the analyzer read-out. This is accomplished by locating two thermocouples in the tube, one inch apart and equidistant from the temperature peak. Any movement of the peak due to a change in fuel octane rating is sensed by a differential temperature controller and causes a compensating change in reactor tube pressure to restore the temperature peak location.

REFERENCES

1. Nelson, W. L., "How Reid and true vapor pressure vary." *Oil and Gas Journal,* June 21, 1954.
2. Waner, N. S., Lecture notes on Physical Properties Analyzers given at Instrument Society of America Short Courses on Process Analyzers. Published by Instrument Society of America, Pittsburgh, Pa.
3. Huffman, H. C., Hass, R. H., O'Brien, N., Unzelman, G. H., and Jones, J. T., "An Analyzer for On-Line V/L Control." Presented at American Petroleum Institute Meeting, Houston, May, 1970.
4. McLaughlin, J. H., and Bajek, W. A.: "An Instrument to Continuously Monitor the Octane Quality of Gasoline." Presented at the Instrument Society of America Conference, Houston, October, 1969.

1.10 TRENDS AND ADVANCES IN ONSTREAM PROCESS ANALYSIS

According to the predictions of the Stanford Research Institute, by 1975 the use of analyzers will double the 1970 level. This growth rate will change the nature of the analyzer industry as it presently exists, because for the first time in process analysis large production runs will replace the custom-built units and funds will become available for large scale research and development activities. This may also change the present practice in which some new analyzers are developed by the users, who have the ability to pay the cost.

The demand for more and better analyzers is reinforced by the growing concern about environmental pollution and the associated need to enforce safety standards by detecting the concentrations of various pollutants (see Chapter V for details). The chemical industry will benefit from the technological fall-out resulting from this effort just as it benefited from the space industry by inheriting improved solid state electronics, telemetering systems for long distance transmission and digital and CRT information display techniques.

Another reason why more and better analyzers are needed is the increasingly competitive and sophisticated nature of the petrochemical industry. A plant may operate at a profit or at a loss depending on its ability to increase the efficiency (by only a few percentage points) of converting feeds into products. This can be done only if the operators know what is flowing in the pipelines in addition to being aware of the flow rates. This represents a higher level of sophistication than was exhibited by the earlier operating philosophies when management was still satisfied to know how much was being produced without knowing what it was or when after the fact laboratory analysis of grab samples was considered to be sufficient.

Another factor in more and better analyzers is the new generation of control systems, replacing the regulatory and feedback concepts with adaptive, feedforward and optimizing modes of control (see Chapter VII in Volume II). For these improved modes of control a complete understanding of the process model (static and dynamic) is needed; and analyzers play an important part in furnishing the data for these process models.

That there is a need for more and better analyzers does not mean that their application is or ever will be a simple routine task. An analyzer system will fulfill its expectations only if careful planning and evaluation precede its purchase, and if the users realize that if an expensive analyzer is worth purchasing it is also worth calibrating and maintaining properly after installation. Another important factor is the operator's acceptance which depends largely on training and familiarization. What follows is a discussion of selection and application in addition to a brief state-of-the-art review.

Analyzer Selection

Some of the more important factors that must be weighed by the instrument engineer are noted in Table 1.10a, in which both the desirable and undesirable features are explained. Unfortunately, no single analyzer combines all the desirable features of providing onstream, specific, continuous, unattended high sensitivity readings without drift or noise or need of a sampling system. Therefore, the selection is always a compromise, which is likely to give satisfactory results only if it is preceded by careful evaluation.

Two main areas of onstream analysis will be discussed including recent developments in analytical detectors, and sample handling considerations.

The review of recent developments is made in accordance with the order in which the analytical sensors are listed in Table 1.10b.

Radiant Energy Sensors

This family of sensors operates on either absorption or reflection principles. If radiation at different wave lengths (Table 1.10c) is passed through a process material, the amount of absorption is an indicator of sample identity or composition. The output of these analyzers is frequently non-linear, because Beer's law of radiation absorption is logarithmic:

$$A = abc = \ln\frac{Ie}{Il} \qquad 1.10(1)$$

where A = absorbence
a = molar absorption
b = sample path length
c = sample concentration
Ie,Il = radiation intensity entering and leaving sample

The microwave analyzer operates in both the absorption and the reflection modes. One of its applications is in the measurement of moisture in solids without requiring physical contact with the sample.

Infrared (IR) was one of the first analyzers to be moved from the laboratory to the pipeline and is available for composition detection of gas, liquid or solid streams. In the absorption mode of operation on liquid samples, the path length has to be very short. The non-contacting back-scatter designs

Table 1.10a
FACTORS AFFECTING ONSTREAM ANALYZER PERFORMANCE

Instruction: Use code to identify analytical instruments in Table 1.10b

	Code	*Desirable*	*Code*	*Undesirable*
Selectivity	2,3,4,5,8, 11,12,14, 16,23,31, 32,33,34, 35,36,37, 38,40,41	Responds only to specific component of interest; is unaffected by or can be compensated for pressure, temperature, other factors.	6,7,9,13, 15,17,20, 21,22,24, 25,26,27, 28,29,39	Responds to a process property which is affected by both the process conditions and the presence of components other than the one of interest. Applications limited to compensated binary systems.
Operation	All except 23,31,32, 33,34	Continuous, unattended	23,31,32, 33,34	Semicontinuous or intermittent
Quality	2,3,6,8, 11,12,16 31,32,33, 35,36,37, 38,39,40,41	Good sensitivity, low drift, high signal to noise ratio.	1,4,5,7, 9,13,14, 15,17,20, 21,22,23, 24,25,26, 27,28,29, 34	Low sensitivity, substantial noise or drift.
Sample phase	3,4,8,21, 22,25,29	Handles both liquid and gas samples.	All except 3,4,8,21, 22,25,29,41	Limited to either liquid or gas samples.
	2,3,9,17, 20,21,22	Can analyze solids.	All except 2,3,9,17, 20,21,22,41	Not applicable to solid samples
Sample handling	2,9,17, 21,22	No sampling system required. Sensor is not in contact with process stream.	4,5,8,11, 23,24,25, 26,27,28, 31,32,33, 34,35,36, 37,38,39, 40,41	Sample must be withdrawn from process and taken to analyzer. Results: transportation lag, possible need for altering.
	3,6,7	Sensor requires a window but is not in contact with process.		
	1,12,13, 14,15,16, 17,22,29	Sensor has retractable probe or can be cleaned ultrasonically by air or by liquid jets.		This could involve flashing, condensing, filtering, diluting, drying, or other manipulation that could affect representativeness of sample.

Table 1.10b

A LISTING OF ONLINE COMPOSITION ANALYZERS

Instruction: Use code to locate analytical instrument characteristics in Table 1.10a

Code	Sensor	Reference	Code	Sensor	Reference
	Type: Electromagnetic radiant energy				
1	Ultrasonic	pp. 54–60, pp. 519–521, vol. I	22	Dielectric constant	pp. 856–873, pp. 39–50, vol. I
2	Microwave	pp. 873–881, vol. I	23	Flash, melt, pour point*	Section 1.9
3	Infrared	pp. 772–784, vol. I	24	Flow index	pp. 637–642, pp. 892–913, vol. I
	Visible		25	Molecular weight	pp. 637–642, pp. 892–913, vol. I
4	Colorimeter	pp. 784–791, vol. I	26	Plastometry	pp. 637–642, pp. 892–913, vol. I
5	Flame photometer	p. 786, vol. I	27	Thermal conductivity	pp. 791–797, vol. I
6	Refractive index	pp. 797–806, vol. I	28	Vapor pressure	pp. 892–913, vol. I
7	Turbidity	pp. 881–886, vol. I	29	Viscosity	pp. 569–642, vol. I
8	Ultraviolet	pp. 762–772, vol. I		**Type: Combination, miscellaneous**	
9	Radiation (gamma, neutron)	pp. 63–75, pp. 290–294, pp. 873–881, vol. I		Chromatography	
	Type: Electrochemical		31	Adsorption	pp. 725–762, vol. I
11	Electrolytic cells	pp. 856–873, vol. I	32	Partition	pp. 725–762, vol. I
12	Galvanic cells	pp. 806–821, vol. I	33	Gel-permeation	pp. 892–913, vol. I
13	pH	pp. 821–842, vol. I	34	Total carbon analyzer	Section 1.15
14	Ion selective electrodes	Section 1.13		Ionization (mass spectrometry)	
15	ORP (Red-Ox)	Section 1.14	35	Thermal & flame	pp. 744–747, vol. I
16	Polarographic cells	pp. 806–821, vol. I	36	Electrostatic	p. 817, vol. I
17	Conductivity	pp. 913–919, vol. I	37	Radiation	p. 182, vol. I
	Type: Inferential property		38	Paramagnetic	pp. 806–821, vol. I
17	Conductivity-resistivity	pp. 913–919, vol. I	39	Catalytic combustion	pp. 1039–1050, vol. I
20	Consistency	pp. 886–892, vol. I	40	Piezoelectric	856–873, vol. I
21	Density	pp. 265–305, vol. I	41	Wet chemistry	

Note: Codes 10, 18, 19, and 30 unassigned

* Physical properties sensors

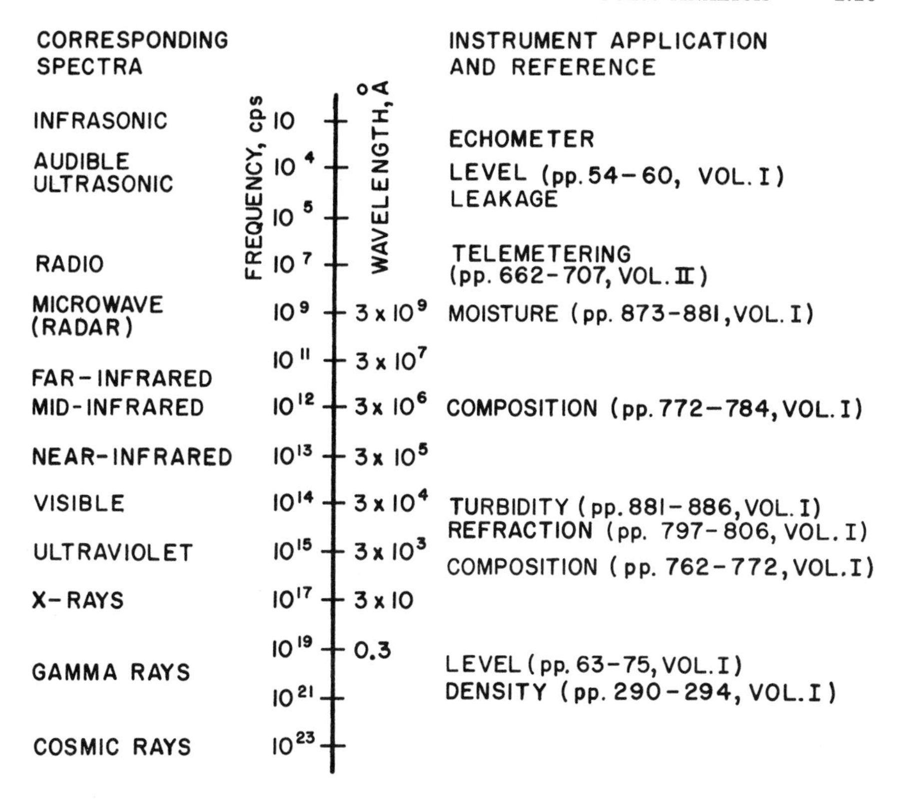

Table 1.10c Radiation spectrum pertaining to analytical instrumentation

detect the moisture in solids. The more recent probe version of the IR analyzer eliminates the need for a sampling system.

The refractive index (RI) is a unique property of a chemical compound and can therefore be used for composition determination in binary systems. Snell's law of refraction expresses the relationship between RI, the angle of incidence (α) and the angle of refraction (β), as light passes through the interface between two materials:

$$\mathrm{RI} = \sin\alpha/\sin\beta \qquad 1.10(2)$$

Differential refractometers utilize this relationship by keeping α constant and therefore a measurement of β expresses RI. At a critical value of α the light is totally reflected; the measurement of this angle can also be related to RI by the following relationship:

$$\alpha \text{ critical} = \arcsin\left(\frac{\text{variable sample RI}}{\text{fixed prism RI}}\right)$$

This later technique (critical angle refractometry) although less sensitive requires no sampling system and therefore is preferred for onstream applications (Figure 1.10d).

As with refractometers, turbidity sensors also work with visible light, either in the absorption mode or in the reflectance (nephelometry), light scattering mode. The first detects the sum of all light absorbing effects including color variations, whereas the second measures only the concentration of suspended solids.

Ultraviolet (UV) analyzers are less specific but more sensitive than IR and can handle only clean, single phase samples. Recent studies indicate that the UV bands of some chemicals shift with changes in pH, a phenomenon that could also be used for composition analysis.

The absorption or back-scatter of neutron or gamma radiation can also be correlated to composition in binary systems. Neutrons have been used to measure the moisture content of solids in processes in which hydrogen is present only in the free water and is not bonded to the other molecules. Gamma rays can penetrate metallic walls and therefore can give composition

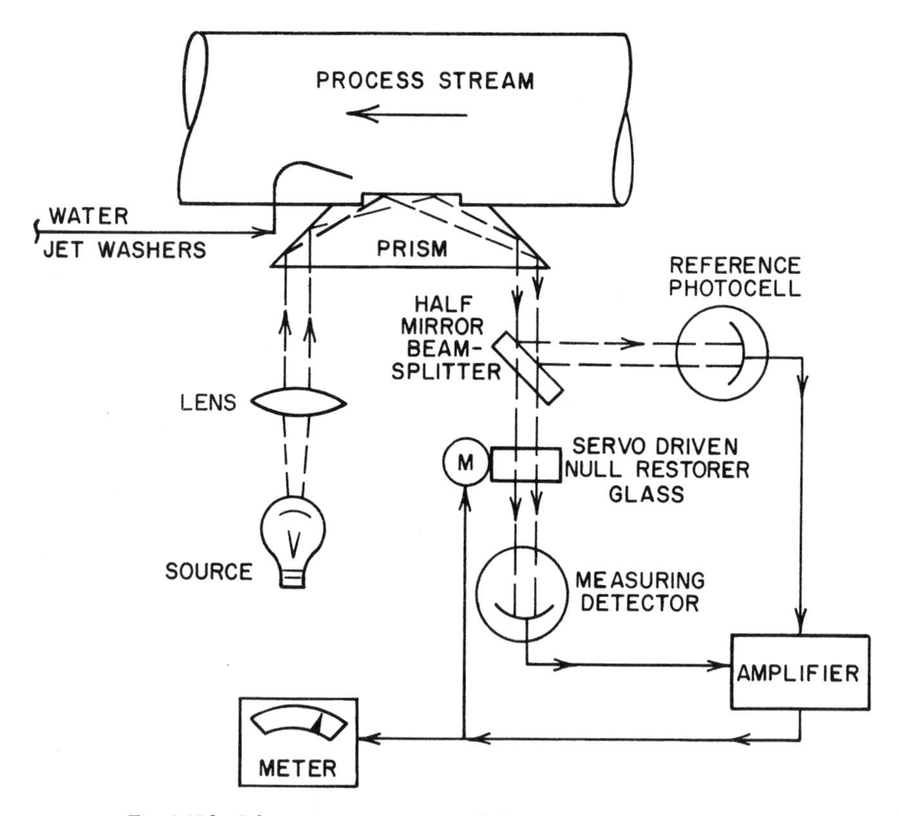

Fig. 1.10d Schematic representation of the critical angle refractometer

data for binary systems without contacting the process stream. Scheduled recalibration (utilizing calibrated absorber plates) is necessary to compensate for source decay and drift.

Electrochemical Sensors

In a recent development, the hydrochromatograph, the chromatographic column separates the water from the sample, and an electrolytic cell generates the electrolysis current, which is proportional to the quantity of water. With this technique it becomes possible to detect the moisture content of unsaturated hydrocarbons, which otherwise could polymerize and plug the electrolytic cell.

In galvanic and polarographic cells in probe type packaging, the electrolyte gel is separated from the process stream only by a membrane, a feature that eliminates the need for a sample system and performs well if the membrane is kept clean. The electrochemical reaction in the electrolyte is either spontaneous or is caused by the polarizing voltage, and the resulting current flow is an indication of composition.

The oxidation-reduction potential (ORP) sensors are also available in probe designs to detect the ratio of reducing agent to oxidizing agent (Section 1.14 in this volume)—an important parameter in effluent treatment controls.

A family of analytical sensors which detects the electrical potential generated in response to the presence of dissolved ionized solids in a solution includes pH, conductivity and ion selective probes. Before discussing each of these devices we will review the principle of their operation, which is based on the Nernst equation

$$E = E_0 + \frac{F}{n} \log (\gamma c + s_1\gamma_1 c_1 + s_2\gamma_2 c_2 \cdots) \qquad 1.10(3)$$

where E = potential difference between sensing and reference electrodes

E_0 = base potential which is a function of the construction and characteristics of the electrodes

F = the Nernst factor which is approximately 60 millivolts if temperature is constant at 25°C

n = the charge on the ion being measured (1 for monovalent, 2 for divalent, and so forth)

γ = activity coefficient of ion being measured

c = concentration being measured

s_1, s_2 = selectivity constants reflecting electrode response to interfering ions

$\gamma_1, \gamma_2, c_1, c_2$ = activity coefficient and concentration of interfering ions to which the electrode is sensitive with the selectivity of s_1 and s_2

Conductivity sensors measure a solution's ability to conduct electricity, which is a function of all dissolved ionized solids in the solution. These detectors are packaged either as probes (with isolating valves for removal, without opening up the process) or in the flow through (inductive) designs shown in Figure 1.10e.

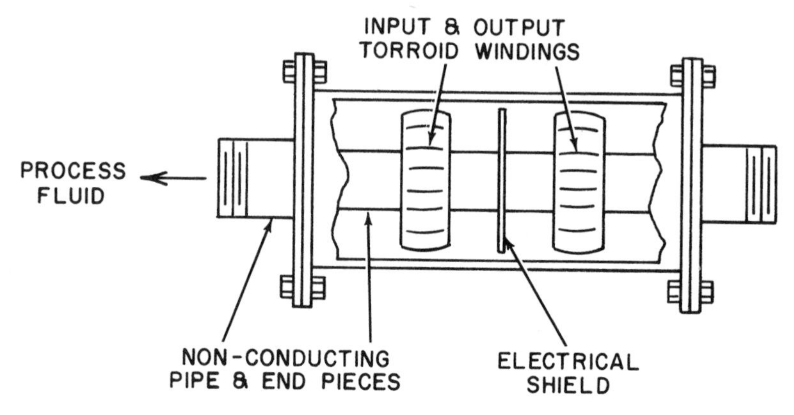

Fig. 1.10e Electrodeless conductivity cell

Ion selective electrodes (Section 1.13) have received much attention recently although the principles (based on the Nernst law) were known and applied (pH) for a long time. If total ionized solids (conductivity) are constant, a correlation can be drawn between the activity of a specific ion and its concentration in the process stream. The ideal reference electrode produces a constant potential, which is independent of the composition of the solution. A "perfect" measuring electrode gives a 60-millivolt change in potential for each tenfold change in the activity of a monovalent ion. It is important to emphasize that it is the *activity* of free ions that the electrodes respond to and *not* concentration. Equally important is it to understand that according to the Nernst equation, concentration (C) can be determined by the measurement of activity (activity $= \gamma C$) *only if* the other variables of the equation (F, s_1, s_2, γ_1, γ_2, c_1, c_2 and so forth) are *constant.* To achieve this involves scrupulous design; occasionally it also requires sample preparation (Section 1.13).

Unless the interferences to which ion selective measurements are subject are recognized and eliminated in the potential installations, misapplications are likely. The available ion selective electrodes are listed in Table 1.10f, where they are also grouped by the type of membrane utilized. Coating or material build-up on the membranes calls for the same degree of maintenance as is required by pH electrodes.

pH sensors are one of the ion selective electrodes, sensitive as they are to the activity of free hydrogen ions in the process stream and thereby

Table 1.10f
ION SELECTIVE ELECTRODES—
TYPES AND APPLICATIONS

Electrode	*Application*
Glass membrane	pH Potassium Sodium
Solid membrane	Bromide Chloride Fluoride Iodide Silver Sulfide
Liquid ion exchange membrane	Calcium Chloride Cupric Divalent metals (water hardness) Nitrate Perchlorate
Silicone rubber membrane (N.I.L.-Pungor)	Bromide Chloride Iodide Sulfide

reflecting acidity or alkalinity of the sample. pH hardware has been improved (Section 1.8) in various ways recently. Local, integral preamplifiers became available, reducing drift, instability and transmission distance limitations. New reference electrodes with "non-flowing junctions" reduced maintenance and eliminated pressurization. Other improvements included the combination of the measuring and reference electrodes into a single probe and the development of various electrode cleaning devices, such as air and water jets and mechanical and ultrasonic cleaners. Despite all this activity, pH remains a difficult measurement in which installation of standby spare sensors and scheduled periodic maintenance are likely to be necessary.

Inferential Property Sensors

Concerning property sensors and combination analyzers, the 1960s represented the decade of consolidation and improvement without major breakthroughs in the state of the art. Some of the more recent improvements and trends are noted below.

The detection of the dielectric constant has been used to measure the composition (moisture content) of samples. Now instruments can measure the dielectric loss[2] of moving materials with all components external to the piping and with increased sensitivity. The dielectric loss tangent of moist materials, for example, is frequently two orders of magnitude greater than the loss tangent of dry materials.

In plastometry, flow index or intrinsic viscosity measurement (the determination of molecular weight distribution and plastic behavior), the main development is that some of the Mooney, Kneader and capillary extrusion plastometers are now available for continuous unattended onstream service in polymer plants.

As with the plastometers, gel permeation chromatographs can also be equipped with automated sampling systems for onstream measurements. The main drawback is the lengthy analysis; the advantage is that it furnishes data on molecular weight distribution in addition to molecular weight averages.

The piezoelectric effect is responsible for the phenomenon that an electric charge appears across crystals when they are exposed to a deforming force or that the frequency of crystal oscillation is affected by material deposits on the surface of the crystals. The first phenomenon has been used to detect pressure, acceleration, temperature, force and thickness or to generate ultrasonic waves; and the second phenomenon has recently been applied to measure the moisture content of gases. The technique is moderately accurate because each microgram of moisture deposited in the hygroscopic coating of the crystal results in a frequency change of 2,000 Hz.

In connection with pollution controls, one of the parameters of interest is the chemical oxygen demand (COD) of the effluent (Section 1.16). The automatic total carbon analyzer (Section 1.15) gives close correlation to COD and is therefore an important analytical tool. The organic and inorganic carbon bearing compounds in the sample are reacted catalytically (or otherwise) to form carbon dioxide and water. Carbon dioxide is then detected by an IR analyzer, and the entire cycle is completed in a few minutes.

The most powerful onstream analyzer is the chromatograph, and its widespread use is reflected in the fact that 30% of all analyzer outlay is for this type of instrument. The chromatograph operates in two distinct steps. First, it separates the component(s) of interest from the rest of the sample; and second, through elution by a carrier gas it permits the use of binary sensors. These sensors are mainly of the thermal conductivity type; flame ionization detectors are the second most frequently used, owing to their higher cost but superior sensitivity. In moisture measurement, for example, the phosphorus pentoxide cell (hydrochromatograph)[3] has been used successfully. Other chromatograph-detector combinations utilize the specific wavelength absorption characteristics of IR, visible and UV analyzers, the piezoelectric effect or, when both quantitative and qualitative analyses are desired, the ionization sensors such as the mass spectrometer.

Improvements in chromatographic technology include parallel columns and programmed multiple temperature zones which contribute to the reduction of analysis time; signal storing peak pickers which close the control loop with an otherwise discontinuous analysis signal; solid state circuitry which improves reliability; and new sampling valves. One improvement is in liquid sampling, in which peak tailing and baseline separation have been minimized by preventing the unvaporized (adsorbed) portion of the sample from entering the chromatographic column.

The cost of the chromatograph itself can be less than half (sometimes only 30%) of the total installation cost, and if the expense of maintenance is also included, the cost is reduced accordingly. This consideration has contributed to the emerging of two schools of thought, as have the reports indicating that for every successful chromatograph installation in the chemical industry, one has failed and been abandoned.

Adherents of the first school favor simple inexpensive chromatographs, monitoring one component in a single sample so as to increase reliability. Proponents maintain that an overburdened chromatograph is more a liability than an asset and that even when it is operating, the volume of information generated is likely more to swamp than to assist the operator. Therefore, their target is to achieve a degree of standardization and simplicity similar to that found in flow or temperature detectors. It is reported[4] that these simple chromatographs can accommodate the majority of present applications. The experience of about one hundred installations, including closed loop control systems, indicates that these standardized interchangeable units are easy to operate and maintain without the need for specialists.

Proponents of the second point of view feel that the answer lies in the opposite direction. They do not suggest reducing the number of samples or components analyzed but propose attacking complex problems with the tools of advanced technology. This means the use of dedicated mini-computers (Section 2.1), each handling six or more chromatographs, at a total computer hardware-software package cost of roughly $60,000. For the most critical analyses, a standby manual mode of operation is frequently available to prevent plant shutdown during the expected annual downtime period of five days. The computer can accurately integrate the chromatographic peaks, log or rearrange the measurement data, calculate control models or control functions, program all sequential or logic steps in the analysis cycle, optimize and thereby reduce cycle time and detect malfunctions. More important, it can anticipate maintenance needs before they occur so as to predict failures (sample valve leakage for example) and increase reliability by preventive maintenance.

Sampling Systems

No discussion of onstream analyzers is complete without a look at sample handling (Section 1.12). As shown by Table 1.10a, the most powerful analyzers are also those that require sampling systems. The installation price for the sampling system frequently exceeds the cost of the analyzer, but its importance is even greater than what this economic consideration implies, because a second class analyzer can still furnish useable data if it operates within an efficient sample system, whereas a poor sample invalidates the entire measurement.

Therefore, the most important criterion requires keeping the samples representative, both in time (short sample lines guarantee minimum transportation delays) and in composition. Whenever possible, the sample should not be tampered with, because the steps of sample preparation (drying, vaporizing, condensing, filtering and diluting) always degrade the representativeness of the sample. If there is no sampling system, the integrity of the sample is automatically guaranteed and therefore preference should be given to those sensors which either are external to the process pipe or penetrate it with a retractable, cleanable probe. Probe type sensors (solid or membrane) require periodic cleaning, which can be done manually (withdrawing the probe through an isolating valve so that the process is not opened when the electrode is cleaned) or automatically. Automatic probe cleaning devices may be pressurized liquid or gas jets, or thermal, mechanical or ultrasonic cleaning and scraping instruments.

When a sample system is unavoidable, it should be made as simple as possible, with the minimum number of components. Even a well designed multi-stream sampling system, such as that shown in Figure 1.10g, will most likely become expensive to maintain owing to the number of components.

If the process is not pressurized, solid teflon aspirators activated by water or air jets can withdraw the samples. They are easier to maintain than are sample pumps. Similarly, the ball valves with block-bleed-block arrangements perform better than solenoids in sampling systems.

A frequent problem is the plugging of sampling systems. If the material to be removed to prevent plugging is dust, the self-cleaning bypass filter (Figure 1.10h) with automatic blowback constitutes a potential solution. In some instances liquid scrubbing or electrostatic precipitation should be considered, as well as cyclone type separators. In this last device (Figure 1.10i), the process stream enters tangentially to provide a swirling action and the cleaned sample is taken near the center. Transportation lag can be kept to less than one minute, and the unit is applicable to both gas and liquid samples. This type of centrifuge can also separate sample streams by gravity into their aqueous and organic constituents.

For the removal of small amounts of polymer dust in vapor samples there are melt filters with removable, heated metallic surfaces which melt and collect the polymer dust from the sample.

When the impurities in the process gas stream are both solids and liquids, such as in particulate matter and mist—carry-over problems in chlorine

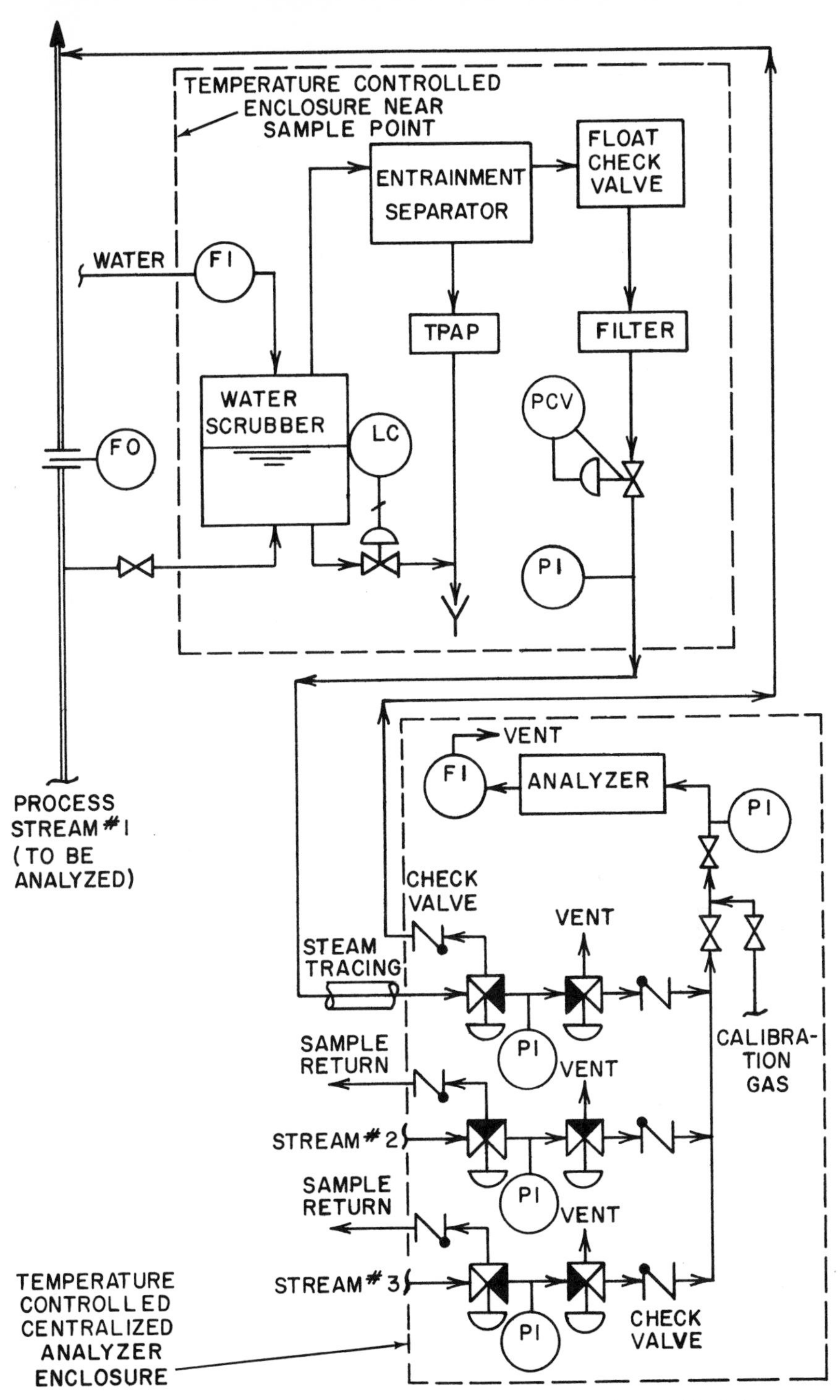

Fig. 1.10g Multistream analyzer sampling system with cleaning section and block, bleed and back-purge features

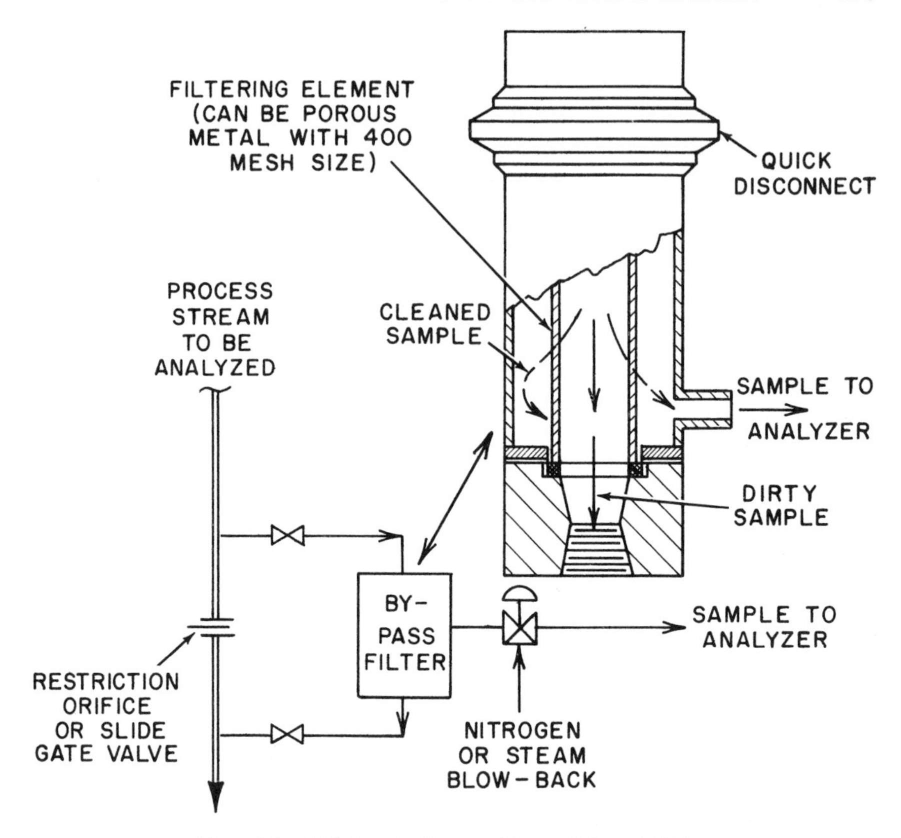

Fig. 1.10h Self-cleaning by-pass filter and its installation

plants—the fiber mist eliminator[5] (Figure 1.10j) should be considered. The liquid particles form a film on the fiber surface, and the drag of the gas moves this film and the dissolved solids radially, while gravity causes them to move downward, resulting in self-cleaning action.

For dissolved solids or polymer forming compounds in a process stream, which would leave a residue and eventually plug the liquid sample valve if not removed, the logical solution is to force the residue formation to take place in a controlled area, such as the fiber-glass filter in the spray stripping chamber shown in Figure 1.10k. If polymers represent a substantial portion of the process stream, the need for filter replacements becomes excessive and therefore impractical. A better technique is to vaporize the unreacted monomers through pressure reduction while keeping the polymers in a molten state through heating. This technique (Figure 1.10 l) not only discharges polymers continuously, but also provides a useable vapor sample.

Where it is necessary to keep clean the windows on the various photometers operating on gas samples, a 2 SCFH warm air purge can be used, keeping

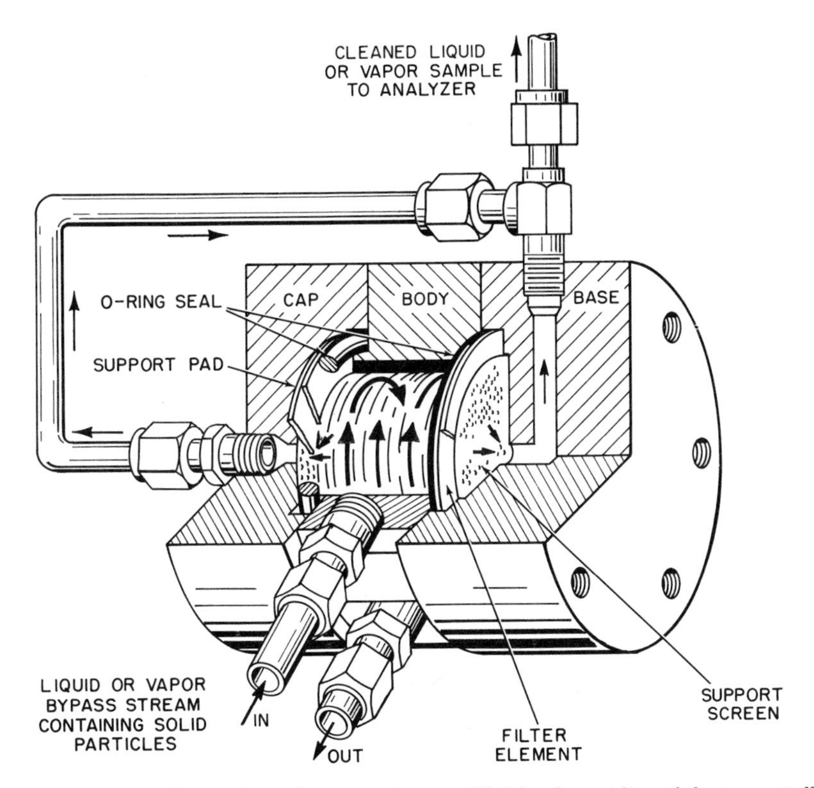

Fig. 1.10i By-pass filter with its cleaning action amplified by the swirling of the tangentially entering sample

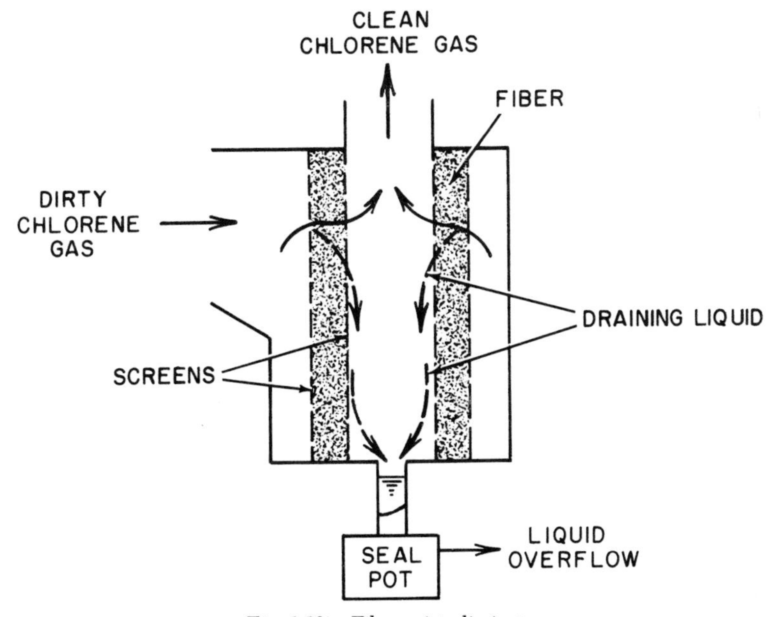

Fig. 1.10j Fiber mist eliminator

Table 1.10m
ORIENTATION TABLE FOR ANALYZERS

Manufacturers of Online Analyzers:

Grouped by the method of analysis applied

Code	*Sensor Type*	Agricultural Control Systems, Inc.	American Instrument Co.	Anacon, Inc.	Applied Automation, Inc.	Automated Environmental Systems	Automation Mfg. & Service Co.	Bacharach Instrument Co.	Bailey Meter Co.	Barton ITT	Beckman Instruments, Inc.	Bendix Corp.	Bristol Div., ACCO	Brookfield Engineering Labs., Inc.	Cambridge Systems, Inc. (EG&G)	CEC (Bell & Howell)	Corning Scientific Instruments Co.	Customline Control Panels, Inc.	Dasie, Corp.	Davis Instrument Co.	Delta Scientific Corp.	DeZurik Corp.	E. I. du Pont de Nemours & Co.	Electron Machine Corp.	Electronic Associates, Inc.	Farrand Optical Co.	Fischer & Porter Co.	The Foxboro Co.	GAM-RAD, Inc.	General Electric Co.	Hach Chemical Co.	Hallikainen Instruments	Hartman & Braun	Hays Corp.	Honeywell, Inc.	Houston Atlas, Inc.
	Electromagnetic Radiant Energy																																			
1	Ultrasonic																																			
2	Microwave																																			
3	Infrared			✓							✓	✓											✓							✓			✓			
	Visible																																			
4	Colorimeter		✓	✓							✓										✓		✓			✓	✓				✓	✓	✓			✓
5	Flame photometer																				✓		✓													✓
6	Refractive index			✓																				✓												
7	Turbidity	✓	✓						✓		✓										✓		✓				✓	✓	✓		✓	✓	✓		✓	
8	Ultraviolet		✓	✓				✓	✓		✓												✓			✓							✓			
9	Radiation (gamma, neutron)																													✓			✓			
	Electrochemical																																			
11	Electrolytic cells															✓																				
12	Galvanic cells			✓					✓												✓													✓		
13	pH										✓						✓				✓							✓					✓		✓	
14	Ion-selective electrodes										✓						✓				✓							✓						✓	✓	
15	ORP (Red-Ox)																				✓							✓							✓	
16	Polarographic cells																										✓								✓	
17	Conductivity										✓										✓							✓					✓		✓	

Table 1.10m
ORIENTATION TABLE FOR ANALYZERS (Continued)

Grouped by the method of analysis applied

Code	Sensor Type / *Manufacturers of Online Analyzers:*	Ikor, Inc.	Industrial Nucleonics Corp.	Infrared Industries, Inc.	Jacoby-Tarbox Corp.	Kay-Ray, Inc.	Kollmorgen Corp.	Leeds & Northrup Co.	Lockwood & McLorie, Inc.	Microwave Instruments Co.	Mine Safety Appliances, Inc.	Monsanto Research Corp.	Norcross Corp.	Nuclear Chicago Corp.	NUS Corp.	Ohmart Corp.	Olkon Corp.	Panametrics, Inc.	Permutit Co.	Precision Scientific Development Co.	Princo Instruments, Inc.	Process Analyzers, Inc.	Research Appliance Co.	Seiscor Div., Seismograph Service Corp.	Taylor Instrument Process Control Div.	Technicon Corp.	Teledyne Analytical Instruments, Inc.	Thermo-Lab Instruments, Inc.	UGC Industries, Inc.	Union Carbide Corp.	Vap-Air Div., Vapor Corp.	Waters Associates, Inc.	Westinghouse Electric Corp.
	ELECTROMAGNETIC RADIANT ENERGY																																
1	Ultrasonic														✓																		
2	Microwave									✓																							
3	Infrared			✓				✓			✓														✓		✓						
	Visible																																
4	Colorimeter						✓																				✓						
5	Flame photometer																																
6	Refractive index								✓																								
7	Turbidity				✓																						✓			✓			
8	Ultraviolet																										✓						
9	Radiation (gamma, neutron)		✓			✓								✓		✓		✓					✓		✓								
	ELECTROCHEMICAL																																
11	Electrolytic cells								✓													✓								✓			✓
12	Galvanic cells								✓													✓					✓	✓					
13	pH							✓																						✓			
14	Ion-selective electrodes							✓																									
15	ORP (Red-Ox)							✓																						✓			
16	Polarographic cells																																
17	Conductivity							✓																						✓			

Table 1.10m
ORIENTATION TABLE FOR ANALYZERS (Continued)

Grouped by the method of analysis applied

Code	Sensor Type / *Manufacturers of Online Analyzers:*	Agricultural Control Systems, Inc.	American Instrument Co.	Anacon, Inc.	Applied Automation, Inc.	Automated Environmental Systems	Automation Mfg. & Service Co.	Bacharach Instrument Co.	Bailey Meter Co.	Barton ITT	Beckman Instruments, Inc.	Bendix Corp.	Bristol Div., ACCO	Brookfield Engineering Labs., Inc.	Cambridge Systems, Inc. (EG&G)	CEC (Bell & Howell)	Corning Scientific Instruments Co.	Customline Control Panels, Inc.	Dasie, Corp.	Davis Instrument Co.	Delta Scientific Corp.	DeZurik Corp.	E. I. du Pont de Nemours & Co.	Electron Machine Corp.	Electronic Associates, Inc.	Farrand Optical Co.	Fischer & Porter Co.	The Foxboro Co.	GAM-RAD, Inc.	General Electric Co.	Hach Chemical Co.	Hallikainen Instruments	Hartman & Braun	Hays Corp.	Honeywell, Inc.	Houston Atlas, Inc.
	INFERENTIAL PROPERTY																																			
17	Conductivity-resistivity										√										√							√					√		√	
20	Consistency													√								√					√	√								
21	Density, thickness								√			√																√				√				
22	Dielectric constant																															√				
23	Flash, melt, pour point																															√				
24	Flow index																																			
25	Molecular weight										√	√																								
26	Plastometry																																			
27	Thermal conductivity		√						√		√	√																					√	√		
28	Vapor pressure																															√				
29	Viscosity											√		√													√	√				√				
	COMBINATION, MISCELLANEOUS																																			
	Chromatography																																			
31	Adsorption				√		√				√	√				√																				
32	Partition				√		√				√	√				√																				
33	Gel-permeation																																			
34	Total carbon, hydrocarbon																				√															
	Ionization (mass spectrometry)																																			
35	Thermal & flame										√	√				√																				
36	Electrostatic															√									√											
37	Radiation											√																								
38	Paramagnetic											√																						√		
39	Catalytic combustion							√												√														√		
40	Piezoelectric																						√													
41	Wet chemistry (auto-titrators)																				√										√					

Table 1.10m

ORIENTATION TABLE FOR ANALYZERS (Continued)

Grouped by the method of analysis applied

Code	Sensor Type (Manufacturers of Online Analyzers:)	Ikor, Inc.	Industrial Nucleonics Corp.	Infrared Industries, Inc.	Jacoby-Tarbox Corp.	Kay-Ray, Inc.	Kollmorgen Corp.	Leeds & Northrup Co.	Lockwood & McLorie, Inc.	Microwave Instruments Co.	Mine Safety Appliances, Inc.	Monsanto Research Corp.	Norcross Corp.	Nuclear Chicago Corp.	NUS Corp.	Ohmart Corp.	Olkon Corp.	Panametrics, Inc.	Permutit Co.	Precision Scientific Development Co.	Princo Instruments, Inc.	Process Analyzers, Inc.	Research Appliance Co.	Seiscor Div., Seismograph Service Corp.	Taylor Instrument Process Control Div.	Technicon Corp.	Teledyne Analytical Instruments, Inc.	Thermo-Lab Instruments, Inc.	UGC Industries, Inc.	Union Carbide Corp.	Vap-Air Div., Vapor Corp.	Waters Associates, Inc.	Westinghouse Electric Corp.
	INFERENTIAL PROPERTY																																
17	Conductivity-resistivity							✓																						✓			
20	Consistency							✓					✓																				
21	Density, thickness		✓			✓								✓	✓	✓			✓		✓	✓			✓				✓				
22	Dielectric constant																	✓															
23	Flash, melt, pour point																			✓													
24	Flow index																																
25	Molecular weight											✓												✓									
26	Plastometry											✓												✓									
27	Thermal conductivity							✓			✓											✓					✓						
28	Vapor pressure																			✓													
29	Viscosity											✓	✓				✓							✓									
	COMBINATION, MISCELLANEOUS																																
	Chromatography																																
31	Adsorption										✓											✓											
32	Partition										✓											✓											
33	Gel-permeation																															✓	
34	Total carbon, hydrocarbon																					✓								✓			
	Ionization (mass spectrometry)																																
35	Thermal & flame										✓																✓						
36	Electrostatic										✓																						
37	Radiation										✓			✓																			
38	Paramagnetic										✓																						
39	Catalytic combustion										✓																						
40	Piezoelectric																																
41	Wet chemistry (auto-titrators)																									✓				✓			

Table 1.10m
ORIENTATION TABLE FOR ANALYZERS (Continued)

Sensor Type (Grouped by component for which analysis is being made) / *Manufacturers of Online Analyzers:*	Agricultural Control Systems, Inc.	American Instrument Co.	Anacon, Inc.	Applied Automation, Inc.	Automated Environmental Systems	Automation Mfg. & Service Co.	Bacharach Instrument Co.	Bailey Meter Co.	Barton ITT	Beckman Instruments, Inc.	Bendix Corp.	Bristol Div., ACCO	Brookfield Engineering Labs., Inc.	Cambridge Systems, Inc. (EG&G)	CEC (Bell & Howell)	Corning Scientific Instruments Co.	Customline Control Panels, Inc.	Dasie, Corp.	Davis Instrument Co.	Delta Scientific Corp.	DeZurik Corp.	E. I. du Pont de Nemours & Co.	Electron Machine Corp.	Electronic Associates, Inc.	Farrand Optical Co.	Fischer & Porter Co.	The Foxboro Co.	GAM-RAD, Inc.	General Electric Co.	Hach Chemical Co.	Hallikainen Instruments	Hartman & Braun	Hays Corp.	Honeywell, Inc.	Houston Atlas, Inc.
Chlorine																				✓						✓									
Combustibles							✓	✓			✓								✓														✓		
Dissolved oxygen																				✓													✓	✓	
Flame								✓																											
Humidity		✓										✓		✓																					
H_2S									✓		✓																								✓
Moisture in gas or liquid samples			✓							✓	✓	✓			✓							✓									✓	✓			
Moisture in solid samples			✓																													✓			
Oxygen			✓				✓	✓		✓	✓									✓												✓	✓		
Ozone																		✓		✓						✓									
Sampling systems						✓	✓			✓	✓						✓														✓		✓		
Smoke and dust density					✓			✓														✓													
SO_2 and SO_3					✓				✓							✓			✓																
Water quality monitoring packages					✓															✓														✓	

Table 1.10m
ORIENTATION TABLE FOR ANALYZERS (Continued)

	Sensor Type / *Manufacturers of Online Analyzers:*	Ikor, Inc.	Industrial Nucleonics Corp.	Infrared Industries, Inc.	Jacoby-Tarbox Corp.	Kay-Ray, Inc.	Kollmorgen Corp.	Leeds & Northrup Co.	Lockwood & McLorie, Inc.	Microwave Instruments Co.	Mine Safety Appliances, Inc.	Monsanto Research Corp.	Norcross Corp.	Nuclear Chicago Corp.	NUS Corp.	Ohmart Corp.	Olkon Corp.	Panametrics, Inc.	Permutit Co.	Precision Scientific Development Co.	Princo Instruments, Inc.	Process Analyzers, Inc.	Research Appliance Co.	Seiscor Div., Seismograph Service Corp.	Taylor Instrument Process Control Div.	Technicon Corp.	Teledyne Analytical Instruments, Inc.	Thermo-Lab Instruments, Inc.	UGC Industries, Inc.	Union Carbide Corp.	Vap-Air Div., Vapor Corp.	Waters Associates, Inc.	Westinghouse Electric Corp.
Grouped by component for which analysis is being made	Chlorine																																
	Combustibles			✓							✓																						
	Dissolved oxygen																													✓			
	Flame																																
	Humidity																	✓													✓		
	H_2S																					✓						✓					
	Moisture in gas or liquid samples							✓			✓							✓				✓		✓			✓	✓			✓		
	Moisture in solid samples					✓				✓	✓		✓			✓									✓								
	Oxygen							✓	✓													✓					✓	✓					✓
	Ozone																																
	Sampling systems								✓		✓													✓									
	Smoke and dust density	✓			✓																		✓							✓			
	SO_2 and SO_3							✓			✓											✓				✓							
	Water quality monitoring packages																									✓				✓			

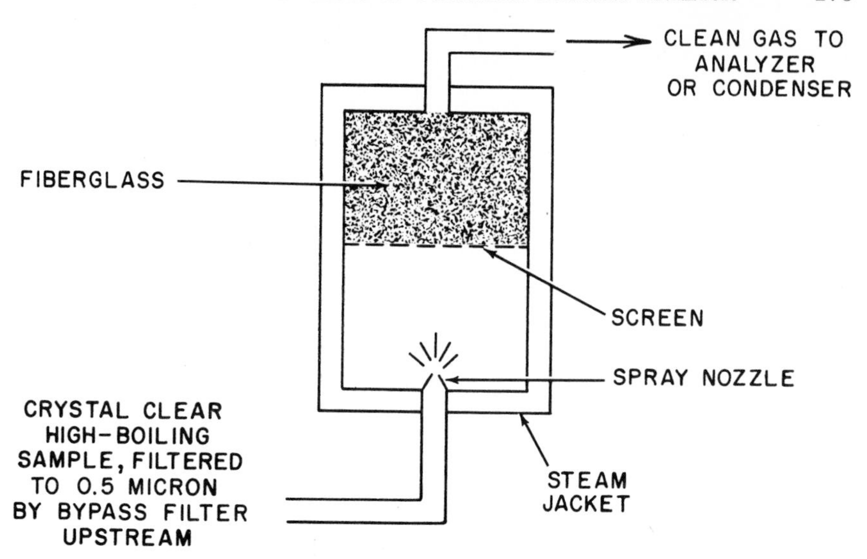

Fig. 1.10k When dissolved inorganic solids or polymer forming compounds are present, stripping the liquid sample may be the answer

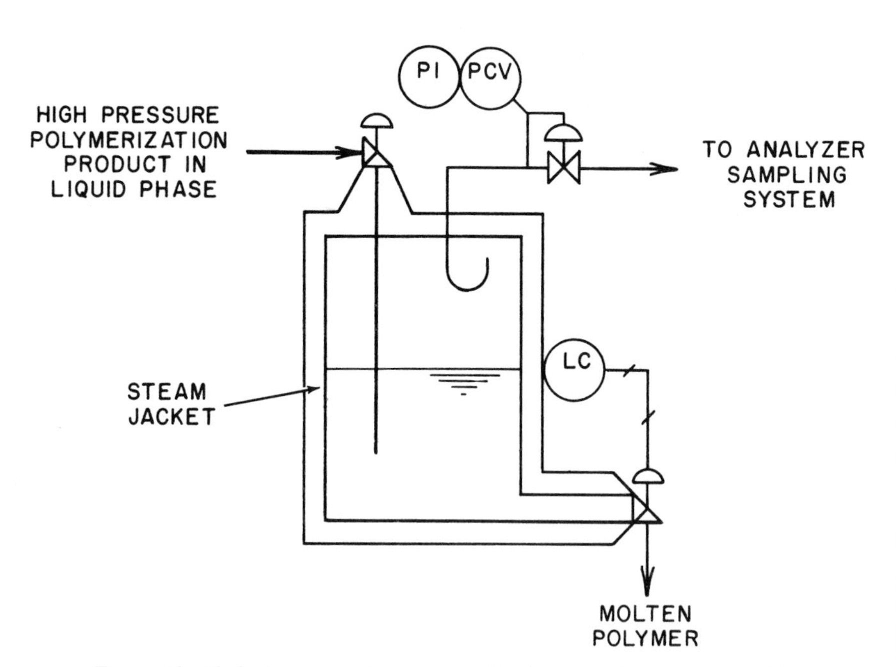

Fig. 1.10l Flash chamber makes the analysis of unreacted monomers possible

the window compartment isolated from the sample. The introduction of this air stream into a 900 SCFH sample stream can usually be tolerated.

Orientation Table

The orientation table (Table 1.10m) summarizes the availability of online analysis hardware. These devices are grouped (by the method of analysis) into electromagnetic, electrochemical, inferential and combination sensors. At the end of the orientation table are some analyzers and systems grouped by the component being analyzed or by the function they fulfill. Consequently, a single detector may be referred to twice in the table. For example, a phosphorus pentoxide type water analyzer is listed as a galvanic cell and as a moisture-in-gas sensor in the bottom part of the table. Obviously, the listing in the lower part of the table contains only the most frequent applications and is far from complete.

REFERENCES

1. "Gamma Backscatter Technique for Level and Density Detection," J. A. McConnell and W. W. Smuck, *Chemical Engineering Progress,* Aug. 1967.
2. Wood, H. H., "An Instrument for On-Line Moisture Measurement Utilizing the Principle of Dielectric Loss," ISA 16th Analysis Symposium, May 1970.
3. Penther, C. J. and Notter, L. J., "Hydro-Chromatography," Anal. Chem., Feb. 1964.
4. "A New Approach to Chromatograph Systems for Process Control," W. H. Topham, ISA 25th Conference, Oct. 1970.
5. Nichols, J. H. and Brink, J. A. Jr., "Use of Fiber Mist Eliminators in Chlorine Plants," Electrochemical Technology, July–Aug. 1964.

1.11 ANALYZER APPLICATION CONSIDERATIONS

Justification for an Analysis System

The reasons for purchasing process analyzers include cost and safety considerations as well as improving product quality or quantity, reducing by-products, decreasing analysis time, tightening specifications and monitoring contaminants, toxicants or pollutants. By continuing to ask how to accomplish these objectives, we arrive at the specific measurement requirement—to measure material A in the presence of material B, or to measure property X of material C.

Additional requirements and characteristics of the measurement should also be defined, such as the frequency of analysis, sample availability and the like. To distinguish between the actual needs and the desirable characteristics can save both time and investment cost.

A form is helpful in gathering data on needs and desires (Table 1.11a). In selecting an analysis system we must ask whether the need for the measurement justifies the cost of determining what type of analysis system is required. If the answer is yes, the study progresses until several types of analyzers are selected, at which time the question of cost is again raised, and some analyzers may be eliminated from further consideration. After the complete system has been defined, the estimated costs can be compared with the expected return and a decision reached.

Method of Analysis

If the problem has been defined as one requiring the measurement of one material in the presence of another, it is necessary first to look for a unique property of the material to be measured. Usually, a first step in determining the suitable methods of analysis is to investigate laboratory methods to determine the desired property. ASTM has established certain test methods for the determination of various properties and materials (see Section 1.9) as have several other organizations. Suppliers of the material to be analyzed are possible sources of information for methods of analysis as are suppliers of online analyzers.

Properties[1] which can be utilized for process composition analysis are shown in Table 1.11b. Similar information is provided in Table 1.10b.

Table 1.11a
ANALYZER SPECIFICATION FORM

Project:.. Date: ....
Specification No.: .... Code: ....
Information Compiled By: ....

A. GENERAL INFORMATION:

1. Plant: .... 2. Unit: ....
3. Process: ....
....

B. CONDITIONS AT PLANNED ANALYZER LOCATION:

1. Ambient Temperature Range: .... to .... Normal .... °C ☐ °F ☐
2. Protected From Weather: Yes ☐ No ☐
3. Unusual Ambient Conditions: ....
(Corrosive or Explosive Atmosphere, Excessive Moisture, Dust, etc.)
4. Power Available: .... Volts .... Hertz
   (a) Voltage Variation: .... to .... Volts
   (b) Frequency Variation: .... to .... Hertz
   (c) Grounding Facilities Available: Yes ☐ No ☐
5. Lighting Level: *Good* *Average* *Poor*
   (a) Front of Instrument ....
   (b) Back of Instrument ....
   (c) Direct Sunlight Will Strike Instrument: Yes ☐ No ☐
6. Steam Lines Near Location: ....
7. Instrument Air Available:
   (a) Pressure Range: From .... to .... Normal .... PSIA
   (b) Temperature Range: From ...... to ...... Normal ...... °C ☐ °F ☐
   (c) Contaminants: ....
   (d) Size of Header: .... Volume: .... ft^3/min

(*Use Separate Page for Each Stream To Be Analyzed*)

C. SAMPLING INFORMATION

1. Form of Sample: Gas ☐ Liquid ☐ Other ....
2. Temperature Range: From .... to .... Normal .... °C ☐ °F ☐
3. Pressure Range: From .... to .... Normal .... PSIA
4. Dew Point: .... °C At .... PSIG
5. Quantity Available: .... Per Hour
6. Low Pressure Return Line Available:
   Yes ☐ No ☐ Back Pressure .... PSIG
7. Specific Gravity ....
8. Contaminants in Sample: Oil ☐ Wax ☐ Solids ☐ Particle Size......
(Identity and concentration to be included in list of components below)
9. Corrosive Nature: Acid ☐ Basic ☐ Other ....

10. Other Data (Viscosity, Unusual Surges, etc.): ....
11. Materials of Construction Which May Be Used in Contact with Sample: ....
12. Distance From Sample Tap to Analyzer Location: .... ft
13. Size of Tap at Process Line if Any: ....
14. Concentration Ranges of All Components (even if only traces) in Stream: (Specify unit of measurement: % by volume, % by weight or ppm for each component)

Components to be Analyzed					*Other Components*				
	Max	*Min*	*Normal*	*Unit*	(liquid) Water (vapor)	*Max*	*Min*	*Normal*	*Unit*

15. The Above Stream Composition Information Is Considered Proprietary: Yes ☐ No ☐
16. Method of Lab Analysis Used to Measure Sample: ....
17. Desired Response Time of Analyzer: .... Minutes .... Seconds
18. Accuracy Required: % of full scale reading

D. INSTALLATION REQUIREMENTS:

1. Type of Installation: Permanent ☐ Temporary ☐ Portable ☐
2. Type of Mounting: None ☐ Rack ☐ Panel ☐
   Other ....
3. Electrical Code: Class .... Group .... Division ....
4. Recorder or Indicator Required: ....
   (a) To Be Supplied with Analyzer: Yes ☐ No ☐
5. Location of Recorder: At Analyzer ☐ Distance from Analyzer .... ft
6. Recorder Mounting: None ☐ Rack ☐ Panel ☐
7. Accessories:
   (a) Alarms: High ☐ Low ☐
   (b) Controls: ....
   (c) Others: ....
8. Date Required: ....
9. Sketch of System Indicating Sample Points and Distances to Analyzer:

Table 1.11b
PROPERTIES USED AS THE BASIS FOR COMPOSITION ANALYSIS

Absorption	Heat of reaction
Ultraviolet	Ionization phenomena
Visible (color)	Molecular weight
Near infrared	Nucleonics
Infrared	Paramagnetism
X-ray	pH and other selective ion electrodes
Sound	Refraction
Boiling point	Thermal conductivity
Chemical reactivity	Viscosity
Combustibility	
Density	
Dispersion	
Electrical capacitance	
Electrical conductivity	
Freezing point	

Analyzer Hardware Selection

Once prospective methods of analysis have been selected, the search for appropriate hardware begins. Manufacturers' names and addresses can be obtained from this handbook or from Buyers' Guides and the Thomas Register. Ordinarily, details of the desired measurement are given to the potential analyzer suppliers and estimates of performance are obtained. The analysis system can often be purchased on a guaranteed performance basis. For proprietary installations the user himself must determine how the general analyzer specifications apply to his needs and since the user is buying hardware only, the supplier will only guarantee the quality of materials and workmanship.

Specificity (selectivity)

Specificity is the characteristic of responding only to the property or component of interest. The specificity of the analyzer will not exceed that of the analysis. Specificity is often a function of the range to be measured, the sample background and the process conditions (solid, liquid or gas; pressure; and temperature).

Accuracy and Repeatability

Frequently, absolute accuracy cannot be established owing to the lack of a suitable calibration standard. For this reason, other terms (such as repeatability) take on added significance. Often, the terms do not have an industry-wide significance, and the definition should be resolved with each supplier. (For a detailed discussion of terms like accuracy, see pp. 1 to 13 in Volume II.)

Repeatability (for the purpose of this discussion) is defined as the ability of an analyzer to produce the same output each time the sample contains the same quantity of component or property being measured. Stability, reliability and reproducibility are sometimes used synonymously with repeatability. A repeatable analysis system, properly calibrated, thus also becomes precise (Figure 1 in Volume II). Lack of repeatability may be caused by the analyzer, the sample system or the effects of temperature, voltage, composition, pressure and flow rate. Repeatability and accuracy are normally expressed either as a percentage of the full measurement range or as a percentage of the actual reading.

Calibration

The ability to calibrate an analyzer properly usually depends on the availability of a reliable reference sample, or on the ability to perform reliable laboratory analysis on the actual sample which is entering the analyzer. This ability to calibrate has been the subject of many articles and is also discussed in Section 1.9 in this volume.[2]

Interferences with the analysis can be defined as physical or chemical effects which cause a deviation in the analyzer output. If the effect of the interfering substance remains constant, a compensation factor can usually be applied in the calibration procedure. During checkout of the installed system one should verify the effect of interference by calibrating the analyzer for each suspected substance.

When the interferences cannot be predetermined, it may be practical to purchase the instrument subject to a plant test or to send known samples to prospective suppliers for evaluation. In any case, known and suspected interferences and their concentrations should be made known to prospective suppliers when requesting quotations.

Whether to believe the analyzer or the calibration standard can become a "chicken or the egg" proposition. Both the method of calibration and the accuracy of the calibration procedure should be established before purchasing a process analyzer.

Analysis Frequency

One should first determine whether continuous analysis, automatic-repetitive batch sampling or an occasional "spot check" is required. The information from the analyzer and the rate of the dynamic changes in the sample are the main factors of consideration.

The need for process control suggests either a continuous type analyzer or an analyzer with a continuous output signal because a continuous analyzer provides a better chance for the system to reach and maintain equilibrium with the sample. Also, the mechanical design of a continuous analyzer ordinarily is less complicated than that of a discontinuous system. However, the discontinuous analyzer may be more attractive if automatic zero checks are frequently needed, if reagents are blended with the sample or if the sample is corrosive. Some analyzers, e.g., chromatographs, are inherently discontinuous.

The rate at which the measured variable changes in the process is an important factor in determining the frequency of discontinuous analyses. If the sample has to be withdrawn and transported to the analyzer, the time lag factor must also be considered. If several different samples are to share the analyzer, additional time allowances are required.

Sample Systems

For a detailed discussion of sample systems, see Section 1.12 in this volume. Sample systems are rarely duplicated and thus each system must be "debugged" as a new entity. In deciding the design of a sampling system, the following questions should be raised:

1. Will the sample be adversely affected by sample transport and conditioning?
2. From what stage in the process will the sample be taken?
3. How can sample be transported to the analyzer?
4. Is the sample solid, liquid, gas or a mixture?
5. In what phase must the sample be for analysis?
6. Must the sample be altered (filtered)?
7. Is sufficient sample available?
8. What will be the time lag introduced by transporting the sample?
9. Where is the excess sample returned?
10. Will the analyzer be shared by one or more samples?

These considerations are applicable both to the continuously flowing sample systems and to the less frequently used grab samples. Grab sampling is limited by variations in cleanliness of the sample containers, changes in method of collecting the sample, delays in transporting the sample and deviations in withdrawing the sample from the sample container.

The piped-in sample system usually includes the hardware for calibrating the analysis system. It also provides a tap for obtaining samples for laboratory testing. Sample conditioning systems are usually costly to maintain.

Analyzer Location

Location for the analyzer must be selected after considering availability of utilities, ambient requirements, available space, read-out location, safety considerations, access for maintenance and analyzer response time.

An inferior quality utility can cause inaccuracies in the output or degradation of the analysis system. Problems are frequently caused by varying voltage, changes in electrical frequency, transients in the voltage, changing

Table 1.11c
ELECTRICAL AREA CLASSIFICATIONS

Flammable Gas Class I		*Combustible Dust Class II*		*Ignitable Fibers Class III*	
Group A: Acetylene		Group E: Metal dust, aluminum, magnesium, etc.		No Groups Listed	
Group B: Hydrogen or mfd. gas		Group F: Carbon black, coal or coke dust.			
Group C: Ethyl-ether, ethylene, cyclopropane		Group G: Flour, starch, grain dust			
Group D: Gasoline, hexane, naphtha, benzene propane, butane, alcohol, acetone, benzol, lacquer solvent, natural gas					
Div. 1	Div. 2	Div. 1	Div. 2	Div. 1	Div. 2

CLASS I FLAMMABLE GASES

Division 1 Locations	*Division 2 Locations*
Locations in which: (1) Hazardous concentrations exist continuously, intermittently or periodically under normal operating conditions. (2) Exist frequently because of repair, maintenance, operations or leakage. (3) Breakdown or faulty operation might release gases which might also cause simultaneous failure of electrical equipment.	Locations in which: (1) Gases are handled, processed, or used, but will normally be confined to closed containers or systems from which they can escape only in accidental rupture or abnormal operation of equipment. (2) Gases are normally prevented by positive ventilation but might become hazardous through failure of ventilating equipment. (3) Adjacent to Class I division 1 locations in which gases might occasionally be communicated.

pressure in vents and drains and oil or moisture, or both, in plant or instrument air. The temperature and humidity of the atmosphere surrounding the analyzer may also contribute to inefficient performance.

Analyzers may be purchased in one housing or in several modules. A modular approach can often reduce space requirements, provide remote read-out, improve safety, provide easy maintenance access and reduce sample transport time.

The major safety consideration is in the area classification for fire and explosion hazards. Table 1.11c summarizes these classifications. The inability of laboratory analyzers to meet electrical area requirements limits their use in online processing. Often, however, purging an analyzer that is not explosion-proof allows its use in a hazardous area of the plant. Samples or stored reagents containing toxic, flammable or noxious substances constitute a safety hazard. For a detailed discussion of electrical safety considerations, see Sections 10.10 and 10.11 in Volume I.

Maintenance

Analyzer hardware is likely to receive better care if it is accessible and housed in pleasant surroundings, such as in air-conditioned buildings. Spare parts and special testing components should be ordered at the same time as the system so as to avoid delay at a future time; the availability and most effective use of qualified maintenance personnel are also factors of major importance.

An analyzer already in the plant is helpful in determining maintenance requirements. Major subassemblies not stocked as spare parts can also be interchanged, and help to obviate waiting for parts for the duplicated system.

Cost

Table 1.11d lists the elements of expense in analysis systems.

Table 1.11d
ANALYSIS SYSTEM COSTS

Engineering study	Spare parts
Analyzer	Startup and checkout in the plant
Recorder or other display hardware	Utilities and reagents
Sample system	Training repairmen
Startup and checkout in the laboratory	Maintenance
Calibration standard	
Installation costs	

REFERENCES

1. Mock, J. A., "Physical properties and tests—A to Z," *Materials Engineering 67:* 82, June 1968.
2. Gray, T. A. and Kuczynsk, E. R., "Calibration of SO_2 monitoring instruments," *ISA Transactions 7:* 327, 1968.
3. Kuller, B. E., Interpretation and Application of Article 500 of the National Electrical Code. Presented at the Symposium on Limiting Electrical Losses. Sixty Third National Meeting of the American Institute of Chemical Engineers.

1.12 ANALYZER SAMPLING SYSTEMS

Partial List of Suppliers: Automation Mfg. & Service Co., Beckman Instruments, Inc., Bendix Corp., Customline Control Panels, Inc., Elliot Bros., Ltd., Greenbrier Instruments, Inc., Hallikainen Instruments, Hays Corporation, Leeds and Northrup, Lockwood and McLoire, Inc., Mine Safety Appliances Co., Process Analyzers, Inc., Seiscor Div., Seismograph Service Corp., W. G. Pye and Co., Ltd.

The purpose of a process control system is to optimize safety of personnel and property, produce quality, process efficiency, process equipment maintenance and pollution control.

The contribution of the analytical system is to satisfy the requirements or specifications for the analysis of the components of the process streams.

The sampling system should fulfill some or all of the following functions: obtain a representative sample, condition the sample, transport the sample, accomplish stream switching, check calibration, dispose of analyzed samples and alert the operator if sample flow or pressure is abnormal.

Sample Data Requirements

A comprehensive listing of the characteristics of each sample stream for the complete processing cycle is desirable. The process control system may require compositional data from pre-startup and materials charging to process shutdown and standby for recharging. Abnormal stream conditions should be included in the listing for all constituents including solids, oils and the line. This body of data is usually available for existing processing units but is not available for new plants. Calculations based on pressure, temperature, and composition ranges reflect the abnormal sampling conditions to be considered. Table 1.12a is a form to be used in summarizing the characteristics of the sample.

Table 1.12a
SAMPLE DATA FORM

Sample phase entering the analyzer:
Liquid .... Gas ....

Sample temperatures:
Maximum .... Minimum ....

Recommended temperature .... Tolerance ± ....

Sample pressures:
Maximum .... Minimum ....

Recommended pressure .... Tolerance ± ....

Sample flowmeters and analyzer speed of response:
Maximum flow rate .... Analyzer response ....
Minimum flow rate .... Analyzer response ....
Recommended flow rate .... Tolerance ± ....

Analyzer internal temperatures:

Maximum .... Minimum ....

Recommended temperature .... Tolerance ± ....

Analyzer accuracy required:
Component ranges .... Cross sensitivity ....

Analyzer environment:
Safety classification .... Ambient temperature ....
Barometric or pressure venting .... Calibration method ....

The type of process to be monitored may allow some freedom in selecting sample points. A sample point should be taken that meets the following conditions: process chemical reactions are completed; single phase sample is available; levels of particulate or liquid contaminants are low; sample is homogeneous; pressure and temperature levels are compatible with easy sample removal; and access for equipment installation and maintenance is convenient.

Inline probe monitors may not require sampling systems although means for removing the probes from the process for periodic inspection and calibration will still be required.

Sample Conditioning

Sample conditioning begins with the sample probe and includes all equipment used in bringing the sample to the analyzer. It is permissible to treat and condition the sample as long as the analyzed sample provides a direct indication of the actual process stream composition.

A simple block diagram showing the sample points, properties and flows is useful in selecting and sizing sample conditioning equipment and transport and by-pass lines.

Ideally, the sample at the selected sample point will require little or no conditioning prior to being analyzed. Sample streams from liquid distillation and extraction systems approach the ideal in which the conditioning requirements are limited to a sample probe or "thief" in the liquid stream, sample line to the analyzer, filter and flow and pressure controls.

Sampling processes that are still reacting chemically may require reaction quenching in the sample probe by cooling or dilution, or both, and backflush with an inert gas or liquid solvent for periodic cleaning of the sampling system (Figure 1.12b).

Some sampling may require a filter and flow restriction in the sample probe to reduce stream pressure.

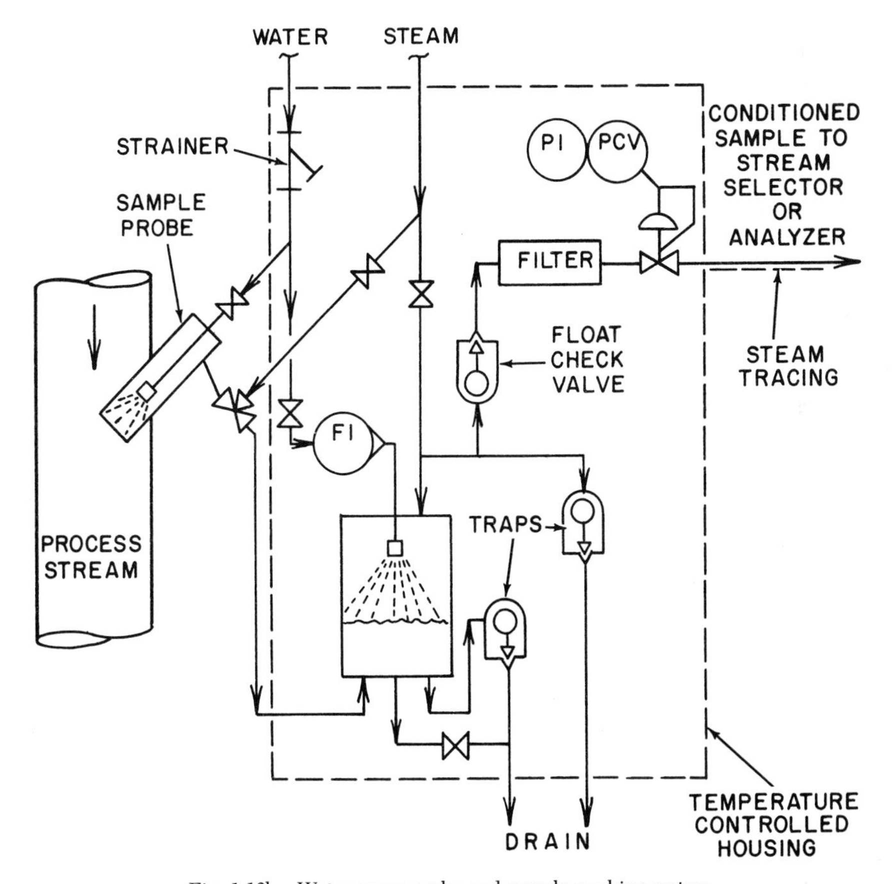

Fig. 1.12b Water spray probe and sample washing system

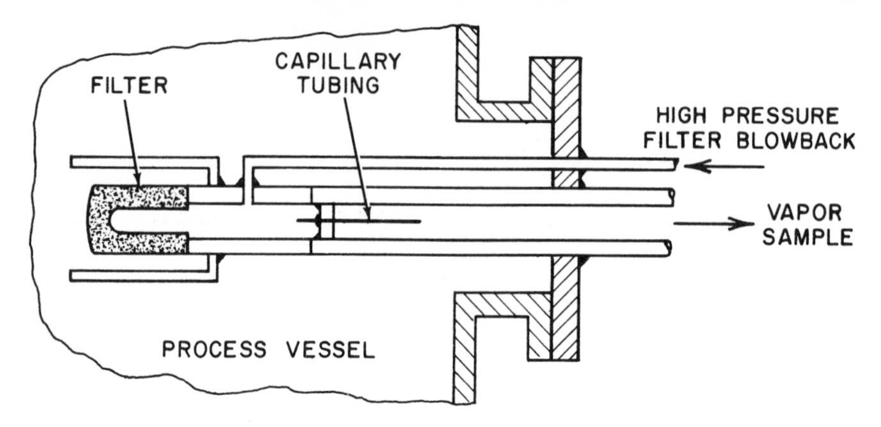

Fig. 1.12c Sample probe with inline filter and flow restriction

A potential application for the probe illustrated in Figure 1.12c is a high pressure liquid stream that will vaporize completely at ambient temperatures. The heat required for vaporization comes from the liquid process stream around the flow restriction sample probe.

Another application is a high pressure-high temperature gas stream that will partially condense at a temperature slightly lower than the process stream temperature. A flow restriction probe lowers the sample pressure, allowing low temperature heat tracing to prevent sample condensation between the sample point and the analyzer.

Sampling from a vent stack may require samples from several locations across the stack if the process gas is not homogeneous. The point most representative of the total sample mix across the stack is usually chosen for continuous sample removal.

Sample Washing, Condensing and Vaporization

Sample washing is usually limited to very dirty particle-laden streams because the sample composition will be altered by the solubility of the components in the liquid used to wash the sample. All flows and temperatures must be closely controlled in order for the original and diluted samples to maintain a predetermined relationship.

Upon washing, the sample will be saturated with vapors of the wash liquid, a step that may necessitate heat traced lines to the analyzer. Some systems take the sample through a cooler to reduce the number of condensibles. Allowing the sample to cool to ambient temperature at the process pressure may be adequate for removing condensables.

Water cooling is used when air cooling is not sufficient or to precool the sample prior to entry into a refrigeration cooler.

The more volatile liquids can be vaporized across heated pressure reducing regulators, fixed restrictors (capillary tubing or small orifices) or needle valves.

Partial vaporization of the sample should be avoided by sufficiently maintaining the pressure of the sample entering the expansion zone and by having sufficient heat available for instant vaporization. Most heated pressure reducing regulators are flow limited. A liquid by-pass stream may be required for rapid changeout of the upstream liquid.

Capillary tubing with upstream and downstream pressure control can maintain at a constant level the vaporization heat requirement. Figure 1.12d shows insulated capillary tubing extending downward into a heated chamber, providing flash vaporization for the removal of non-volatile materials.

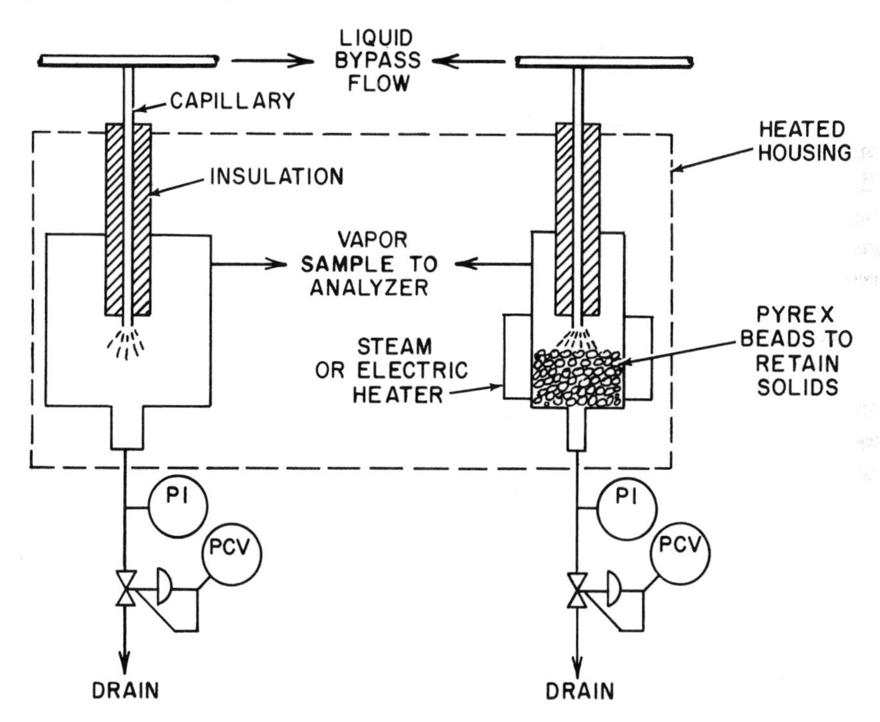

Fig. 1.12d Two versions of capillary type sample vaporizers

Coalescing and Bubble Removal

Coalescers remove mists or entrained droplets from gas streams. The droplets merge as they pass through the coalescing (hydrophilic) element and the collected liquid is removed by liquid traps.

The removal of free water from a hydrocarbon stream is usually accomplished by passing the liquid through a hydrophilic element, causing the water droplets to merge. The hydrocarbon stream is then passed through a second element (hydrophobic) that rejects the water, removing it from the bottom with the hydrocarbon by-pass stream.

Bubbles can be effectively removed from liquid streams by by-pass filters or by passing the liquid through a packed bed where the small bubbles merge to form larger bubbles (Figure 10.12e). The large bubbles continue to flow upward while the by-pass stream and the liquid to be sampled flow out the bottom. A rise in temperature or a sample pressure reduction cause more bubbles to form, and a bubble removal device may be required in the analyzer, especially if the latter is operated above ambient temperature.

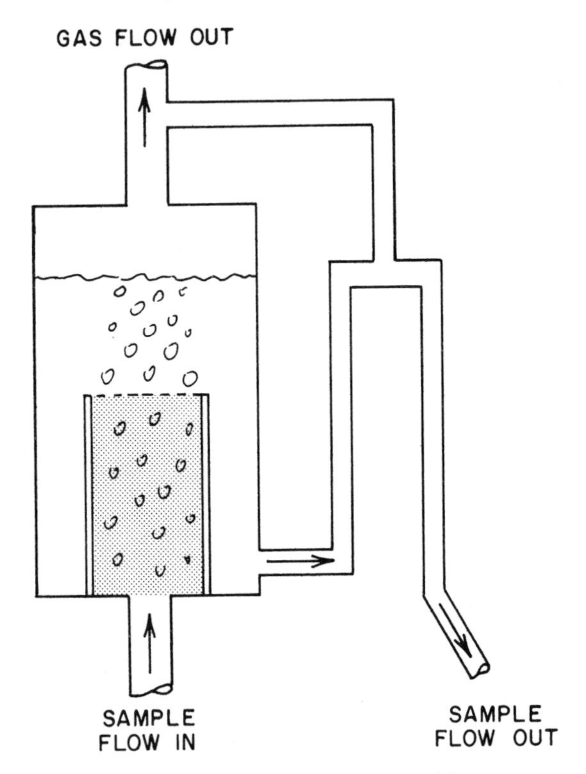

Fig. 1.12e Bubble removal assembly

Adsorption, Diffusion and Dew Point Effects

The materials of construction in sampling systems, if present in the ppm range, can appreciably affect the analysis results.

Adsorption of sample components on the walls or surfaces will affect the analysis until equilibrium is established—when the number of molecules reaching and leaving a surface are in balance. A low ppm water vapor sample will reach equilibrium much faster in a stainless steel sample line than in copper or plastic tubes. The diffusion of gases through permeable materials can also affect the analysis as is illustrated by the fact that a ppm water vapor analyzer using a 2,000 PSIG sample cylinder and a pressure reducing regulator to lower the sample pressure to 20 PSIG gives a decrease in its read-out from 20 ppm to 10 ppm when the regulator is by-passed.

Small leaks can also alter the sample composition because when a leak is so small that the mass flow velocity through the leak is less than the molecular velocity, gas molecules move in both directions.

The net flow of any gas depends on the relative partial pressure of that gas on each side of the leak. For example, if a sample is flowing to a ppm oxygen analyzer through a stainless steel line at 18 PSIG and a tube fitting is loosened just enough to build a soap bubble, the analyzer will respond with a full scale reading.

The instrument engineer should consider the following design criteria for trace analysis sampling and for high vacuum systems:

1. Stainless steel should be given preference over other materials. The materials should be inert to the sample stream and smooth surfaced with low porosity.
2. When low flows are used, the wall area to tubing volume can be critical in small bore tubing due to the adsorption-desorption phenomena. A rule-of-thumb is to use tubing one size smaller than that which would be necessitated by response or lag time.
3. All tubing, fittings and regulators should be thoroughly cleaned of oils, grease or other contaminating materials.
4. Only metal, unplasticized dense fluorocarbon or vinylidene chloride (Saran) diaphragm pressure regulators should be used in contact with the sample stream. The metal diaphragm is the regulator of choice. The use of all regulators in the main flow stream should be kept to a minimum to reduce the added wall adsorption effects and longer cleanout times.
5. A continuous unbroken run of seamless metal tubing is recommended for piping. The use of screwed fittings should be kept to a minimum. Flare type tube fittings are preferred to reduce vibration initiated leaks. Pipe dope should be used only when tinning the male pipe threads with soft solder is not permissible. The tinning should afford thorough coverage of the threads, and the joint must be taken up a sufficient number of turns to provide normal thread engagement.
6. Shutoff valves should be brass, bronze, stainless steel or other non-ferrous corrosion resistant alloys. Packless valves of the diaphragm or bellows type are recommended.
7. The use of filters in the sample stream should be carefully con-

sidered. Metal filters are preferred, and the use of non-metal filter inserts, gaskets and drip pots is to be avoided. The presence of liquid in the sampling system should be viewed with suspicion. It is desirable to have high flow rates through porous or sintered metal filter inserts. Liquids flowing or being forced through the filter may be "atomized" and deposited farther along in the system thereby adding considerable adsorption-desorption time to the system. Chemical reaction with the liquid coated sample system is also a possibility.

8. Condensation of the sample after removal from the process should be prevented by heat tracing of the sample lines. If condensation is permitted, the conditions for it should be closely controlled and the condensate completely and quickly removed from contact with the remaining gas sample stream. A heat traced system may be required to prevent condensation farther downstream in the system.

The sample flow and temperature should be constant in order to promote equilibrium between the flowing sample stream and the components in the sampling system. Precise temperature control of long sample leads is not practical. The flow rate through the sampling system should be high (but constant) so that over-all equilibrium is rapidly established.

The dew point in the sampling system determines the minimum temperature allowable to prevent sample condensation. Heated sample lines or lower sample transport pressures may be required to prevent condensation.

Partial pressure is the product of the total system pressure and the mole fraction (or volume percent) of the condensible component:

$$PP = (M)(P) \qquad 1.12(1)$$

Sample Pumps, Ejectors and Aspirators

When mechanical sample pumps must be used, the pump analysis system and materials of construction, must be compatible. Some types of pumps expose the sample to lubricants and shaft seal leakages. The lubricant can create serious adsorption-desorption problems in addition to coating the sample valves and other analyzer components (especially photometric type detectors) thus keeping the cost of maintenance at a high figure.

Diaphragm type pumps solve the lubricant and shaft seal leakage problems and are recommended, especially for low range analytical measurement systems. Diffusion and adsorption-desorption effects can usually be reduced by judicious pump material selection and by using high flow rates through the pump to a sample by-pass.

Liquid and gas activated ejectors or aspirators are used extensively on water wash vent stack sampling systems or when highly corrosive samples preclude mechanical pumps. An aspirator pulls the sample through the analyzer, and its exhaust can be returned to the process if the aspirating fluid does not contaminate it.

Stream Selection Systems

Stream selection systems are required in multi-sample installations. Their design should minimize or prevent sample mixing and limit sample pressure or flow changes. For example, a percent component range analyzer usually does not require purging of all dead endline volumes, whereas a ppm component range analyzer or a dual high-low range analyzer requires such features.

A simple block diagram should be prepared, showing the sample stream parameters and the types of analyzers to be furnished for each stream. Moving the sample takeoff can sometimes simplify the multi-sample–multi-component analysis systems. If a computer is available, other stream compositions may be used to calculate the composition of a stream not directly analyzed.

In the design for a low range (ppm and ppb) sampling system (Figure 1.12f), critical flow orifices and back pressure regulators maintain constant pressure and flow conditions throughout the three systems. The filter at the sample point (especially if it is not easily accessible) can be sized to pass particles as large as one-half the orifice diameter to reduce filter pluggage. Most of the pressure gauges, safety pressure relief valves and backpressure regulators are on by-pass streams and thus will not introduce time lag or equilibrium problems. The sample selector valves should be "fullflow"—ball or rotary plug type valves.

Solenoid valves are sometimes used on stream switching systems. In such case, in addition, to maintenance the temperature rise in the valves also should be considered. The temperature rise in a typical energized solenoid valve (neither insulated nor traced) with and without 300 cc per minute air flow through it is tabulated in Table 1.12g.

Heated Sampling Systems

A steam or electrically heated sampling system may be required for samples containing condensible components. The dew points for the pressure levels anticipated in the system should be calculated and noted on the block diagram.

Heated housings for the sample conditioning equipment should be located at the sample takeoff points if it is probable that periodic maintenance will be required, or if the sample should be kept at the lowest practical pressure. Low dew points reduce the heating requirements.

When required, it is essential that all parts of the system be heat traced. The heat capacity of a gas is very low, and a short section of unheated tubing can lower the sample temperature by several hundred degrees.

Bulkhead type sample line fittings should not be used because they act as heat sinks, and the heat tracing and insulation of the lines should extend into the heated enclosures.

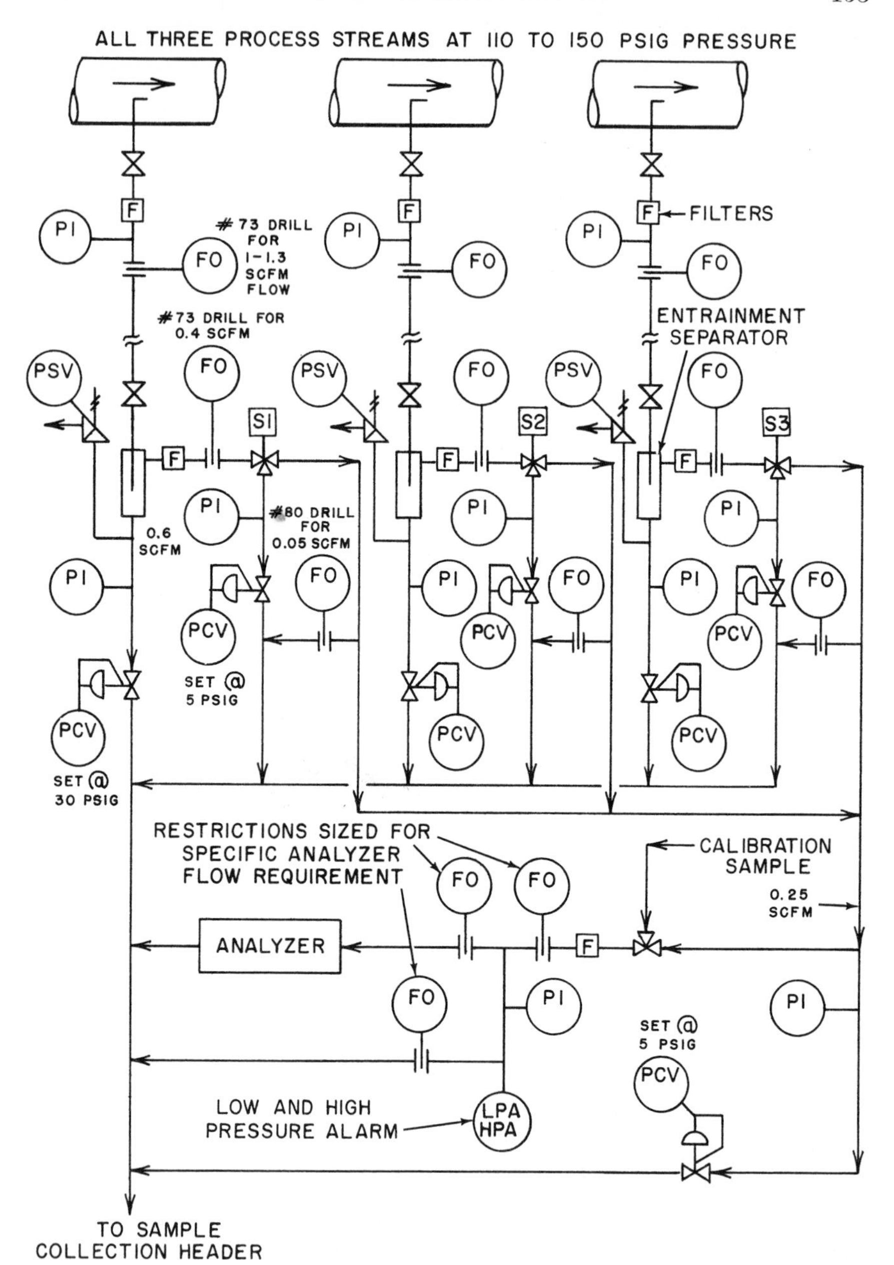

Fig. 1.12f Multipoint sampling system for trace quantity analysis

Table 1.12g
TEMPERATURE RISE IN
ENERGIZED SOLENOID VALVE*

Time Period of Energization (minutes)	*Actual Air Temperature (°C) at Valve Inlets and Outlets*					
	Inlet		*Outlet*			
			Normally Closed		*Normally Open*	
	No Flow	*Flow*	*No Flow*	*Flow*	*No Flow*	*Flow*
0	25	25	25	25	25	25
5	29	32	30	31	45	48
10	35	37	36	38	56	58
15	42	41	43	44	67	64
20	47	45	49	48	74	70
25	52	49	54	52	79	75
30	56	52	58	56	83	78
35	58	54	61	57	86	81
40	60	54	63	58	88	83
45	62	54	65	58	90	83
50	63	54	66	58	92	84
55	64	54	68	59	93	84
60	66	55	69	59	94.5	84

* Solenoid uninsulated in 25°C still air, with 300 cc per minute air flow through it when open and energized by a 110VAC, 60Hz, 0.174 ampere holding current.

When several steam-heated sample lines are to run to a central analyzer location, a steam distribution header may be used for grouping the sample lines around it. If heat sensitivity of certain samples requires a lower transport temperature, these lines can be run between two layers of insulation as shown in Figure 1.12h. On long steam-heated lines the steam line may have to be trapped several times to assure constant heating.

Sample Flow and Back Pressure Requirements

Most analyzers require constant flow and pressure for optimum operation. Figure 1.12i shows the effects of vent pressure changes on several types of gas analyzers.

Flow control regulators are required in the more critical cases.

Equipment and Analyzer Locations

Samples from high pressure-low temperature volatile liquid streams are usually vaporized near the sample point. The volume on the liquid side of the vaporizer is kept as low as practical for fast liquid sample changeout. Heat tracing may not be required after the sample is vaporized.

Less volatile liquids are usually vaporized at the analyzer by taking a sidestream from a fast flowing liquid by-pass. If a by-pass stream is unavailable,

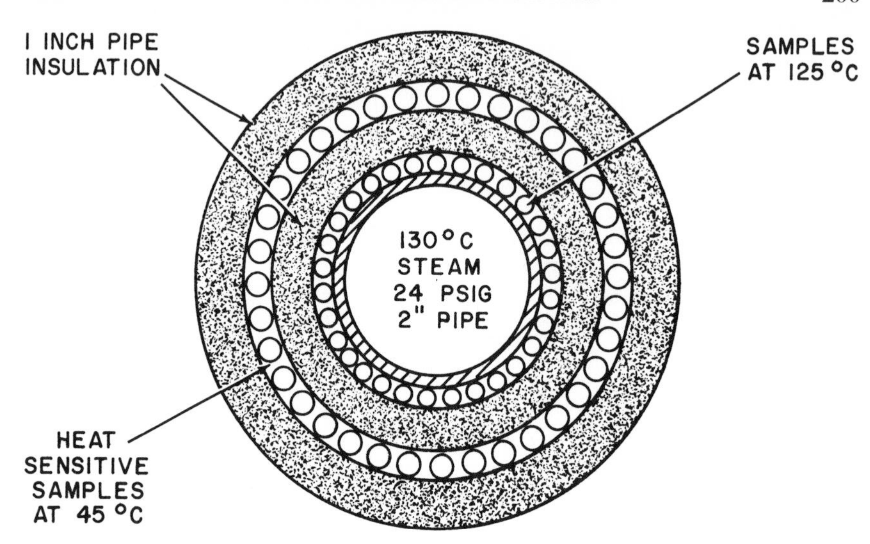

Fig. 1.12h Sample lines run on steam header

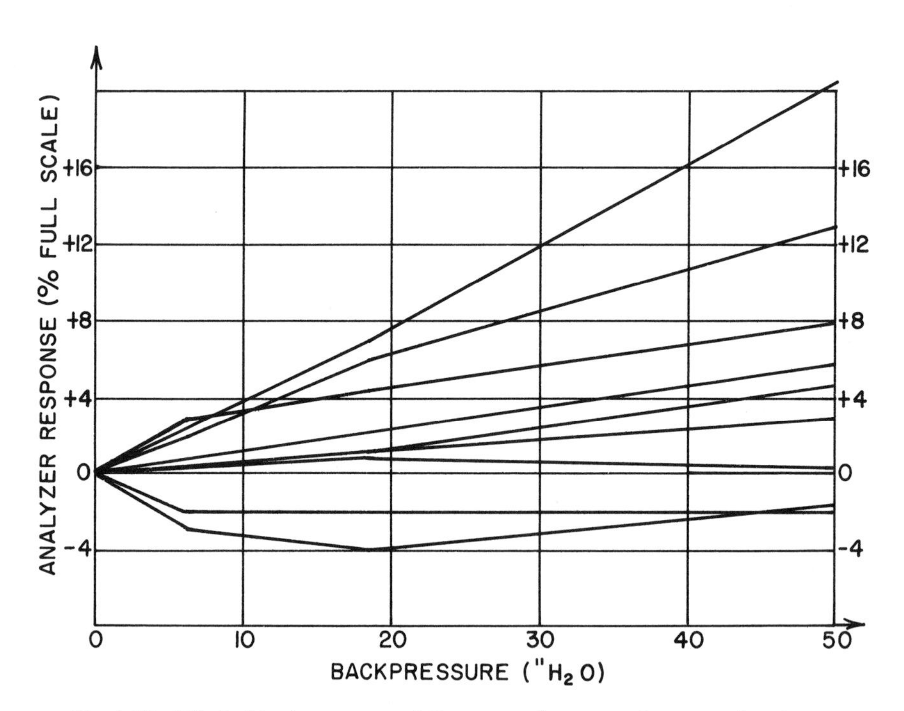

Fig. 1.12i Effect of back pressure variations on performance of a group of analyzers

vaporization at the sample point becomes necessary so as to avoid excessive sample time lags in the liquid line; heat tracing of the vapor sample line is usually required.

Pressure reducers for gas samples are commonly located near the sample tap in order to reduce time lags. The flow required for sample purge out is directly proportional to the absolute line pressure.

Washing and cleaning equipment is usually located near the sample point to minimize the probability of sample line plugging.

The location of the analyzer is determined by evaluating sampling difficulties (such as sample transport and conditioning), equipment maintenance, utility requirements, sample disposal, communications or signal transmission, area safety classification, analyzer environmental control and personnel safety. When several analyzers are being considered, economics also becomes a factor.

Environmental control becomes a necessity for close tolerance analytical measurements. Figure 1.12j shows temperature stability curves for several analyzers. No analyzer is unaffected by ambient temperature variations, but it can perform within very close tolerance specifications if the environment is controlled.

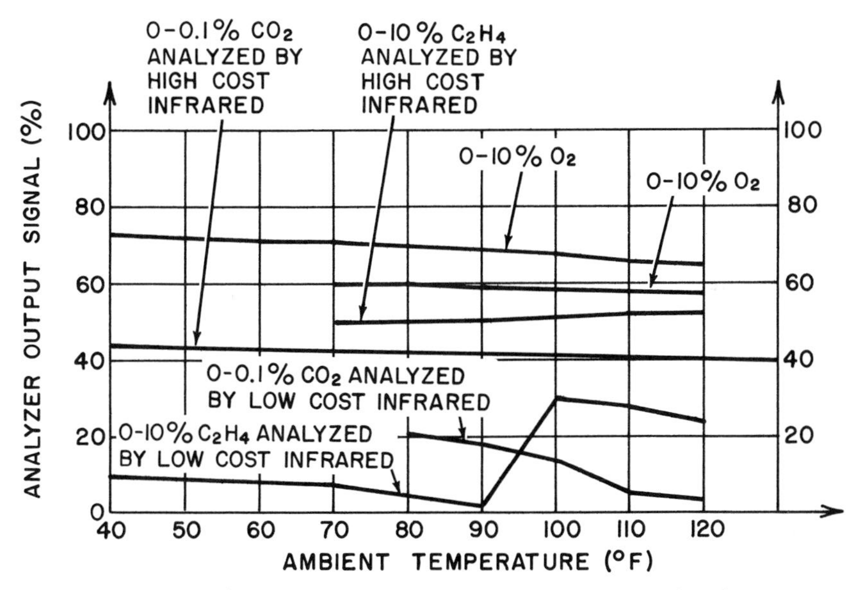

Fig. 1.12j Ambient temperature effect on performance of several analyzers

Calibration Techniques

Analyzer calibration techniques include:

1. Insertion of a screen or partially opaque translucent filter to block or absorb part of the emitted photometric energy. This method does not insure that the sample flow will be adequate.

2. Injection of pure components in smaller volumes than the normal sample inject volume (as in a chromatograph).
3. A calibration check scanning analyzer. A single chromatograph can monitor the status of several other analyzers on single stream services.
4. Electrolysis generators for low ppm oxygen calibration.
5. Permeation devices using the molecular flow of gases through plastic membranes into a flowing stream for ppm and ppb gas mixtures.
6. Calibration samples prepared by partial pressure, measured volume or weight or flow proportioning.

The frequency of the analyzer calibration checks may vary from minutes with a scanner analyzer to days or weeks with other methods of calibration. In some systems the panel controls or the computer may initiate the calibration checks.

Startup and Documentation

The field installation of sampling systems and analytical equipment should be closely supervised. The methods of tube or pipe cutting for joining, pipe dope and heat trace applications, line slopage, leak checking and line cleaning should be reviewed in detail. Small sectional sketches of the sampling system are useful for on-site discussions. Sections of the sampling systems may be assembled and inspected prior to field installation.

The use of gas cylinders for leak checks also permits one to determine the sample transport lag times and to make the proper adjustments before charging the system from the process.

Some analyzers are checked on calibration samples prior to system startup. It may be desirable to by-pass the analyzers during final checkout of the sampling systems, utilizing the process sample to prevent contamination of the analyzer if there is dirt or liquid in the sample lines which normally would not be present.

Documentation should include a written description of the functioning of each part of the system. Sketches should be available to show the flow rates, temperatures, dewpoints, bubble points, directions of flows and other pertinent information such as electrical wiring diagrams and voltage check points.

All field changes should be incorporated in the updated drawings. All costs, startup and unexpected maintenance and installation problems should be recorded and reviewed at the completion of the project.

1.13 ION SELECTIVE ELECTRODES

Electrode Types:	Glass, solid state, silicon-rubber matrix, liquid ion exchange.
Design Pressure:	Generally dictated by electrode holder. 0 PSIG for liquid ion exchange type, 0 to 100 PSIG for most electrode types and over 100 PSIG for solid state designs.
Design Temperature:	0 to 50°C for liquid ion exchange type, −5°C to +80°C for most others, with 100°C intermittent exposure being permittable.
Range:	From low ppm to concentrated solutions.
Relative Error:	For direct measurements in process applications, ±1.0 millivolt. This in % relative terms means ±5% for monovalent ions and ±10% for divalent ions; for end point detection or batch control, ±0.25% or better is possible; for expanded scale commercial amplifiers, better than ±1% of full scale.
Cost:	Similar to those of pH installations (see P. 821 in Volume I). Electrodes: $50 to $300. Systems: $1,000 to $3,000.
Partial List of Suppliers:	Beckman Instruments, Inc., Cambridge Instrument Co., Coleman Instruments, Corning Scientific Instruments, Delta Scientific Corp., Electronic Instruments Ltd., Foxboro Co., Hays Corp., Honeywell, Inc., Kero-Test Manufacturing Co., Leeds

and Northrup Co., Orion Research Inc., A. H. Thomas, Inc., Universal Interlock Co.

Ion selective electrodes constitute a class of primary elements used to obtain information on the chemical composition of a process solution. Electrochemical transducers which generate a millivolt potential when immersed in a conducting solution containing free or unassociated ions to which the electrodes respond, the magnitude of their potential is a logarithmic function of the activity of the measured ion (not the total concentration of that ion) as expressed by the Nernst equation [equation 1.13(5)]. The familiar pH electrode for measuring hydrogen ion activity is the best known of the ion selective electrodes and was the first to be available commercially (see Section 8.8 in Volume I and Section 1.8 in this volume for more on pH measurement). With few exceptions, notably the silver-billet electrode for halide measurements and the sodium-glass electrode, the pH electrode was the only satisfactory electrode available to the process industry prior to 1966. Today, more than two dozen electrodes are suitable for industrial use. Table 1.13a gives a list of the electrodes for which process applications have been reported.

Advantages and Disadvantages

Compared with other composition measuring techniques such as photometric, titrimetric, chromatographic or automated-classic analysis, the ion selective electrode has an impressive list of advantages. An electrode measurement is simple, rapid, non-destructive, direct and continuous, and is therefore easily applied to closed-loop process control. It is similar to using a thermocouple for temperature control. Electrodes can also be used in opaque solutions and in viscous slurries. Also, the electrodes measure the free or active ionic species in the process (under process conditions) and, thus, measure the status of a process reaction.

Together with these advantages, however, are several disadvantages, one of which is that the electrodes do not measure the total concentration of ions—frequently the very information required. This occurs because prior to the introduction of electrodes, concentration information was the only information available from the chemist due to his "classic" measurement techniques. Control laboratory chemists and process engineers are not accustomed to thinking in terms of activity, even when making pH measurements (Section 8.8 in Volume I). This disadvantage may well disappear as the new ion selective technique becomes more widely used.

Occasionally, however, concentration is a desirable measurement, e.g., for material balance calculations or for pollution control. Knowing the material balance allows a prediction of where a process reaction will be. This

Table 1.13a
ION SELECTIVE ELECTRODES FOR PROCESS APPLICATIONS

Electrode Class	*Electrode Designed For*	*Process Application*
Glass	Hydrogen (pH)	See Section 8.8, Vol I and Section 1.8 in this volume
	Sodium	Pulp and paper, food industry, ion-exchange unit breakthrough
	Ammonium	Chemical industry
	Potassium	Chemical industry, food industry
	Carbon dioxide*	Food industry, chemical industry
	Ammonia*	Chemical industry
Solid state or silicon-rubber matrix	Fluoride	Potable water monitoring, HF leaks in petroleum industry, chemical industry, aluminum smelting
	Chloride Bromide Iodide	Food industry, pharmaceutical and photographic industries
	Cyanide	Waste treatment
	Lead Copper Cadmium	Plating solutions, industrial wastes, and refining
	Sulfate** Phosphate**	Water pollution, pharmaceutical industry, chemical industry
Liquid ion exchange	Calcium/water hardness	Power industry, water treatment, chemical industry
	Divalent ions	Metals industry, plating industry, waste treatment

* Based on permeable membrane enclosed pH electrode.
** Not yet proven in industrial applications.

information is necessary if a process is to be controlled by introducing changes that will counter those predicted. In pollution control, it is generally accepted that many ions, even in the combined state, are detrimental to life forms. Fluorides, cyanides and sulfides, to name a few, are harmful to fish and human beings in many combined forms, in which state, however, they are not detected by ion selective electrodes. Consequently, pollution control agencies usually require data on concentration, which the electrodes can supply if they are calibrated with solutions matching the process. If this technique is unsatisfactory, an electrode in one line control can separate grab samples analyzed by other procedures in order to supply the regulatory information on concentration.

Another disadvantage arises from a misunderstanding about precision or accuracy, or both (see introduction to Volume II for discussion of precision and accuracy). Many classic analytical techniques are accurate to $\pm 0.1\%$ relative error. Ion selective electrodes are accurate to ± 4 to 8% relative error. In terms of pH, this accuracy is equivalent to making a pH measurement to ± 0.02 pH units—normally considered a satisfactory measurement. When used with some degree of understanding, ion selective electrodes can supply satisfactory composition information and allow closed-loop control, a previously unattainable approach. When in doubt, the user should consult electrode manufacturers or his own analytical chemists.

Types of Electrodes

Ion selective electrodes are classified according to the type of sensing membrane employed.

Glass Electrodes

Glass electrodes are constructed from specially formulated glass that responds to ions by an exchange of mobile ions within the membrane structure. The membrane is fused to a glass body so that the outer surface makes contact with the sample or process stream, whereas the inner surface makes contact with an internal filling solution containing a constant activity of the ion for which the membrane is sensitive (Figure 1.13b). A stable electrical contact

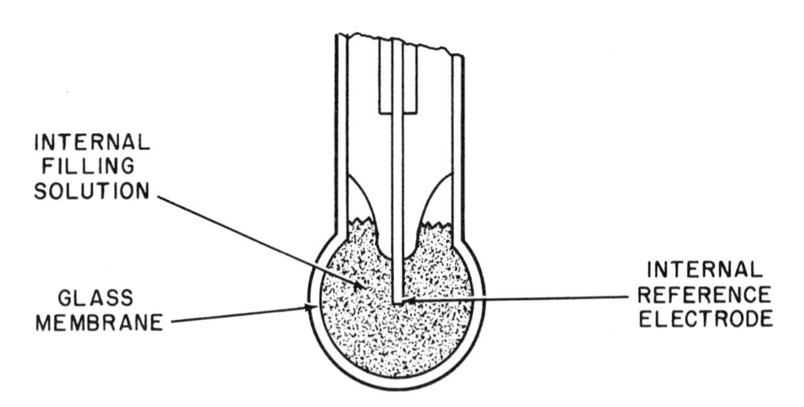

Fig. 1.13b Conventional glass pH electrode

is made with the internal solution by a silver wire coated with silver chloride. Other internal contacts have been used (mercury-mercurous chloride or thallium amalgam thallous chloride) but the silver-silver chloride is the most common. The internal filling solution must contain a constant chloride ion activity and be saturated with silver chloride so that a stable potential is maintained at the metal salt-solution interface (Figure 1.13c). In the conventional glass electrode the internal solution is buffered at a pH of 7 and contains a chloride level similar to that in the external reference electrode. (See Section 1.8 in this volume.) Other glass electrodes finding process use are the sodium electrode, ammonium electrode and potassium ion electrodes.

The sodium electrode, in addition to the construction already described,

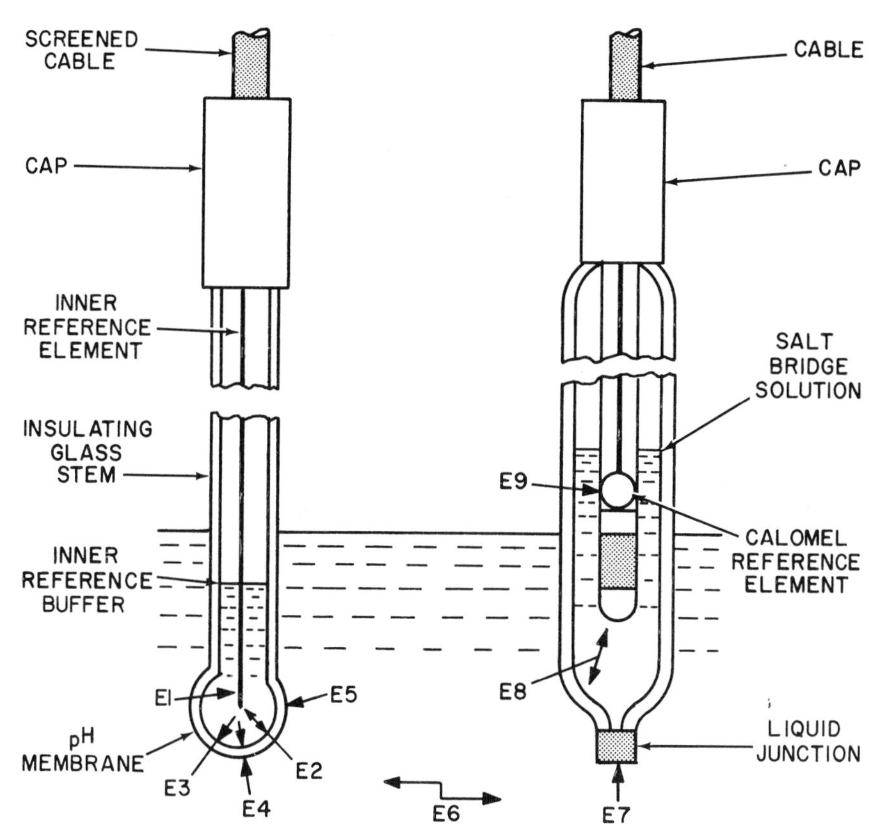

Fig. 1.13c pH electrodes

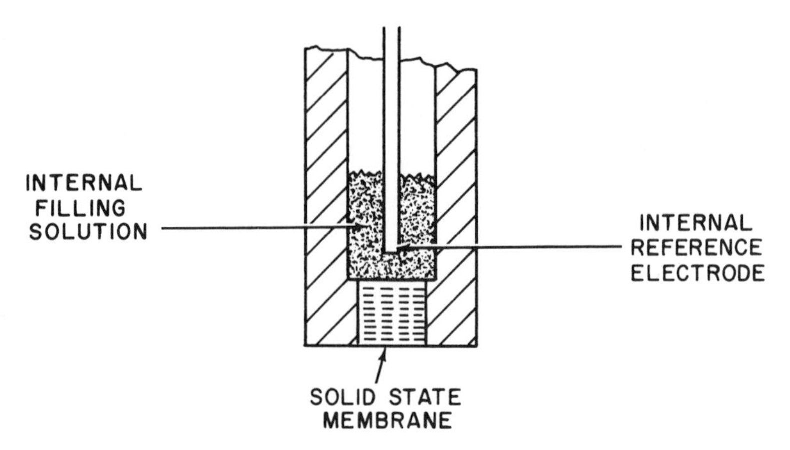

Fig. 1.13d Solid state membrane electrode

can be prepared by slicing a thin section from a rod of sodium sensitive glass and cementing it to an epoxy body (Figure 1.13d), which eliminates the familiar glass body of the pH electrode. The epoxy construction is not yet available for pH measurement due to difficulties inherent in cementing pH glasses to epoxy.

A carbon dioxide and an ammonia electrode can be made from a pH electrode by covering the membrane with a permeable membrane sac filled with pH buffer. The gas in solution will selectively diffuse in or out of the permeable membrane, causing a pH change. The latter is dependent on the activity of the gas in the process solution.

Solid state

Solid state electrodes are made of crystalline membranes with exacting requirements regarding the size and charge of the mobile ions within the membrane. The composition of the membrane varies as a function of the required measurement. The fluoride electrode, for instance, has a single crystal of lanthanum fluoride as a sensing membrane. The silver and sulfide membranes are pressed pellets of insoluble silver sulfide. The solubility of silver sulfide is such as to prevent the coexistence of silver and sulfide ions (except in extremely small amounts), and the electrode can measure either of these ions. Like the sodium electrode already mentioned these membranes are sealed in epoxy bodies (Figure 1.13d).

Table 1.13e lists some of the commercially available solid state electrodes and the composition of their sensing membranes.

Some pressed pellets and the single crystalline silver-salt membranes can have a metal deposited on the surface and an electrical lead connected to the metal deposit (Figure 1.13f). A solid connector permits the electrodes to be in any position without breaking electrical continuity. Also, there are no

Table 1.13e
SOLID STATE ELECTRODES AND THEIR MEMBRANE COMPOSITION

Electrode	*Membrane*	*Form*
Fluoride	LaF_3	Single crystal
Silver/sulfide	Ag_2S	Pressed pellet
Chloride, bromide or iodide	AgX^* $AgX\text{-}Ag_2S$	Single crystal Pressed pellet
Cyanide	$AgI\text{-}Ag_2S$	Pressed pellet

* X = Cl, Br or I

internal solutions to deteriorate with time or temperature. Figure 1.13g shows a conventional silver-wire or silver-billet electrode which behaves identically to its corresponding solid state electrode. However, minor imperfections in the silver halide coating expose free silver metal to the process solution, thereby developing variable oxidation-reduction potentials (Section 1.14), and these electrodes have not found extensive use in industry. When placed in clean controlled environments such as the inner filling solution of ion selective electrodes, they produce stable reference potentials.

Silicon-Rubber Matrix

Many of the solid state electrodes, especially those made from insoluble precipitates, can be constructed by supporting the precipitate in a silicon-rubber matrix and sealing the membrane in a glass body (not shown). These electrodes were developed in Hungary[1] and have not been used very much in the United States. However, as the state of the art advances and as the

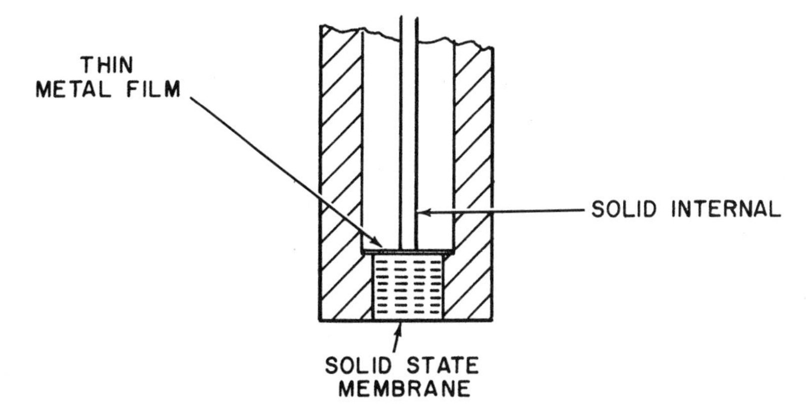

Fig. 1.13f Solid state membrane electrodes with solid internals

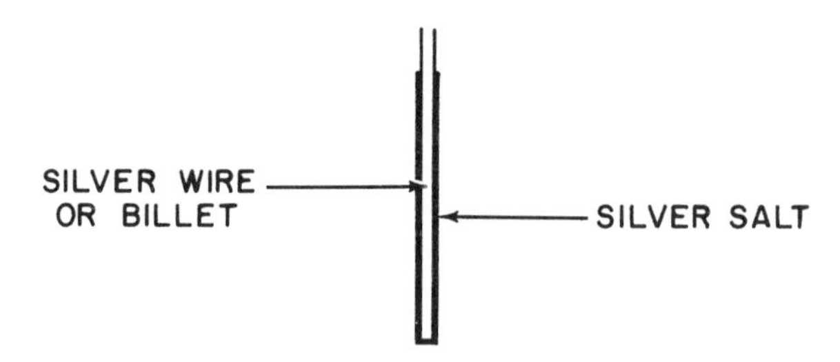

Fig. 1.13g Conventional silver-silver salt electrode

efficacy of new electrodes is proved (especially the sulfate and phosphate electrodes), these electrodes will be used more often.

Liquid Ion Exchange

There are many ions for which no glass or crystalline membrane can be found that is suitable for process measurements. Fortunately, chemistry is a versatile discipline and by using techniques familiar in ion exchange and solvent extraction technology, electrodes can be built for some of these ions. An inert hydrophobic membrane, such as a treated filter paper, can be made selective to certain ions by saturating it with an organic ion exchange material dissolved in organic solvent. This feature requires construction of the electrode as shown in Figure 1.13h, which is a cross-section of the tip of a liquid-ion exchange electrode.

This electrode has two filling solutions, an internal aqueous filling solution in which the silver-silver chloride reference electrode is immersed and an ion exchange reservoir of a non-aqueous water immiscible solution, which "wicks" into the porous membrane. The membrane serves only as a support for the ion exchange liquid and separates the internal filling solution from the unknown solution in which the electrode is immersed. In effect, there is a "sandwich," with the bottom layer being the unknown process solution, the

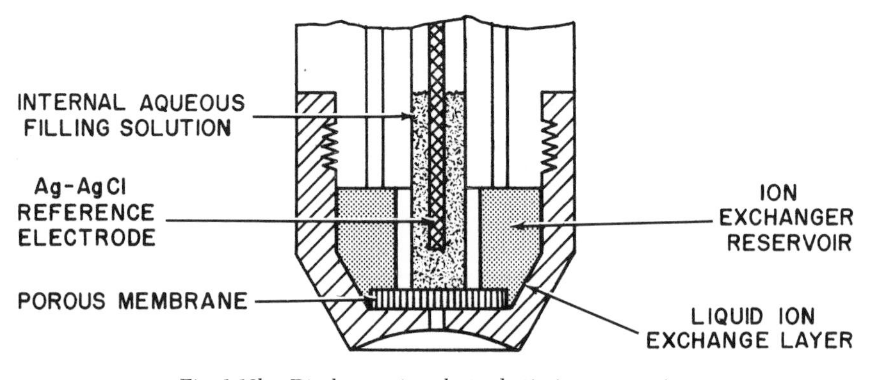

Fig. 1.13h Divalent cation electrode tip in cross-section

filling being the non-aqueous liquid ion exchange solution and the top layer being the internal aqueous solution. If the liquid-ion exchanger is selective for calcium, for example, a potential across the membrane is created by the difference in calcium activity between the internal filling solution and the process solution.

The electrode is designed so that the liquid-ion exchanger, used as a sensing element, has a very small positive flow into the process stream. Liquid ion-exchange membrane electrodes, therefore, require recharging with ion exchanger. Liquid-ion-exchange membrane electrodes come in kit form, and the kit contains an electrode body and sufficient ion exchanger, internal filling solution and membranes to recharge the electrode many times. In a properly designed system a single recharging should last several months. Unlike the solid state or glass electrodes, liquid-membrane electrodes cannot be used in non-aqueous solutions since the latter would dissolve the liquid-ion exchanger. The body of the electrode is a chemical-resistant plastic.

Measurement Range

The upper limit of detection for ion selective electrodes is the saturated solution. However, due to the problems of making measurements with reference electrodes having large liquid-junction potentials (Section 1.8), the electrodes are specified as having an upper limit of 1 M. If the problems of large liquid-junction potentials are solved, measurements can be made in saturated or almost saturated solutions. The lower limit of detection is usually determined by the solubility of the solid state sensing element or the liquid-ion exchanger; the solution pH occasionally determines the lower limit of detection. Some dilute solutions are unstable, but activity measurements may be made if the solution is buffered with respect to the ion being measured, i.e., if the free ion is in equilibrium with a relatively large excess of complexed ion, such as occurs when measuring free silver in photographic emulsions or sulfide, cyanide or fluoride in acid solutions.

Interferences

All ion selective electrodes are similar in principle and operation. They differ only in the details of the process by which the ion to be measured moves across the membrane and other ions are kept away. Therefore, a discussion of electrode interferences will be in terms of the membrane materials.

The glass electrodes and the liquid-ion exchange electrodes both function by an exchange of mobile ions within the membrane, and ion exchange processes are not specific. Reactions will occur among many ions with similar chemical properties, such as the alkali metals, alkaline earths or transition elements. Thus, a number of ions may produce a potential when a given ion selective electrode is immersed in a solution. Even the pH glass electrode will respond to sodium ions at a very high pH (low hydrogen ion activity).

Fortunately, an empirical relationship can predict electrode interferences, and a list of selectivity ratios for the interfering ions can be obtained by consulting the manufacturers' specifications or the chemical literature. Selectivity constants have been described in connection with equation 1.10(3).

Solid state and silicon-rubber matrix electrodes are made of crystalline materials, and interferences resulting from ions moving into the solid membrane are not expected. Interference is usually by a chemical reaction with the membrane, one of which is observed with the silver halide membranes (for chloride, bromide, iodide and cyanide activity measurements) and involves reaction with an ion in the sample solution, such as sulfide, to form a more insoluble silver salt. Specific data of electrode side reactions can be found in the manufacturers' specifications and chemical literature.

A true interference is one that produces an electrode response which can be interpreted as a measure of the ion of interest. For example, the hydroxyl ion (OH^{1-}) causes a response with the fluoride electrode at fluoride levels below 10 ppm. Also, the hydrogen ion (H^{1+}) creates a positive interference with the sodium ion electrode. Often, an ion will be thought of as interfering if it reduces the activity of the ion of interest through chemical reaction. It is true that this reaction (complexation, precipitation, oxidation-reduction and hydrolysis) results in an activity of the ion which differs from the concentration of the ion by an amount greater than that caused by ionic interactions. However, the electrode is still measuring the true activity of the ion in the solution.

An example of solution interference demonstrates this point. Silver ion in the presence of ammonia forms a stable silver-ammonia complex ion which is not measured by the silver electrode—only the free, uncombined silver ion is measured. The total silver ion may be obtained from calculations involving the formation constant of the silver-ammonia complex and from the fact that the total silver is equal to the free silver plus the combined silver. Alternatively, a calibration curve can be drawn relating the total silver (from analysis or sample preparation) to the measured activity. The ammonia is not an electrode interfering ion.

Most confusion comes from the fact that analytical measurements have been in terms of concentration without regard to the actual form of the material in solution, and electrode measurements often do not agree with the laboratory analyst's results. However, the electrode reflects what is actually taking place in the solution at the time of measurement. This may be far more important in process applications than the more classic information. With some of the techniques suggested, the two measurements are often reconciled.

Calibration Solution

Calibration solutions for ion selective systems are normally not buffered to resist changes as are the standard solutions for pH systems. They are affected by dilution, evaporation or contamination with foreign matter from the process fluid and air oxidation. Thus, more care must be taken in preparing and handling these solutions than is generally needed in a typical pH application. Attention should be paid to eliminate carry-over from one test solution to another or from distilled water rinses.

Calibration solutions should be prepared by a competent laboratory in accordance with accepted principles of analytical chemistry. Many common chemical standards are available as stock solutions from laboratory supply houses. Generally, only solutions at a reasonably high concentration level (greater than 0.01 M or 100 ppm) should be made for storage. Serial dilutions of these stock solutions should be made at the time of use since very dilute solutions are particularly liable to lose some of their ions by absorption on the walls of the storage vessels. Use of a satisfactory grade of plastic storage bottle is recommended.

Table 1.13i lists solutions frequently used to check the performance of ion selective measuring systems. For solutions in which the ion background is held constant (sulfide, chloride, cyanide and pH), the potential difference between two solutions follows Nernst's law (Table 1.13j). For the others, the potential differences should be calibrated to decade changes in activity.

To achieve the utmost in accuracy and useful measurements, the ion selective measuring system should be standardized (and thus made as effective as possible) in a solution which has been carefully chosen to be chemically similar to the process solution at the point of prime interest. This solution should be at a stable temperature near the actual process temperature ($\pm 2°C$). A grab sample of the process solution checked in the laboratory may be the best and most convenient standard to use.

The Nernst Equation

The potential developed across an ion selective membrane is related to the ionic activity as shown by the Nernst equation

$$E = \frac{2.3\ RT}{nF} \log \frac{a_1}{a_{int}} \quad 1.13(1)$$

Where E is the potential developed across the membrane, a_1 is the activity of the measured ion in the sample or process and a_{int} is the activity of the same ion in the internal solution. The term 2.3RT/nF is the Nernst slope (slope of the calibration curve) and is a function of the absolute temperature T and the charge on the ion being measured, n. R is the gas law constant. Table 1.13j shows how the Nernst slope changes with temperature and the charge on the ion. When the ratio of the two activities is unity, the potential across the membrane is zero.

Equation 1.13(1) assumes that the membrane has identical selectivity properties on both sides. If for some reason this is not true, the equation is

Table 1.13i
CALIBRATING SOLUTIONS FOR ION SELECTIVE ELECTRODES

Electrode	*Chemical Composition*	*Ionic Concentration*	*Approximate Ion Activity*	*Approximate emf vs 1.0M KCl, AgCl, Ag Reference at 25°C*
Hydrogen (pH)	0.05 M KH phthalate	4.008 pH buffer at 25°C	$10^{-4.008}$ M H^{1+}/liter	+143 (+178)°
	0.025M KH_2PO_4 + 0.025M + Na_2HPO_4	6.86 pH buffer at 25°C	$10^{-6.86}$ M H^{1+}/liter	−35 (0)
	0.01M Borax	9.18 pH buffer at 25°C	$10^{-9.18}$ M H^{1+}/liter	−149 (−114)°
Fluoride	22.10 mg NaF/liter	10.0 mg F^{1-}/liter	9.8 mg F^{1-}/liter	−59
	2.21 mg NaF/liter	1.0 mg F^{1-}/liter	1.0 mg F^{1-}/liter	0.0
Chloride	1.00×10^{-2}M KCl in 1.00M KNO_3	1.00×10^{-2}M Cl^{1-}	0.61×10^{-2}M Cl^{1-}	+118
	1.00×10^{-3}M KCl in 1.00M KNO_3	1.00×10^{-3}M Cl^{1-}	0.61×10^{-3}M Cl^{1-}	+177
Silver	1.00×10^{-2}M $AgNO_3$	1.00×10^{-2}M Ag^{1+}	0.90×10^{-2}M Ag^{1+}	+443
	1.00×10^{-3}M $AgNO_3$	1.00×10^{-3}M Ag^{1+}	0.96×10^{-3}M Ag^{1+}	+385
Sulfide	1.00×10^{-1}M Na_2S in 1.00M NaOH	1×10^{-1}M S^{2-}	0.15×10^{-1}M S^{2-}	−860
	1.00×10^{-3}M Na_2S in 1.00M NaOH	1×10^{-3}M S^{2-}	0.15×10^{-3}M S^{2-}	−800
Cyanide	1.00×10^{-3}M NaCN in 1.00×10^{-1}M NaOH	1.00×10^{-3}M CN^{1-}	0.76×10^{-3}M CN^{1-}	−192
	1.00×10^{-4}M NaCN in 1.00×10^{-1}M NaOH	1.0×10^{-4}M CN^{1-}	0.76×10^{-4}M CN^{1-}	−133
Water Hardness	1.00×10^{-2}M $CaCl_2$	1.00×10^{-2}M Ca^{2+} or 1000 mg/liter as $CaCO_3$	0.55×10^{-2}M Ca^{2+} or 550 mg/liter as $CaCO_3$	+34
	1.00×10^{-4}M $CaCl_2$	1.00×10^{-4}M Ca^{2+} or 10 mg/liter as $CaCO_3$	0.92×10^{-4}M Ca^{2+} or 9.2 mg/liter as $CaCO_3$	−18

° vs 4M KCl, AgCl, Ag reference electrode at 25°C.

written

$$E = E_{asy} + \frac{2.3\ RT}{nF} \log \frac{a_1}{a_{int}} \tag{1.13(2)}$$

Where E_{asy} is the asymmetry potential and amounts to a few millivolts. This equation is simplified by the fact that the a_{int} term is fixed by the internal structure of the electrode, giving

Table 1.13j
NERNST SLOPES

Electrode Temperature		*Millivolts per Decade of Activity**	
°C	*°F*	n = ±1	n = ±2
0	32	54.19	27.10
10	50	56.17	28.08
20	68	58.17	29.08
25	**77**	**59.16**	**29.58**
30	86	60.15	29.58
40	104	62.15	29.58
50	122	64.12	32.03
60	140	66.10	33.05
70	158	68.09	34.04
80	176	70.07	35.04
90	194	72.05	36.02
100	212	74.04	37.02

* The slopes are positive for cations and negative for anions.

$$E = E^{\circ\prime} + \frac{2.3\ RT}{nF} \log a_1 \qquad 1.13(3)$$

where $E^{\circ\prime}$ is a new constant.

In operation, the potential of a single electrode cannot be measured by itself but only in conjunction with a reference electrode and a high input impedance voltmeter (see Figure 1.13k). The latter is necessary to prevent current flow through the electrode, which would tend to cause electrochemical reactions in the solution phase around the membrane. The potential read on the voltmeter is equal to the algebraic sum of the potentials developed within the system (see Section 8.8 in Volume I). That is, the observed meter potential is the sum of the potentials developed by the measuring electrode (E), the reference electrode (E_{ref}) and a small but important liquid-junction potential (E_j).

$$E_{meter} = E - E_{ref} + E_j \qquad 1.13(4)$$

Under normal operating conditions, the reference electrode is assumed to be constant as is the liquid-junction potential. This is not always the case (see Section 1.8 in this volume).

Substituting equation 1.13(3) into 1.13(4), and combining constant terms (including E_{ref} and E_j), we obtain the general form of the Nernst equation

$$E_{meter} = E^{\circ} + \frac{2.3\ RT}{nF} \log a_1 \qquad 1.13(5)$$

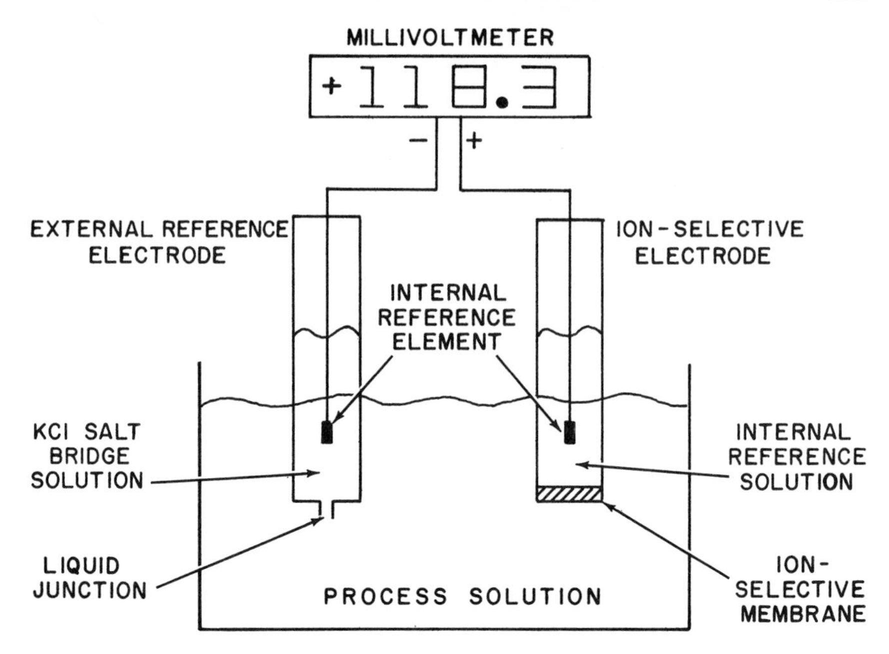

Fig. 1.13k Ion selective electrode measuring system

where E° is a constant for a given electrode system at a specific temperature. It depends on the choice of reference electrode and includes the liquid-junction potential.

The Nernst equation for the electrode pair can be written as an instrument input-output type equation

$$\text{Output} = A + B \log (\text{Input}) \qquad 1.13(6)$$

Where the output is a millivolt signal to a meter, A is a zero adjustment and B is a span or slope adjustment around a temperature independent point (see P. 826 in Volume I). The input to the electrodes is the composition of the solution in terms of activity.

Equation 1.13(5) states that the output of an electrode pair is linear with respect to the logarithm of the activity of the ion being measured (see Figure 1.13 l). The slope of the curve relating E_{meter} to log a_1 is 59.16 mv (at 25°C for n = 1) or 29.58 mv (at 25°C for n = 2). Ignoring the effects of chemical reactions which would complex ions, the activity of the ions is related to the analytical concentration, C, as follows:

$$a = \gamma C \qquad 1.13(7)$$

where γ is the activity coefficient and is a measure of the interaction among ions in solution. It can be thought of as an empirical factor to explain the difference between the actual behavior of ions in solution and the ideal behavior. At zero ion concentration, i.e., no ionic interaction, γ is taken as unity, and the activity is equal to concentration. As the concentration increases, γ at first decreases, passes through a minimum and then rises, often to values greater than unity in very concentrated solutions. The activity coefficient is constant when the ionic composition of the solution is constant.

Substituting equation 1.13(7) into equation 1.13(5) gives

$$E_{meter} = E^\circ + \frac{2.3\ RT}{nF} \log (\gamma C) \qquad 1.13(8)$$

or, at constant total ionic conditions

$$E_{meter} = E^\circ + \text{Constant} + \frac{2.3\ RT}{nF} \log C \qquad 1.13(9)$$

where the constant term is RT/nF log γ. Equation 1.13(9) is linear with respect to the concentration term (see Figure 1.13 l). However, if the ionic background of a solution varies, as in the preparation of a series of standards by dilution, the activity coefficient is no longer constant and equation 1.13 (8) is non-linear (see Figure 1.13m).

Equation 1.13(5) can predict the change in potential to be expected from a given change in activity or concentration (constant activity coefficient). For

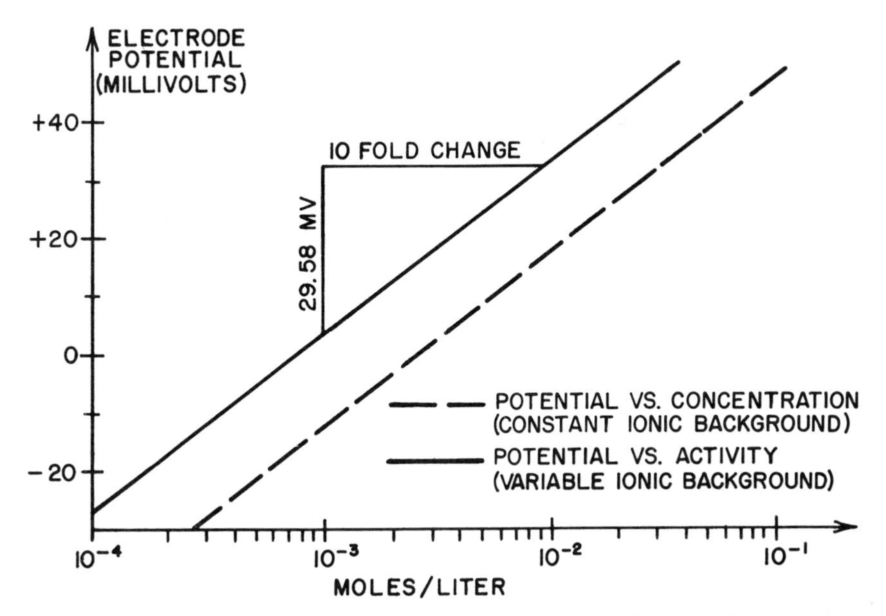

Fig. 1.13l Electrode potential for calcium chloride solutions as a function of concentration and activity

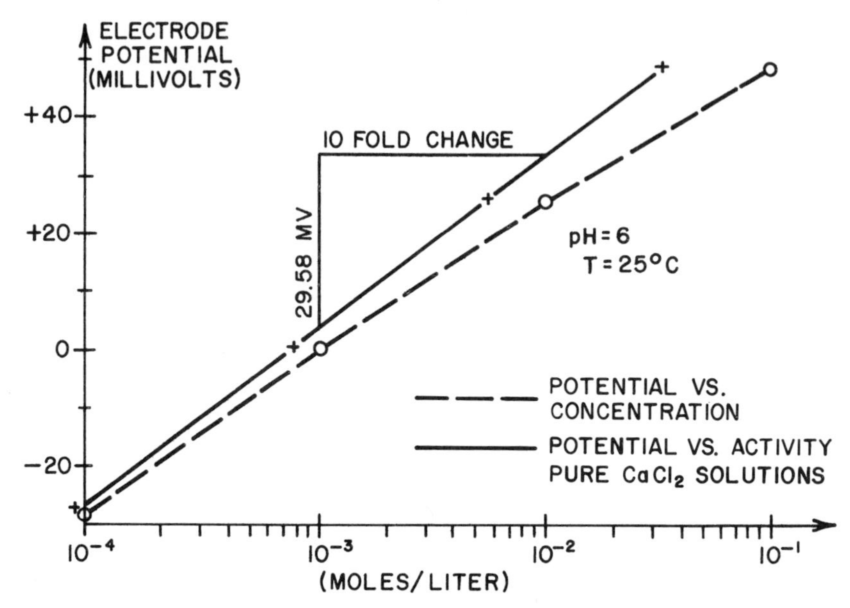

Fig. 1.13m Concentration and ion activity vs electrode potential

instance, if the activity changed twofold (100% change) we obtain

$$E_{meter} = E^{\circ} + \frac{2.3\ RT}{nF} \log 2a_1 \qquad 1.13(10)$$

Subtracting equation 1.13(5) from 1.13(10) gives

$$\Delta E = \frac{2.3\ RT}{nF} \log 2 \qquad 1.13(11)$$

or 18 mv at 25°C for n = 1. A similar argument would show that for the same sample an 18 mv decrease would be observed if the initial activity was cut in half (50% change). This change is not dependent on the magnitude of a_1 and is the same whether measuring fluoride at the ppm level, chloride in 4% salt solutions or pH of a 15% sulfuric acid solution.

Table 1.13n lists the changes in potential to be expected for as much as a tenfold change in activity. Column 1 is the ratio of the original activity, a_1, to the final activity, a_2. The last column shows the equivalent pH change if the hydrogen ion was being measured. This body of data indicates that for precise measurements of small changes in activity it is necessary to use an expanded scale meter. A span of 60 mv, for instance, would allow a tenfold change to be measured, using the full scale of the measuring instrument.

Temperature Effects

There are three temperature effects on ion selective measurements. They are the T term in the Nernst equation (Table 1.13j), the thermal characteristics

Table 1.13n
CHANGES IN METER POTENTIAL
FOR CHANGES OF ACTIVITY
(See Equation 1.13(11).)

$\frac{a_1}{a_2}$	ΔE *in mv* (25°C)		*Equivalent pH change* (n = 1)
	n = 1*	n = 2*	
.1	−60**	−30**	−1.0
.25	−36	−18	− .6
.5	−18	−9	− .3
.79	−6	−3	− .1
1.0	0	0	0
1.26	+6	+3	+ .1
2.0	+18	+9	+ .3
4.0	+36	+18	+ .6
10.0	+60	+30	+1.0

* Data are for positive ions. For negative ions, sign should be reversed.
** Values rounded off from 59.16 and 29.58 (see Table 1.13j).

of the electrodes and the thermal characteristic of the solution. The T term in equation 1.13(5) states that the potential produced by the electrode system is a function of temperature as well as of ion activity. This effect can be compensated for manually or automatically by manipulating the input signal to the converter to indicate the true activity at the measured temperature. Temperature can also be accommodated by designing the electrode pair so that there is a zero temperature error at a particular ion activity. Figure 1.13o shows this effect for the fluoride electrode used to control the fluoridation of public water supplies. The point at which the temperature curves intersect, 1 mg F per liter, is the control level for fluoridation. This point of intersection is called the isopotential point, and temperature effects are negligible on either side for a 10° to 15°C (18° to 27°F) change in temperature. The isopotential point for pH measuring systems is normally around pH 7 and has a potential value of zero millivolts (see Types of Electrodes in this section and Figure 8.8b in Volume I). The activity coordinate of the isopotential points of the solid state electrodes is fixed during manufacture, and the millivolt coordinate is also dependent on the choice of reference electrode. However, due to the construction of the liquid-membrane electrode, the isopotential points can be changed to fit the process.

The second effect associated with temperature is one created by the different internal thermal characteristics of the measuring and reference electrodes. This effect can be minimized if the internal elements of the measuring electrode are matched to the reference electrode. Most commercial pH systems employ matched internal elements for both electrodes.

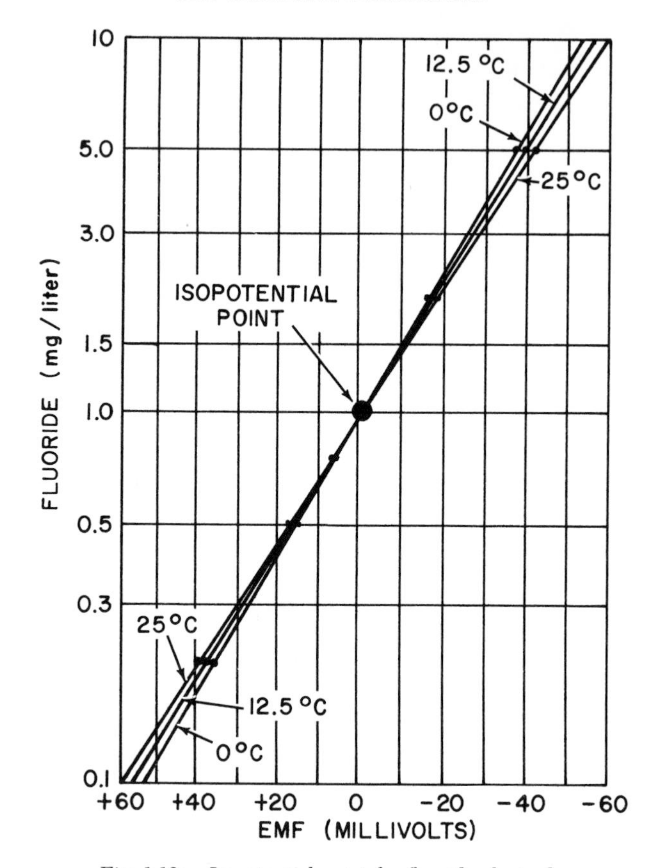

Fig. 1.13o Isopotential point for fluoride electrode

The temperature effect on the chemistry of solutions is the third factor that can create an apparent error in measurement. This effect is difficult to measure but can be offset by calibrating the system with pre-analyzed process samples at the temperature of the process measurement. It should be noted that this is not a system error. The electrode reflects the true activity as a function of temperature changes. As long as the status of the process is what is required, the solution temperature effect is not important since the true activity is the quantity desired. This effect is usually not compensated for by the measuring instruments.

It is important to remember that when electrodes are changed from process samples to standard samples at a different temperature, or when sudden and wide variations in process temperature are encountered, time is required for a new state of thermal equilibrium to be reached (approximately 30 minutes for a 10° or 15°C change). During this time the potential of the electrode system will drift. The duration of the drift depends on the particular electrode system and the magnitude of the temperature change. It is, therefore, important either to avoid changes in temperature during calibration or to allow thermal equilibrium to be established.

System Accuracy

The accuracy of measurements derived from an analytical system is a composite of all contributing variables. These variables for ion selective measuring systems are the measuring electrode, the reference electrode including the liquid-junction potential, the selective ion potential converter, recorder, the temperature and solution errors.

The relationship between over-all emf errors (ΔE) and ionic activity a may be derived from the Nernst equation 1.13(5). Substituting the values of the thermodynamic constants R and F

$$E = E^\circ + \frac{0.1984T}{n} \log a \qquad 1.13(12)$$

and taking the first derivative of Equation 1.13(12) with respect to activity, we obtain

$$\Delta E = (0.2568/n)\ (\Delta\ a/a)\ (\text{at } 25^\circ C) \qquad 1.13(13)$$

$$\Delta E = (0.2568/n)\ (RE) \qquad 1.13(14)$$

where RE represents relative error in the activity.

A plot of equation 1.13(14) is given in Figure 1.13p. The relative error in measuring activity is dependent only on the absolute error in the emf and is independent of the activity range and of the size of the sample being measured. This is similar to an equal-percentage valve wherein equal incremental changes in valve opening (electrode potential) produce equal-percentage changes in flow (RE in activity) for all valve openings—assuming constant differential pressure (constant temperature). Reflecting a logarithmic function, an electrode gives a constant precision throughout its dynamic range. Concentrated solutions can be analyzed with the same accuracy as dilute solutions.

Laboratory instruments for ion selective electrode measurements with an accuracy of ± 0.1 millivolt are commercially available. It is possible to make laboratory pH measurements with a relative accuracy of ± 0.002 pH unit (equivalent to ± 0.12 mv). Similarly, under carefully controlled conditions, ion selective electrodes may be made repeatable within 0.1 millivolt. The accuracy attained to date in process instruments has been limited by the reference electrodes rather than by the ion selective electrodes.

Ion selective measurement systems for process applications are repeatable to ± 1 millivolt. For an electrode responding to univalent ions, an over-all error of 1 millivolt corresponds to a 3.9 percent error in activity; for an electrode responding to divalent ions, the relative error is 7.8 percent per millivolt. This means that they are roughly 5 percent (of value in activity)

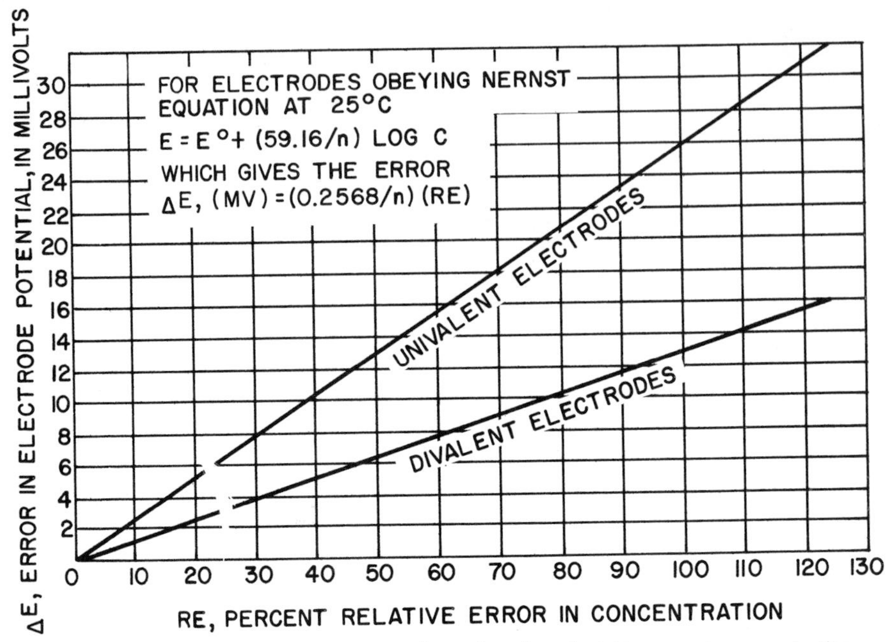

Fig. 1.13p Theoretical error in potential as a function of relative error in concentration

devices when measuring univalent ions, including H^{1+} in acidic or basic solutions or 10 percent (of value in activity) devices when making divalent ion measurements.

The figures mentioned apply only to direct electrode measurements. When electrodes are used as end point detectors in titrations, batch reactions or differential systems, relative errors of 0.1 percent are possible.

Conclusions

It is evident from the previous discussion that ion selective electrode measurements and pH measurements using the glass electrode are identical in theory and also similar in practice. The electrodes are generally the same size and fit the pH holder assemblies. Electrical insulation of ion selective electrodes is just as important as for glass electrodes. In addition, the electrodes are subject to the same fouling by oils and slimes as glass electrodes are, and can in general be cleaned by methods already proven for the pH electrodes. (See also Section 1.8 concerning ultrasonic cleaning.)

These similarities make discussion on the application of ion selective electrodes unnecessary. The reader is referred to the sections on pH measurement in Section 8.8 of Volume I and in Section 1.8 of this volume for further details.[5]

REFERENCES

1. Pungor, E., Tóth, K., and J. Havas, "Silicone rubber membrane electrodes." *Instruments and Control Systems 38:* 105, 1965.
2. Butler, J. W., "Ionic Equilibrium." Reading, Mass.: Addison-Wesley, 1964.
3. Ringbom, A., "Complexation in Analytical Chemistry." New York: Interscience, 1963.
4. Frankenthal, R. P., "Handbook of Analytical Chemistry." New York: McGraw-Hill, 1963.
5. Durst, R. A., (ed.) "Ion-Selective Electrodes." Washington, D.C.: U.S. Government Printing Office, 1969. Chapter 10.

1.14 ORP PROBES

Design Pressure:	0 to 100 PSIG, dictated by electrode holder.
Design Temperature:	$-5°$ to $+110°C$.
System Accuracy:	Difficult to define since measurement represents a ratio; similar to pH and ion selective electrodes.
Range:	±2000 millivolts.
Cost:	Electrodes: $30 to $150. Systems: less than $1,000.
Partial List of Suppliers:	Beckman Instruments, Inc., Cambridge Instrument Co., The Foxboro Co., Kero-Test Manufacturing Co., Leeds and Northrup Co., Universal-Interloc.

Oxidation-reduction potential (ORP) electrodes are primary elements that, like the pH electrodes (Section 8.8 in Volume I) and other ion selective electrodes (Section 1.13 in this volume), produce a millivolt signal as described by the Nernst equation [equation 1.14(1)]. However, unlike the electrodes that measure the activity of selected ions, the ORP electrodes are inert and measure the ratio of the activities of the oxidized to the reduced species in a redox reaction (Section 10.14 in Volume II).

$$E_{meter} = E° + 2.3 \frac{RT}{nF} \log \frac{(Ox)}{(Red)} \qquad 1.14(1)$$

where E_{meter} is the potential read on a millivolt meter when an ORP probe and a reference electrode are placed in a process solution (Figure 1.13k). $E°$ is a constant that depends on the choice of reference electrode; R and F are thermodynamic constants; T is the temperature in degrees Kelvin; and n is the number of electrons involved in the reaction (Table 1.14a). 'Ox' and 'Red'

Table 1.14a
REDUCTION POTENTIALS OF SOLUTIONS OF INTEREST IN ORP MEASUREMENT

	Reduction equation	*E° (volts)***
1.	$O_3 + 2H^{1+} + 2e^{1-} = O_2 + H_2O$	2.07
2.	$HClO + H^{1+} + e^{1-} = \frac{1}{2}Cl_2 + H_2O$	1.63
3.	$Cl_2 + 2e^{1-} = 2Cl^{1-}$	1.359
4.	$Cr_2O_7^{2-} + 14H^{1+} + 6e^{1-} = 2Cr^{3+} + 7H_2O$	1.33
5.	$ClO^{1-} + 2H_2O + 2e^{1-} = Cl^{1-} + 2OH^{1-}$	0.89
6.	$Fe^{3+} + e^{1-} = Fe^{2+}$	0.771
7.	$Hg_2Cl_2 + 2e^{1-} = 2Hg + 2Cl^{1-}$ (sat. KCl)	0.244*
8.	$AgCl + e^{1-} = Ag + Cl^{1-}$ (1M KCl)	0.235*
9.	$AgCl + e^{1-} = Ag + Cl^{1-}$ (4M KCl)	0.199*
10.	$SO_4^{2-} + 4H^{1+} + 2e^{1-} = H_2SO_3 + H_2O$	0.17
11.	$2H^{1+} + 2e^{1-} = H_2$ (SHE)	0.000*
12.	$CNO^{1-} + H_2O + 2e^{1-} = CN^{1-} + 2OH^{1-}$	−0.97
13.	$2CO_2 + N_2 + 2H_2O + 6e^{1-} = 2CNO^{1-} + 4OH^{1-}$	(0.4 estimated)

* Reactions for reference electrodes.
** It should be pointed out that by themselves, E° values have little meaning to a real process except to indicate reagents that are strong oxidizing agents and those that are strong reducing agents. Unless otherwise noted, values are for unit activity relative to the standard hydrogen electrode at 25°C.

in the equation represent the activities of the oxidized and reduced forms of the ions being measured. (For a more detailed discussion of the Nernst equation see Section 1.13 in this volume.)

An example of an oxidation-reduction reaction is the air oxidation of ferrous iron (Fe^{2+}) to ferric iron (Fe^{3+}) in steel pickling liquors. When an ORP electrode and a suitable reference electrode (Section 1.8) are immersed in this solution the potential developed is given by

$$E_{meter} = E^\circ + \frac{2.3\,RT}{F}\log\frac{(Fe^{3+})}{(Fe^{2+})} \qquad 1.14(2)$$

E_{meter} is dependent on the ratio term of equation 1.14(2) and on the choice of a reference electrode; it is not dependent on the ORP electrode, which is inert.

Electrodes

The inert electrode serves only as an electrical conductor. The conducting material is usually metal (platinum, gold, nickel) but carbon can also be used. The catalytic properties of platinum may prohibit its use in the presence of strong reducing agents in which hydrogen ions are reduced at the platinum surface, resulting in a mixed potential.[1]

Gold electrodes in an alkaline cyanide solution (a strong reducing agent) are used presumably for this reason, but the reason does not appear to be documented. In waste treatment applications in which the cyanide levels are low, either platinum or gold electrodes should give satisfactory results.

ORP electrodes are fabricated from wire, foil or billets sealed in a suitable body material—glass or epoxy (Figure 1.14b). They are generally the same size and shape as pH and other ion selective electrodes and can be used in the same electrode chambers. In addition to the conventional designs they can also be formed in different shapes to suit particular applications. Figure 1.14c shows a cylindrical arrangement as part of a sample line. The platinum cladding is employed to conserve the noble metal.

Applications

Table 1.14a lists some of the oxidation-reduction reactions pertinent to the process industry including reactions for several reference electrodes (equations 7 to 9). Also given are values of E° relative to the standard hydrogen electrode (SHE). Values of E° related to other reference electrodes can be calculated by subtracting the corresponding E° for the reference electrode from the value given in the table. For example, E° for the iron reaction (equation 6) is 0.771 volt with respect to the SHE and 0.536 (0.771 minus 0.235) with respect to a 1 M KCl silver-silver chloride reference electrode. Likewise, the potential measured with any reference electrode can be converted to the SHE scale by adding the E° for the reference electrode, i.e., 0.536 plus 0.235 = 0.771 volt.

The values of E° given in Table 1.14a were determined under ideal

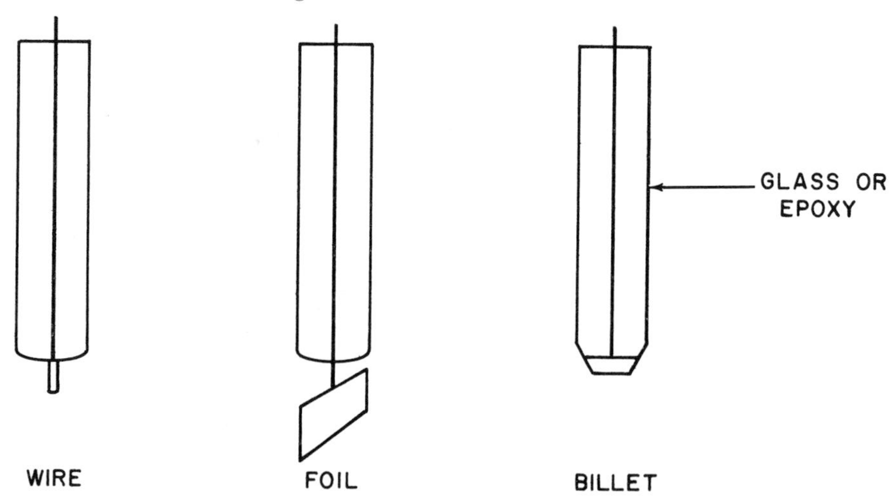

Fig. 1.14b Types of metallic ORP electrodes

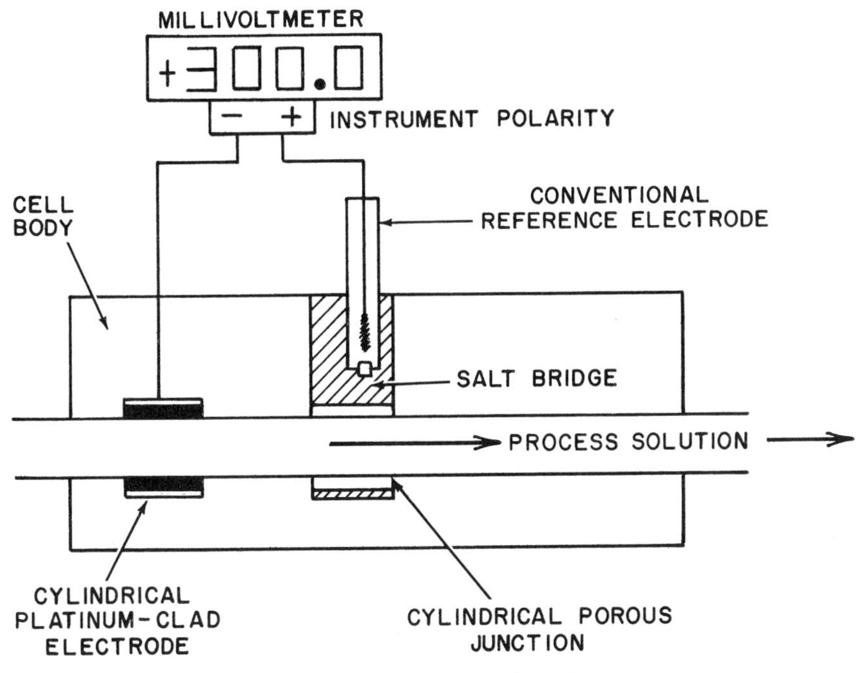

Fig. 1.14c Cylindrical ORP electrode cell

laboratory conditions and may not represent values obtained in complexed waste treatment solutions. However, they can predict the limiting potentials to be expected in oxidation-reduction systems and the magnitude and direction that the potentials will change during process upsets. For instance, the reduction of hexavalent chromium by sulfur dioxide was discussed in Section 10.14 in Volume II. It was stated that the potential level for hexavalent chromium was 700 to 1000 mv, whereas that for the reduced form (excess sulfur dioxide) was 200 to 400 mv. The controller was set at approximately 300 mv. This information can be obtained from the Nernst equation and equations 4 and 10 of Table 1.14a. The potential of an inert platinum electrode with respect to a 1M KCl, silver-silver chloride reference electrode ($E^\circ = 235$ mv), both in a solution containing hexavalent chromium, is given by equation 1.14(3).

$$E_{meter} = 1.095 + \frac{2.3\ RT}{6F} \log \frac{(Cr_2O_7^{2-})(H^{1+})^{14}}{(Cr^{3+})^2} \qquad 1.14(3)$$

where 1.095 is equal to E° (Cr^{6+}, Cr^{3+}) minus E° (1 M reference electrode) or 1.33 minus 0.235. The activity of water is assumed to be unity. Using the value of 0.060 for 2.3 RT/F at 25°C (actually 0.05916; see Table 1.13j) and factoring out the hydrogen ion term gives

$$E_{meter} = 1.095 + \frac{0.060 \times 14}{6} \log (H^{1+}) + \frac{0.060}{6} \log \frac{(Cr_2O_7^{2-})}{(Cr^{3+})^2} \quad 1.14(4)$$

or

$$E_{meter} = 1.095 - 0.14\,pH + 0.1 \log \frac{(Cr_2O_7^{2-})}{(Cr^{3+})^2} \quad 1.14(5)$$

The potential is pH dependent (p. 1541 in Volume II). Most of the reactions in Table 1.14a involve H^{1+} or OH^{1-} and are therefore pH dependent. Chromium reduction is normally carried out at a pH of 2 to 2.5 (for kinetic reasons).

The potential of a dichromate solution at pH 2 which is 10% reduced is given by equation 1.14(6). If the dichromate is 10% reduced, 90% of it is in the oxidized form. In fractions of the total, these values are 0.1 and 0.9 respectively. Because two trivalent chromiums (reduced form) are produced for every dichromate ion (oxidized form) that is reduced, the concentration of the trivalent chromium is twice that of the dichromate or 0.2. According to equation 1.14(5) this term is squared, which results in the $(0.2)^2$ value.

$$E_{meter} = 1.095 - 0.280 + 0.10 \log \frac{0.9}{(.2)^2} \quad 1.14(6)$$

Note: $1\ Cr_2O_7^{2-} = 2\ Cr^{3+}$

The potential calculated from equation 1.14(6) is 950 mv versus the 1 M KCl, silver-silver chloride reference electrode, which is in agreement with the potentials given earlier. If the chromium is reduced with sulfurous acid (SO_2 in H_2SO_4), the potential of the solution after complete reduction would be determined by the excess SO_2 according to equation 1.14(7).

$$E_{meter} = -0.065 + \frac{0.06}{2} \log \frac{(SO_4^{2-})(H^{1+})^4}{(H_2SO_3)} \quad 1.14(7)$$

where -0.065 is equal to E° (SO_4^{2-}, SO_3^{2-}) minus E° (1 M reference electrode) or 0.17 minus 0.235.

Note that the numerator of the log term in equation 1.14(7) includes the sulfate ion which is a common ion between the oxidized form of sulfur dioxide and the acid used for controlling the pH. Solving equation 1.14(7) for a pH of 2 and assuming a ratio of $(SO_4^{2-})/H_2SO_3$ of 100 gives 295 mv which also agrees with the potentials listed earlier.

This assumption is an arbitrary but reasonable one for a sulfuric acid solution (10^{-2} M) with a slight excess of H_2SO_3 (10^{-4} M or 8.2 ppm). The set point on an ORP controller for the chemical destruction of hexavalent chromium then would be a weighted average of the extremes[2] with the weight toward excess reducing agent. In normal operations this is close to 295 mv for a 1 M KCl, silver-silver chloride reference electrode.

A similar treatment can be used to obtain potential ranges for the chlorine oxidation of cyanide wastes or for hypochlorite bleach applications.

The use of ORP probes in hypochlorite bleach is an interesting application which incorporates the pH and ion selective electrodes in addition to ORP. Using equation 5 of Table 1.14a the potential for an ORP measurement at 25°C is

$$E_{meter} = E^\circ + \frac{0.060}{2} \log \frac{(ClO^{1-})}{(Cl^{1-})} - 0.060 \log (OH^{1-}) \qquad 1.14(8)$$

where the last term in equation 1.14(8) is the pH. In bleach production, chlorine gas is absorbed in sodium hydroxide to produce equal concentrations of hypochlorite and chloride ions; i.e., the ratio is always one and the second term of equation 1.14(8) is always zero. A conventional ORP measurement is nothing more than a pH measurement. By using a chloride ion selective electrode (Section 1.13) instead of a reference electrode the potential is given by equation 1.14(9).

$$E_{meter} = E^{\circ\prime} + \frac{0.060}{2} \log \frac{(ClO^{1-})}{(Cl^{1-})} + $$
$$+ 0.060 \log (Cl^{1-}) - 0.060 \log (OH^{1-}) \qquad 1.14(9)$$

where $E^{\circ\prime}$ includes the E° value for the chloride electrode. Or by combining terms

$$E_{meter} = \frac{0.060}{2} \log (ClO^{1-})(Cl^{1-}) - 0.060 \log (OH^{1-}) + E^{\circ\prime} \qquad 1.14(10)$$

If (ClO^{1-}) is equal to (Cl^{1-}) the potential can be related to (ClO^{1-}) and pH.

$$E_{meter} = E^{\circ\prime} + 0.060 \log (ClO^{1-}) - 0.060 \log (OH^{1-}) \qquad 1.14(11)$$

If the bleach is used for oxidizing, the relationship between (ClO^{1-}) and (Cl^{1-}) is no longer one to one and equation 1.14(11) becomes more complicated.

Conclusions

An ORP measurement is not a selective measurement. The measured potential is the sum of each contributing reaction including that from organic species (oils, dyes and microbes). Each application should be evaluated on its own merits. It should not be assumed that a real process will behave as predicted unless the prediction was based on data obtained from the process itself.

REFERENCES

1. Latimer, W. M., "The Oxidation States of the Elements and Their Potentials in Aqueous Solutions," Englewood Cliffs, N.J.: Prentice-Hall, 1952.
2. Butler, J. N., "Ionic Equilibrium," Reading, Mass.: Addison-Wesley, 1964.

1.15 CARBON ANALYZERS

Methods of Detection:

(a) Non-dispersive infrared analyzer.
(b) Trace oxygen analyzer.
(c) Hydrogen flame ionization detector.
(d) Thermal conductivity detector.
(e) Specific conductance detector.

Note: In the feature summary below the letters (a) to (e) refer to the listed analyzer types.

Sample Pressure:

Generally atmospheric since most of these analyzers are laboratory types; however, several models are available as process analyzers and these will accept sample streams at pressures of 5 to 10 PSIG.

Sample Temperature:

32° to 200°F since most samples are injected as liquids.

Sample Flow Rate:

Generally grab samples are brought to the analyzer but for those using flowing sampling systems, flow rates vary from 3 cc per minute to 200 cc per minute.

Materials of Construction:

Samples normally come in contact only with corrosion resistant materials such as glass, quartz, Teflon, hastelloy, stainless steel, polypropylene and polyethylene.

Accuracy:

Generally in the ±1% to ±2% of full scale range when detecting concentrations above 50 ppm and in the ±3% to ±5% of full scale when measuring concentrations less than 50 ppm.

Ranges:	0 to 2 ppm carbon as high as 0 to 5,000 ppm carbon without dilution. Higher concentrations may be measured by prior dilution.
Response:	Most analysis cycles are in the two-minute to five-minute range, but the response time for the continuous analyzer can vary from five to 15 minutes for full scale response, depending on flow parameters.
Cost:	Basic analyzer systems range from $2,400 to $9,800. With special sample handling equipment such as filters, multi-stream selectors or remote recorders, prices may be as high as $15,000.
Partial List of Suppliers:	Beckman Instruments, Inc. (a), Bendix Corp. (c), Delta Scientific Corp. (e), Envirotech Corp. (c), Hewlett-Packard (d), Ionics, Inc. (b), Mine Safety Appliances Co. (a and c), Oceanography International Corp. (a), Precision Scientific, Inc. (a), Raytheon Co. (a), Union Carbide Corp. (a and c).

During the past ten years more than a dozen carbon analysis systems have been developed for laboratory analysis and for onstream monitoring. The major reason for this developmental activity is the recent interest in environmental protection. A measurement of ppm carbon in waste waters or streams is one of the most reliable indications of pollution from process industries and sewage treatment plants. Also, some manufacturing operations, such as in the semiconductor industry, are very sensitive to trace levels of carbon in the water used in the manufacturing processes.

For fifty years, dating from the mid-1800s, the only method for determining pollution in water was by biological assay techniques which resulted in biochemical oxygen demand (BOD) readings. Around the turn of the century analytical chemists developed a wet chemical technique—chemical oxygen demand (COD)—involving high temperature oxidation of the sample in the presence of a strong acid and excess potassium dichromate. The excess dichromate is backtitrated to determine the amount of dichromate consumed by reaction with oxidizable material. (See Section 1.16 in this volume for BOD and COD analysis systems.)

Early in the 1900s Pregl and co-workers perfected microanalytical techniques for determining the ratios of carbon, hydrogen and nitrogen (C,H,N analysis) in unknown samples of organic materials. The work proved to be very important in the recent development of carbon analyzers, since the latter has to a large extent been a process of mechanizing the techniques developed more than fifty years ago by Pregl and his co-workers.

Although both the BOD and COD tests were long accepted as standard methods of evaluating pollution loads or as means of measuring the efficiency of waste treatment plants, they were too slow for actual control of plant operations. The BOD test originally required 20 days (later shortened to five); the COD test required roughly three hours. The search for a more rapid method of analysis led to the development of methods to measure the carbon content of waste water samples.

Late in the 1950s, researchers at Dow Chemical and Union Carbide, working independently, began the development of carbon analysis systems. Both groups approached the problem in much the same manner, i.e., high temperature conversion of organic carbon to carbon dioxide and subsequent detection with a non-dispersive infrared analyzer. The major difference in the two systems was that the Dow analyzer was basically a laboratory device, whereas the Union Carbide analyzer was a process monitoring instrument.

Carbon Analysis of Liquid Samples

In the following discussion all systems involve high temperature conversion of carbonaceous material to carbon dioxide, carbon monoxide or methane unless otherwise noted.

Non-Dispersive Infrared Detectors (a)

There are 5 major carbon analyzer systems in use today that utilize the non-dispersive infrared analyzer as a detector. Figure 1.15a is a generalized diagram of four of these systems. A, B, C and D in Figure 1.15a refer to specific components in the various analytical systems and will be described as the individual systems are discussed. Infrared analyzers are treated in Section 8.3 in Volume I.

SYSTEM 1 (AIR AS CARRIER GAS)

The first system utilizes oxygen or pre-purified air as a carrier gas. Depending on the range of analysis, a 20 to 200 μl water sample is injected into the flowing stream of carrier and is swept into a catalytic combustion tube containing a cobalt oxide-impregnated asbestos packing. The combustion tube is enclosed in an electric furnace with its thermostat set at 950°C. The water is vaporized, and all carbonaceous material is oxidized to carbon dioxide. The air flow carries the cloud of steam and carbon dioxide out of the furnace where the steam is condensed and removed at point D. The carbon dioxide

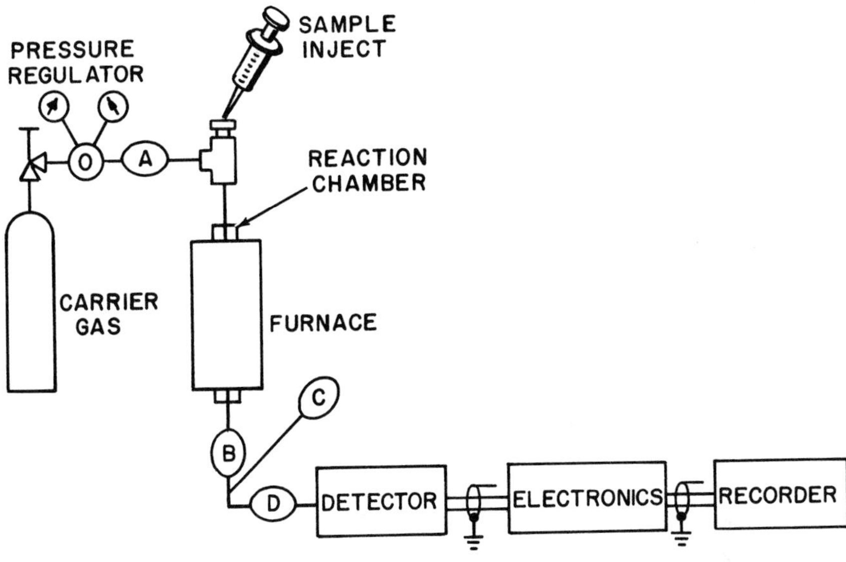

Fig. 1.15a Carbon analyzer systems

is swept into the non-dispersive infrared analyzer which is sensitized to measure carbon dioxide.

The transient carbon dioxide peak is recorded on a strip chart recorder. The peak height is a measure of carbon dioxide present, which is directly proportional to the concentration of total carbon in the original sample (this includes organic carbon, inorganic carbon and carbon dioxide dissolved in the sample). By using standard solutions, the analyzer can be calibrated to read milligrams total carbon per liter of sample; ranges are from 0 to 10 to 0 to 4,000 mg per liter.

Addition of a second sample injection system and a new furnace at point C in Figure 1.15a converts the total carbon analyzer to an organic carbon analyzer by the following operation. A similarly sized sample is injected into the flowing stream of carrier and is swept into the second reaction chamber which contains quartz chips wetted with 85% phosphoric acid. This tube is enclosed in an electric heater with thermostat controlled at 150°C, which is below the temperature at which organic matter is oxidized. The acid-treated packing and the operating temperature cause the release of carbon dioxide from inorganic carbonates and the water to be vaporized. The air flow carries the cloud of steam and carbon dioxide out of the furnace where the steam is condensed and removed. The carbon dioxide is swept into the infrared analyzer and is recorded on the strip chart recorder. This peak height is proportional to the concentration of inorganic carbonates plus carbon dioxide dissolved in the original sample. By using standard solutions, the analyzer can be calibrated in milligrams of inorganic carbon per liter of sample.

Subtracting the results obtained in the second operation from the results of the first operation yields a measure of total organic carbon (TOC) in milligrams TOC per liter of sample. With high inorganic backgrounds and low organic levels, the accuracy of this approach may be poor.

The time from sample injection to peak read-out for each portion of the system is about two minutes. Therefore, a complete analysis of total carbon and total inorganic carbon may be obtained from water or waste water samples in less than five minutes.

If the syringe injection port shown in Figure 1.15a is replaced by an automatic liquid sample injection valve (Figure 1.15b), the total carbon analyzer can be converted to an automatic analyzer and (when properly packaged for inplant operating conditions) may be used as an online process monitoring instrument.

The main advantage of the carbon analyzer is the short analysis time—roughly two minutes for a single analysis. Two major disadvantages are high cost, ranging from $6,000 to $10,000, and poor correlation of test results (except in special cases) with either BOD or COD; and these are the numbers most pollution control technicians understand.

SYSTEM 2 (NITROGEN AS CARRIER GAS)

The second type of carbon analyzer uses nitrogen as a carrier gas with, ordinarily, a precision sample injection valve (Figure 1.15b) instead of a

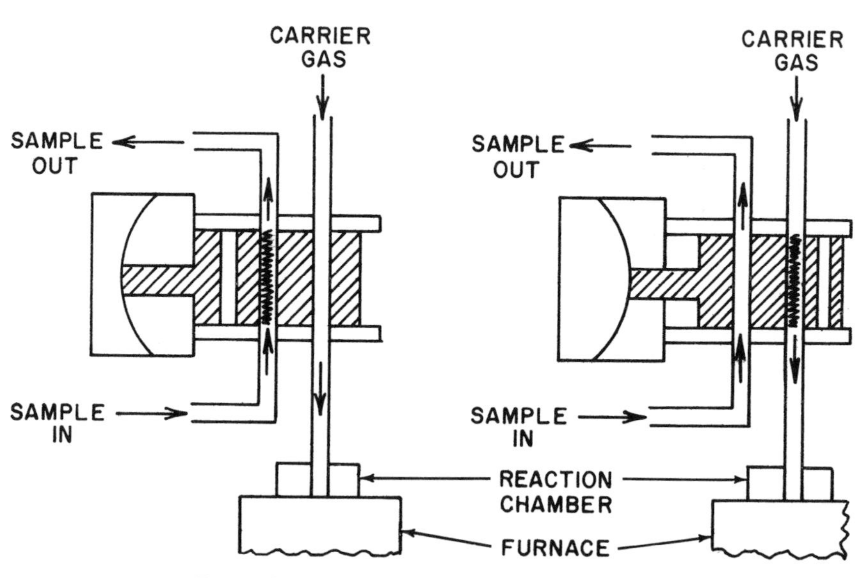

Fig. 1.15b Automatic liquid sample injection valve

hypodermic syringe. Typically, a 40 μl sample of water is injected into the hastelloy reaction chamber where it is heated to about 850°C. A palladium catalyst converts carbon bearing compounds into carbon dioxide and water. The reaction products are then carried into the gas liquid separator (B in Figure 1.15a) where the water is removed, and the vapor is swept into a non-dispersive infrared analyzer sensitized to measure carbon dioxide. The infrared analyzer can be calibrated to reflect carbon dioxide levels with a full scale range from 50 ppm to 3,000 ppm total carbon. Since many waste water samples contain dissolved salts or other solids, a rinse cycle is incorporated in the sample injection valve to help keep the catalytic system clean. When the sample injection valve is returned to the sample-taking position, a 100 μl sample of distilled water is injected into the reaction chamber to clean out residues.

Because the carrier gas is inert, the analyzer is much safer to operate in the often hazardous ambient conditions of a modern industrial plant than are the analyzers using oxygen or air as carrier. The inert carrier gas also reduces the maintenance requirements on the high temperature combustion system.

This carbon analyzer can also be converted to measure organic carbon by adding a low temperature system at point C as previously discussed. The major difference between this low temperature system and the one previously described is that an amphoteric salt, such as zinc chloride, is used in place of phosphoric acid to assist in the decomposition of carbonates to carbon dioxide. The salt permits the low temperature system to be operated for a much longer time before it needs to be reactivated.

The second carbon analysis system has the same advantages and disadvantages listed for system 1.

SYSTEM 3 (CARBON DIOXIDE AS CARRIER GAS)

The third type of carbon analyzer uses carbon dioxide as the carrier gas. A purifying carbon furnace (A in Figure 1.15a) converts traces of oxygen present in the feed gas to carbon monoxide. This level of carbon monoxide is the background or zero level for the analyzer. A 20 μl water sample is injected by hypodermic syringe and swept into the reaction chamber by the mixture of carbon dioxide-carbon monoxide from the carbon furnace. A platinum catalyst oxidizes the pollutants to carbon monoxide and water vapor. The water is stripped out in a drying tube and the reaction products are then passed through a second platinum catalytic treatment. The carbon monoxide concentration is then measured by an integral non-dispersive infrared analyzer sensitized for carbon monoxide. The resultant reading is converted to COD (milligrams per liter) by a calibration chart.

Unlike the first two analyzers described, which give results in total carbon or total organic carbon, the third carbon analyzer yields a body of data that correlates well with the classic COD test. Unfortunately, this instrument is only available as a laboratory analyzer.

SYSTEM 4 (OXYGEN AS CARRIER GAS)

Unlike those previously described, this unit is a continuous analyzer, utilizing oxygen as the carrier gas. At point A in Figure 1.15a, a continuous sparging system removes dissolved carbon dioxide and inorganic or carbonate carbon. A continuous sample of water is mixed with 0.5 N HCL and is continuously sparged by nitrogen to remove carbon dioxide. A metering pump feeds a continuous liquid sample to the reaction chamber, which is a gas-fluidized bed of 180 mesh aluminum oxide particles heated to 850°C.

Oxygen is supplied to the reactor at ten times the stoichiometric requirement in order to oxidize the maximum amount of organic carbon present. The oxygen and the vaporization of the sample provide the gas requirement for fluidization of the reactor bed. In the reactor, the organic carbon is oxidized to carbon dioxide. The effluent is a gas stream of water, carbon dioxide and oxygen, plus oxidation products of other carbonaceous compounds in the water sample. A two-stage condenser located at point B (Figure 1.15a) cools the effluent stream to less than 50°C.

The carbon dioxide in the effluent gas is measured in a non-dispersive infrared analyzer, and is proportional to the total organic carbon in the sample stream; typical measurement ranges are from 0 to 50 and 0 to 4,000 milligrams per liter.

If the acid sparging system is eliminated from point A in Figure 1.15a the analyzer operates as a total carbon analyzer. If a measure of both total carbon and total organic carbon is desired, it is possible to modify the design so that the sample is alternately fed through the sparging system (to measure total organic carbon) and through a by-pass (to measure total carbon).

The major advantage of this carbon analyzer is that it operates continuously, rather than monitoring discrete samples. However, it is complex in design and is not a fully tested, commercially available instrument. Also, for some types of chemical wastes, the sparging system will strip out volatile organics and cause the organic carbon reading to be low.

SYSTEM 5 (NO CARRIER GAS)

This instrument (Figure 1.15c) is a manually operated, batch-type analyzer which does not utilize a flowing carrier gas to transport the sample to the reaction chamber. It consists of two units: a glass ampule purging and sealing unit, and a carbon analyzer unit. The method of analysis is based on the wet oxidization of organic matter in the sample by potassium persulfate dissolved in dilute phosphoric acid.

The purging and sealing unit uses purified oxygen to sparge the acidified sample (which has been placed in the glass ampule) to remove the inorganic

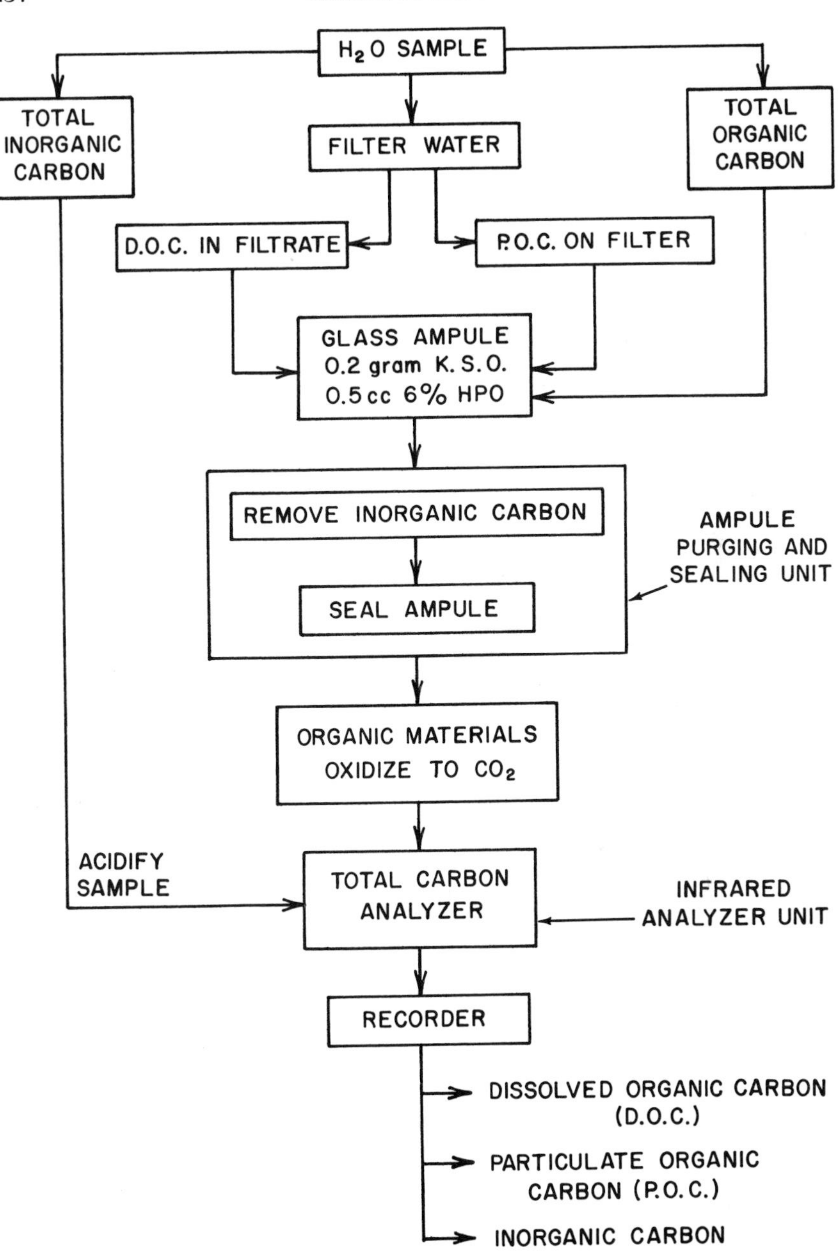

Fig. 1.15c Analysis flow chart

carbon. The ampules are sealed by the same unit with an oxygen-propane microburner. The neck of the ampule is held by a clamping assembly, and the lower portion is twisted manually after the neck becomes molten. Purging cones prevent carbon dioxide contamination from the microburner during the sealing operation. The organic material contained in the sealed glass ampule is converted to carbon dioxide by the potassium persulfate at elevated temperature and pressure.

The ampules are placed into the carbon analyzer and after the ampule stem is broken, a bubbling tube is inserted into the vial, and the gaseous contents of the ampule are flushed through the infrared analyzer by a nitrogen stream. The amount of carbon dioxide recorded is proportional to the organic carbon in the sample.

Total inorganic carbon may be determined by by-passing the oxidation step and allowing the sparging gas to flow directly into the infrared analyzer. If the sample is filtered, it is possible to determine dissolved organic carbon (DOC) by analyzing the filtrate or to measure particulate organic carbon (POC) by analyzing the material retained on the filter.

The major advantage of this carbon analysis system is its ability to measure low ppm concentrations (0.2 to 2 ppm) of carbon with a relatively high degree of precision ($\pm 5\%$). It is used only in the laboratory because of the need for considerable manipulation of the sample and apparatus by the analyst.

Trace Oxygen Analyzer Detector (*b*) (*System 6*)

One of the carbon analyzers uses a trace oxygen fuel cell as detector. This sixth system has a nitrogen carrier gas that contains approximately 200 ppm oxygen as a component. Referring to Figure 1.15a, point A, a short length of elastomeric tubing inserted in the carrier line allows a relatively constant amount of atmospheric oxygen to diffuse into the flowing nitrogen stream.

A 20 μl aqueous sample is swept into the reaction chamber by the nitrogen carrier gas which contains trace amounts of oxygen. At the surface of a 900°C platinum catalyst within the reaction chamber, two reactions occur, including one in which the impurities in the sample are oxidized, thereby partially depleting the oxygen adsorbed on the platinum surface; and one in which the oxygen equilibrium on the catalyst surface is restored by the oxygen in the carrier gas stream. The second reaction results in a momentary depletion of the oxygen in the nitrogen stream, which is measured in a platinum-lead fuel cell and recorded as a negative oxygen peak on the recorder. The total oxygen demand (TOD) for the sample is obtained by comparing the recorded peak height to a calibration curve.

The major advantage of this analyzer is that it detects the oxygen consumed by materials other than carbon, e.g., hydrogen, nitrogen and sulfur, but does not detect inorganic carbon. Hence, the analytical results correlate better with the ultimate oxygen demand exerted in a natural water system. Although the minimum full scale range is 0 to 40 ppm TOD, at low TOD levels the amount of dissolved oxygen in the sample can introduce significant errors.

This analyzer is designed primarily for use in the laboratory, although the manufacturer indicates that an analyzer will shortly be introduced ". . . designed for continuous monitoring and control of industrial effluents. . . ."

Hydrogen Flame Ionization Detectors (*c*)

The hydrogen flame ionization detector (HFID) has a significantly greater sensitivity than any of the other detectors currently in use (p. 744 of Volume I).

SYSTEM 7 (PYROGRAPHIC)

A nitrogen carrier flows through a manually operated sampling valve and when the valve is actuated, the aqueous sample is swept into the pyrolysis chamber (Figure 1.15a), which is maintained at 800°C; however, since there is no catalyst present, the organic molecules are not oxidized but are thermally cracked into low molecular weight fragments. Exiting from the pyrolysis chamber, the carrier and organic fragments plus water vapor enter a flow stabilizer at point B. The flow stabilizer prevents the detector flame from being extinguished after sample injection.

A hydrogen stream is introduced from point C and the organic fragments are detected by the HFID within one minute after injection. Results are reported as organic load in mg per liter.

Major advantages of the pyrographic technique include

1.) high sensitivity with wide range—up to 3,000 mg per liter without dilution; 2.) detection of organic molecules only without being affected by inorganic carbon or carbon dioxide; and 3.) reasonably satisfactory correlation with BOD_5 on raw sewage and primary sewage treatment plant effluents. However, secondary sewage treatment (biological treatment) preferentially removes biodegradable organics and, thus, the analyzer gives high (twice BOD_5) results on secondary effluent samples.

The major disadvantages of this system are the unequal response of the HFID to different molecules and the fact that this instrument is still in the development stage and is not commercially available.

SYSTEM 8 (CATALYZED REACTION)

Another system utilizing the HFID is a combination of a very sensitive air pollution monitor and the second carbon analysis system described earlier. This eighth system uses high purity nitrogen as a carrier gas (Figure 1.15a). A 40 μl liquid sample is flushed from the sample injection valve and carried into the palladium-catalyzed reaction chamber. All carbonaceous material is converted to carbon dioxide. At point D is a nickel-catalyzed reaction chamber where the carbon dioxide (with a small flow of hydrogen gas) is converted to methane at a temperature of approximately 400°C. The methane concentration, which is directly proportional to the total carbon content, is then measured by the HFID.

As already described, a low temperature reaction system may be correlated with the high temperature system (at point C), and the analyzer can be programmed to record total carbon, inorganic carbon or both inorganic and organic carbon. The value for organic carbon may be recorded by electronically subtracting the value of the inorganic carbon peak from the total carbon peak.

High sensitivity and wide range of this system are major advantages; furthermore, the conversion to carbon dioxide followed by methanization eliminates the major objection to the HFID, namely the unequal response to different types of organic molecules. The need for three gaseous utilities to operate this analyzer (nitrogen, air and hydrogen) is a disadvantage if these gases are not readily available.

SYSTEM 9 (CHROMATOGRAPHIC)

This ninth analysis system is not, strictly speaking, a carbon analyzer; instead a temperature-programmed chromatographic oven takes the place of the furnace in Figure 1.15a, and the reaction chamber is replaced by a chromatographic column. Some of the organic molecules are separated and—without conversion—measured by the HFID.

The main advantage of the chromatographic system is that specific compounds can be identified. When used in an industrial plant to check sewers or plant outfalls, the exact source of a pollutant may be identified by knowing what the organic compound is.

There are, however, several limitations to this system:

1.) There is no single column that will separate all compounds. 2.) The user must know what compounds will be in the water before the system can be programmed. 3.) Unlike the other systems, each installation must be custom engineered since (in general) no two users will have the same compounds to be measured. 4.) Analysis time is long—often an hour or more.

Thermal Conductivity Detector (d)

SYSTEM 10 (CHROMATOGRAPHIC)

The tenth system is also a special type of chromatographic analyzer. Unlike those previously described, this system is not designed to analyze aqueous samples, but is a modern version of the classic C,H,N analysis system.

Referring to Figure 1.15a, the C,H,N analysis system uses high purity helium as the carrier gas. A pre-weighed sample of the solid or liquid is placed in a combustion boat (catalyst and oxidant—WO_3 and MnO_2—are added), and the boat is inserted into the combustion furnace (adjustable between 750° and 1050°C). The carrier gas does not flow through the furnace until the timed combustion period has elapsed. After combustion, the carrier gas sweeps the hot gases through hot copper oxide packing where any carbon monoxide is oxidized to carbon dioxide.

At point B is a reducing furnace (400° to 600°C) packed with reduced copper, in which any excess oxygen gas formed during combustion is removed ($2\ Cu + O_2 \xrightarrow{heat} 2\ CuO$) and any oxides of nitrogen are reduced to nitrogen gas. The carrier gas continues to move the combustion products along, and at point C is the chromatographic column which separates the sample into three component peaks—nitrogen gas, carbon dioxide and water—whose areas are directly proportional to the amount of nitrogen, carbon and hydrogen (respectively) in the original sample. A thermal conductivity detector senses the three peaks and the entire system is calibrated with compounds of known composition and purity.

Although this system is not used for pollution measurements, the description has been included here because of the relationship between classic C,H,N analytical techniques and the development of the carbon analyzer systems.

SYSTEM 11 (LOW SENSITIVITY)

The eleventh carbon analysis system uses air or oxygen as a carrier gas and a thermal conductivity cell as the detector. At point A in Figure 1.15a is a gas purifying unit that removes traces of carbon dioxide from the oxygen or air.

A sample, usually 40 μl, is injected into a combustion tube maintained at 950°C. The carbon dioxide formed is swept through a drying tube located at point B and into a collection trap at point D. After a pre-determined collection time (which can be adjusted), the trap is heated and helium (introduced at point C) drives the carbon dioxide into the thermal conductivity detector. The output signal can be read directly in mg per liter, and samples containing as much as 10,000 mg per liter of carbon can be analyzed without dilution.

The only significant advantage of this system is the simplicity of the detector; some users may have difficulty maintaining the more sophisticated detectors, such as the infrared or the HFID. However, the system's sensitivity is relatively poor. The minimum full scale range is about 500 mg per liter.

Specific Conductance Detector

SYSTEM 12 (*e*)

The water sample to be analyzed is injected into a combustion chamber maintained at 850°C. All carbon is converted to carbon dioxide and the carbon dioxide-steam-carrier gas stream is passed through an air-cooled condenser, located at point B in Figure 1.15a. Following the condenser, the mixture passes into a dilute acid chamber (point D) where alkaline gases are removed and the condensed water vapor is disentrained.

The remaining pre-purified air carrier gas and carbon dioxide are bubbled through a fritted tube into a carbon dioxide-absorbing solution in the detector chamber. A conductivity cell compensated for temperature measures the decrease in conductance of the solution caused by the reaction with the carbon dioxide. This reading is directly related to the total carbon content of the sample.

By pre-treatment of the sample (precipitation or acidification and sparging with carbon dioxide-free gas) to remove carbonates and dissolved carbon dioxide, the analyzer can be calibrated to reflect the content of organic carbon. In some cases these techniques have yielded poor results due to coprecipitation or loss of volatile constituents.

Two significant features of this carbon analyzer are the low price (less than $3,000) and the ability to inject large samples without overloading the detector, thus permitting the user to analyze samples containing suspended carbon without blending or filtering the sample. Minimum and maximum ranges are 0 to 100 ppm and 0 to 5,000 ppm. Although this analyzer has just recently been introduced, it should perform well either in the laboratory or for spot checking in the field. However, as a continuous monitor, it would suffer from the same disadvantages as the Capuano analyzer (US Patent No. 3,322,504), namely:

1.) Frequent replacement is required of chemicals and reagents used to treat the carrier gas or remove interfering compounds formed in the combustion chamber, or both. 2.) Unlike the Capuano system, it is not corrected for interferences from other acid gases (HCl, SO_2 and SO_3) that will be formed if the sample contains measurable quantities of chlorides or sulfates. 3.) Recalibrating the analyzer after changing detector reagent is necessary.

Sampling Considerations

Analyzer sampling systems are discussed in greater detail in Section 1.12 in this volume. However, a few thoughts concerning carbon analyzer sampling systems and sample handling are included here.

The organic compounds in aqueous solutions are subject to oxidation or bacterial decomposition. Care should therefore be taken to insure that the time between sample collection and the start of analysis is minimal. Collected samples should be kept in a cool environment and protected from atmospheric oxygen; quick freezing has proved to be satisfactory. Do not store in plastic containers and do not use mercury or acid as preservatives.

If the carbon analyzer is an onstream instrument, consideration should be given to the following factors:

1.) Lag time or transport time should be kept to a minimum if the analyzer is used to monitor spills. Part of the sample will probably have to by-pass the instrument. 2.) The sample may be filtered to remove solids if the amount of dissolved carbonaceous matter is required. If the solids to be filtered are extremely fine, such as silt, it may be necessary to use either a back-washed filter of fine mesh screens or even a sand filter. 3.) If suspended matter is to be analyzed, large volume sampling valves and reaction chambers may be required, as may an inline blender or homogenizer if the solids are fibrous. 4.) If the analyzer is to monitor potable water, or if drinking water is used to flush the system, check the local sanitation code because most local codes require that an air-gap device be used to connect the analyzer to the drinking water system.

If the analyzer is in a laboratory, first let samples containing appreciable amounts of suspended solids settle, before filtering or homogenizing. Otherwise most sampling valves and syringes will either become plugged or deliver non-repeatable samples. An ordinary household blender is often adequate for the job. Samples containing very high concentrations of carbonaceous matter, high salt content or strongly basic or acidic components may require dilution in order to obtain optimum results.

Carbon Analysis of Gaseous Samples

Most carbon analyzers have been developed to measure aqueous samples because of the need to control water pollution (see Section 5.1). However, several of these analyzers can be adapted to monitor gaseous samples (particularly air) since they all utilize a carrier gas (often air or oxygen), and most of them oxidize the sample to carbon dioxide before detection. Thus, if a continuous sample of air is pumped through the analyzer, many of the analyzers already described can monitor the carbon level in the air stream.

Owing to the high cost of the apparatus, this approach is justifiable only

1.) to measure the actual concentration of carbon, rather than (say) total hydrocarbons; and 2.) if the user already has a carbon analyzer and the gaseous samples are to be analyzed on a short-term basis, to obviate the purchase of another analyzer.

The most economical way to measure normal levels of carbon in gaseous samples is to use either the non-dispersive infrared analyzer or the HFID as a total hydrocarbons analyzer. (Both of these analyzers are described in greater detail in Section 5.2.) However, both these analyzers suffer from the same limitation, i.e., they are not equally sensitive to all carbon-containing molecules. Therefore, if it is necessary to know the actual number of carbon atoms present in the sample, all carbonaceous material must be converted to a common form, e.g., carbon dioxide, and this quantity measured.

Non-dispersive Infrared Detector

In 1957, research workers at Air Products began development of a hydrocarbons analyzer to monitor liquid oxygen processes for traces of hydrocarbons. Such an instrument was needed to detect the build-up of hazardous levels of hydrocarbons in the firm's air separation plants.

The carrier gas is eliminated because the vaporized liquid oxygen sample serves as the carrier gas (Figure 1.15a). The sample inject system is also eliminated, since the analyzer is continuous. A by-pass loop is connected between points A and D, and the sample continuously flows through the reaction chamber where copper oxide at a temperature of roughly 850°C converts all of the hydrocarbons to carbon dioxide. A filter at point B stops the particles of catalyst that would eventually befoul the infrared cell. The carbon dioxide-oxygen stream flows through the infrared analyzer which is sensitized to measure carbon dioxide. The system is calibrated by introducing standard samples at point C.

By switching the three-way valves at points A and D, unreacted sample may be passed through the detector to measure the level of carbon dioxide in the process stream, which is also an important operating parameter in air fractionation plants[1].

Hydrogen Flame Ionization Detectors

Carbon analyzers using HFID to monitor gaseous streams are examined in greater detail in Section 5.2. The two instruments most commonly used are the total hydrocarbons analyzer (THA) and the air monitors reflecting levels of carbon monoxide, methane and total hydrogen.

The THA is a continuous analyzer that contains only the HFID, gas flow control equipment and electronic hardware to condition the detector signal. The air monitor has two chromatographic columns. The first separates carbon monoxide, methane and air from the other components in the sample; the second separates carbon monoxide and methane. It also has a methanizer to convert carbon monoxide to methane (see Section 1.15, System 8), gas flow control equipment, HFID and electronic hardware. The system can be modified to measure trace quantities of carbon monoxide and carbon dioxide in gaseous hydrocarbon streams by the proper choice of chromatographic columns.

Carbon Analyzers Selection

The following considerations should aid the reader in arriving at the correct selection for a particular application:

1.) Must the data correlate with either BOD or COD? If yes, this immediately rules out all systems except 3, 6 and 7.) Is the analyzer primarily for laboratory use or online monitoring? Only system 9 was originally designed as an onstream instrument and has been proved in a wide variety of field installations. 3.) Is continuous read-out essential? If yes, only system 4 meets this requirement. 4.) Is high sensitivity (less than 10 ppm full scale) required? Only systems 5, 7 and 8 are suitable for this service. 5.) Must the cost be minimal? If yes, only system 12 currently falls within this category. 6.) Does the analyzer have to identify specific compounds? Only system 9 is capable of identification, although some experimental work has been done with the pyrographic technique to attempt to "fingerprint" specific compounds; 7.) Is there a need to measure particulate carbon? Only system 5 provides this body of data.

Although classic wet chemical and bioassay (COD and BOD) analytical techniques are still in widespread use, they are neither sufficiently accurate, specific nor rapid to permit real time control of waste treatment plants and pollution control systems. Unfortunately, most of the rapid instrument systems described do not correlate with either of these historical tests. As Professor Davis has observed[2] . . . "Established aerobic systems reaeration kinetics have been statistically proven to be directly dependent on and measurable by BOD values. As yet COD, TOC, and TOD values have not been applied specifically for the total evaluation of each of these criteria . . . Finally, the question must be raised again as to the meaning of these types of analyses . . . It appears then that dependence on the established standard tests for biochemical and chemical oxygen demand will continue until either as rapid but less expensive tests for carbon are developed or the full meaning of the carbon analysis results are set forth. . . ."

Some of the analyzers are more specific than others; some are more sensitive than others; and some are more accurate than others; however, if properly applied, calibrated and maintained, all of them will provide the user with valuable data.

REFERENCES

1. Ent's, W. L., "Automatic Continuous Measurement of Oxygen," Annuals of the N.Y. Academy of Sciences, Vol. 91, Article 4, pp. 888–900, June 2, 1961.
2. Davis, Ernst M., "BOD vs COD vs TOC vs TOD," Water & Wastes Engineering, Vol. 8, February, 1971, pp. 32–34 and 38.

1.16 BOD, COD AND BIOMETER SENSORS

Methods of Detection:	BOD DETECTION (a) Chemical analysis (a1) Wet chemical (Winkler) analysis (a2) Dissolved oxygen probe (b) Physical observation (b1) Manometric pressure change COD DETECTION (c) Chemical oxidation (c1) Titrimetric (c2) Colorimetric (c3) Thermal conductivity (d) Catalytic oxidation (d1) Infrared BIOMETER DETECTOR (e) Luminescence *Note:* In the feature summary below, the letters (a) to (e) refer to the listed analyzer types.
Sample Pressure:	Essentially atmospheric; nearly all are manually operated and are used to analyze grab samples.
Sample Temperature:	Near ambient
Sample:	Samples are usually grab samples.
Materials of Construction:	Glass, quartz, Teflon, polyethylene, polypropylene, polyvinyl chloride, tygon tubing.
Accuracy:	BOD ±15 to 20%. COD ±8%.
Ranges:	BOD—0.1 mg per liter and up. COD—5 mg per liter and up.

	Biometer—10^{-5} to 10^{-1} μg ATP per ml.
Response:	BOD—5 days to 20 days. COD—2½ hours to 3 hours. Biometer—0.5 second after sample injection.
Cost:	BOD—$200 to $20,000. COD—$100 to $20,000. Biometer—$4,860.
Partial List of Suppliers:	Delta Scientific Corporation (a1, a2, c1 and c2), E. I. du Pont de Nemours (e), Hach Chemical Co. (b, c1, and c2), Horiba, Ltd. (a2), Axel Johnson & Co., HAB (c2), Lockwood & McLorie, Inc. (c3), Nalco Chemical Co. (c1), Precision Scientific Co. (d), Technicon Industrial Systems (c2), J. M. Voith GmbH (b), Weston and Stack, Inc. (a2), Yellow Springs Instrument Co. (a2).

One of the most significant indicators of water pollution is the amount of organic material present. It is important because organic compounds in a waste water outlet or potable water supply can result in unacceptable taste, color and odor, and can cause toxicity. In the management of waste disposal facilities, the measurement of organic material furnishes an index of the historical efficiency of the treatment processes. In streams and lakes, organic compounds can reduce dissolved oxygen and make the water unacceptable for aquatic life and recreational use. In industrial wastes, the detection of high concentrations of organic matter may signal improper process control or a major spill or process leak.

Prior to the development of total organic carbon analyzers, two methods of analysis to evaluate the organic loading of a water or waste included:

1. *The Biochemical Oxygen Demand* (BOD) test. This method measures the quantity of biologically consumable substances in the liquid by allowing microorganisms to digest or stabilize the decomposable organic matter under controlled incubation conditions. The estimate is made by measuring the amount of oxygen consumed during 5 to 20 days.

2. *The Chemical Oxygen Demand* (COD) test. This method was developed in an attempt to devise a more rapid approach, using chemical oxidizing agents to stabilize the decomposable organic matter instead of the oxidizing ability of the microorganisms. The COD test takes 2½ to 3 hours.

For more details on the chemistry and calculations for the BOD and COD tests, see References (1), (2), and (3).

Most of the analyzers described in this section are of the laboratory type; in fact, most are completely dependent on the expertise of the analyst. However, since a few of the devices are automatic or semiautomatic, the discussions on both the BOD and the COD are divided into manual systems and automatic systems.

BOD Analysis

The BOD test is one of the basic analytical procedures used by environmental chemists and engineers to evaluate the effect of domestic sewage and industrial wastes on treatment plants and receiving bodies of water. It is the only test that measures the amount of biologically oxidizable organic matter present. It can determine the rates at which oxidation occurs. Despite its importance, the BOD test is incompletely understood and its quantitative data are often misinterpreted.

The test is essentially a bioassay measurement of oxygen consumed by living organisms (mainly bacteria) while decomposing the organic matter in a waste, under conditions as close as possible to those that occur in nature. Being a bioassay, it is extremely important that environmental conditions are suitable for the living organisms to function in an unhindered manner at all times. This means that toxic substances must be absent and that all accessory nutrients needed for bacterial growth, such as nitrogen, phosphorous and certain trace elements, must be present.

The BOD test can be best understood by studying oxygen utilization versus time curves. The typical BOD curve (Figure 1.16a) consists of three phases: lag, log growth and endogenous metabolism. The duration of the lag phase is predominately dependent on the bacteria population (Figure 1.16b), the acclimation of the bacteria seed (Figure 1.16c) and the degradability of the organic waste (Figure 1.16d).

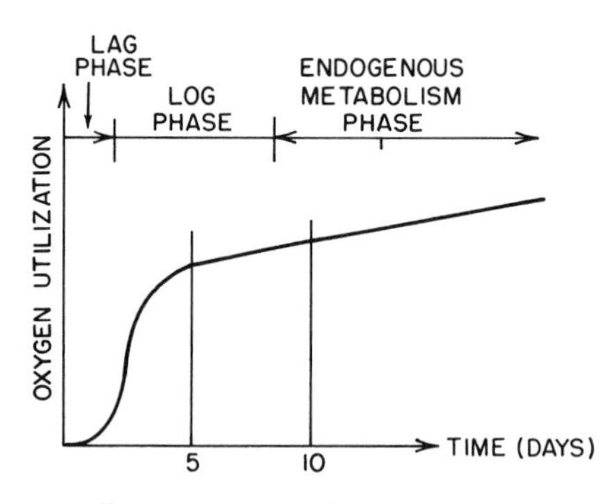

Fig. 1.16a Typical BOD curve

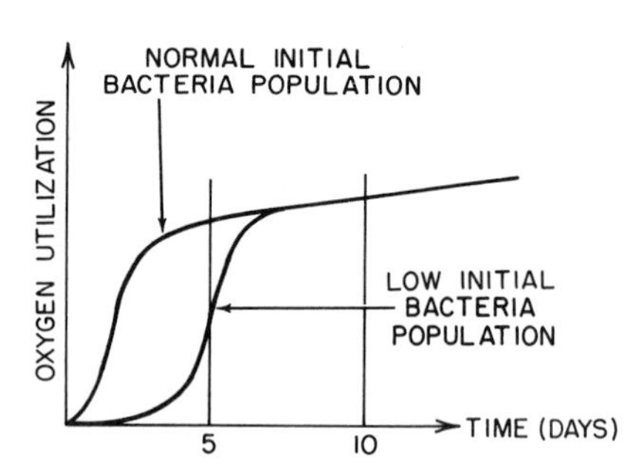

Fig. 1.16b Effect of low initial bacteria population on BOD test

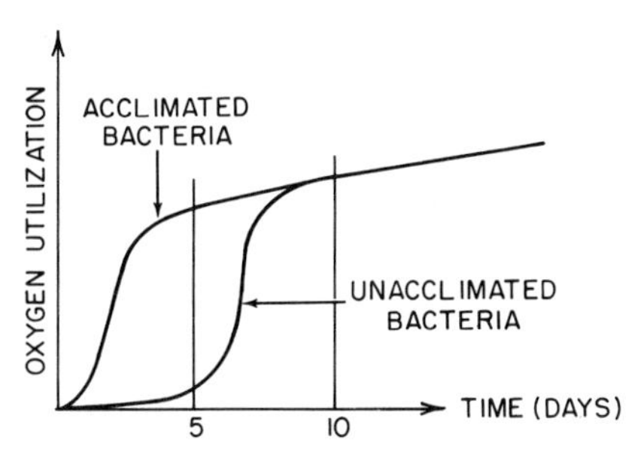

Fig. 1.16c Effect of unacclimated bacteria seed on BOD test

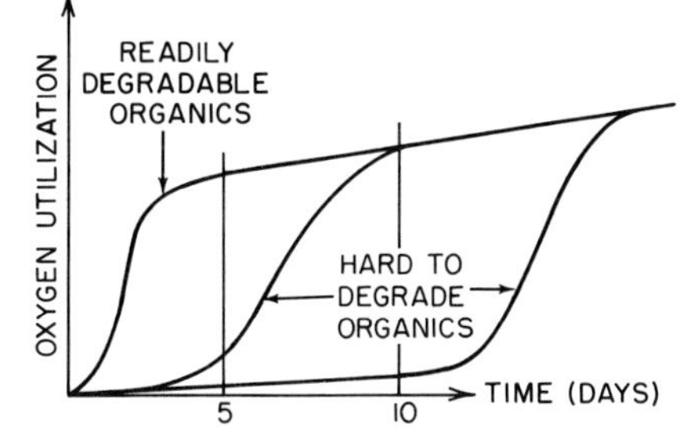

Fig. 1.16d Effect of hard-to-degrade organics on BOD test

The log growth phase follows the lag phase and must be virtually completed within the prescribed incubation period if reliable BOD results are to be obtained.

The amount of oxygen utilized during the endogenous metabolism phase can be either determined by direct measurement over a longer incubation period (if nitrification is inhibited) or approximated by calculations based on short-term, e.g., 5 days BOD data. The sum of oxygen consumed in all three phases represents the ultimate BOD. Because of the length of time in achieving ultimate BODs (20 days), environmentalists will usually settle for shorter time periods (5 days) and try to extrapolate the data to ultimate BOD values. There are many interpretation problems if the incubation period is too short or if too few points are measured. The analyst should be conscious of these problems before proceeding with the BOD test.

Interferences can adversely affect the results unless they are taken into account when conducting the BOD test. The immediate oxygen demand of organic or inorganic compounds, or both, that react with the oxygen, e.g., sulfite, should be satisfied prior to the BOD test. The oxygen demand of the biological seed (endogenous metabolism) must be distinguished from the oxygen demand of the waste water. Samples that are strongly alkaline or acidic must be neutralized to the pH range of 6 to 8 before BOD examination. Residual chlorine must be removed. The potential presence of other toxic materials must be tested for.

The dilution water should contain neither organic nor inorganic compounds that will introduce an oxygen demand. The dissolved oxygen consumed by nitrification (the bacterial oxidation of ammonia nitrogen to nitrites and nitrates) should be distinguished from carbonaceous oxygen demand (Figure 1.16e). Nitrification normally occurs between the fifth and seventh day of incubation, thereby contributing only a minor error in the normal five-day BOD test. However, on tests longer than five days, a serious interference occurs. Interference can be reduced by inhibiting the nitrifying bacteria with thiourea.

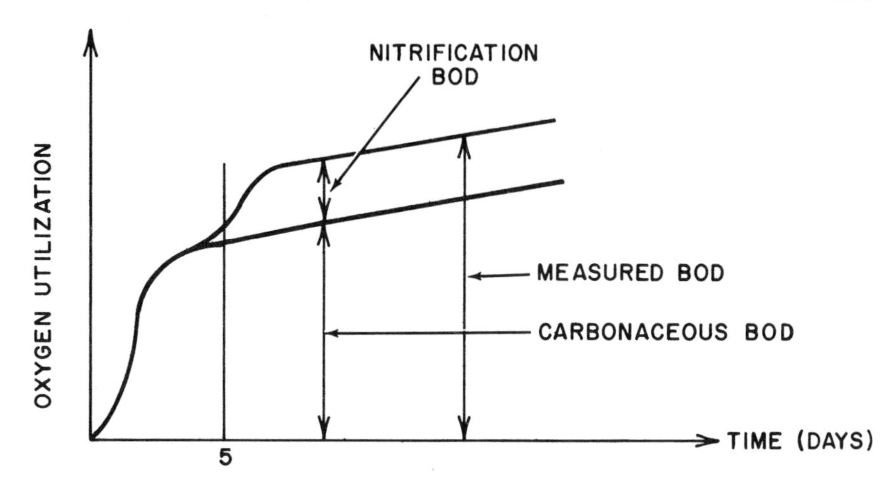

Fig. 1.16e Effect of nitrification on BOD test

Despite the complexity of the BOD test, reliable results can be obtained by a trained analyst. BOD data, based on an abundant number of points obtained during an incubation of longer than five days, furnish both a better understanding of the test samples and more reliable data.

Figure 1.16f illustrates the basic approaches to BOD analysis to be discussed. In both manual and semiautomatic systems the analyst has a choice between chemical and physical measurements to quantitate the amount of oxygen utilized during the incubation period.

Manual Systems

CHEMICAL ANALYSIS (*a*)

The publication *Standard Methods for the Examination of Water and Wastewater* published by American Public Health Association, Inc.,[1] describes the dilution technique with a Winkler titration as the referee method of

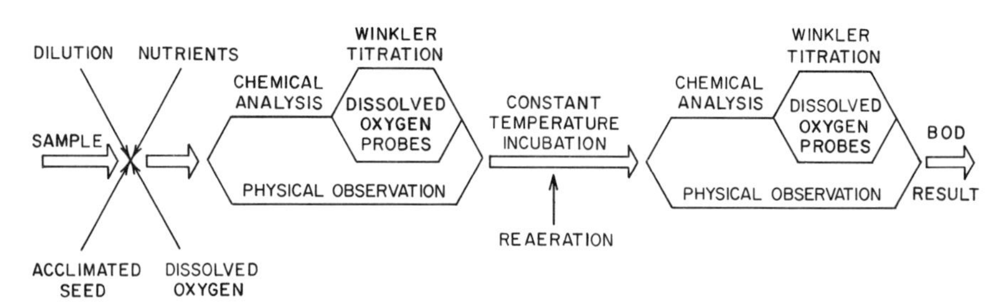

Fig. 1.16f Basic approaches to BOD analysis

measuring the dissolved oxygen concentration. Laboratory hardware that facilitate the test measurement include micropipets for rapid addition of reproducible quantities of Winkler reagents, specifically sized volumetric flasks for measuring precise quantities of sample from the BOD bottle to the titrating flask and semiautomatic titration equipment that dispenses the amount of titrant needed to reach the equivalence point. The Winkler titration method is still the most frequently used analytical check on BOD results obtained by other more advanced techniques.

The use of dissolved oxygen probes is gaining acceptance. Several manufacturers now fabricate oxygen probes that fit into BOD bottles with a minimum displacement of sample. Some probes are even equipped with their own stirrer so as to provide sufficient agitation to insure accurate dissolved oxygen analysis. The advantages of the probe include analysis that is easier and quicker than the Winkler titration, more important, the sample being analyzed is not destroyed as it is during the Winkler titration. Consequently, additional dissolved oxygen measurements can be obtained during incubation which will lead to a better understanding of the sample being tested. In addition, a larger amount of sample (less dilution) can be used since the samples can be periodically re-aerated. Winkler titrations necessitate the reduction of the organic material in a BOD dilution to a concentration which will not exhaust dissolved oxygen during incubation. Dilutions to a low concentration create problems of amplified background consumption of oxygen by dilution water impurities and biological seed. The background problem is alleviated if a larger quantity of organic material (lower dilution factor) is employed, but to do so a re-aeration is required. Re-aeration is by bubble aeration, using a fritted glass diffuser or by a coupled bottle shaking technique (Figure 1.16g).

PHYSICAL OBSERVATION (*b*)

Manometric methods for the determination of BOD have been used for many years but have never been completely satisfactory. They require as much skill to carry out as does the chemical analysis and their cost is considerably higher.

The principle of operation involves a measured sample of sewage or waste water being placed in a brown glass mixing bottle (Figure 1.16h). The bottle is then connected to a closed-end mercury manometer. Above the sewage or water sample is a quantity of air containing 21% oxygen. Bacteria in the sewage use up the dissolved oxygen to oxidize organic matter in the sample. The air in the closed sample bottle replenishes the dissolved oxygen, which results in a drop in air pressure. This pressure drop is registered on the mercury manometer and is read directly as ppm BOD.

During the test period (usually five days) the sample is continually agitated by a magnetic stirring bar. Carbon dioxide is produced by the oxidation of

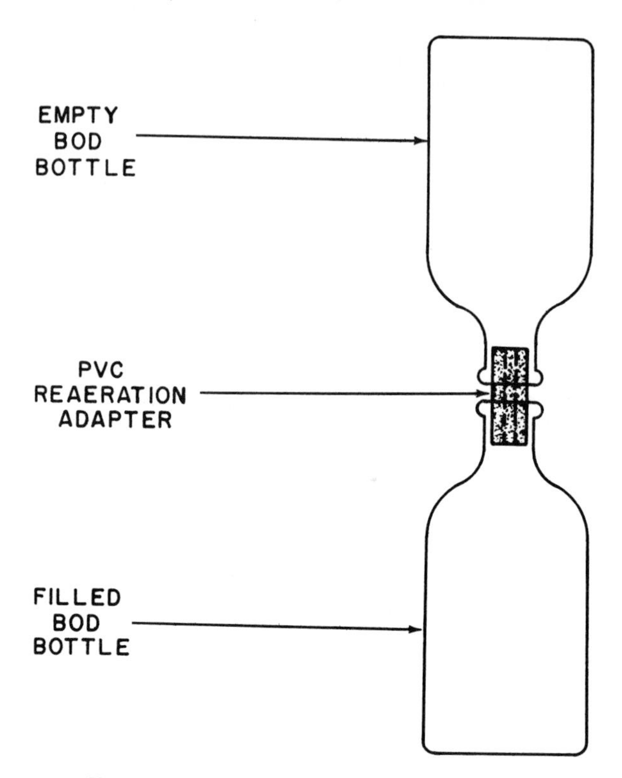

Fig. 1.16g Coupled bottle re-aeration apparatus

organic matter and must be removed from the system so as not to develop a positive gas pressure which would interfere with the measurement. This is accomplished by the addition of a few drops of potassium hydroxide solution in the seal cup in each sample bottle.

New hardware allows both trained and untrained technicians to determine the results with ease and accuracy. Higher precision is obtained because large representative samples are used. Closed-end manometers make the measurement independent of barometric changes. A physical change being observed, chemical laboratory analysis is not required to measure oxygen utilization. Since direct readings can be made from the manometer, numerous data points can be plotted to construct a BOD curve of oxygen utilization versus time. The large amount of information obtainable from a BOD curve makes the extra effort worthwhile.

Comparison of the manometric method with the chemical analysis (using a Winkler titration under controlled laboratory conditions), demonstrates that results and precision were the same.

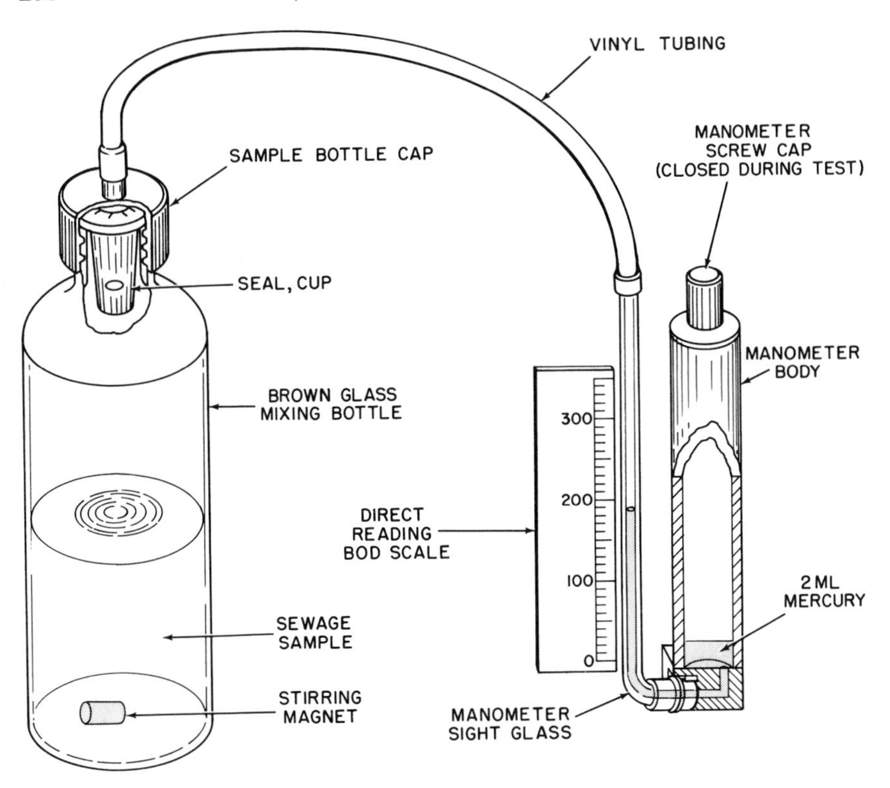

Fig. 1.16h Manometric BOD apparatus (showing one cell)

Semiautomatic Systems

CHEMICAL ANALYSIS (*a*)

A relatively recent development is an instrument that automatically measures the BOD of as many as 11 samples. Before the automatic features can be initiated, the samples have to be placed manually on the turntable and the instrument controls set. Automation is achieved both by automatic re-aeration at pre-set intervals in order to supply the sample with required oxygen and automatic dissolved oxygen measurements with a probe. The instrument incorporates all the advantages of the manual methods already discussed as well as automatically sensing (at pre-set intervals of 4 to 12 hours) the need for re-aeration of the sample to a pre-determined and measured dissolved oxygen level. The ability to re-aerate the sample automatically when the oxygen reserve is low eliminates the need for dilution. Therefore, improved precision and accuracy are obtained and direct BOD readings can be recorded.

The instrument consists of the measuring unit and the control unit. The measuring unit (Figure 1.16i) is a measuring station equipped with a dissolved oxygen probe, aerator water sealing mechanism, unplugging mechanism, sample bottles and turntable. They are housed in a thermostatic oven, maintained at 20°C. All steps are controlled by a pre-determined program. To take measurements, sample water is first poured into a bottle and placed over the turntable which accommodates as many as 11 bottles. When the measurement is initiated, the turntable rotates 30° and maintains that position for a timer-controlled interval.

When a sample is directly below the measuring position (Figure 1.16j), the water sealing and the unplugging mechanisms automatically unseal the bottle and remove the plug. Next, the dissolved oxygen probe is lowered into the bottle to take the first measurement of dissolved oxygen. The aerator then enters the bottle and aerates for five minutes, following which the sensor is immersed in the sample water for a second measurement, and the sensor transmits the measurement signal to the recorder for one minute. The unplugging and the water sealing mechanisms are actuated to lower the plug and seal the mouth of the bottle. These maneuvers are completed in 20 minutes, and the turntable rotates an additional 30° to bring the next sample bottle to the measuring position.

Thus, within 4 to 12 hours, these maneuvers are repeated, and results

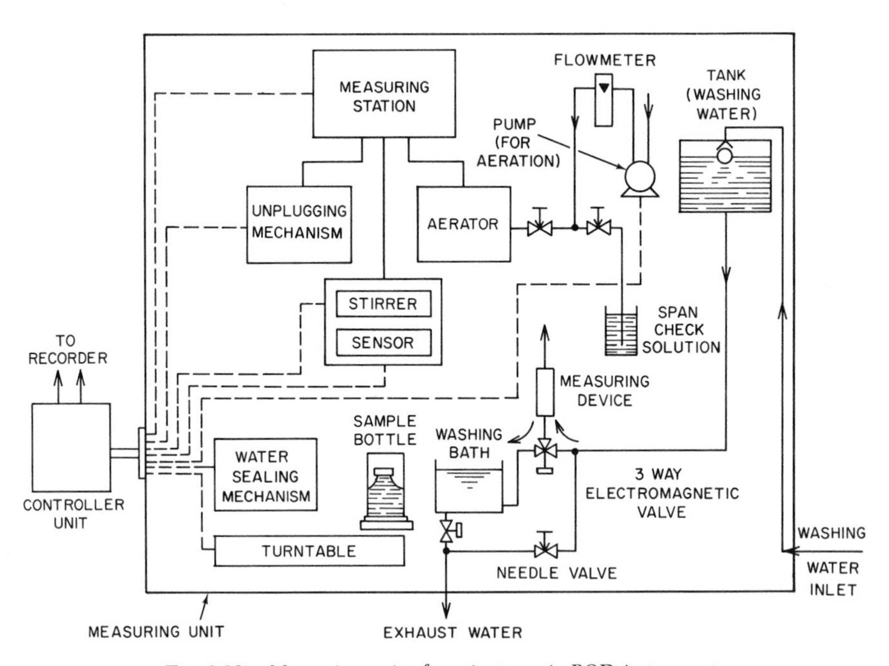

Fig. 1.16i Measuring unit of semiautomatic BOD instrument

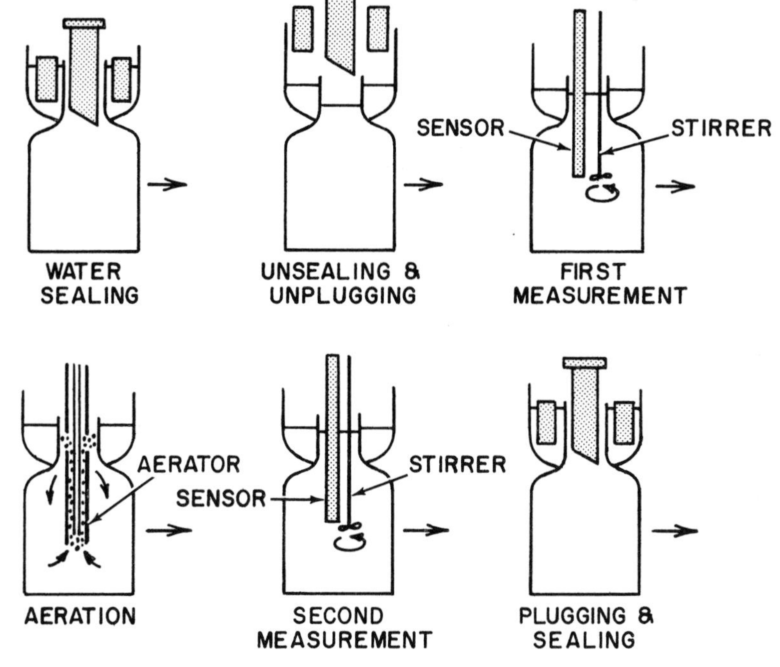

Fig. 1.16j Measuring program of semiautomatic BOD instrument

of dissolved oxygen measurements are recorded for each sample. BOD results are obtained by calculating the oxygen consumption during the incubation period from the difference in the measured values of dissolved oxygen.

A cleaning mechanism (Figure 1.16i) in the measuring station cleans the sensor and the aerator automatically after each sample is measured.

PHYSICAL OBSERVATION (*b*)

The samples are manually placed on the turntable. Automation is initiated by a differential pressure switch on each reaction vessel that continually monitors the addition of oxygen to the reaction vessel. The BOD results are obtained by measuring the time during which the oxygen generator is actuated to maintain a constant pressure in the reaction vessel. The differential pressure switch and oxygen generator on each reaction vessel maintain uniform dissolved oxygen levels. The uniformity of dissolved oxygen in the reaction vessels during the incubation period reflects high accuracy.

Each measuring unit (Figure 1.16k) includes a reaction vessel, an oxygen generator and a differential pressure switch, interconnected by plastic hoses. The sealed measuring system is not affected by barometric air pressure fluctuations. The sample is agitated in the reaction vessel by a magnetic stirrer so

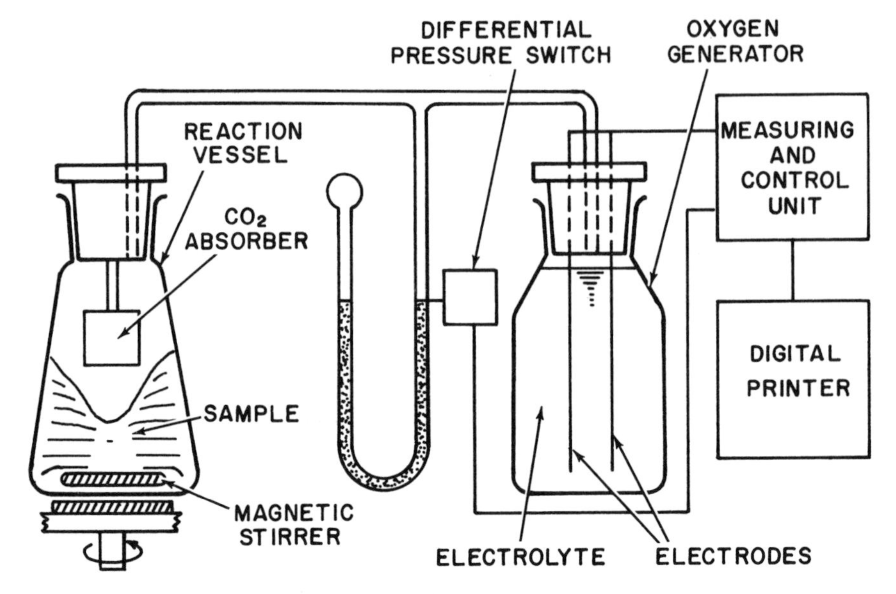

Fig. 1.16k Measuring unit of semiautomatic BOD instrument

that the oxygen required to decompose the organic matter may penetrate the substrate from the gaseous atmosphere above the sample. The carbon dioxide released as a result of the metabolic processes of the microorganisms enters the vapor space and is absorbed by the carbon dioxide absorber. This absorption creates a vacuum, a condition that is detected by the differential pressure switch. The switch actuates an amplifier which initiates the generation of oxygen until pressure is balanced.

The operating time of the oxygen generating set is measured and the quantity of the sample, the impulse sequence frequency and the amperage are chosen so that the counter directly indicates the BOD of the analyzed sample in mg per liter. The measuring time is preselected, at the end of which the unit shuts down automatically.

Orientation Table

Table 1.16 l summarizes the advantages of each BOD system.

COD Analysis

This test measures the pollution of domestic and industrial wastes in terms of the total quantity of oxygen required to oxidize pollutants to carbon dioxide and water, and is based on the fact that most organic compounds can be oxidized by strong oxidizing agents in acid solutions.

Organic matter is converted to carbon dioxide and water regardless of

Table 1.16l
ORIENTATION TABLE FOR
MANUAL AND SEMI-AUTOMATIC BOD DETECTION SYSTEMS

Advantages	*Manual*			*Semi-automatic*	
	Chemical		*Physical*	*Chemical*	*Physical*
	Winkler Titration	*Dissolved Oxygen Probe*	*Manometric*	*Dissolved Oxygen Probe*	*Manometric*
Sample not destroyed		√	√	√	√
Significant dilution not required		√	√	√ *	√ *
Intermediate oxygen utilization values can be obtained		√	√	√	√
Decomposition rates can be measured	**	**	√		√
Toxicity can be measured		**	**	√	√
No chemical analysis			√		√
Independent of barometric changes	√	√	√	√	√
Direct reading of BOD values			√ ***	√	√
Independent of weekend and holiday considerations				√	√
Unlimited number of samples can be tested at one time	√	√			
Best potential precision and accuracy				√	√
Inexpensive hardware	√	√	√		

* Sample is automatically re-aerated; no dilution required..
** Decomposition rates and toxicity determined with more difficulty than with the semiautomatic systems.
*** Assuming no dilution is required, but because there is no automatic re-aeration it may be necessary to dilute samples.

the biological assimilability of the substances. For example, glucose (rapidly biodegradable) and lignin (slowly biodegradable) are both completely oxidized. As a result, COD values are higher than BOD values, especially when significant amounts of biologically resistant organic matter is present. The inability to differentiate between biologically oxidizable and biologically inert organic matter is a serious limitation of the test in studying the effect of a certain waste on the environment.

The COD procedure oxidizes certain inorganic materials (nitrites, ferrous iron and sulfides) which are not affected by the BOD procedure, and does not oxidize certain organic materials (benzene, toluene and pyridine) which are readily oxidized by the BOD procedure. Chlorides can be oxidized in the COD test and thus can cause erroneously high results. Fortunately, this interference can be reduced by mercuric sulfate.

The major advantage of the COD test is the shorter time (3 hours vs. 5 days for BOD) for evaluation. Consequently, it is often a substitute for the BOD test. However, the fundamental differences between the two tests often make their data difficult to correlate.

Figure 1.16m illustrates the two basic approaches to COD analysis. In both manual and automatic systems the analyst has a choice among several detection measurements.

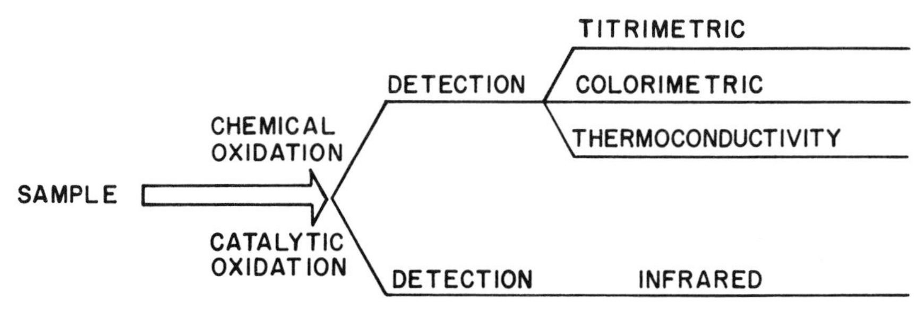

Fig. 1.16m Basic approaches to COD analysis

Manual systems

CHEMICAL ANALYSIS (*c*)

Standard Methods for the Examination of Water and Wastewater[1] recommends as the referee method for measuring COD a dichromate-sulfuric acid oxidation [equation 1.16(1)] with backtitration (ferrous ammonium sulfate) of the excess dichromate [equation 1.16(2)]. Maximum oxidation is achieved by heating the strongly acidic dichromate mixture in a flask with an attached reflux condenser. The condenser prevents the loss of volatile organic compounds.

Oxidation:

$$C_nH_aO_b + cCr_2O_7^{2-} + 8cH^{1+} \longrightarrow nCO_2 + \frac{a + 8c}{z} H_2O + 2cCr^{3+} \quad 1.16(1)$$

where $c = \frac{z}{3} n + \frac{a}{6} - \frac{b}{3}$

Titration:

$$6Fe^{2+} + Cr_2O_7^{2-} + 14H^{1+} \longrightarrow 6Fe^{3+} + 2Cr^{3+} + 7H_2O \quad 1.16(2)$$

Available laboratory hardware makes the test simpler to conduct. Automatic pipets for rapid addition of reagents in reproducible quantities, reflux hardware, and semiautomatic titration equipment (to dispense the exact amount of ferrous ammonium sulfate) are all available.

Field kits also simplify the COD test. The hardware is usually miniaturized for portability and the reagents come in easy-to-dispense packages.

The COD can be measured by titrating the remaining chromate or by colorimetrically measuring the green chromic ion present. The former method is more accurate but also more tedious than the latter. The latter method, when performed with an accurate photoelectric colorimeter or spectrophotometer, is more rapid, easier and of sufficient accuracy for most applications. Some field kits eliminate the reflux condenser and curtail reaction times to less than two hours. These analytical shortcuts may be appropriate in exceptional cases but ordinarily their results are inaccurate.

CATALYTIC OXIDATION (*d*)

Catalytic oxidation with oxygen, air or nitrogen as the carrier gas has been used for several years to measure the total carbon (TCA) or total oxygen demand (TOD) of liquid samples. (see Section 1.5). A relatively new instrument (similar to TCA and TOD instruments) based on the catalytic conversion of organic compounds, uses carbon dioxide as the carrier gas and a non-dispersive infrared analyzer sensitized for carbon monoxide.

The measurement is based on the fact that the number of moles of carbon monoxide produced is the same as the number of reactant moles of oxygen. Therefore, instrument readings of the carbon monoxide formed can usually be directly related to the referee determination for COD. Occasionally, the direct relationship may be weakened because some organic compounds are difficult to oxidize by the COD method, e.g., aromatics, but are readily oxidized catalytically.

This system uses a purifying carbon furnace to convert traces of oxygen in the carrier gas to carbon monoxide (Figure 1.16n). This level of carbon monoxide is the background or zero level for the analyzer. A 20 μl water sample is injected by hypodermic syringe and swept from the carbon furnace into the reaction chamber by the carrier gas. A platinum catalyst oxidizes the pollutants to carbon monoxide and water vapor. The water is stripped out in a drying tube and the reaction products then undergo a second platinum catalytic treatment. Carbon monoxide concentration is then measured by an integral non-dispersive infrared analyzer. The result can normally be converted to COD (milligrams per liter) by a calibration chart.

The major advantages are rapid results (complete test in 2 minutes), ease of operation, wide concentration range (10 to 300 mg per liter; higher levels by sample dilution), oxidation of chemically hard-to-degrade organic com-

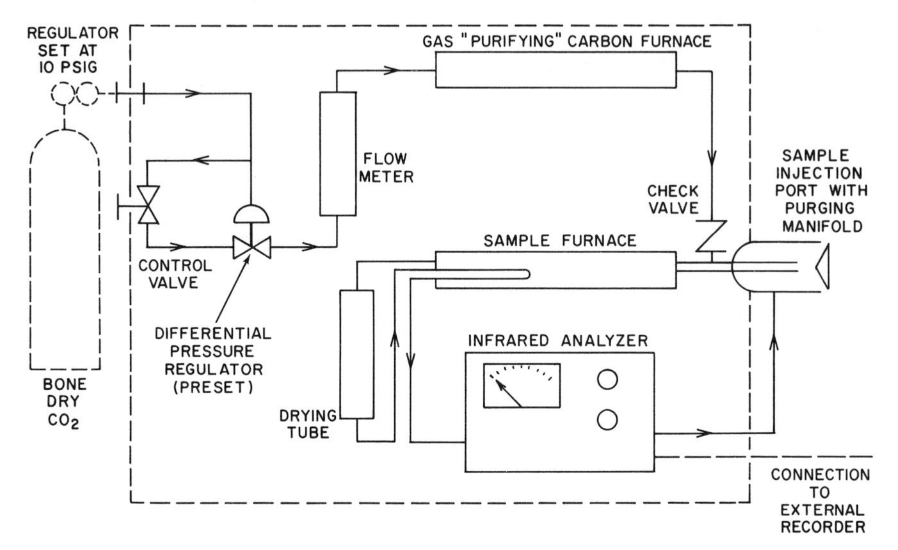

Fig. 1.16n Flow diagram of COD instrument based on catalytic oxidation and infrared detection

pounds, and the analysis can be performed without having to handle potentially dangerous chemicals (e.g., concentrated sulfuric acid).

Automatic Systems

CHEMICAL OXIDATION (*c*)

Analytical systems are available that automate the chemistry of the referee method outlined in *Standard Methods for the Examination of Water and Wastewater.*[1] The hardware (Figure 1.16o) includes the sampler, proportioning pump, mixing coils, heating bath, cooling coil, onstream colorimeter and recorder.

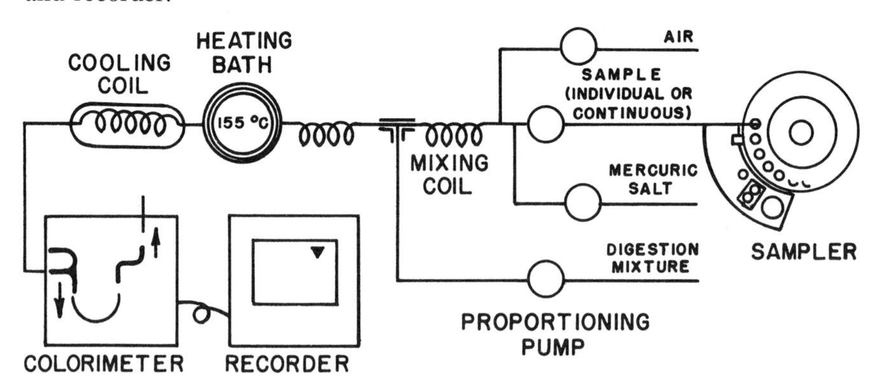

Fig. 1.16o Flow diagram of COD instrument based on chemical oxidation and colorimetric detection

A few changes from the referee method are needed to make the system reliable. A colorimeter (600 mμ) measures the concentration of green chromic ion instead of the (more precise) titrimetric procedure that measures the remaining chromate. Digestion occurs at a lower temperature and for a significantly shorter time ($<$ 15 minutes) than in the referee method (2 hours). These changes may affect the accuracy of the COD results with samples that contain hard-to-degrade organic compounds.

These automated instrumental systems have the advantage of rapid analysis (20 samples per hour) with a response time of only 15 minutes from sampling to reading. The manpower requirements are commensurately reduced as compared to the conventional COD analysis. The disadvantages are similar to those outlined under manual systems and include high cost of maintenance.

Another automatic system utilizes a thermal conductivity detector instead of a colorimeter. The oxidation still involves a heated mixture of dichromate-sulfuric acid. The advantages are similar to those previously discussed but maintenance, reagent consumption and cost are all higher.

Conclusions

The referee method for measuring COD involves a dichromate-sulfuric acid oxidation with backtitration of the excess dichromate. When portability is a requirement, field kits are available to simplify the COD test and still afford sufficient accuracy for most applications, and the analyst has a choice between titrimetric and colorimetric detection. The best laboratory system for rapid analysis and accuracy is based on catalytic oxidation followed by infrared detection. Occasionally, the cost of this system is prohibitive.

The automated systems based on chemistry identical to the manual referee method yield similar COD results. In automating the instruments, however, some compromises were made that may affect COD accuracy. The major advantage of the automatic analyzer is the speed of analysis, thus allowing the instrument to be used for control purposes. A major potential disadvantage is the high cost of maintenance.

Biometers

The inadequacies of biological control parameters have long been recognized by waste treatment research workers and plant operators. Monitoring methods inappropriate to measuring biodegradation and toxicity result in unit operations unacceptable to legislation requiring high quality effluent and incompatible with a high level of efficiency.

Current biological control (or monitoring) tests in biological waste treatment are divided into three categories including

1.) unit operation efficiency: BOD and COD reduction; 2.) biological population density: mixed liquor volatile suspended solids (MLVSS); and 3.) biological oxidative activity: sludge age, sludge yield, rate of BOD exertion as indicated by the first order rate constant

The time required to run the standard BOD test presents the most obvious difficulty for using it as a parameter in biological control. Also, BOD measurements only approximate unit efficiency between influent and effluent because of the different rate constants for the influent and effluent. The COD circumvents some of the disadvantages of the standard BOD test, but is considered unacceptable by those who question the relationship between material oxidizable under strong acid and high temperature conditions and material oxidizable by microorganisms under physiological circumstances.

Monitoring by MLVSS is generally accepted in the waste treatment field, although it is recognized as an indirect and incomplete measure of the viable sludge floc. A major problem of MLVSS measurements is that rapid reduction in the active biomass, e.g., from a toxic sludge, is reflected only slowly by changes in MLVSS values.

Sludge age is at best only a relative measure of sludge activity. Sludge yield reflects synthesis activity, but is a poor indicator of over-all activity since it does not measure the oxidative activity of the biomass. The problem in looking at the first order BOD rate constant is the difficulty of distinguishing between the combined effects of the organic material being degraded and the oxidative ability of the organisms to utilize the organic material. Also, there is little reason to expect biological activity in long-term, dilute and quiescent BOD tests to reflect sludge activity in short-contact time, highly concentrated and turbulent activated sludge units.

Owing to the inadequacies of current biological control tests, a significant amount of research is being conducted in an effort to develop a better control test. Test measurements currently being considered include dehydrogenase, oxygen uptake and adenosine triphosphate (ATP). Presently, a commercial instrument is available for the ATP measurement only.

Luminescence (e)

The instrument is an extremely sensitive photometer (detects 1×10^{-13} grams of ATP per 10 μl injection), specifically designed to afford increased capabilities in research-oriented and routine measurement of bioluminescent and chemiluminescent reactions. ATP reacts with luciferin and luciferase (extracted from firefly lanterns) to yield a complex that subsequently reacts with oxygen, forming an excited state followed by light emission. The number of light quanta emitted is directly proportional to the initial ATP content of the reaction system.

Because of its fundamental role in cellular processes, ATP seems to be an attractive parameter to study the response of cells to their environment.

Results indicate that the ATP analysis is a quantitative measurement of microbial biomass, as well as of activity. The approximation of cell populations is based on the proportionality between the quantity of extractable cellular ATP and the number of cells of a given size, e.g., bacteria. The extraction and measurement can be done within minutes after sampling, and the process is therefore, indicative of the quantity of life in the sample, not of the life that can be cultured under artificial conditions. The measurement does not detect dead cells. The ATP is a specific indicator of cell viability; for example, 15 minutes after addition of a lethal dose of mercuric chloride, no ATP was present.

Conclusions

The measurement of unit operation efficiency, biological population density and biological oxidative activity by conventional tests is difficult, inaccurate, time consuming and occasionally impossible. Much current research is devoted to developing better tests. Tests under consideration include a dehydrogenase measurement, an oxygen uptake measurement and an ATP measurement. An instrument commercially available quantitatively measures ATP with an extremely sensitive photometer. Measurement of cellular ATP is simple and rapid; the described technique is highly sensitive and precise. The biological significance of ATP is that it can measure the quantity and activity of living matter.

REFERENCES

1. *Standard Methods for the Examination of Water and Wastewater,* 13th Edition, American Public Health Association, Inc., New York, 1970.
2. *Annual Book of ASTM Standards: Water; Atmospheric Analysis,* Part 23, American Society for Testing and Materials, Philadelphia, 1970.
3. *FWPCA Methods for Chemical Analysis of Water and Wastes,* FWPCA Analytical Quality Control Laboratory, Cincinnati, 1969.

1.17 MEASUREMENT OF ELECTRIC QUANTITIES

Types of Meters and Accessories:

(a) Permanent magnet, moving coil meter
(b) Rectifier type meter
(c) Moving iron vane meter
(d) Electrodynamic meter
(e) Current transformer
(f) Shunt
(g) Potential transformer
(h) Resistor

Note: In the summary below, the letters (a) to (h) refer to the listed equipment. A table with additional information will be found at the end of this section.

Meter Selection for Various Measurement Requirements:

AC current (c,b), high range AC current (c,b with e).
DC current (a), high range DC current (a with f).
AC voltage (c,b), high range AC voltage (c,b with g).
DC voltage (a), high range DC voltage (a with h).
1 or 3 phase AC power (d with e,g).
DC power (d), high range DC power (d with h)

Accuracy:

±0.25% of rating (f).
±0.3% to ±0.6% of secondary rating (h).
±0.6% to ±1.2% of secondary rating (e).
±1% to ±2% of full scale (a,c,d).
±3% of full scale (b).

Approximate Cost:

$15 (e,f), $20 (a,c), $30 to $40 (b,h), $70 (g), $200 to $300 (d)

Partial List of Suppliers: API Instruments Co. (a,b,c,d), Bitronics Inc. (a,b,c,d,f), Compagnie Générale de Metrologie (a,b,c), Dowa Trading Co., Ltd. (a,b,c,d,f), Esterline-Angus Div., Esterline Corp. (All), Ferranti Ltd. (All), General Electric Co. (All), Hickok Electrical Instruments Co. (a,b,c,d,f), Parker Instrument Corp. (a,b,c,d,e,f,g), Simpson Electric Co. (All), Singer Co., Instrumentation Div., (All), Triplett Electrical Instrument Co. (a,b,c,d,f), Voltron Products Inc. (a,b,c,d), Westinghouse Electric Corp., Relay and Instrument Div. (All), Weston Instruments Division, Weston Instruments Inc. (All).

Basic Meter Movements

Four types of meter movements that commonly measure basic electric quantities include 1.) permanent magnet moving coil; 2.) rectifier; 3.) moving iron vane; and 4.) electrodynamic.

Although details of construction vary widely among the different meter movement designs, essentially the same principle underlies them all, i.e., rotation of the pointer is caused by a current-induced magnetic field.

Permanent Magnet Moving Coil Movements

More commonly known as the D'Arsonval type, these movements consist of a pointer attached to a coil of fine wire suspended between the poles of a permanent magnet (Figure 1.17a). Current through the coil creates a magnetic field which reacts with the field of the permanent magnet, causing a deflection of the coil proportional to the amount of current. The coil shaft is usually mounted on jeweled bearings, and hairsprings provide the force necessary to restore the pointer to zero. An alternate method is to support the coil by flat metal bands attached to a supporting framework. The bands carry current to the coil and furnish the necessary restoring force. Permanent magnet, moving coil meters are accurate and sensitive, consume small amounts of power, possess linear display scales and are widely used to measure DC current and voltage.

Rectifier Type Movements

These movements consist of a permanent magnet, moving coil type sensor combined with silicon rectifiers arranged in a full wave bridge for use on AC

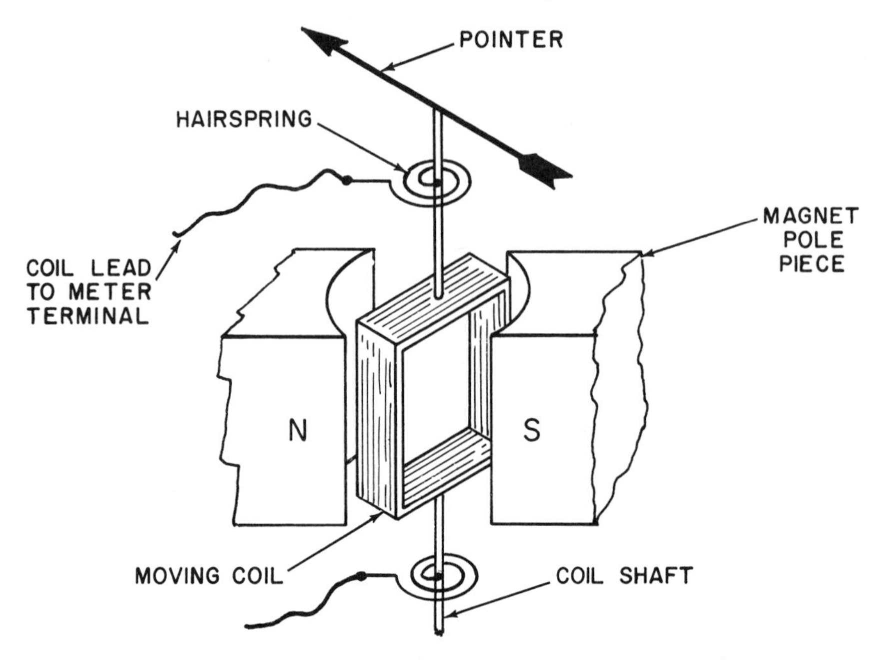

Fig. 1.17a Internal construction of a permanent magnet moving coil meter

circuits (Figure 1.17b). Rectifier current ratings limit these movements to relatively low ranges of AC current and voltage measurements. A wide band of frequencies can be accommodated because accuracy is not affected by frequency. Scales are essentially linear with some crowding at the lower end due to changes in rectifier values at very low currents. Meters must be calibrated to read effective or root mean square values since the movement responds to average AC values. Sensitivity is very high; power consumption is very low.

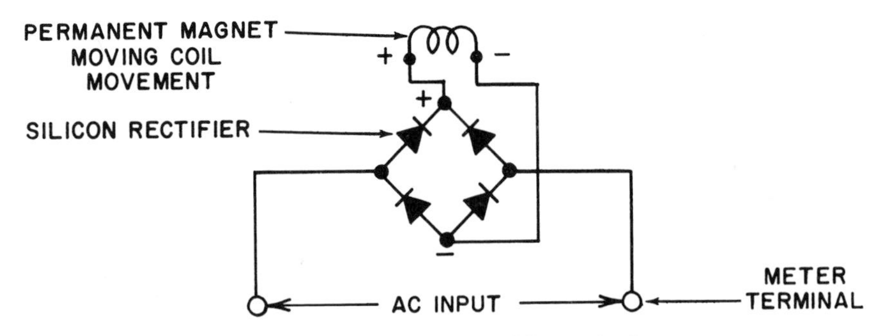

Fig. 1.17b Rectifier movement with full wave bridge circuit

Moving Iron Vane Movements

These movements consist of two cylindrical soft iron vanes mounted within a fixed current carrying coil (Figure 1.17c). One vane is held immobile and the other is free to rotate, carrying with it the pointer shaft. Current in the coil induces both vanes to become magnetized, and repulsion between the similarly magnetized vanes produces a proportional rotation. A hairspring provides restoring force. Only the fixed coil carries current and therefore the movement may be constructed so as to withstand high current flows. Moving

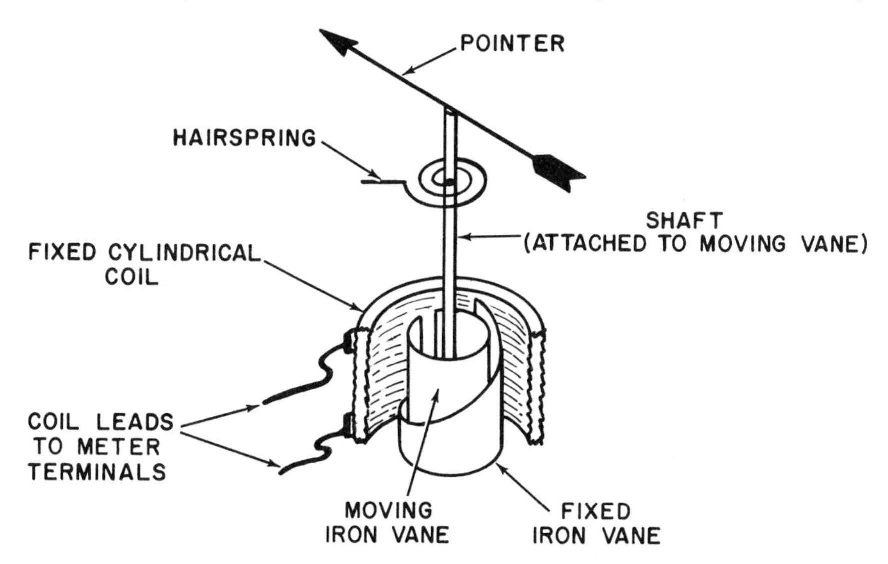

Fig. 1.17c Internal construction of a moving iron vane type meter movement

iron vanes may be used for DC current and voltage measurement, but small DC errors due to residual magnetism cause them to be more widely utilized for AC current and voltage detection even though sensitivity is low and power consumption is moderately high. These movements indicate effective or root mean square AC values and are subject to minor frequency errors only. Scales are non-linear and somewhat crowded in the lower third, since pointer deflection is approximately proportional to the square of the coil current.

Electrodynamic Movements

These movements are similar to the permanent magnet, moving coil type elements except that the magnet is replaced by two fixed coils which produce the magnetic field when energized (Figure 1.17d). These movements can measure AC or DC current, voltage and power. Cost and performance compared with other movements restrict the use of this design to AC and DC power measurement. When used for power measurement its scale is linear

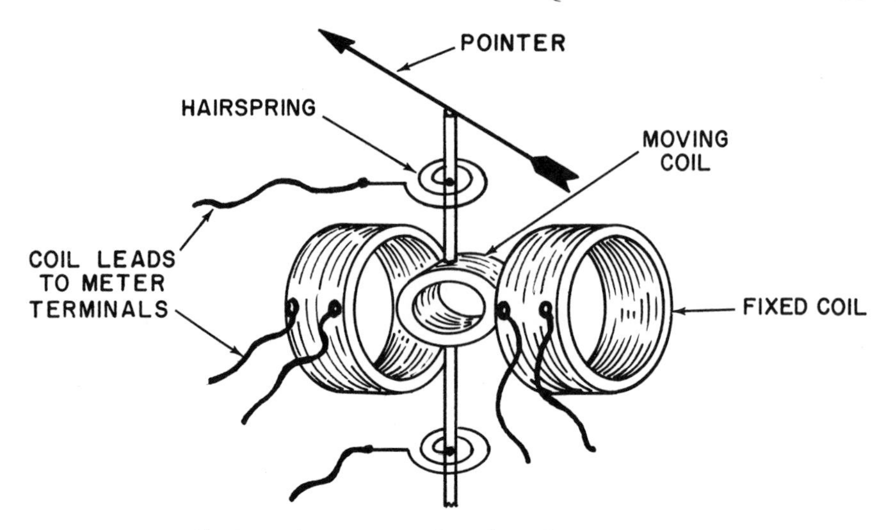

Fig. 1.17d Internal construction of an electrodynamic meter movement

(calibrated in average values for AC). Its accuracy is high but its sensitivity is low.

Additional Meter Movements

These are less commonly used meter movements or those that are restricted to specific applications.[1-5]

Induction type movements depend for their operation on the reaction between the magnetic field of a fixed coil and the current induced in a solid moving disk or cylinder. These movements may be used for current, voltage or power measurement.

Electrothermic movements usually consist of one or more dissimilar metal thermocouple junctions in contact with a heater. Current through the heater causes the thermocouple to produce a DC voltage which may be measured with a permanent magnet moving coil meter. These instruments are suitable for measurements of current, voltage and power.

Electrostatic movements utilize the forces of attraction or repulsion between electrically charged parts to deflect a pointer and measure voltage only.

Electronic meters with vacuum tube or transistor circuitry utilize a wide variety of circuit arrangements to permit measurement of current, voltage and power.

Transducers permit the use of the permanent magnet moving coil meter for a broad range of AC and DC current, voltage and power measurements. Transducer designs vary widely; however, they generally are one of the following types: balanced bridge circuits, analog converters or multipliers, hall effect transducers, saturable reactors or resistance divider networks. (See Sections 2.1, 6.2 and 6.3 in Volume II.)

Current Measurement

AC Current

AC current indicators are almost exclusively of the moving iron vane type owing to their wide range and low sensitivity to frequency variations (less than the electrodynamic type) and waveform errors.

This meter movement in ranges from 1 to 50 amperes full scale can be used for higher current ranges by the addition of external current transformers. Operating voltage is limited to approximately 750 volts. Above this level a current transformer must be used even if the current is within the meter rating. Meters calibrated for 60 Hz are usable from 25 Hz to 400 Hz with an additional error of only ½% full scale and may be recalibrated for use at levels as high as 1,000 Hz. For measurement of small currents at frequencies above 400 Hz or, when variable frequencies must be measured, the rectifier type meters should be used. This type of meter movement is available in ranges from 500 microamperes to 20 milliamperes full scale and may be used from 20 Hz to 10,000 Hz without loss of accuracy on sine wave circuits.

Current Transformers

For current measurement above the range of moving iron vane meters a current transformer should be inserted in the circuit in order to provide the proper ratio between the meter movement and the measured current (Figure 1.17e). Most current transformers are designed to deliver 5 amperes at the secondary terminals when full primary current is flowing. The primary winding can be the current carrying conductor passed through the center of the secondary winding in the form of a single turn coil, which design requires no mechanical connection or break in insulation.

Selection of the current transformer ratio should be based on the closest standard primary rating over the maximum current to be measured. Because there are large intervals between standard primary ratings for current transformers, it is often desirable to modify the transformer ratio by passing the primary conductor through the center of the secondary winding two or more times. A typical current transformer with a rating of 100 to 5 amperes has a normal ratio of one primary turn to 20 secondary turns or 1 to 20. Passing the primary conductor through the center of the secondary winding twice (Figure 1.17f) will increase the primary turns to two and the ratio will become 1 to 10, giving a new rating of 50 to 5 amperes. By taking additional primary turns the rating may be modified to correspond more closely with the desired scale.

The new primary rating is determined by dividing the original primary

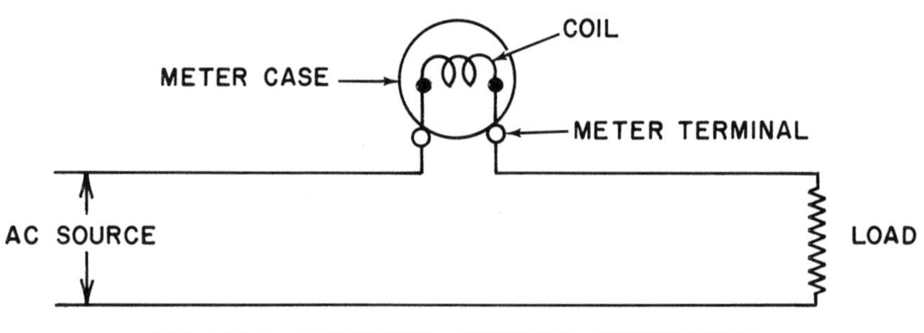

(1) SELF-CONTAINED AMMETER CONNECTION

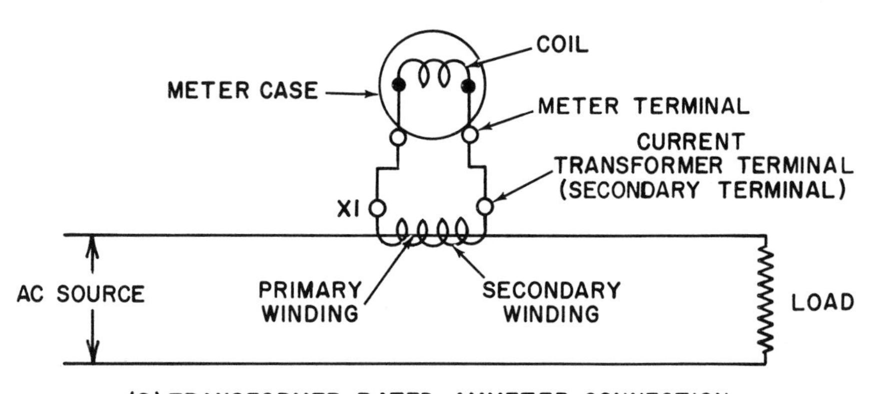

(2) TRANSFORMER RATED AMMETER CONNECTION

Fig. 1.17e AC ammeter connections

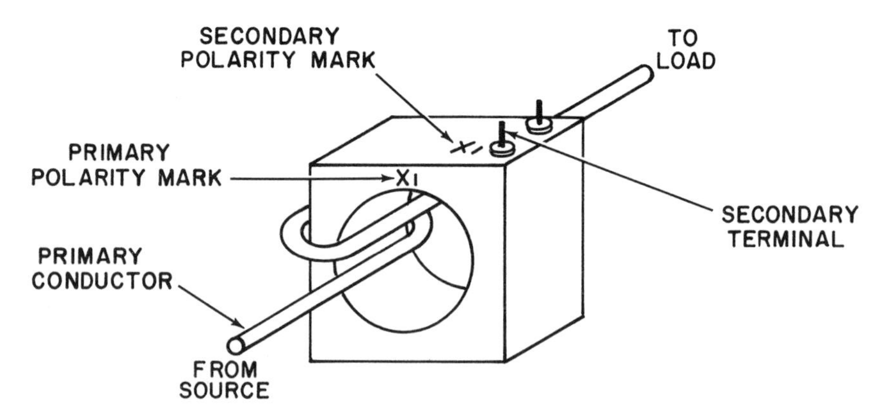

Fig. 1.17f Current transformer with two primary turns added

rating by the number of turns taken. Small adjustments to the transformer ratio may be made by passing one of the secondary leads through the center of the secondary winding so as to effectively add or subtract secondary turns (Figure 1.17g). A current transformer with a 100 to 5 ampere rating has 20 secondary turns. Therefore, the addition of one secondary turn changes the turns ratio to 1 to 21, for a new rating of 105 to 5 amperes. By adding or subtracting the required number of secondary turns, the rating may be adjusted to almost any value with the new primary rating being determined by multiplying the total number of secondary turns by the secondary rating which is 5 amperes in all cases. When both primary and secondary turns are modified, the new primary rating may be calculated by equations 1.17(1) and 1.17(2).

$$\text{Original secondary turns} = \frac{\text{Original primary turns}}{\text{Secondary rating}} \tag{1.17(1)}$$

$$\text{Adjusted primary rating} = \frac{(\text{Total secondary turns})(\text{Secondary rating})}{\text{Primary turns}} \tag{1.17(2)}$$

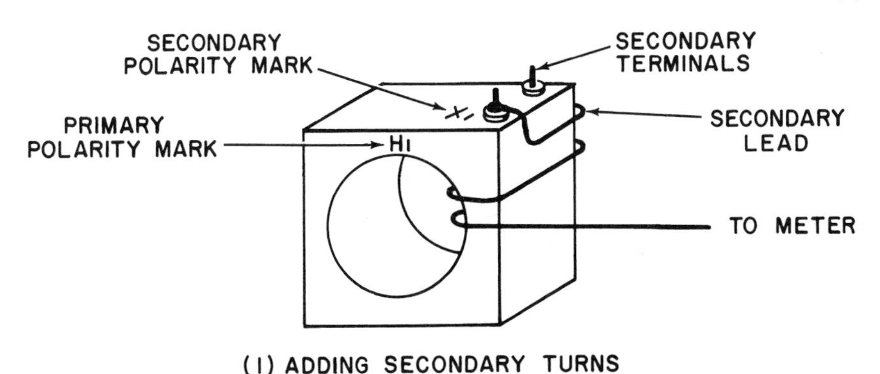

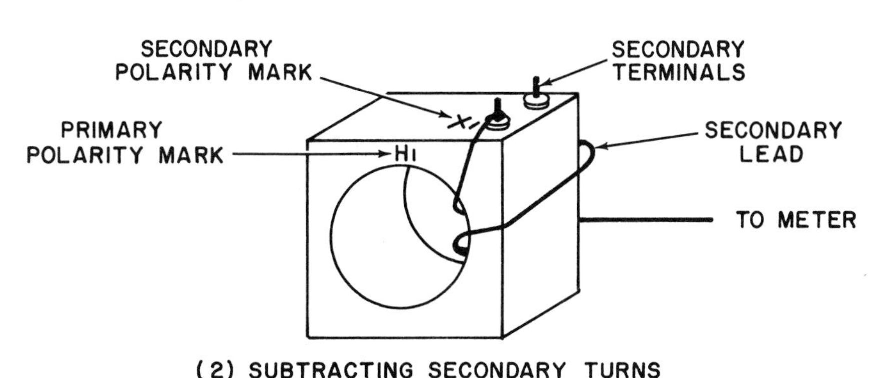

Fig. 1.17g Modifying the number of secondary turns on a current transformer

Primary and secondary polarity markings are provided on current transformers with the primary identified as H1 and the secondary terminal as X1. To add secondary turns one should connect a lead to terminal X1 and pass it through the center of the secondary winding from the side opposite the H1 polarity mark as indicated in Figure 1.17g. To subtract secondary turns, connect a lead to terminal X1 and pass it through the secondary winding from the same side as the H1 polarity mark. Current transformer accuracies are expressed as the maximum error for a particular load class as specified by USASI. Although load accuracy data are available for the meter manufacturers, it is seldom necessary to check the meters because if they are listed as transformer rated they are also designed so that current transformer accuracy will be maintained when connected with #14 AWG leads at lengths up to 150 ft and #12 AWG leads at lengths up to 250 ft.

Current transformers are generally insulated for use on systems up to 600 volts, but 600 volt transformers may be used at higher voltages if the primary conductor is fully insulated. Current transformers designed for ammeters are available with 0.6% and 1.2% accuracy and in the following standard primary ratings: 100, 150, 200, 250, 300, 400, 500, 600 and 800 amperes.

DC Current

DC current indicators are usually of the permanent magnet moving coil type to take advantage of the low power consumption and inherent linear scale factor while avoiding the residual magnetic errors of the iron vane types and the slower response of electrodynamic movements. Current carrying capacity of the internal connections to the coil limit the self-contained meters to a range of 20 microamperes to 1 ampere full scale. So-called self-contained meters are available for up to 50 amperes; however, they are actually low range movements with an internal shunt. Above this range external shunts are employed. Meters are suitable for use on up to 600 volt circuits. Above this value other types of meter movements or transducer rated meters are generally used.

Ammeter Shunts

DC ammeter ranges can be greatly extended by connecting a shunt of the proper resistance in parallel with the meter movement so that a specific proportion of the current passes through the meter movement, the remainder being carried by the shunt (Figure 1.17h). Meters designed for shunts are rated either at 50 millivolts or at 100 millivolts, depending on the voltage drop across the meter terminals at full scale deflection. When a shunt is used it must have a corresponding millivolt rating. The actual resistance value of the proper shunt does not have to be known since shunts are rated on the basis of full scale meter current and are available from 5 amperes to 20,000 amperes,

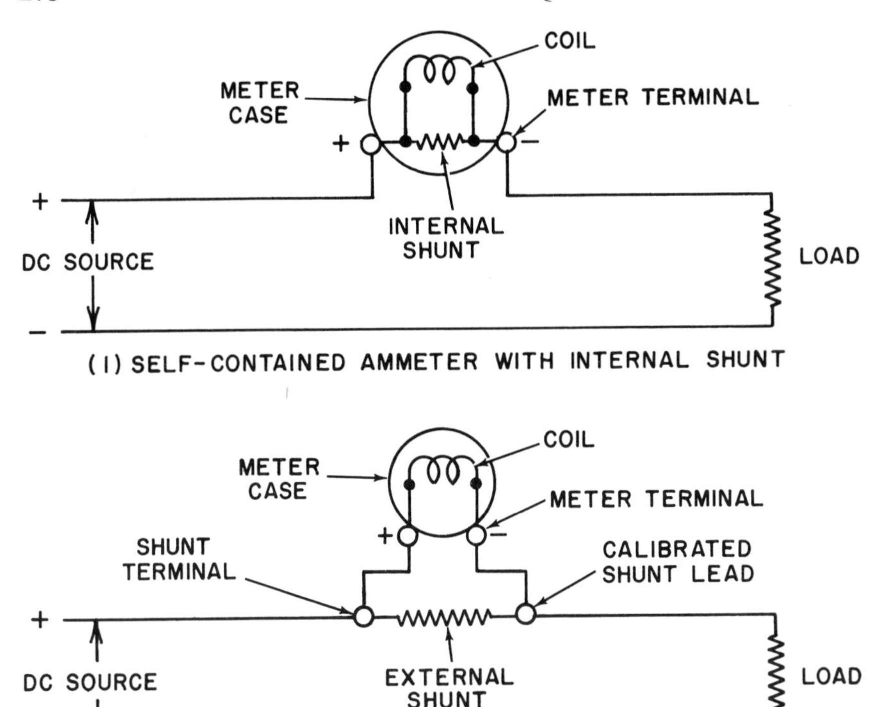

Fig. 1.17h DC ammeters with internal and external shunts

calibrated to ±0.25% accuracy. Shunt rated ammeters are calibrated for a particular shunt lead resistance, factors which vary from one manufacturer to another, and the correct calibrated shunt leads should be used for maximum accuracy or the meter recalibrated for use with different leads. Non-calibrated shunt leads of #14 AWG wire have an additional error of ½% for 50 millivolt meters at distances up to 25 ft or for 100 millivolt meters at distances up to 40 ft.

Voltage Measurement

AC Voltage

Indicators for AC voltage measurement may be of the moving iron vane type or the rectifier type since each movement has certain advantages for a particular voltage range. Moving iron vane voltmeters have the advantage of high accuracy and wide range although their sensitivity is poor. Voltmeter sensitivity is determined by the coil current required for full scale detection and is expressed in ohms per volt since the coil current is inversely proportional to the total resistance. A series resistor is generally installed within the meter case to provide the total resistance required by the movement so that full voltage equals full scale meter current. Self-contained moving iron vane voltmeters are available in ranges from 3 volts to 600 volts full scale with corresponding sensitivities ranging from 4 ohms per volt to 250 ohms per volt. Meters are usually calibrated for 60 Hz; however, they may be used in circuits from 25 Hz to 1,000 Hz when calibrated for a specific frequency.

Rectifier type meters have an advantage in that their high resistance coils permit greatly increased sensitivity although they are not as accurate as the moving iron vane type, owing to varying rectifier losses. They are available in ranges from 3 volts to 800 volts full scale with standard sensitivity of 1,000 ohms per volt at all ranges. Calibrated for 60 Hz these meters may be used from 20 Hz to 10,000 Hz without loss of accuracy, although they show considerable errors if used in circuits with wave shapes other than sine wave.

Potential Transformers

For measurements on circuits above 600 volts or on lower voltage circuits in which isolation is desirable, a moving iron vane meter is used with a potential transformer to provide the proper ratio between circuit voltage and meter movement (Figure 1.17i). Meters used with transformers are usually provided with 150 volt movements having 100 ohms per volt sensitivity. The corresponding potential transformers are always furnished with a ratio based on a 120 volt secondary to permit indication of small over-voltages. For this reason potential transformers must be selected on the basis of maximum circuit voltage and not on full scale meter reading. Potential transformer accuracies are specified by the USASI in a manner similar to that of current transformer accuracy. Checks are seldom required because accuracy is not affected by leads as long as several hundred feet. Potential transformers with primary ratings from 120 volts to 14,400 volts are available with 120 volt secondaries and with accuracies of either 0.3% or 0.6% at frequencies from 50 Hz to 400 Hz.

DC Voltage

Almost all meters for DC voltage measurement are the permanent magnet moving coil type because of its high sensitivity, linear scale and wide range. Self-contained DC voltmeters are available in ranges from 1 volt to 600 volts full scale with a standard sensitivity of 1,000 ohms per volt in all ranges. Above this range self-contained meters are not available because the internal resistors would become very high and resistor losses would result in excessive heating. External resistors can be used on circuits above 600 volts to extend the measurement range of the permanent magnet moving coil meter. Circuit

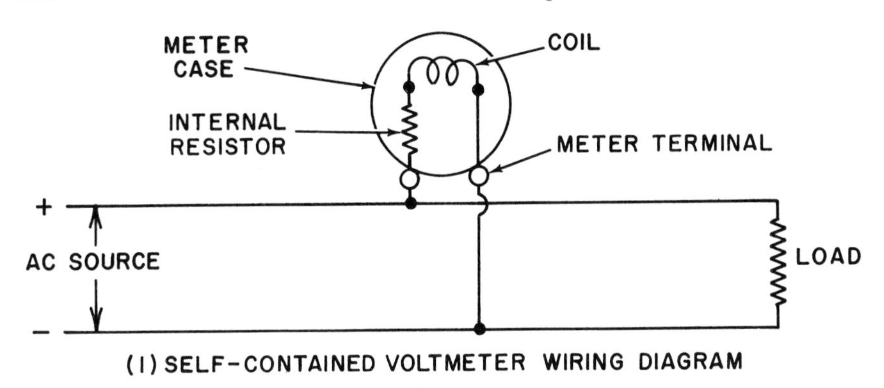

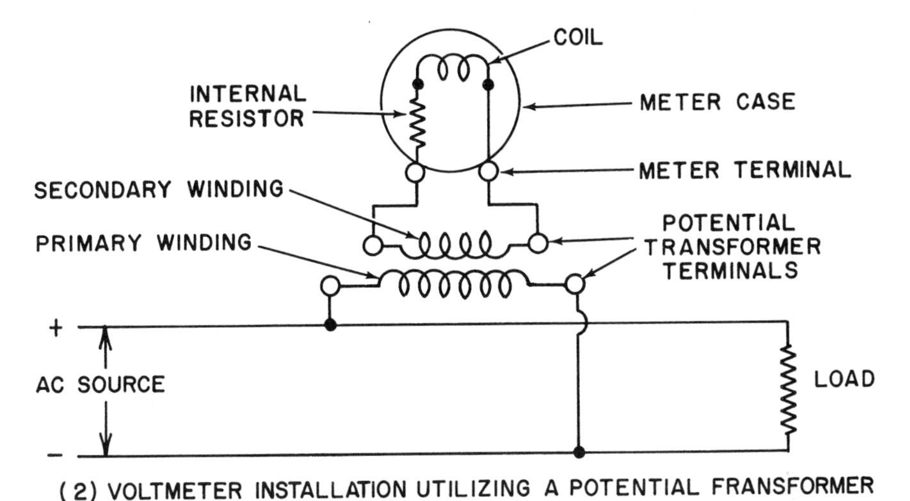

Fig. 1.17i AC voltmeter installations

isolation is not possible in this arrangement and the meter will be damaged if the external resistor is accidentally shorted.

Movements with sensitivities up to 20,000 ohms per volts are available, although not in all ranges.

Voltmeter Resistors

Inserting an external resistor of the proper value in series with a permanent magnet moving coil voltmeter movement (Figure 1.17j) permits measurement of higher voltages or increased sensitivity at low ranges. Meters with external resistors generally have 1 milliampere movements and 125 ohms internal resistance. Standard resistors designed to afford a sensitivity of 1,000 ohms per volt may be selected on the basis of full scale meter reading without the necessity of calculating resistance. Standard resistors are available

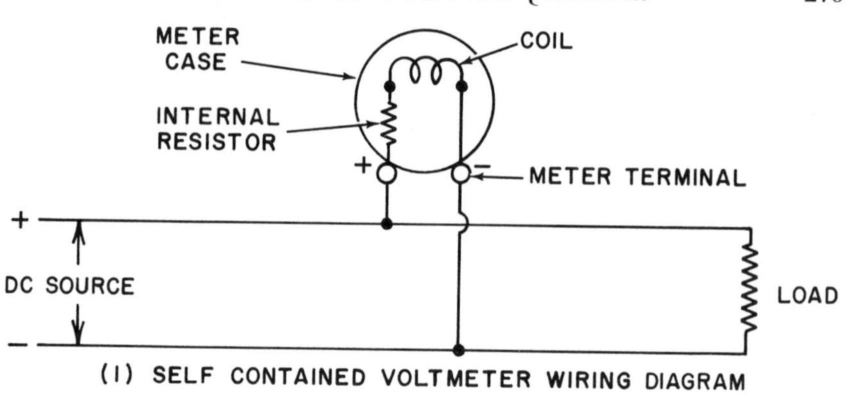

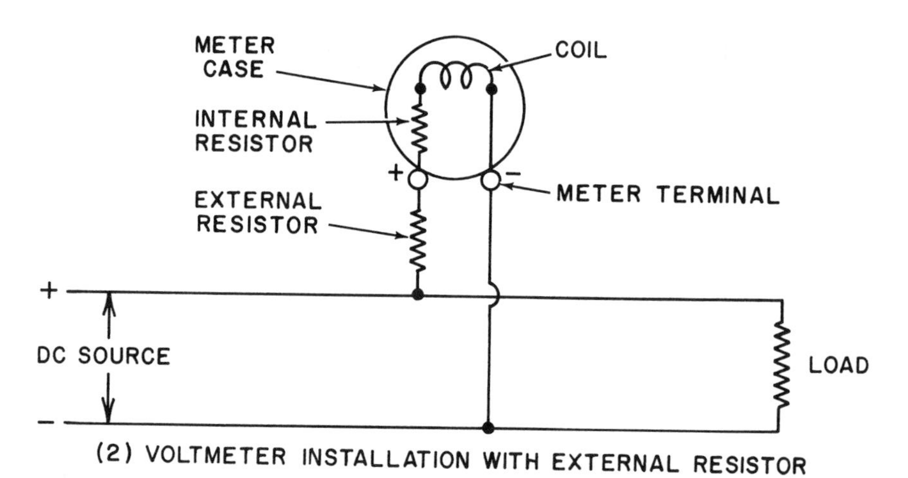

Fig. 1.17j DC voltmeter installations

in voltage ratings from 250 volts to 30,000 volts, and resistors for non-standard voltages may be calculated using equations 1.17(3) and 1.17(4)

$$\text{Total Resistance} = \frac{\text{Full scale meter reading}}{\text{Full scale meter current}} \qquad 1.17(3)$$

$$\text{External resistor value} = \text{Total resistance} - \text{meter resistance} \qquad 1.17(4)$$

Movements requiring less current for full scale deflection permit greater sensitivity and once the total resistance has been determined for a particular voltage range the new sensitivity may be calculated from equation 1.17(5)

$$\text{Sensitivity} = \frac{\text{Total resistance}}{\text{Full scale meter reading}} \qquad 1.17(5)$$

Accuracy is not noticeably affected by lead length because the value of the external resistor is usually very high and calibrated accuracies of $\pm0.25\%$ are generally available.

Power Measurement

AC Power

Indicators for AC power measurement are usually constructed with electrodynamic movements because the separate fixed and moving coils permit two different types of input signals for the same movement. The fixed coils are usually connected to measure currents, and the moving ones are connected to monitor voltages (Figure 1.17k). Connection of the coils in this way causes a deflection of the moving coil proportional to the instantaneous product of the circuit current and voltage. Inertia prevents the moving coil from responding quickly to current and voltage variations in AC circuits above 25 Hz and the pointer will indicate the average AC power regardless of wave shape. The way wattmeters are wired introduces a certain amount of error which can be minimized by altering the wire connections as a function of the current and voltage ranges being measured.

In the wattmeter shown in Figure 1.17k the current in the current carrying coils is the true load current, but the voltage across the potential coil is higher than the load voltage by an amount equal to the voltage drop across the current coils. This arrangement will result in a positive error in

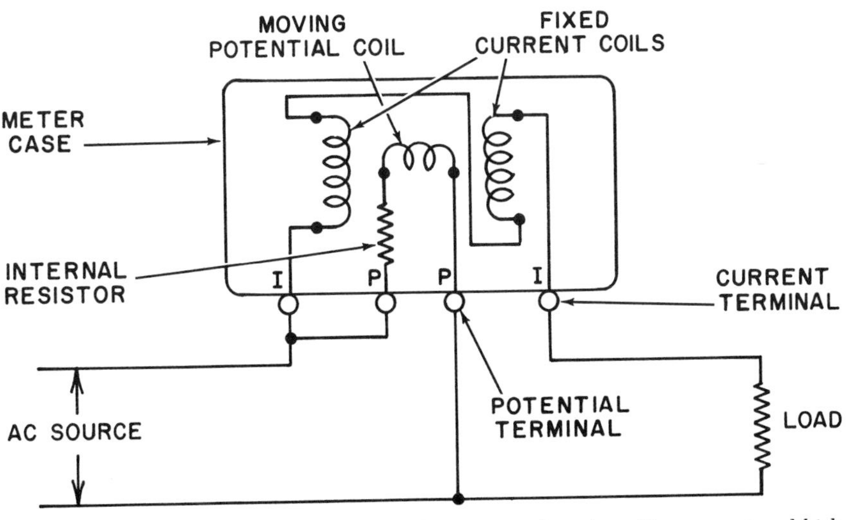

Fig. 1.17k Single element wattmeter, wired for accurate detection of low current and high voltage loads

power indication by an amount equal to the power consumed by the current coils. This error will be lowest when the wattmeter is used on high voltage circuits with low current loads. If the external wiring is changed as shown in Figure 1.17 l, the voltage across the potential coil will be the true load voltage but the current drawn by the potential coil will also pass through the current coil and the wattmeter reading will therefore be high by an amount equal to the power consumed in the potential coil. The error will be lowest when the meter is used in low voltage circuits with high current loads.

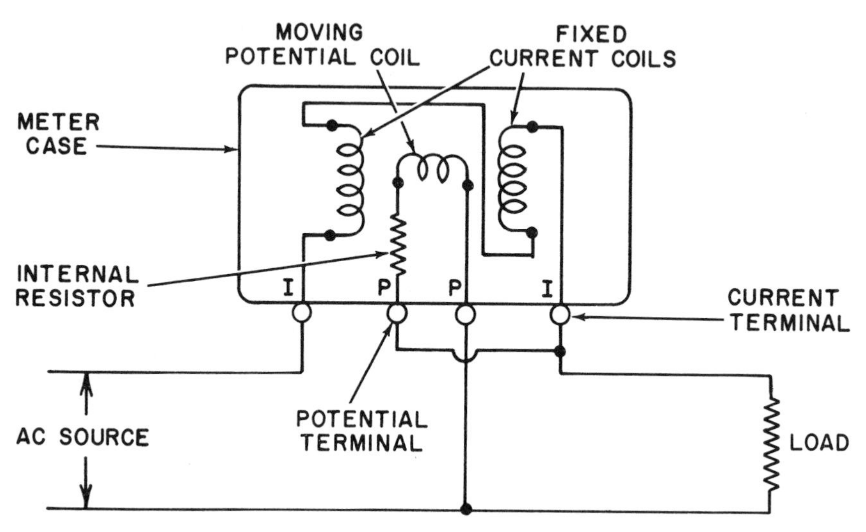

Fig. 1.17 l Single element wattmeter, wired for accurate detection of high current and low voltage loads

Meter scales are linear, accuracy is high, but low sensitivity is the result of the high power consumption. Calibration is usually for 60 Hz and recalibration is necessary for use at any frequency up to 400 Hz. Wattmeters are rated for maximum current and voltage in addition to maximum wattage because circuits with large phase angle differences between current and voltage (circuits with low power factor) can overload the current or voltage coils without causing large wattage readings. Power measurement in single phase, two-wire circuits requires a single element wattmeter consisting of two fixed current coils and one moving potential coil.

The fixed current coils are wound with heavy wire as required by the meter current rating, whereas the moving coil is usually wound with fine wire and is provided with a series resistor to insure a high resistance-low inductance circuit for potential measurement. Self-contained meters have ranges from 125 watts to 1,000 watts full scale with current ratings from 1.25 amperes to 10 amperes at 120 volts. Transformer rated meters are also available, with ratings of 5 amperes at 120 volts and scale ranges from 1,000 watts to above 100 megawatts, depending on current transformer and potential transformer ratios.

Power measurement in three-phase, three-wire circuits necessitates a two-element wattmeter consisting of two single element movements with the moving coils attached to a common shaft that is wired as shown in Figure 1.17m. Each of the elements measures a portion of the power drawn by the load and adds to the common moving coil shaft a proportional torque so that the pointer will indicate total power.

Multi-element wattmeters are almost always transformer rated with 5 ampere current coils and 120 volts to 600 volts potential coils. Meter ranges from 5,000 watts to more than 100 megawatts are common although ranges as low as 1,000 watts are also available.

Power measurements can be made in three-phase, four-wire circuits by a three-element wattmeter. Such a design is seldom utilized because the two element wattmeter movement can be modified to permit measurement on three-phase, four-wire systems by reconnecting one of the fixed coils for each element (Figure 1.17n). Meters of this type are known as $2\frac{1}{2}$ element wattmeters and will correctly indicate power for three-phase, four-wire loads as long as the line to neutral voltages are balanced for all three-phases. Meter ranges from 10,000 watts to more than 100 megawatts are available in this design with ratings of 5 amperes at 120 volts or 240 volts.

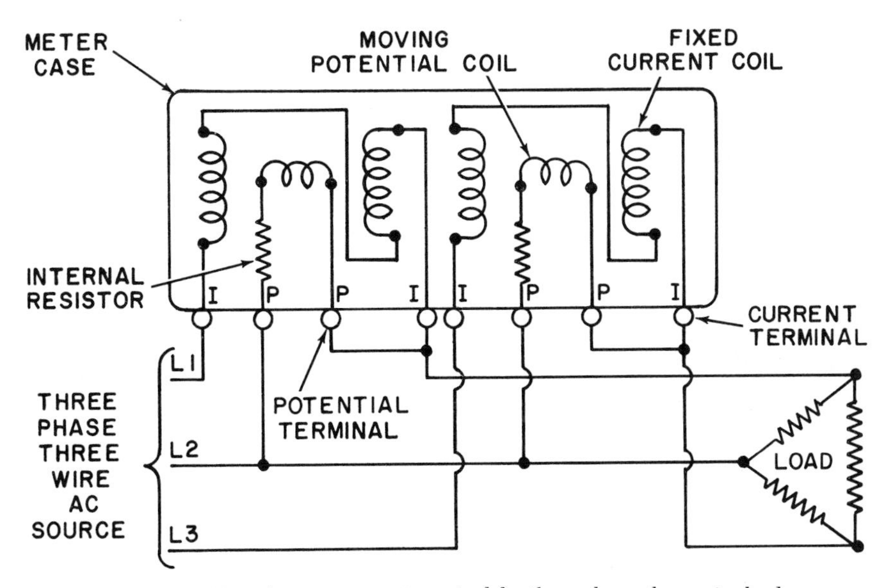

Fig. 1.17m Two element wattmeter, wired for three phase, three wire loads

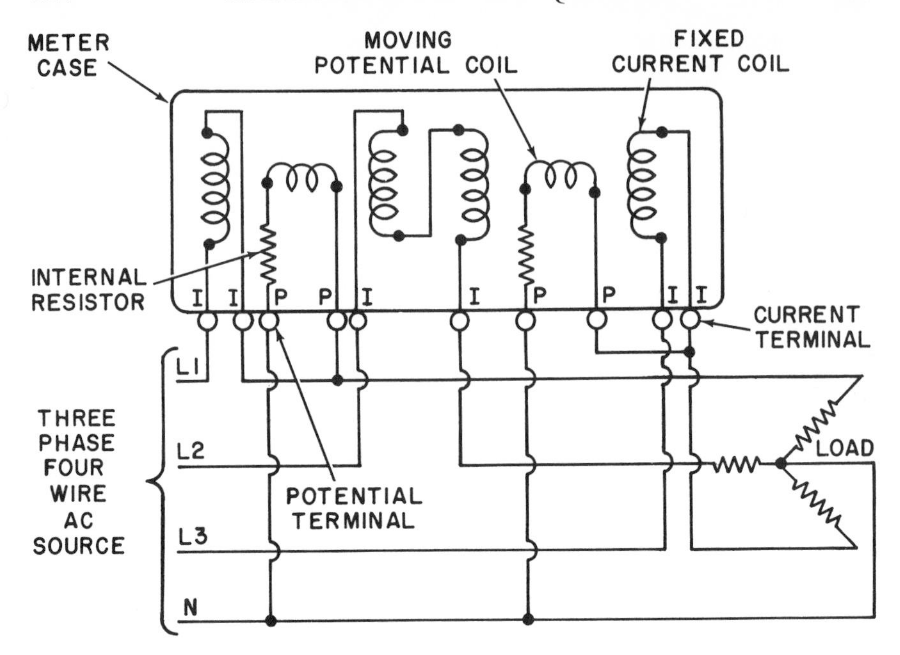

Fig. 1.17n $2\frac{1}{2}$ element wattmeter, wired for three phase, four wire loads

DC Power

DC wattmeters utilize the same movement as single element AC wattmeters but must be calibrated for use on DC circuits. Ranges from 100 watts to 2,000 watts are available with ratings from 1 ampere to 20 amperes at 120 volts. Higher ranges may be accommodated by adding an external resistor to the potential circuit for use on higher voltages although this introduces additional errors. Therefore, the higher ranges are generally measured by a DC watt transducer.

Meter Scales

In general, the best choice of meter scale range is one in which the maximum anticipated value of current, voltage or power falls to 80% of the full scale reading. Scale selection on this basis provides reasonable utilization of the meter scale, furnishing good visibility with the capability to indicate moderate overloads.

When many meters are grouped in a small area it may be desirable to have all normal readings at the mid-scale position in order to permit easier identification of an abnormal condition. The mid-scale pointer position on meters with non-linear scales indicates a value of approximately 65% of full scale, whereas on linear displays it corresponds to a value of 50% of full scale.

Table 1.17o
ORIENTATION TABLE FOR AMMETERS, VOLTMETERS AND WATTMETERS

Features / *Meters*	*Type of Meter Movement*	*Accessories Required*	*Full Scale Meter Range (A-Amperes V-Volts W-Watts)*	*Permissible Overload in Multiples of Full Scale and for Noted Time Duration*	*Recommended Applications*
Ammeters AC	Rectifier	None	$0.5–20 \times 10^{-3}$A	*For meters:* 1.2×: 8 hr 100×: 1 sec	Low range, high frequency
	Moving iron vane	None	1–50 A		General use up to 750 volts
	Moving iron vane	Transformer	10–8,000 A	*For transformer:* 50×: 2 sec	High range, over 750 volts, long meter leads
DC	Permanent magnet moving coil	None	0.02×10^{-3} – 50 A	1.2×: 8 hr	General use
	Permanent magnet moving coil	Shunt	20–20,000 A	100×: 1 sec	High range
Voltmeters AC	Rectifier	None	3–800 V	*For meters:* 1.2×: continuous 100×: 1 sec	Low range, high frequency
	Moving iron vane	None	3–600 V		General use
	Moving iron vane	Transformer	150–18,000 V	*For transformer:* 1.1×: continuous 1.25×: 1 min	High range, circuit isolation
DC	Permanent magnet moving coil	None	1–600 V	1.2×: continuous 100×: 1 sec	General use
	Permanent magnet moving coil	Resistor	250–30,000 V		High range, high sensitivity

Table 1.17o (Continued)

Meters \ Features	Type of Meter Movement	Accessories Required	Full Scale Meter Range (A-Amperes V-Volts W-Watts)	Permissible Overload in Multiples of Full Scale and for Noted Time Duration	Recommended Applications
Wattmeters Single Phase AC	One element electrodynamic	None	125–1,000 W	*For current:* 1.5×: continuous 10×: 1 min	Low power, single phase 2 wire circuits
	One element electrodynamic	Transformer	1,000 – 100 × 10^6 W		General use, single phase circuits
Three Phase AC	2 element electrodynamic	Transformer	1,000 – 100 × 10^6 W	*For voltage:* 1.2×: continuous 10×: 1 min	General use, three phase, three wire
	2½ element electrodynamic	Transformer	1 × 10^4 – 1 × 10^8 W		General use, three phase, four wire
DC	One element electrodynamic	None	100–2,000 W		General use, low power
	Permanent magnet moving coil	Transducer	400–100 × 10^6 W	Varies with design	High power

REFERENCES

1. Drysdale, C. V. and Jolley, A. C., "Electrical Measuring Instruments," 2 vols., Ernest Bern, Ltd., London, 1924.
2. Edgcumbre, K. and Ockenden, F. E., "Industrial Electrical Measuring Instruments," Sir Isaac Pitman and Sons, Ltd., London, 1933.
3. Golding, E. W., "Electrical Measurements and Measuring Instruments," Sir Isaac Pitman and Sons, Ltd., London, 1948.
4. Harris, F. K., "Electrical Measurements," John Wiley and Sons, Inc., New York, 1952.
5. Laws, F. A., "Electrical Measurements," McGraw-Hill Book Co., Inc., New York, 1938.

Chapter II

NEW DEVELOPMENTS IN COMPUTERS AND DATA TRANSMISSION

J. R. Copeland
S. P. Jackson
V. A. Kaiser

CONTENTS OF CHAPTER II

2.1 MINICOMPUTERS

Cost:

For a 4K core memory, $4,000 to $20,000. A "typical" $10,000 minicomputer package includes $4,000 for its 4K memory, $4,000 for its central processing unit and input-output hardware and $2,000 for its panels and mechanical assembly. The following extras may apply:

- Additional 4K of 8-bit memory, $3,000.
- Additional 4K of 16-bit memory, $5,000.
- Other related extras include parity check, $600; power-fail restart, $600; real time clock, $700; and I/O expansion chassis, $1,000.
- Teletypewriter, $1,500 to $3,500.
- Tape deck, $4,000 to $20,000.
- Disk, $8,000 to $25,000.
- Line printer, $5,000 to $28,000.
- Card punch, $10,000 to $22,000.
- Cathode ray tube display, $5,000 to $25,000.
- Paper tape puncher, $1,000 to $4,000.
- Paper tape reader, $1,000 to $4,000.

The above figures allow one to estimate the hardware cost for the installation. Software expenses may be twice the total figure. Programmable controllers without I/O hardware cost from $3,000 to $7,000.

Size of Memory:

4K to 32K capacity (1K to 4K for programmable controllers).

Memory Cycle Time: About 1.0 microsecond.

Word Lengths: 6 to 24 binary digits (see Table 2.15 for more information).

Partial List of Suppliers

MINICOMPUTER Atron Corp., Cincinnati Milacron, Compiler Systems, Inc., Computer Automation, Computer Logic Systems, Control Data Corp., Datacraft Corp., Data General Corp., Datamate Computer Systems, Inc., Digital Computer Controls, Inc., Digital Equipment Corp., Elbit Computers, Ltd., Electronic Processors, Inc., Electronic Products International, Inc., EMR Computer, Ferranti, Ltd., Friden Division, Singer Co., Fujitsu, Ltd., General Automation, Inc., General Electric Process Computer Products, GRI Computer Corp., Hewlett-Packard Co., Honeywell Information Systems, IBM Corporation, Information Technology, Inc., Interdata, Inc., Laben Division of Montedel, Lockheed Electronics Co., Matsushita, Microdata Corp., Modular Computer Systems, Motorola Instrumentation and Control, Inc., Multidata, Inc., Nuclear Data, Inc., Philips Corp., Raytheon Computer Corp., Redcor Corp., ROLM Corp., Scientific Control Corp., Selenia S.p.A., Spiras Systems, Inc., Systems Engineering Laboratories, Inc., Tempo Computers, Inc., Texas Instruments, Inc., Unicomp, Inc., Varian Data Machines, Varisystems Corp., Wang Laboratories, Westinghouse Electric Corp., Xerox Data Systems.

PROGRAMMABLE CONTROLLERS Allen-Bradley Co., Digital Equipment Corp., General Electric Co., Modicon Corp., Reliance Electric Co., Square D Co., Struthers-Dunn, Inc., Unicom, Inc.

A glance at the list of minicomputer manufacturers should dispel any feeling of paucity. In a few cases the list is redundant since some units are sold under several labels. Nevertheless, about 100 units are available around the world.

One of the principal characteristics of the minicomputer is the size of its word length. Typical sizes range from 8 to 18 bits. Figure 2.1a illustrates the relationship between word length and precision. Generally expected accuracies of various types of instruments are superimposed for cross-referencing.

For an indepth treatment of the many aspects of process computers see Chapter VIII in Volume II.

Computer Terminology[*]

A brief glossary of terms is presented for the reader's convenience. A more complete listing of computer terms can be found on P. 882 in Volume II.

A/D Analog to digital.

A/D Converter Device that encodes an analog input into a corresponding digital signal.

Assembler Software that operates on symbolic input codes to produce machine instructions on a one-to-one basis.

Bit Binary digit.

Bootstrap Hardware or software program that provides for the entry of operating software, application programs or data.

Bulk Storage Disks, drums or magnetic tape units that store large quantities of data or information.

Bus Conductor, shared by several devices, which transmits power or data.

Byte Sequence of adjacent bits, less than a word, operated on as a unit.

Clock Hardware that generates periodic signals to synchronize internal computer operation.

Compiler Software that generates machine language instructions from a high-level, user-oriented program; the symbolic statements and over-all logic structure of the source program are considered, and multiple machine instructions are typically generated by each symbolic statement.

Core Programmable, random access working memory of a computer.

CRT Cathode ray tube.

Cycle Time Interval between requesting and receiving data from memory; used as the basic speed unit of the computer.

D/A Digital to analog.

[*] This glossary is from *Instruments and Control*, March, 1971, p. 102.

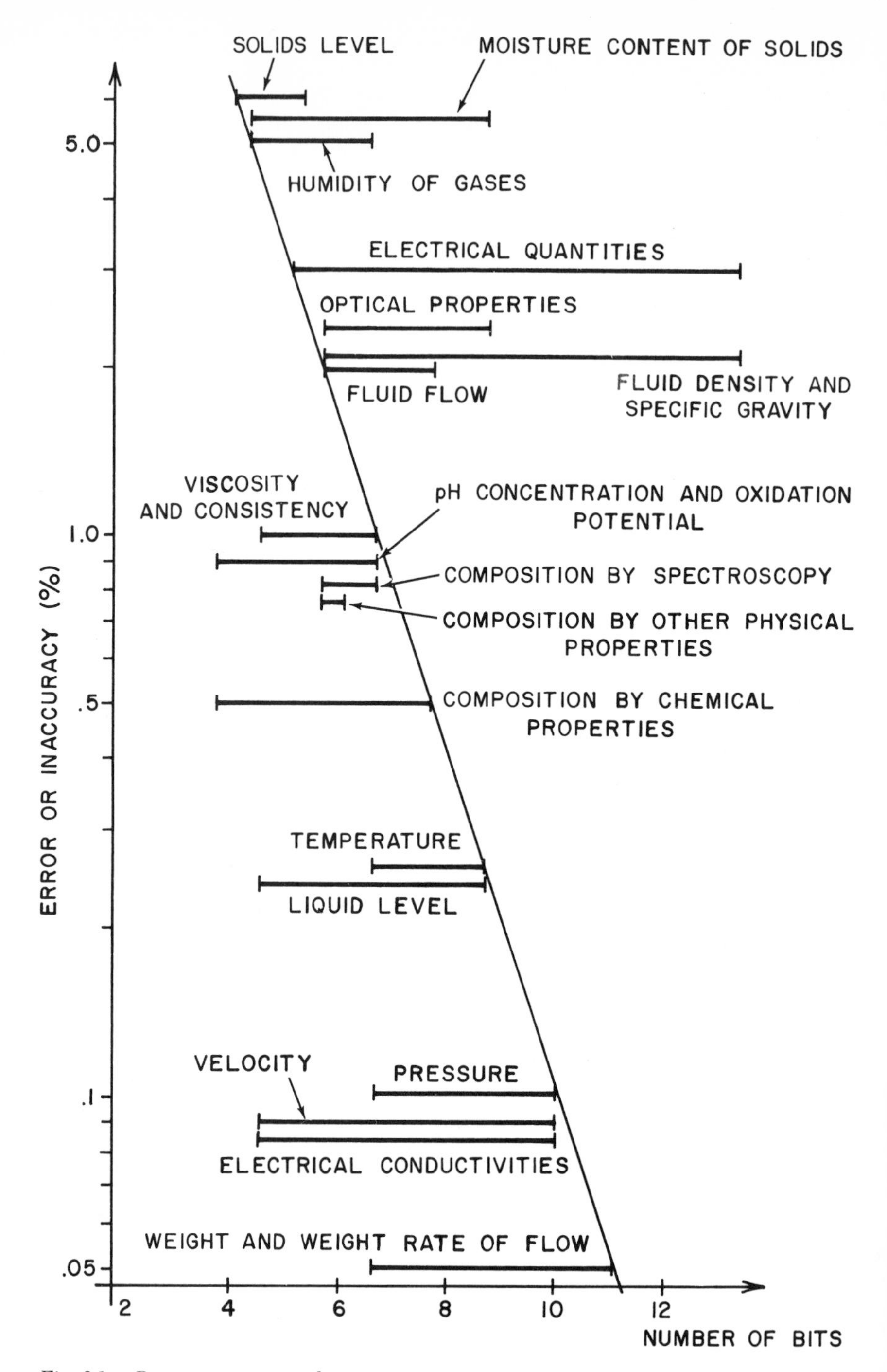

Fig. 2.1a Process instrument element action. (Generally expected accuracy of typical range of accuracy derived from Chemical Engineering 1969 Guide to Process Instrument Elements, *Chemical Engineering Magazine*, June 2, 1969.)

D/A Converter Device that translates digital outputs to analog signals for compatibility with final control elements.

DDC Direct digital control, a mode in which the computer calculates control action from inputs and algorithms, and generates outputs to actuate control elements.

Debug Software which enables programmer to check and correct programs.

DMA Direct memory access.

Direct Memory Access The ability to store or retrieve data from memory at the machine cycle speed.

Disk A device capable of storing data as a magnetic pattern on a rotating plate.

Drum A device capable of storing data as magnetic patterns on a rotating cylinder.

Input-Output Data channels that feed information to and receive data from the mainframe.

Instruction Symbolic code which causes the computer to perform prescribed hardware functions.

Instruction Repertoire Set of codes provided with the computer.

I-O Input-output.

Machine Language Numerical codes used directly by a computer to manipulate data internally.

Main Frame Central processing unit, in which all arithmetic and logic functions are performed.

Microprogramming Programming instructions consisting of several machine language commands; usually stored in read only memory (ROM) to provide increased speed for common operations.

Modem (MOdulator-DEModulator) A device that modulates and demodulates signals over a communications channel.

Multiplexer A device that samples input and output channels and interleaves signals in frequency or time to provide apparently simultaneous communications.

Multiprogramming Concurrent execution of multiple programs.

Parallel Processing Concurrent or simultaneous execution of two or more operations.

Parity Extra bits added to words or bytes so that even or odd number of ones or zeroes will be present; helps to ensure that no errors are made when reading from memory.

Peripherals Hardware distinct from the mainframe which provides external communication functions.

Read Only Memory Permanent storage device in which data cannot be dynamically altered.

Registers Hardware devices capable of storing active data; used for arithmetic, data manipulation, indexing and temporary storage.

ROM Read only memory.
Serial Processing Sequential execution of multiple operations.
Supervisory Control A mode in which a computer regulates the action of a dedicated control device.
TTY Teletypewriter
Word Series of bits considered as a unit by the computer.

The Process Control System

All computer control systems are composed of a combination of the five elements shown in Figure 2.1b, which is deceptively simple. Each of the five elements can be selected from several possible options, but the selection is not independent. A given choice in one block may not be compatible with options for other elements.

For a given set of system specifications, the designer's goal is to achieve maximum reliability at minimum cost. Frequently, lower cost is synonymous with higher reliability—if fewer high quality components are used.

To provide a background for system specification, each of the five elements will be discussed. An attempt will be made to show the interdependence of some of the element selections and the effect on cost and system performance of these selections.

Operator Console

In many systems, performance is largely determined by the selection of the operator console options. The speed at which important data can be written will frequently determine system performance. If operator interaction is important, the form of data presentation must be carefully chosen. If the

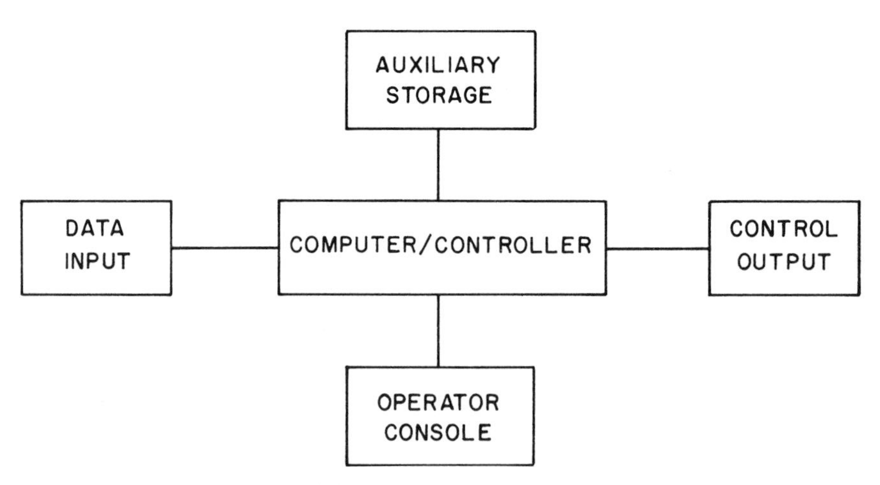

Fig. 2.1b Basic computer control system

application requires logging of large quantities of data, the speed and cost of the writing device must be considered with respect to the rate at which data are being taken and the amount of data storage being accumulated.

Table 2.1c is a list of operator console elements. Omitted are specially designed switches, lights and controls which are designed for specific projects. The price for the most frequently specified options or a range of prices is given. No attempt has been made to correct for quantity or functional discounts.

Operator console costs versus writing speeds have been plotted in Figure 2.1d. Apparent is the wide variation in cost at the slower speeds owing to 1.) the cost of options available, and 2.) the frequency and amount of maintenance required. Special cabinets, pilot lights, push buttons, registers and other read-out or control devices are excluded; these items cost extra.

Data Input

The options for data input are many, and data sources may be of four basic types: switch closure, pulse train or frequency, analog and coded digital. Accuracy of source data may change from point to point. The number of data points also changes with application and time. Special interface equipment may be required to eliminate undesirable transients or isolate potential levels. Data may either be taken in a desired sequence or directed on a priority basis. Reliability begins with the data input section (detectors, transmitters, starters, switches and limit switches). From this point, the system can only

Table 2.1c
OPERATOR CONSOLE OPTIONS

Item	*Function*	*Speed*	*Cost*
Teletype (ASR33–35)	Program Input Data Output	10 characters/second	$ 800–3,500
IBM Selectric	Program Input Data Output	14.7 characters/second	1,500–2,500
Printer (Execu Print I)	Data Output	11 characters/second	2,150
Cathode Ray Terminal (Beehive)	Program Input Data Output	10 to 500 characters/second	3,500
Teletype (Inktronic)	Data Output	120 characters/second	5,450
Printer (Gulton)	Program Input Data Output	30 characters/second	5,500
Printer (A.B. Dick)	Data Output	250 characters/second	7,000
Line Printer (Nortec)	Data Output	200 lines/minute	7,000

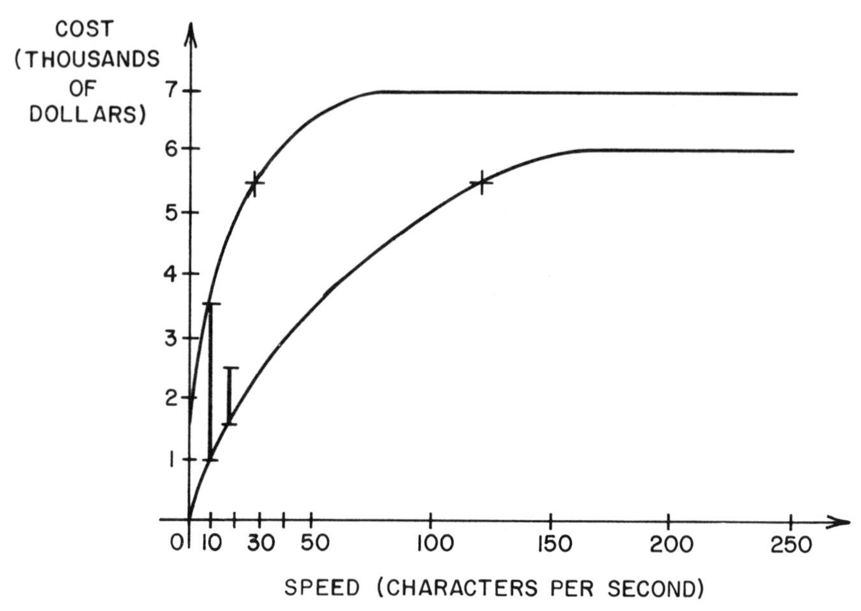

Fig. 2.1d Operator console cost writing speed

manipulate. Thus, careful attention to the input sources is extremely important.

SWITCH CLOSURES

Data in the form of switch closures (such as relay points) are the easiest to handle and least expensive. They come in simple binary form and therefore can be efficiently packed into memory in large amounts. Since the loading of this type of information is relatively simple, high speed data acceptance is characteristic.

Switch closures generally offer adequate isolation from equipment-damaging transients, but not from noise. It is often necessary to filter out "contact bounce" so that a clean signal is presented to the computer. A hardware interface is essential for reliable transmission of switch closures to the computer. Typical cost is $25.00 to $50.00 per closure. A frequent oversight in process plant applications is to utilize low quality (high maintenance) limit switches as computer inputs.

PULSE TRAIN OR FREQUENCY DATA

The computer can count pulses or detect frequency. As an example, variable reluctance (gear teeth) type speed indicators supplying output pulses proportional to speed can be entered into the computer directly. The computer then counts the pulses for a given interval, and counts per unit time is proportional to frequency.

The ability to measure time is important in many parts of the process control cycle. It provides a means of integrating pulses over any desired integrating period. The technique may be extended to obtain inexpensive analog to digital conversion.

Also of interest is the computer's ability to use long integrating periods during which sensors responding to long-term chemical process can be handled both accurately and efficiently by the computer. Periods of hours are not uncommon.

ANALOG DATA

Cost of handling analog data in a system depends on 1.) the voltage level of the signal, 2.) the speed at which data must be taken, 3.) the accuracy required and 4.) the number of locations from which data are taken.

For an analog system Figure 2.1e shows the subsystem components. System A uses a single amplifier between the multiplexer and the analog to digital converter. All signals are amplified. In some amplifiers, the gain factor may be controlled by the computer.

System B employs amplifiers on the low level signals only. The multiplexer and analog to digital converter operate at the same "high" signal level. This system is used when the input level varies over a wide range. By adjusting the amplifier gain in each signal line, a normalized voltage level may be presented to the system.

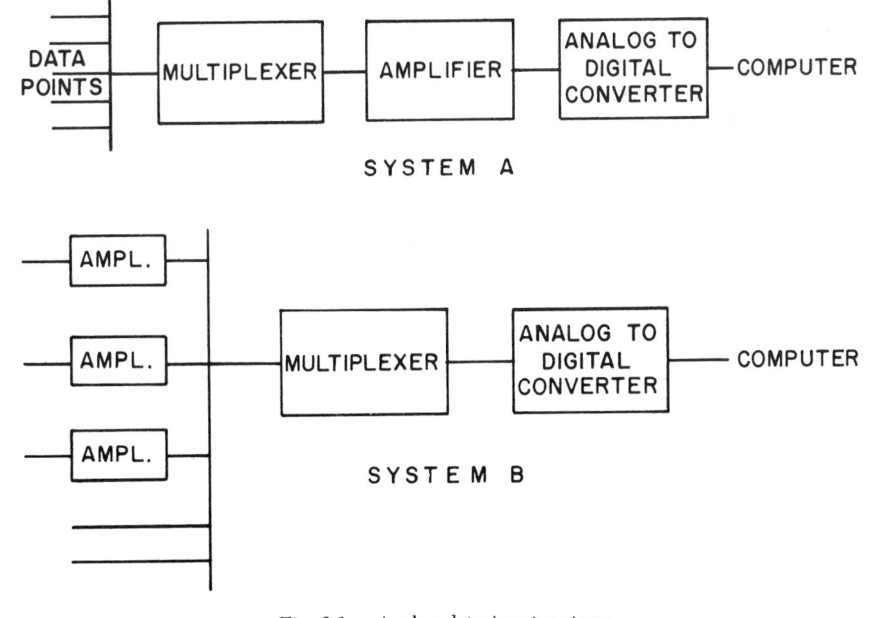

Fig. 2.1e Analog data input systems

For very small systems, multiplexers are occasionally omitted, and an inexpensive analog to digital converter is dedicated to each input. This last has the advantage that all channels can be sampled simultaneously, whereas a multiplexer samples only a single channel at any instant.

System accuracy deteriorates if the word length selected is insufficient to define the quantity measured to the same accuracy as the sensor. Table 2.1f lists the number of bits in the word and gives the corresponding largest decimal number defined by the word and the highest precision. Eight bits are adequate to handle data in the 1% error class, ten bits for 0.25%, and twelve bits for measurements with ±0.1% error limits.

Table 2.1g supplies typical accuracies to be expected in the various sensor categories listed. For more refined information on measurement errors, see Volume I.

The cost of a multiplexer-converter tends to be proportional to the word length-throughput product. Cost is high for long words at high throughput rates (Figure 2.1h). A 15-bit, 100 KHz converter equivalently equipped to those illustrated would cost about $4,000. This body of data is from a single manufacturer and should be considered representative but not illustrative of all possible combinations.

Converters with greater accuracy can also increase the cost of the computer, auxiliary memory and analog inputs.

Table 2.1f
ACCURACY OF DIGITAL CONVERTERS

Number of Bits	*Decimal Number*	*±% Error (or Inaccuracy)*
1	2	50
2	4	25
3	8	12.5
4	16	6.25
5	32	3.125
6	64	1.562
7	128	0.781
8	256	0.390
9	512	0.195
10	1,024	0.098
11	2,048	0.049
12	4,096	0.024
13	8,192	0.012
14	16,384	0.006
15	32,768	0.003
16	65,536	0.0015
17	131,012	0.00075
18	262,024	0.000375

Table 2.1g
PROCESS INSTRUMENT ELEMENT ACCURACY*

Measurement	*Typical Inaccuracy or Error* (±%)
Temperature	0.25–1
Pressure	0.1–1
Force, Tension Compression	0.5–1
Radioactivity Measurements	High
Fluid Flow	0.5–2
Liquid Level	0.25–4
Solids Level	3 or better
Weight and Weight Rate of Flow	0.05–1
Thickness and Displacement	3 or better
Velocity	0.1–4
Fluid Density and Specific Gravity	0.01–2
Viscosity and Consistency	1–4
pH Concentration and Oxidation Potential	1–7
Electrical Conductivity	0.1–1
Thermal Conductivity	High
Caloric Value, Combustible Content and Explosibility	0.25–10
Humidity of Gases	1–5
Moisture Content of Solids	0.2–5
Optical Properties	0.2–2
Composition by Chemical Properties	0.5–5
Composition by Spectroscopy	0.1–2
Composition by Other Physical Properties	1–2
Electrical Quantities	0.01–3

*Generally expected accuracy or typical range of accuracy derived from Chemical Engineering 1969 Guide to Process Instrument Elements, *Chemical Engineering Magazine*, June 2, 1969.

SELECTION OF SENSORS

The minicomputer is directing more attention to the need to reduce sensor costs, while at the same time requiring more accurate sensors. One way of accomplishing these goals is to use the minicomputer as part of the sensor system. (For a detailed discussion of measurement accuracy see pp. 1 to 13 in Volume II.) Error or inaccuracy can be based on the actual reading or on full scale values. The former is based on the rated maximum value shown by the lower and upper lines in Figure 2.1i; the latter is based on the actual value as shown by the shaded area. The sensor is defined as being accurate, based on one of these criteria.

Figure 2.1j shows the performance of a non-linear sensor. One way to improve its accuracy is to store the non-linearity in the computer itself. One method, the polynomial fit, is very compact in its use of memory, and the quantity of storage occupied for this rectification process is proportional to

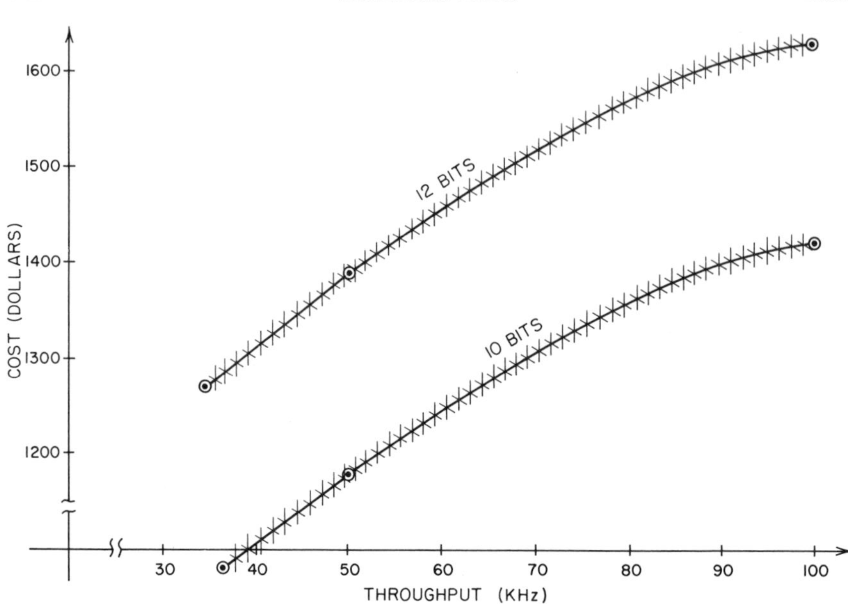

Fig. 2.1h Cost of an eight-channel multiplexer with A/D converter as a function of throughput

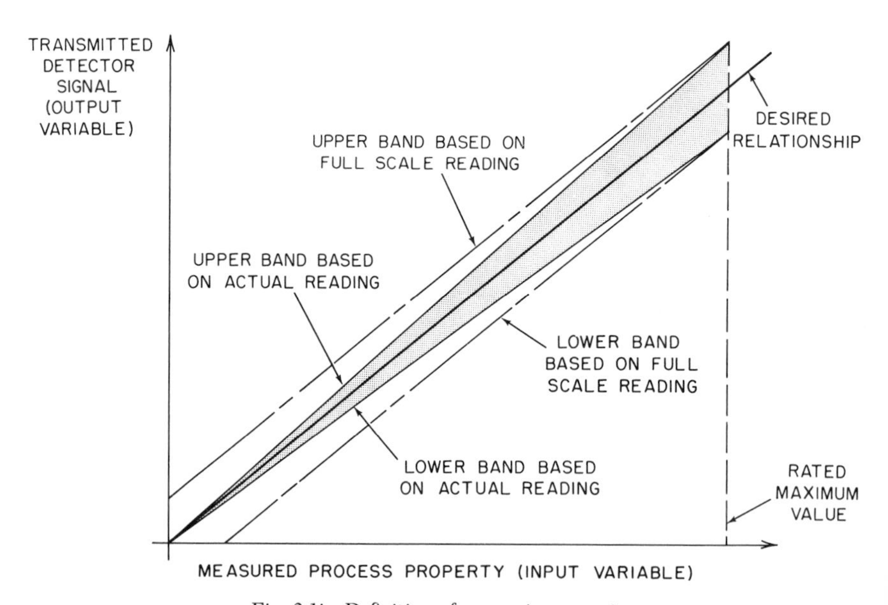

Fig. 2.1i Definition of sensor inaccuracies

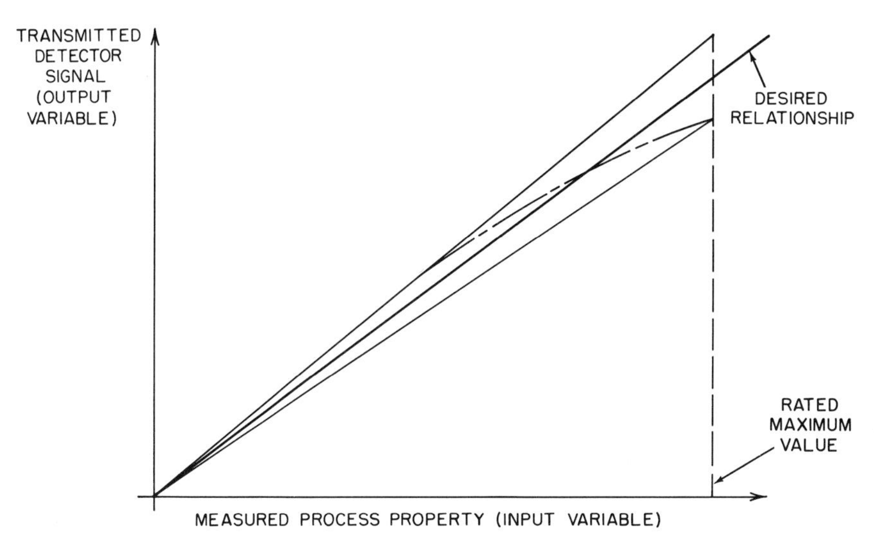

Fig. 2.1j Performance of a nonlinear sensor

the quality of fit desired. By storing the curvature in the computer memory, a low cost sensor might actually achieve greater accuracy.

Figure 2.1k illustrates sensor backlash within the limits of percent full scale error. Backlash can be accommodated economically in the computer itself. It uses about $1\frac{1}{2}$ times the amount of storage required for the rectification of non-linearity.

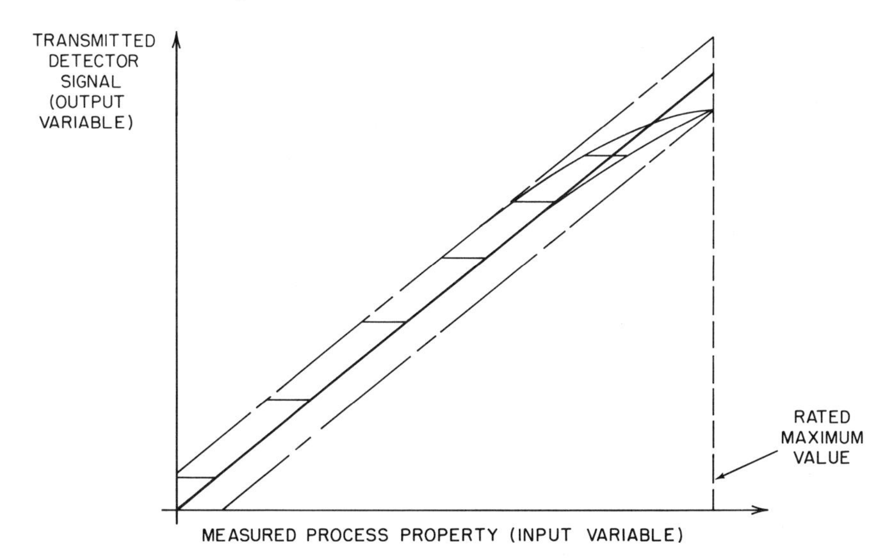

Fig. 2.1k Sensor backlash

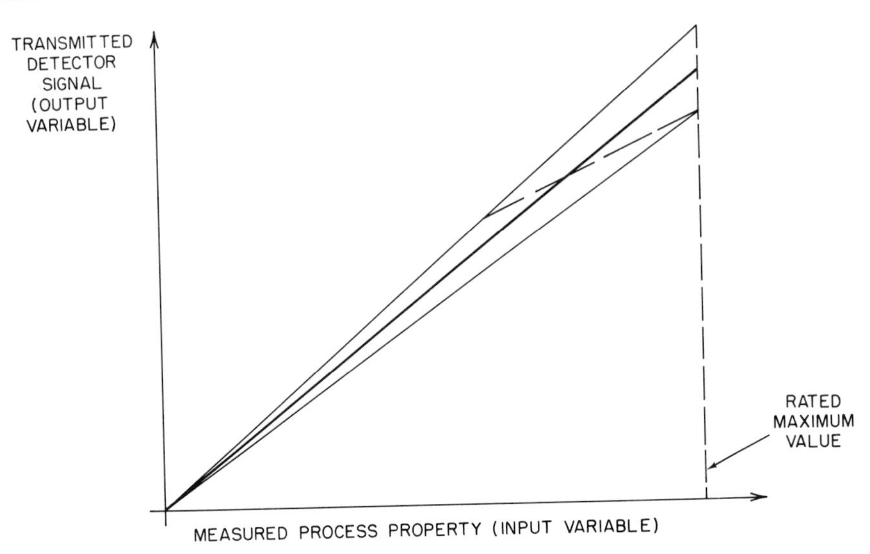

Fig. 2.1 l Sensor performance corrected by linear approximation

Figure 2.1 l shows a non-linearity consisting of two straight lines. This might occur in a mechanical system in which a second spring is engaged at the point of discontinuity. An abrupt change occurs in the rate of change of the variable at the point of discontinuity. This function is shown in Figure 2.1m and can easily be handled in the computer. It requires a linear approximation and a memory space equivalent to that needed to correct for sensor backlash.

In addition to non-linearity, backlash and rate of change corrections, one other possible input signal improvement is in smoothing. It is relatively simple for the computer to determine the average value of a signal for any interval of time (Figure 2.1n). The computer also determines the peak readings and indicates and stores the highest value.

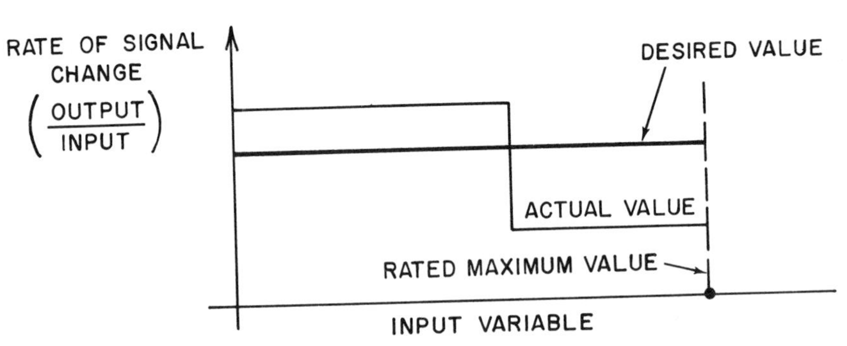

Fig. 2.1m Rate of change in sensor output signal expressed as a ratio

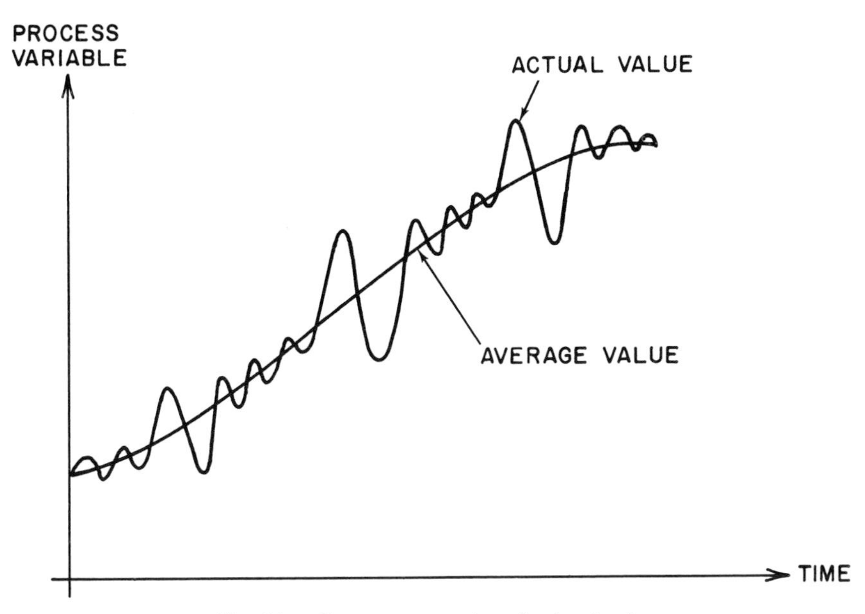

Fig. 2.1n Computer averaging of noisy signals

Many other types of digital filtering and smoothing procedures are possible, such as low-pass or band-pass filtering, and calculation of weighted averages. For additional information see Chapters VIII and IX in Volume II. With these techniques the instrument engineer can change the sensor characteristics of installed systems.

The computer is also valuable (as part of the sensor system) in online calibration of sensors, which is done by programming the computer to bring the process variable to a predetermined level and then compare the sensor output to a stored reference value.

Another task for the computer is the handling of signals with long time constants and the anticipation of their eventual values from early trend data.

The computer can also calculate dependent process properties from the values of easily measured variables or compensate for variations in variables, e.g., compensating for temperature in flow or viscosity measurements. The computer can also check the authenticity of measurement signals, detect sensor failures and guarantee fail-safe operation under all conditions.

In most processes, the location of the sensor has an effect on the measurement value. In Figure 2.1o the solid black line shows the actual variation associated with sensor location.

It would be desirable to add a sensor at point 5. Rather than adding a sensor, however, it may be more feasible to make a model of the relationship between the output variable and length, and thus be able to calculate the

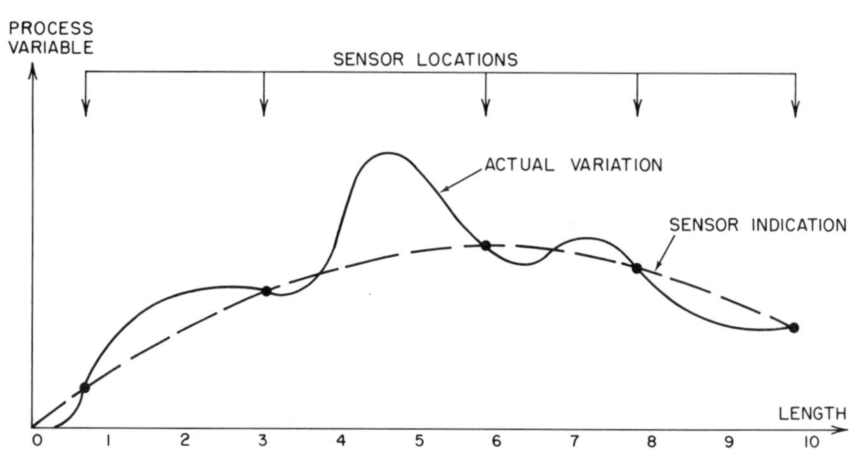

Fig. 2.1o Interpolation of process variable values

output variable as a function of length and base the process control decision on point 5 or any other point. The technique may be expanded to determine the average or RMS value of several readings, or to model and thereby obtain measurements at previously inaccessible sites.

The computers assist in improving sensor performance and in reducing sensor cost. The sensor of the future will probably have a digital output or pulse train or frequency outputs. Thus, the minicomputer is forcing sensor designers to review their design philosophy because the computer can frequently handle tasks that were previously the sensor's; and other design features become desirable because of the computer.

Computer-Controller

Because of the variety of options, prices of computers vary. The lowest priced computer with 4K (4096 words) of core memory is about \$5,000. The next general price range is from \$10,000 to \$12,000.

Neither word length nor memory access time is an infallible indicator of cost. In general, the nebulous "computer architecture" is the parameter most directly related to cost. The problem of defining system value may be approached by describing the first two computer levels in application terms.

The \$5,000 level computer-controllers are specifically designed to handle data acquisition and editing. The heart of a typical machine is an expandable 4096×18 bit core memory which is able to restructure its own data basis. Each word is additionally addressable 3 bits at a time. Use of this capability in conjunction with the decision tables makes possible a computer without the cost of the central processing unit. Memory cycle times of 1 microsecond are typical.

In a complex system, a minicomputer is the first step in a hierarchical array. Multiple minicomputers can be controlled by an executive computer (Figure 2.1p). The latter may be of the "mini" variety but should also be able to multiply and divide. Variable word length and short memory access time are also desirable features. Once confidence is gained, the process control system may be expanded on a modular basis.

Typical of the \$10,000 to \$12,000 class computer is one having a main memory of 4096 words, 8, 12, or 16 bits in length. Memory cycle time is generally 1 microsecond, although speeds from 250 to 400 nanoseconds are obtainable with the semiconductor memories. The unit may be able to vary the word length, have hardware memory protection, provide hardware multiply-divide and offer extensive interrupt ability, in addition to having flexibility in handling I-O devices. Also available is additional memory capacity, frequently as much as eight times (32K) the basic 4096 word memory.

Efficient utilization of these machines requires assembler language for the specific machine. Although many of the minicomputers have higher order languages, they require greater memory capacity and cost more.

The value of the higher order language is to make programming less time-consuming and therefore less costly. If the program being written is infrequently changed, such as occurs in process control, there is little value in higher order languages.

Medium sized computers (\$50,000 to \$60,000 class) have increased word lengths, larger memories and instruction sets. Software packages tend to be more complete and generally include a higher order language compiler.

The first step in selecting the proper computer-controller is to define the

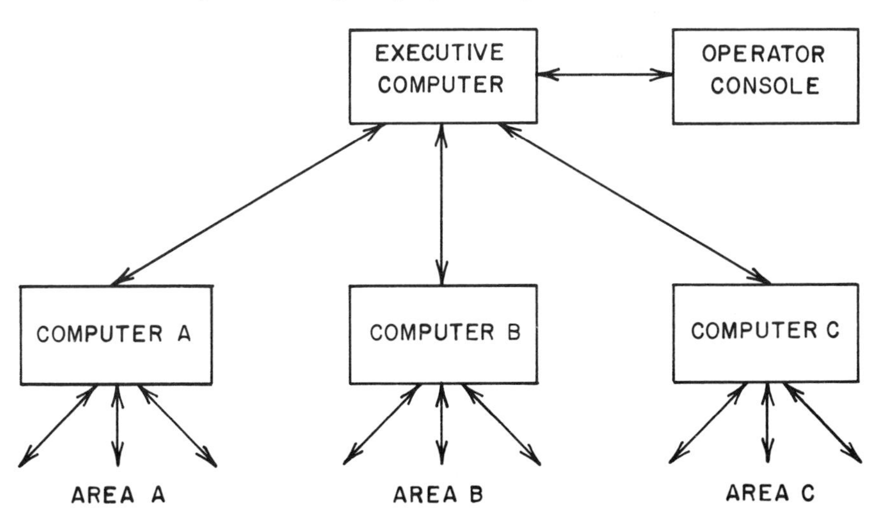

Fig. 2.1p Hierarchical computer array

tasks that it is to perform, including data acquisition (logging, monitoring and direct digital control and an evaluation of the complexity of control for each loop). Signal conditioning must also be reviewed, including averaging, noise elimination, correlation, cancellation, data reduction and alarm condition handling.

If rapid reaction to alarm conditions is important, a computer with fast hardware priority interrupt should be selected. Emphasis on report writing and data logging underlines the need for a unit with efficient character handling. A large system dependent on arithmetic calculation suggests the need for a machine with fast hardware multiply-dividing ability.

Auxiliary Storage

For a detailed treatment, see Section 8.9 in Volume II. The cost of auxiliary storage is largely dependent on size and access time. Core memories are fastest, followed by disk or drums and, lastly, magnetic tapes as shown in Figure 2.1q.

Control Outputs

A detailed treatment of computer interface hardware features is given in Section 8.7 in Volume II.

The least critical of the process control system components are the control

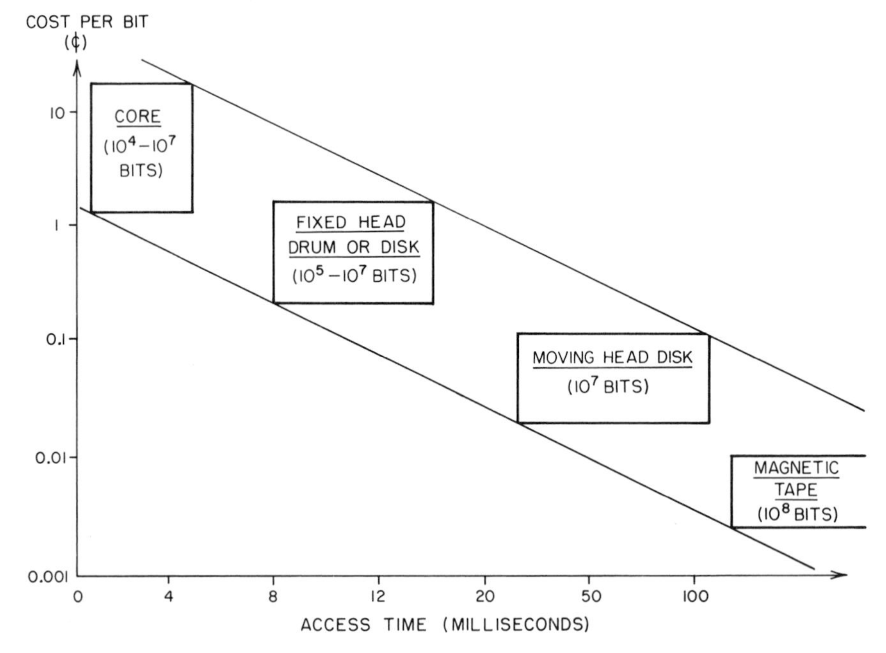

Fig. 2.1q Auxiliary data storage cost as a function of access time

outputs. If the outputs are switch closures, the natural grouping of the relays may have a bearing on computer selection, which affords considerable latitude for creativity if many relay outputs are required.

The cost of digital outputs is \$50 to \$100 each. Thus, a group of 8 relay outputs may cost \$700, including power supplies, enclosures and so on.

Analog output costs depend on accuracy (input word length) and the number of channels. Figure 2.1r illustrates this relationship.

Systems Considerations

If all the steps in the process are known and the relationships among various parameters are defined, it is easy to assure that the steps will occur

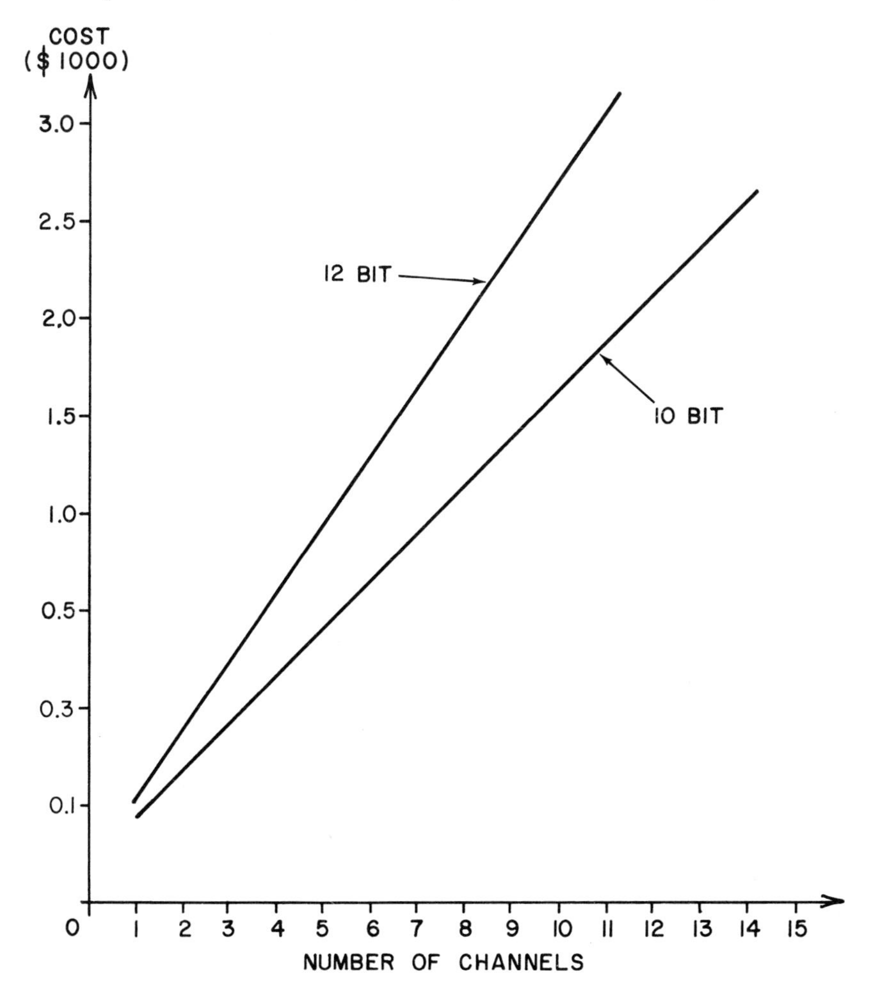

Fig. 2.1r Cost of analog outputs as a function of word length and number of channels

in the proper sequence. If the systems are ill-defined, the process yield may depend on the expertise of the operators, in which event it is first necessary to determine the relationships among the procedural parameters. One approach is to install a data logging system to obtain reliable information on the characteristics of the process. After the process model is determined, the computer may be reprogrammed to control the process.

In another—"minicomputer"—approach the process is subjected to analytical and empirical modeling, and is tested against the actual plant conditions to determine its accuracy. The basis for selection of the computer may be process *control* rather than general data acquisition as was the case in the first approach. Thus, a smaller computer is sufficient.

In developing a working model of the process, a higher level of understanding results which is of substantial value even if the computer is not installed. The model is also a valuable tool in training operators and evaluating process dynamics, and both procedures can take place prior to the construction of the plant or installation of the computer.

Minicomputers

The term "minicomputer" is misleading in that many of the instruments so characterized are greatly expandable. The availability of modules allows the system to satisfy almost any requirements. Optional hardware includes automatic multiply-divide, floating-point arithmetic and high-speed I-O features.

The computer interface with the measuring and control instruments require input-output cards, the cost of which may vary from $150 to $1000 per I-O device, depending on the complexity of the computer. The complexity is largely a matter of the design of the computer I-O and interrupt structure and the need for data buffer registers.

Minicomputers are not necessarily slower than the larger process computers. Cycle time alone does not define the speed of the minicomputer, because the efficiency of the instructions and their compatibility with the particular task play equally important roles. Some instructions are universal and require only a single memory cycle (the instruction "fetch"). The execution of this class of instruction requires the same amount of time as the memory cycle itself. The "load," "store" and "add" class of instructions requires two cycles (one instruction "fetch" and one "data reference" cycle). An unconditional jump may require one or two cycles, depending on whether the jump address is contained in the instruction or in a second word which must be accessed. Other types of instructions relating to input-output or interrupt servicing may require three or more memory cycles to store or to obtain various pieces of data. If indirect addressing is employed, an additional memory cycle is required for each indirect level.

Most minicomputers have parallel binary organization; there are some, however, that deliberately sacrifice processing speed for flexibility in microprogramming. In parallel binary operation, all bits of a single program or data word are stored and retrieved from memory simultaneously, and most logical or arithmetic operations are performed on all bits of the word simultaneously or nearly so.

Component Selection

Keeping up with the continuous evolution of computers is time-consuming. It is rewarding in optimizing a system to know the characteristics and languages of a large number of the available machines. The problem is that if one "keeps up," there is no time for system design. Rapid access to auxiliary memory is expensive. High reliability suggests that working programs can be stored in main memory, whereas data can be stored in disk or tape files.

Computer installations for process control must be regarded as a transient system. The long-range view of the process requirements should be taken into consideration in order to insure that today's decisions do not impede tomorrow's growth. Short-term "savings" in the selection of memory size of software can be very expensive in the long run.

System cost is estimated by first determining the total hardware cost. The resulting figure should be doubled since the average software cost equals that of the hardware.

Unless the system duplicates a previous one, it is probable that the cost of software will represent more than 50% of the total outlay.

Selection of the minicomputer begins with selection of minimum word size; this can be based on the precision of the input signals, on the need for handling text data and on the relative ease of performing multiple-precision arithmetic operations. Accuracy also dictates the word size of A/D and D/A converters. Other computer options such as memory capacity and the number and speed of priority interrupts may also be important in the selection. Maintenance and programming experience with a particular computer can be significant and should always be considered when a new installation is contemplated. In selecting the computer, one should also consider availability of spare parts, ease of delivery and quality of service.

Minicomputer Features

The tabulation of 88 models from 47 manufacturers is included to provide the reader with the general features of available minicomputers (Table 2.1s).

The typical minicomputer has a maximum memory size of 32K of 16 bit words, a memory cycle time of one microsecond, an optional Direct Memory Access (DMA) channel, eight hardware registers, sixty-four priority interrupt levels with a response time to interrupt of six microseconds, a cost (with 4K core) of $9,000 and was first delivered in 1970.

Table 2.1s

DISTRIBUTION AND AVAILABILITY OF MINICOMPUTERS WITH VARIOUS CHARACTERISTICS

(See Reference #2)

Word length (bits)	*6*	*8*	*12*	*16*	*18*	*24*
Number of computer models	1	16	8	55	6	2

Maximum memory capacity	*8K*	*12K*	*16K*	*32K*	*65K*	*110K*	*124K*	*131K*	*263K*
Number of computer models	9	1	9	52	12	1	1	1	1
	← Minicomputers →				← Larger computers →				

Memory cycle time (microsecs.)	*.09*	*.22–.4*	*.75–.96*	*1.0–1.35*	*1.5–1.8*	*2.0–2.66*	*3.0–3.3*
Number of computer models	1	3	27	22	21	11	2

Direct memory access channel	*Optional*	*Standard*	*Not available*
Number of computer models	47	38	3

Total number of hardware registers	*0*	*1*	*2*	*3*	*4*	*5*	*6*	*7*	*8*	*9*	*10*	*11*	*13*	*15*	*16*	*17*	*19*	*21*	*25*	*28*	*32*	*33*
Number of computers	3	2	2	1	3	3	9	12	14	6	3	3	2	1	10	2	1	1	2	1	2	2

Price with 4K (dollars)	*3,120–4,000*	*4,600–6,000*	*6,200–7,990*	*8,400–9,980*	*10,500–13,800*	*14,000–16,795*	*18,000–19,950*
Number of computer models	8	13	12	19	15	13	6

Priority interrupt levels	*1*	*2*	*3*	*4*	*8*	*11*	*12*	*14*	*16*	*18*	*32*	*48*	*61*	*64*	*128*	*256*	*384*
Number of computer models	5	2	2	13	3	1	1	1	16	1	3	3	1	23	2	7	2

Response time to interrupt (microsecs.)	*Not available*	*.8-.9*	*1.1-1.9*	*2-2.7*	*3-3.8*	*4-4.6*	*5-5.25*	*6*	*7.2-8*	*9-9.75*	*10-11*	*12-12.6*	*13.2-13.6*	*14.8*	*17*	*19.4*	*23*	*36*	*91*
Number of computer models	5	4	11	9	8	7	5	11	8	2	5	2	4	1	1	1	1	1	2

Year of first delivery	*Not available*	*1965*	*1966*	*1967*	*1968*	*1969*	*1970*	*1971*
Number of computer models	10	1	0	3	5	20	37	12

In recent years the semiconductor industry has supplied more reliable and more complicated circuit functions at lower cost in integrated circuit form and the computer industry has borrowed proven techniques from the large, general purpose data processing computers offering powerful minicomputers at costs related to the size of the main memory. Each price reduction brings with it a new range of applications with the most adaptable minicomputer designs surviving the competition.

Architecture

Two structures are used in the design of minicomputers. The older multibus architecture features an arithmetic oriented organization. A number of buses interconnect the memory, the registers and accumulators. Operations usually involve accumulating data, performing an arithmetic operation and returning the results. In general its structure requires programming in terms which bear little relationship to the control system functions.

Recently, the bus systems have been simplified. Two buses, the source and the destination, together with the bus modifier, provide a path between any input and output device. Operations are performed as the body of data passes from any source to any destination, including input and output. The main memory can be treated in the same way as any other device, and the resulting modularity tends to decrease system cost.

Software

Software packages from the manufacturers or from systems houses have increased greatly, yet it is seldom that an existing package will fit the user's requirements exactly, and adaptations are common. Higher level languages, such as Indac, Fortran, Basic and Algol with their higher costs and slower speeds are not particularly useful because the minicomputer in process control loses much of its justification if it is handicapped by a compiler. Where the user insists on a higher level language as a matter of convenience, the author believes that an executive with the capability of callout in this language is a better compromise.

Programmable Controllers

As the use of minicomputers increases, it is inevitable that programmers will become scarce. To meet this challenge, a number of manufacturers have made their computers easy to program for those who know relay system design. That a minicomputer is buried in the system is further masked by calling the resulting equipment a "controller." The purpose of this euphemism is to induce the designers of relay systems to learn to use minicomputers. After acquiring an easily attained facility in programming controllers, one might explore the minicomputer language in order to promote maximum utilization of the machine.

Typical controller specifications include:

Memory: 1K expandable to 4K
Word Size: 8, 12, 16 or 18 bits
I-O Lines: 256 or 512
Input Scan Rate: 5 to 40 milliseconds per cycle

Accessory programming panels make it easier to program a relay logic system. Prices for the basic package are $3,000 to $7,000 and do not include input and output interfaces. A complete working controller costs more than $10,000 and generally does not exceed $20,000.

BIBLIOGRAPHY

Bhushan, A. K., "Guidelines for minicomputer selection," *Computer Design 10,* 4: 43, April, 1971.

Jackson, S. P., "Minicomputers and the Process Control System," ISA Show, Philadelphia, October, 1970.

Jackson, S. P. and Copeland, J. R., "Analysis of Small Computers Available for Process Control," ISA Reprint No. 69-510, October 1969.

Kaenel, R. A., "Minicomputers—A Profile of Tomorrow's Component," IEEE Transactions on Audio and Electroacoustics, Vol. AU-18, No. 4, December 1970.

Lapidus, G., "Programmable Logic Controllers—Painless Programming to Replace the Relay Bank," *Control Engineering:* 49, April 1971.

Survey of small computers, *Instruments and Control Systems:* 101, March 1971.

2.2 PRINTERS AND TYPEWRITERS

Printer Types:	(a) IMPACT PRINTERS (a1) tab printer (a2) on-the-fly printer (b) NON-IMPACT PRINTERS (b1) electrostatically guided printer (b2) light sensitive with CRT (b3) light sensitive with photo-optical systems *Note:* In the feature summary below the letters (a) to (b3) refer to the listed printer types.
Reliability and Maintenance:	100 to 1000 hours meantime between failure for group (a), an order of magnitude better for (b).
Cost:	(a1) \$600 to \$3,500, (a2) \$300 to \$61,000, (b1) \$6,500, (b2) \$2,000, (b3) \$10,000.
Throughput:	(a1) 10 to 15 characters per second, (a2) 20 to 100 characters per second, (b) 120 characters per second.
Partial List of Suppliers:	Anelex; Beckman; Bristol; Burroughs; Clary; Codamite; Computer Measurements; Connecticut Technical; Control Data; Data Products; DI/AN Controls; Franklin Electronics; Fatri-Tek, Inc.; General Electric; Hewlett-Packard; Hickok Electrical Instr.; Honeywell; IBM; Kleinschmidt Div. of SCM; Litton Datalog Div.; NCR; Photon, Inc.; Potter Instr.; ISD; Shepard Laboratories; Spectra-Strip REO Div.; Strombert Data Graphics; Teletype; Univac.

Impact printers are available with character type bars, character set bars and continuously moving character sets. Non-impact printers use photo-optical systems, electrosensitive paper or electrostatically guided ink droplets. Throughput speeds range from 10 characters per second in conventional typewriters to 40,000 characters per second in some sophisticated data graphics systems. Prices vary from a few hundred dollars to tens of thousands of dollars. Printing devices provide permanent readable records and can be installed either as separate components or in combination with keyboards to form a terminal.

"Hard copy" data records are prepared in a number of ways. The various techniques can be subdivided into impact and non-impact printers, with the latter type claiming higher speed and lower maintenance.

Impact Printers

In the simplest design, the typewriter style mechanisms are characterized by individual type bars. A single bar is engaged on command to print a given character, and the maximum speed is roughly 10 characters per second.

In "tab" printers a bar contains the entire character set. When a character is actuated, the bar moves the particular character into the print position. Typical examples are the IBM selectric typewriters with the character set on a sphere, and the printer of teletype which uses a cylinder for the same purpose. Most of these printers run at speeds of 10 to 15 characters per second.

The fastest impact mechanism is in the "on-the-fly" printers, in which a continuously moving vehicle presents all the characters to the print position and a particular character is printed by a hammer striking the paper against the given character "on the fly." Precise timing of the hammer is essential to proper operation. The vehicle may be a cylinder, chain or belt. In each case the character set is distributed around the vehicle and kept in continuous motion. Speeds of 40 characters per second per column are typical.

Non-Impact Printers

Non-impact printers use light-sensitive or voltage-sensitive papers or electrostatically guided ink droplets. With electrosensitive papers, a stylus matrix in the print head serves as a source of current, which flows from the appropriately excited matrix through the paper to a ground plate beneath.

Light-sensitive printers require a cathode ray tube or other image forming devices in conjunction with an optical system to transfer the image to the paper. Additional processing of the light-sensitive paper is frequently required to preserve the image.

In the electrostatic deflection system a stream of ink is attracted to the platen from a series of nozzles set in a horizontal plane. The ink stream is then broken into small droplets which form the characters under the influence of the electrostatic guiding system. Speeds of 120 characters per second are typical.

Throughput

The significant properties of a printer are throughput, reliability and cost. Other functional requirements which may also have a bearing on the selection of a particular unit include type-style, paper width, ease of operation and cost of interface.

The speed of a printer is expressed in characters per second or lines per minute. Individual character bar units operate at 10 characters per second. Tab printer speeds range from 10 characters per second to about 60 characters per second. Most of them offer speeds of 10 to 15 characters per second.

Reliability

Impact printer maintenance requirements vary from a complete overhaul every 400 hours to a meantime between failure of 1000 hours or more. No figures are available on the reliability of non-impact units but they are expected to be of a higher order of magnitude than for the impact types.

Cost

Cost varies with throughput and to some extent with reliability. Tab printers cost from about $600 to $3,500, with the higher price reflecting the more reliable units.

"On-the-fly" printers cost between $300 and $61,000. The lower priced units have lower throughputs (about 20 characters per second). The higher priced units have speeds of roughly 2,380 characters per second (Honeywell 222-6).

Electrostatically guided ink printers cost about $6,500 and have speeds of 120 characters per second. The cost of light-sensitive units varies from $2,000 to $10,000. The lower priced units use cathode ray tubes, the higher priced ones use photo-optical systems.

2.3 MULTIPLEXING AND PARTY LINE SYSTEMS

Multiplexing Systems

Types of Systems:	(a) Process control computer (b) Data acquisition-logging (c) Satellite multiplexing (d) General purpose digital communication
Stand-alone Capabilities:	(e) Slave operation only (f) Individual point addressing (g) Scan sequence control (h) Display or recording, or both (i) Data storage
Multiplexer Types:	(j) Electromechanical (k) Solid state (l) Analog (m) Digital (n) Signal
Control and Data Transmission:	(o) Analog (p) Digital-parallel (q) Digital-serial (r) With format (s) Without format (t) Simplex or half-duplex (u) Duplex
Cabling and Communication:	(v) Baseband (w) Amplitude shift keying (ASK) (x) Frequency shift keying (FSK) (y) Phase shift keying (PSK)

(z) Open or twisted pairs
(aa) Coaxial cable
(ab) Modem, voice grade line
Note: In the feature summary below the letters (a) to (ab) refer to the listed multiplexer systems and types.

Cost: Installed wiring cost on a per foot per conduit pair basis: $0.2 to $1.2. Analog multiplexing system cost on a per point basis: $40. Higher if simultaneous operation of several multiplexers is involved.

American Multiplex Systems, Inc. (b,c,f,g,h,i,j,k,l,m,n,q,r,t,v,z), Analog-Digital Data Systems, Inc. (b), Applied Computer Systems, Inc. (a), CompuDyne Controls, Inc. (c,e,k,l, m,q,r,t,x,z), Computer Products, Inc. (c,e,k,l,o,z), Control Data Corp. (a), Datatron, Inc. (b), Digital Equipment Corp. (a), Electronic Modules Corp. (b), Fisher Controls, Inc. (a), Foxboro Co. (a), General Automation, Inc. (a), General Electric Co. (a), Honeywell, Inc. (a), Houston Engineering Research Corp. (b), IBM Corp. (a), I/C Engineering Corp. (c,f,g,h,k,l,m,n,q,r,t,x,y,aa), Larse Corp. (d,f,g,k,m,q,r,t,v-y optional, z-ab optional), Leeds and Northrup Co. (a), Motorola Instrumentation and Control, Inc. (b,f,g, h,i,j,l,m,o,p,t,v,z), Process Automation Co. (a), Redcor, Inc. (a), Scanivalve, Inc. (c,e,j,l,n,o,p,v,z), Systems Engineering Laboratories, Inc. (a), Varian Data Machines (a,b), Vidar (b), Xerox Data Systems (a).
Note: Suppliers who furnish process control computer systems (a) generally also supply data acquisition-logging systems (b); this same equipment can ordinarily be used as satellite multiplexing systems (c). Likewise, data acquisition

systems (b) can also usually be supplied as satellite multiplexing systems (c).

Party Line Systems

Major Types:

(a) Process control systems
Complete systems
(b) Computer systems, including interface
(c) Hardware scan-sequence control capability with optional computer interface
(d) Multiplexing subsystems designed for operation by computers
(e) Stand-alone analog systems with optional computer interface
(g) Digital communication systems

Maximum Analog Scan Speeds:

(h) 1,000 to 5,000 points per second
(i) 100 to 1,000 points per second
(j) 1 to 100 points per second

Communication and Wiring:

(k) Twisted pairs
(l) Coaxial cables
(m) Modem

Monitoring and Control Capabilities:

(n) Analog input
(o) Analog output
(p) Digital input
(q) Digital output

Maximum Resolution of Analog-to-Digital Conversion (bits):

(r) 10 or less
(s) 11 to 13
(t) 14 or more

Partial List of Suppliers:

ACCO Bristol, Datamaster Division (a,c,j,m,n–q,r), American Multiplex Systems, Inc. (a,c,i,k,n–q,r), CompuDyne Controls, Inc. (g,k,n–q), Control Data Corp. (a,d,h,m,n–q,s), Direct Digital Industries, Inc. (g,k,p,q), Electronic Modules Corp. (a,e,j,k,n,p,s), General Electric Co. (a,c,j,m,n–q,s), Houston Engineering Research Corp. (a,e,j,k,n,s), I/C Engineering Corp. (a,c,h,l,n–q,s), International Business Machines

Corp. (a,d,j,k,n–q,s), Leeds and Northrup Co. (a,b,c,j,k, or m, n–q,t), Motorola Instrumentation & Control, Inc. (a,e,j,k,n,p,q).

Multiplexing systems and party line systems share a common characteristic—both enable multiple signals to time-share common equipment, which may include signal conditioners, communication channels and data processing equipment. The sharing is usually for economic reasons, with the tradeoff being a reduction in maximum speed of operations because of the necessity of performing functions sequentially rather than simultaneously. Multiplexers (Figure 2.3a) consist of terminals, to which two or more signals are connected, a switching mechanism or circuit that can route any one of the signals connected at the input terminals to common output terminals and a driver mechanism or circuitry that operates the switching mechanism. Additions to the basic multiplexer may include multiplexer terminal address decoding and shared signal conditioning equipment at the output terminals. A multiplexing system consists of the multiplexer itself and the enhancements provided in the package. The signals switched can be either analog (continuous) or digital (discrete) signals.

The term "party line" ordinarily refers to a method of communication rather than to a piece of equipment. A party line system is one in which a common communication channel is shared by equipment distributed along the length of the communication channel (Figure 2.3b). At each station the equipment sends or receives information, or both, e.g., coded electrical signals,

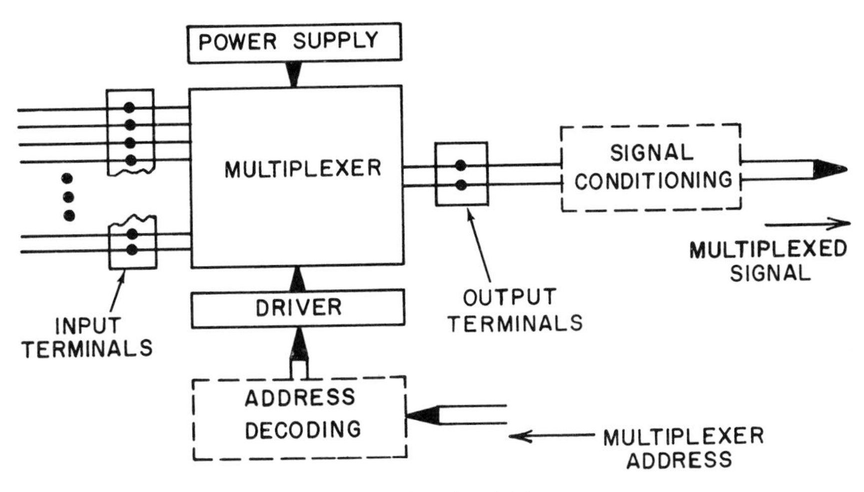

Fig. 2.3a Signal multiplexer

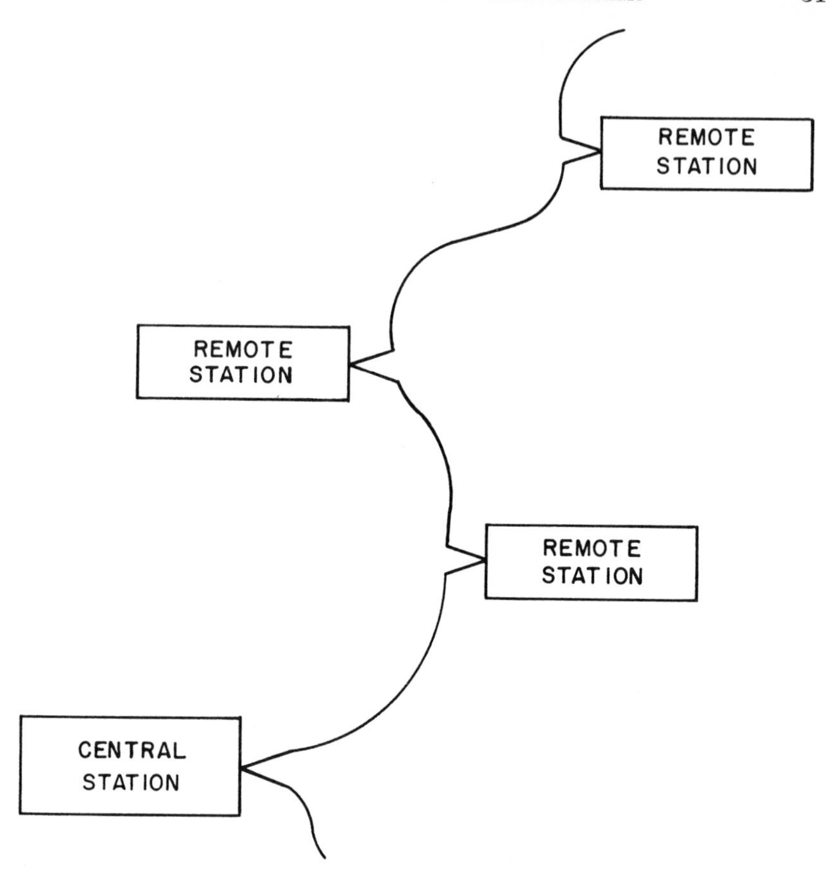

Fig. 2.3b Party line arrangement

from other stations along the party line. Typically, a station contains a local multiplexing system, allowing multiple signals to share the station and the party line. Multiplexers are therefore an integral part of party line systems.

Multiplexing and Scanning

Operating information can be gathered in a variety of ways. Consider measurements of process parameters (temperatures, pressures and flows) taken throughout a plant with instrumentation located at the various points of measurement. As an example, an operator responsible for operation of the plant might periodically tour the plant and take a reading of each measurement. This method of scanning instrument readings has in modern processing plants largely been replaced by one in which the information at the measurement locations is transmitted to the operator in a centralized control room. The information can be scanned and processed directly or indirectly in the central control room, using one of many alternative configurations of equipment.

Each measurement signal can be transmitted from the sensing element in the field to the control room on individual channels, typically pneumatic tubing or electrical conductors. The information can be displayed on individual indicators or recorders, in which case the operator "scans" the readings spatially. Certain signals may share a display device, such as a multipoint indicator or recorder, in which case a scan is affected temporally instead of spatially. These signals are sequentially switched into the common display, making this design financially attractive when the display costs are high relative to the costs of the equipment required to perform signal switching or multiplexing.

The multipoint thermocouple indicator is a simple example of multiplexing. Millivolt signals from thermocouple junctions are switched into circuitry which performs reference junction temperature compensation and linearization, and generates the driving signal for the indicator. The equipment to perform the selection and switching of the individual thermocouple leads is less costly than having separate indications for each thermocouple reading. Temperature readings are scanned either manually or automatically.

In manual operation of a multiplexer, the operator addresses—usually with a pushbutton or rotary switch—one of the thermocouple signals (channels) for temperature indication. Automatic multipoint recorders are typically cyclic; at a given speed (scan rate), the signals and the print character are selected in turn and the temperature level is recorded on the chart. Multiplexer channels can therefore be either randomly or sequentially addressed and can be operated either manually or automatically.

In the data logger or process computer, the individual signals are usually multiplexed in order to time-share amplifiers, analog-to-digital converters and hardware or software for additional processing which may include conversion to appropriate engineering units, checking for alarm conditions, logging operating data and calculating performance and control data. The distinction between this equipment and the multipoint equipment discussed earlier is that the signals are converted to digital signals prior to presentation to the process operator.

Multiplexer Designs

Important characteristics in properly selecting multiplexer equipment for a particular application include 1.) accuracy, or the degree of signal alteration through the multiplexer; 2.) size, or number of channels switched; 3.) input signal requirements such as voltage range and polarity, source impedance and common mode rejection; 4.) channel cross-talk; 5.) driving signal feed-through; 6.) maximum sampling rates; 7.) physical and power requirements, and 8.)

whether one (single-ended), two (differential) or three (including shield) lines are switched for each multiplexer point.

As discussed in Section 8.7 in Volume II, electromechanical switching with motor-driven commutators, crossbar switches or relays are more accurate and less expensive, particularly for low-level (millivolt) signals than are solid state switches. In general, electromechanical switches exhibit open to closed resistance ratios as high as 1×10^{14}, and little feedthrough of the driving signal. Although solid state switches are faster and smaller and consume less power, they are generally more likely to exhibit channel cross-talk and feed-through of the driving signal. The use of optical methods, e.g., light-emitting diodes optically coupled to transistors, and transformer coupling for isolation are becoming increasingly attractive as circuit costs decrease.

Packaged multiplexer systems may include address decoding capability, analog-to-digital conversion, sample-and-hold capability, scan sequence control and power supplies, in which the actual multiplexer is only one component in the system. A remote multiplexer station can therefore range from a multiplexer-only operated by a separate central system to a system with stand-alone, e.g., scanning and alarming, capability and computer interface to a complete, dedicated computer system.

Alternative System Configurations

The simplest system is probably one in which all process signals are brought to a common location and connected to terminals on a multiplexer. The multiplexer can be manually operated—as when an operator randomly selects a channel for a digital reading—or can be driven by other equipment. The multiplexer can be part of the "front-end" equipment of a data logging system or a process computer system, in which case the signals may be randomly addressed or sequentially scanned. Internal storage of the digital data obtained may or may not be a capability of the system.

This configuration (Figure 2.3c) has the advantage that all major equip-

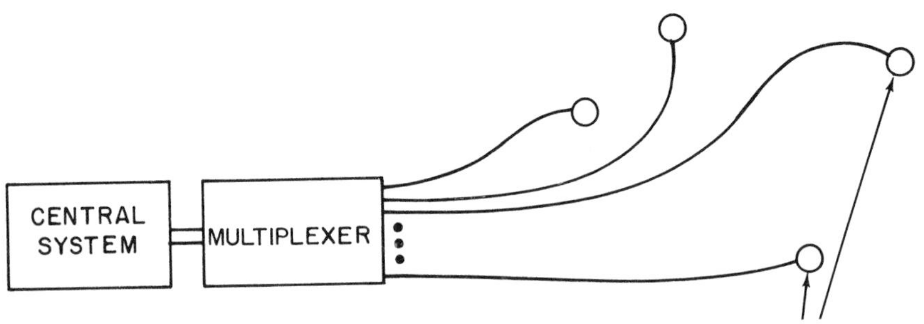

Fig. 2.3c Direct wired system

ment items are in one location, facilitating environmental protection and maintenance, and simplifying signal transmission. A disadvantage is that each signal must be individually transmitted to the multiplexer terminal. If the distances between the signal sources and the multiplexer are great, signal transmission costs may become prohibitive.

If the signals are pneumatic, they must be converted into electrical signals external to the data logger or computer. The conversion can be performed either with individual pressure-to-electrical signal converters (see Section 6.3 in Volume II) or with pneumatic signal multiplexers (Section 8.7 in Volume II). If the signals are electrical, they may be either analog or digital. Following sound wiring practices (Section 8.6 in Volume II), wiring costs range from about 20 cents per foot per conductor pair to as high as (and higher than) $1.00 per foot-pair, including conduit or trays and installation. The costs of transmitting, say, 1,000 electrical signals an average distance of 1,000 feet can easily be more than the costs of the remainder of the multiplexing and data logging or computer equipment associated with the installation.

If the signal sources are concentrated at a location remote from the data logging or computer equipment, significant savings in wiring costs may be obtained by locating the multiplexer near the signals (Figure 2.3d). The signals sequentially selected by the multiplexer can share a single, high quality communication channel to the data logger or computer. The remote multiplexer will typically be a slave of the primary system, which may either address points individually or have start-stop control on sequential or cyclic scans. The remote equipment may merely multiplex the signal to the common line as received, or additional conditioning of the signal may be performed at the remote location prior to transmission.

If the signal sources are concentrated in more than one location, it may be preferable to install more than one remote multiplexer. Each of these remote stations must communicate with the centrally located data logger or computer system. One method is to have the communication to each remote station independent of the other stations, which leads to the "radial" arrangement illustrated in Figure 2.3e. Another method, in which the communication with all stations is done through a common cable, is the party line arrangement

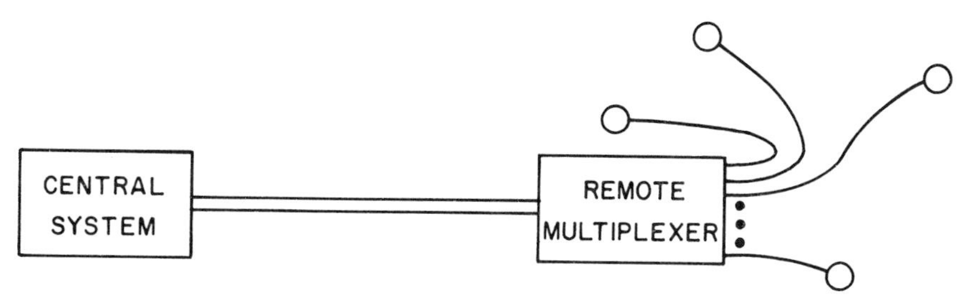

Fig. 2.3d Remote multiplexing

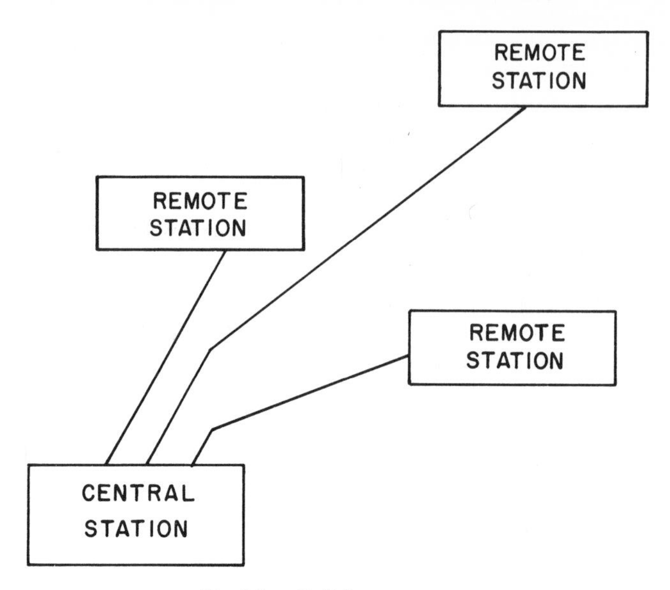

Fig. 2.3e Radial arrangement

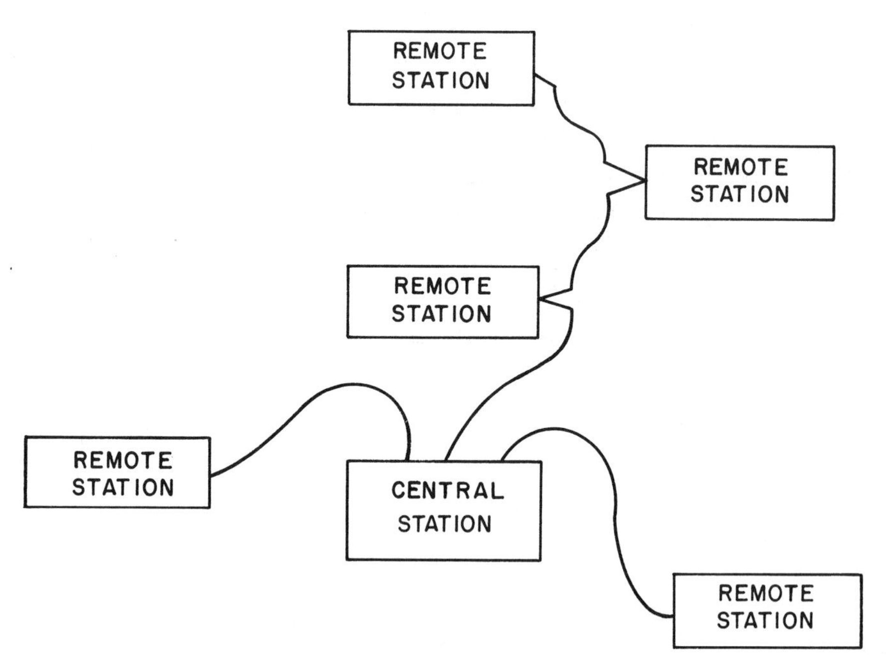

Fig. 2.3f Combination radial and party line arrangement

illustrated in Figure 2.3b. A third arrangement—a combination of the first two—is illustrated in Figure 2.3f.

The radial arrangement has the advantage that the communication links between the central station and the various remote stations are independent. If a link to one remote station is broken, communication with the others is unaffected. As the number of remote stations and their location relative to the central station increase, wiring costs tend to favor the party line arrangement, or at least a combination of radial and party line communication links. Depending on the design, a break in the party line communication link may or may not interrupt communications with all stations on the party line. In one design, for example, the party line is actually a loop, and communication can be in either direction around the loop; in this case a single break interrupts no communications. The choice of configurations depends on wiring costs, security requirements and the required number, capability and distribution of the stations.

Remote Stations

Remote multiplexing systems are characterized primarily by 1.) the amount of self-contained control logic capability, and 2.) the extent of the signal conditioning performed. A rotary electromechanical thermocouple multiplexer, responding only to "start" and "stop" commands, which switches the millivolt signals directly on the common lines to the central station, is an example of minimum remote station capability. Adding the capability to identify and respond to specific and random addressing of individual multiplexer terminals, rather than responding to only a step-to-next-position command, makes a "smarter" remote station. Adding signal conditioning (for example of the amplifier and other circuitry to effect an emf-to-current conversion of the thermocouple millivolt signals so that current can transmit the signal to the central station) is also an extension of the basic remote station capability.

These basic configurations are illustrated in Figure 2.3a for analog multiplexing systems with optional signal conditioning and address decoding. Costs for these systems are about $40 per point for sizes of 100 to 500 points and for basic speeds of 10 to 50 points per second. Simultaneous operation of multiple small multiplexers can increase effective sampling rates to about 100 points per second. Signal conditioning and address decoding capabilities can increase prices from $80 to $100 per point range.

An additional extension of remote station capability is the transfer of the analog-to-digital conversion function from the central station to the remote station. Digital rather than analog signals are then transferred to the central station. The digital communication may be on parallel paths (Figure 2.3g), or the digital data from the analog-to-digital converter may be serialized for transmission along a single path (Figure 2.3h). Typically, terminals on remote

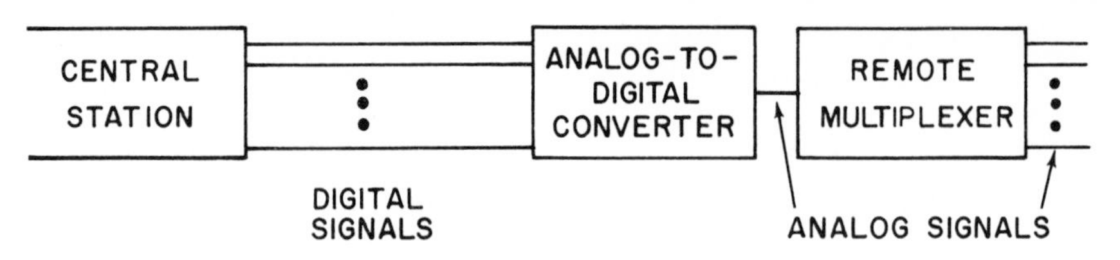

Fig. 2.3g Parallel digital transmission

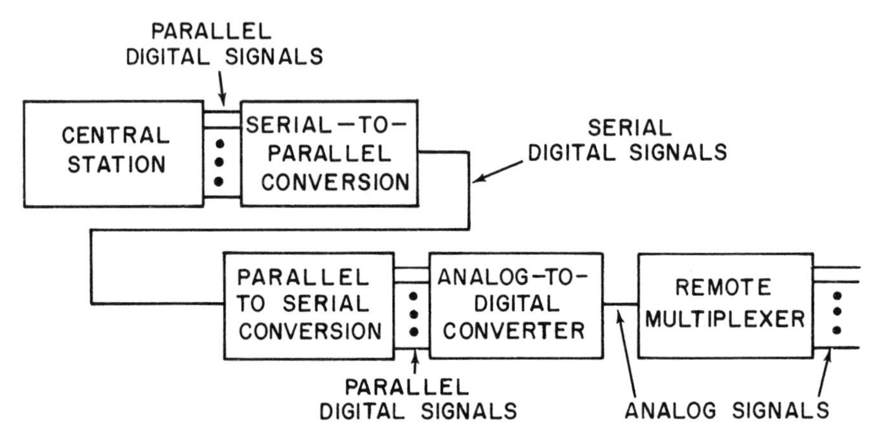

Fig. 2.3h Serial data transmission

multiplexing stations of this type are randomly addressable through serial or parallel transmission of addresses and return to the central station a message containing the requested data, the address selected, some code-checking, (e.g., parity) signals and timing and control signals if required.

The functions of the remote station can be enlarged to include stand-alone capabilities, examples of which include 1.) local control of timing and addressing so that periodic sequential scans can be made; 2.) local (e.g., core) storage of scan data; 3.) local thermocouple reference junction compensation and linearization; 4.) local testing for high-low limit alarms and equipment problems, e.g., open thermocouple detection; 5.) manual addressing and display capability; 6.) logging capability; and 7.) computational capability. A very "smart" remote station would include a digital computer with virtually the same basic scanning, computational and control capability as would be expected of the central computer. Two or more levels of computers result in a hierarchical arrangement (see Section 8.14 in Volume II).

Thus, a very wide range of capabilities is available in remote multiplexing stations from stripped, basic multiplexers with no stand-alone capability to complete, full-scale computer control systems. The efficient and economic allocation of over-all system functions among remote stations and between these stations and the central station is the primary task of the system designer.

Supercommutation and Subcommutation

Hardware for random addressing of multiplexer channels is an extra expense and complexity that is not always a requirement of the application. Scanning systems, which select channels in a fixed sequential or cyclic order, are less expensive and can be used for periodic or continuous scans of all channels.

A scanning system which is incapable of direct addressing of individual channels does not necessarily mean that all signals must be sampled at the same rate. With supercommutation and subcommutation, certain selected signals can be sampled at periods which are integer multiples or divisions of the basic scan cycle. As an example, consider a 100-point multiplexer operating with a basic scan cycle of two seconds i.e., 50 points per second. If one signal were to be connected to every tenth multiplexer terminal, it (supercommutated channel) would be sampled at a rate ten times the basic scan period, or every 0.2 second. The penalty paid for this higher sampling rate is, of course, the use of more than one multiplexer terminal per channel. In the example cited, the supercommutated channel occupies ten multiplexer terminals, leaving only 90 for the remaining signals.

Subcommutated channels are channels which share by means of another level of multiplexing a single terminal of the primary multiplexer. The subcommutated channels are sampled at a lower rate than the channels connected directly to the multiplexer. If, for example, ten subcommutated channels occupied one primary multiplexer channel, each of the ten would be sampled one-tenth as often as the basic scan cycle. Each of the subcommutated channels can in turn be further subcommutated to obtain sub-subcommutated channels. Supercommutation allows selected points to be sampled at a higher rate than that of the basic scan cycle at the expense of multiplexer channel capacity. Subcommutation allows the expansion of channel capacity at the expense of reduced sampling rates and additional submultiplexing equipment. The alternatives and the economics associated with the implementation must be considered for each application.

Digital Signal Transmission

If analog signals are converted to digital signals at the remote locations, communication with the central unit will all be digital. There is a variety of ways in which digital information can be transmitted between the remote stations and the central unit. Data loggers and computers generally communicate internally, using binary or two-state code, with each state represented by a logic-level voltage. Internal communication and communication with local peripherals are usually done a word or byte at a time, with each of the bits that make up the word or byte transmitted simultaneously on parallel paths. The electrical signal on each path is a voltage varying with time between the two possible states. The word transfer rate on the communication channel and the bit transfer rate on each path are equal.

Parallel data transmission over pairs of twisted wires is a common practice when distances are short and required transmission rates are not high. Twisted wire can generally handle a maximum of about 2,500 bits per second reliably. Higher transfer rates are possible with more expensive coaxial cable—500,000 bits per second for distances up to 200 feet and 200,000 bits per second for distances up to 400 feet have been reported for coaxial cable. Beyond 400 feet, signal loss for coaxial cable becomes significant, and the cost of cable becomes a major consideration. Serial rather than parallel transmission must be considered for distances beyond 400 feet.

In serial transmission the bits which make up a word or byte are transmitted sequentially along the same path rather than simultaneously over parallel paths. Word rates are therefore only a fraction of the bit rates. Because data logging and computing systems are generally parallel machines internally, parallel-to-serial and serial-to-parallel interface equipment is required if serial data transmission is used. The interface costs must be related to the cable costs at the distances involved and considered together with the reduced data transfer rates of serial transmission. Serial data can be transmitted with the two level or on-off (baseband) waveform, or the signals can be translated into an alternating current form through a modulation process. If the communication link is to be a modulated alternating current type, modulators and demodulators (commonly called modems) are required at each end of the communication channel.

Three basic modulation alternatives are available as shown in Figure 2.3i. The baseband signal can modulate the amplitude of a sinusoidal carrier; this is amplitude modulation or, in common data transmission parlance, amplitude shift keying or ASK. The two-level signal can control the frequency of the alternating current signal; this is frequency modulation or frequency shift keying, FSK. The third technique is control of the phase of the sinusoidal signal with the two-level signal for phase modulation or phase shift keying, PSK.

FSK is generally preferred for low-speed applications, i.e., when data rates are not greater than about 1,200 bits per second. ASK and PSK are most popular for higher-speed data transmission. Most low-speed data communication systems use binary coding, but more than two states (amplitudes, frequencies and phases) can be used with any modulation technique if the level of noise and other signal distortions are not excessive. With more than two states, each time-frame carries more than one bit of information. The term baud is used for pulses (two or more states) per second; with binary transmission, bits and bauds are equivalent. If R states are used, each baud contains the equivalent information of $\log_2 R$ bits, and the bit rate is R times the baud rate.

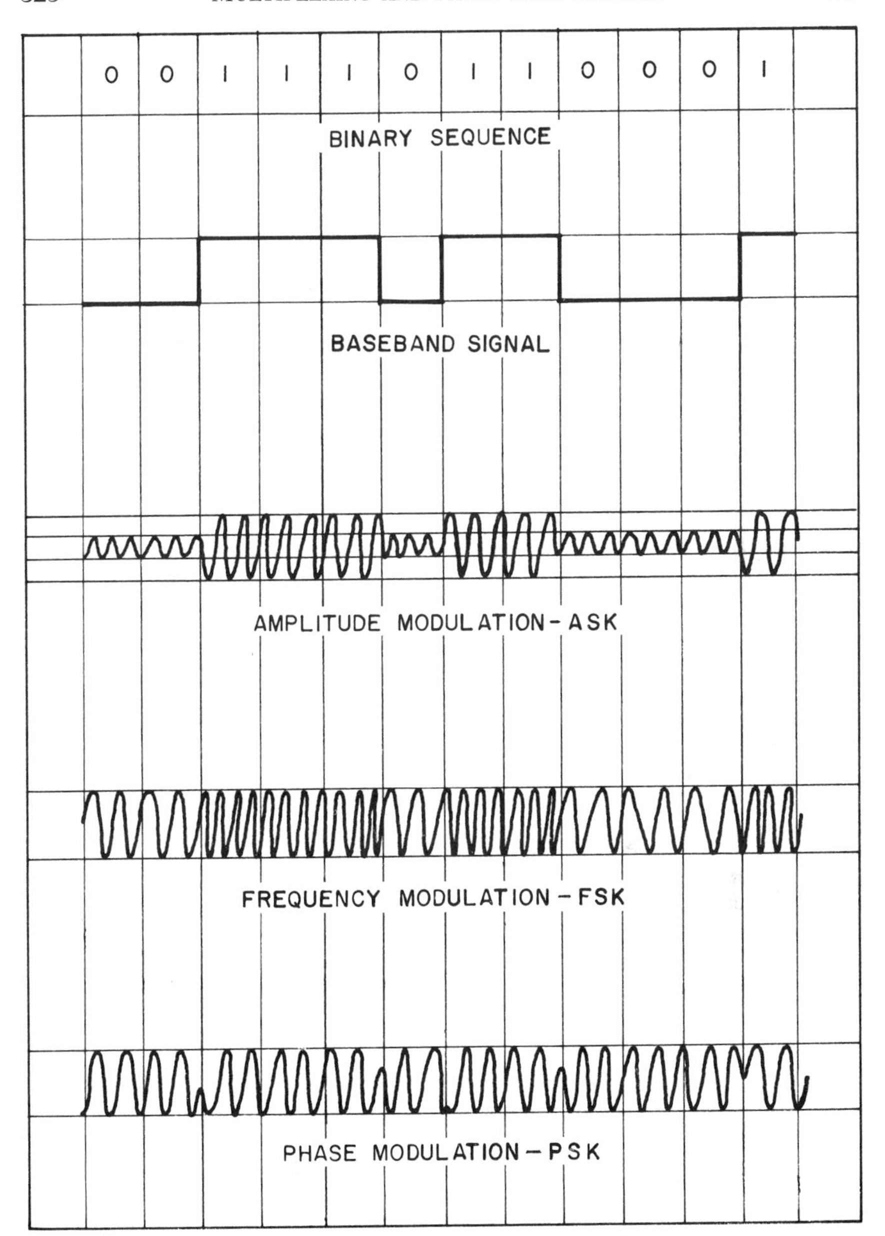

Fig. 2.3i Modulation methods

Communication channels can be multiplexed or shared, so that more than one "message" can be transmitted over the same physical link. Two alternatives are frequency division multiplexing and time division multiplexing. In frequency division multiplexing, the entire channel bandwidth is divided into a number of smaller bandwidth channels (subchannels), and signals are sent in parallel across the subchannels. The carrier frequencies for each channel are spaced across the entire available channel bandwidth so that no interference occurs between subchannels; frequency separation is achieved by band-pass filters. Filters of nearly exact characteristics and highly linear amplifiers are required for FDM. In time division multiplexing, bits or bytes are interleaved in time. The multiplexing can be either synchronous—in which timing is controlled by a master system clock—or isochronous (time-independent synchronous), in which storage buffers instead of a master clock maintain order.

Data concentrators perform functions similar to data multiplexers, with one major difference. Concentrators format (or reformat) the data prior to transmission. Concentrators are generally minicomputers or programmable multiplexers that take character streams and strip away the unnecessary bits and pack the remainder to transmit complete messages with formats. If, for example, the data transmitted consisted of a series of "frames" each containing a number of bits defining a character (or number), plus additional timing, control and parity bits, the concentrator may retransmit the block of character bits (message) with timing, control and parity bits only for the entire block.

The digital information transferred in party line systems may consist of addresses for station and point identification; data, which may define the status of field contacts, control commands or the value of a converted analog signal; self-test and security coding; and system and timing and control bits. In some cases, addresses and data use separate lines; in others, the message frame includes both address and data. Nearly all systems have some sort of security provisions—usually a cyclic (e.g., the Bose-Chandhuri) code or parity bits included in the messages transmitted. Sometimes both horizontal and vertical parity checks are made.

REFERENCES

1. Allen, D. P., How to choose data acquisition systems, *Control Engineering,* 102, Nov., 1969.
2. Arnett, W., Metallic contacts vs. solid state switches, *Control Engineering,* 102, Nov., 1969.
3. Aronson, R. L., Line-sharing systems for plant monitoring and control, *Control Engineering,* 57, Jan., 1971.
4. Ball, J., Tying computers together, *Control Engineering,* 119, Sept., 1966.
5. Blasdell, J. H., Jr., Getting the most out of data links through multiplexing, *Computer Decisions,* 14, Jan., 1971.
6. Buckley, J. E., Telephone carrier systems, *Computer Design,* 10, March, 1971.
7. Cofer, J. W., Saving money on data transmission as signals take turns on party line, *Electronics,* April 15, 1968.

8. Digitized thermocouple compensation yields direct reading for data logger, *Electronics,* Feb. 2, 1970.
9. Feldman, R., Selecting input hardware for DDC, *Control Engineering,* 75, Aug., 1967.
10. Harper, W. L., The remote world of digital switching, *Datamation,* 22, March 15, 1971.
11. Hersch, P., Data communications, *IEEE Spectrum,* 47, Feb., 1971.
12. Hoeschele, D. F., Jr., *Analog-to-Digital/Digital-to-Analog Conversion Techniques,* New York: Wiley, 1968.
13. Insose, F., Takasugi, K. and Hiroshima, M., "A Digital Data Highway System for Process Control," ISA Paper 70–510. Presented at the 1970 ISA Annual Conference, Philadelphia, October 26–29, 1970.
14. Kaiser, V. A., New configurations in computer control, *Instrumentation Technology,* 69, Oct., 1968.
15. Klosky, R. A. and Green, P. M., "Computer Architecture for Process Control," ISA Paper 69–512. Presented at the 1969 ISA Annual Conference, Houston, October 27–30, 1969.
16. Pierce, J. R., Some practical aspects of digital transmission, *IEEE Spectrum,* 63, Nov., 1968.
17. Simon, H., "Multiplex Systems Save Multibucks in Refinery and Chemical Plants," ISA Paper 70–564. Presented at the 1970 ISA Annual Conference, Philadelphia, October 26–29, 1970.

Chapter III

HUMAN ENGINEERING READ-OUTS AND DISPLAYS

N. O. Cromwell, J. A. Gump,
D. W. Lepore, D. D. Russell,
M. G. Togneri, R. A. Williamson

CONTENTS OF CHAPTER III

3.1 HUMAN ENGINEERING

Human engineering, also known as Engineering Psychology, Human Factors Engineering and Ergonomics (in England), is probably the most basic form of engineering because the fashioning of any tool to suit the hand is part of human engineering. The discipline began as a cut-and-dried effort in which modification was conveniently left to future planning rather than being made part of the original design. This approach sufficed when technological progress was sufficiently slow to allow several generations of improvement. The efficiency and cost of tools were such that improper original design was not severely punishing.

Current technology has supplied man with highly efficient, complex and, therefore, expensive tools, the ultimate success of which rests with his ability to use them. Today's tools are required not only to fit the human hand, but to reinforce many physiological and psychological characteristics, such as hearing, color discrimination and signal acceptance rates. The complexity and cost of tools make the trial-and-error method unacceptable.

World War II introduced our first attempt at technical innovation and our first serious effort to treat human engineering as a discipline. Postwar economic pressures have been almost as grudging of critical mistakes as war has been; thus, the role of human engineering has increased with technological progress. A few highlights of human engineering described in this section will aid the instrument engineer in evaluating the impact of this discipline on process plants.

Applicability extends from instruments as sophisticated as those used in the Apollo moon flight to those as prosaic as the kitchen stove.

Man-machine System

The man-machine system is a combination of men and equipment interacting to produce a desired result from given inputs. The majority of industrial control systems are indeed sophisticated man-machine types.

Because the purposes of an instrument and control system are limited, satisfactory definition is difficult. Subdividing the purpose into its constituent goals is an important first step in clarification. Many operators cannot supply much more information than what is presently available and regarded as insufficient. Learning from mistakes, although necessary, is not a desirable engineering approach.

The backbone of a man-machine system is a flow chart of man-related activities. This approach concentrates design efforts on improving operator information and control capability—not just on improving instruments. Proper design criteria are obtained by asking what should be done, not what has been done. The machine should be adapted to the man. Two constraints exist including 1.) the level of ability of the average operator and 2.) the amount of information and control required by the process. Sight, hearing and touch help the operator to control information storage and processing, decision making and process control. Information is processed by man and machine in a similar manner (Figure 3.1a).

The process being similar, the abilities to execute diverse functions vary greatly between man and machine. Man is better suited than machines to process information qualitatively, an aptitude needing experience or judgment. His ability to store enormous quantities of data (10^{15} bits) and to use it subconsciously in decision-making allows him to reach decisions with a high probability of being right, even if all necessary information is not available. Machines are suited for quantitative processing and storage of relatively small quantities of exact data (10^{8} bits) with rapid access. Machine decisions must be based on complete and exact data.

In a process control system the differences are best applied by training the operator in emergency procedures and letting the machine store data on limits that, when exceeded, trigger the emergency. Table 3.1b summarizes man-machine characteristics. Information being the primary quantity processed by a man-machine system, a brief account of informational theory is necessary.

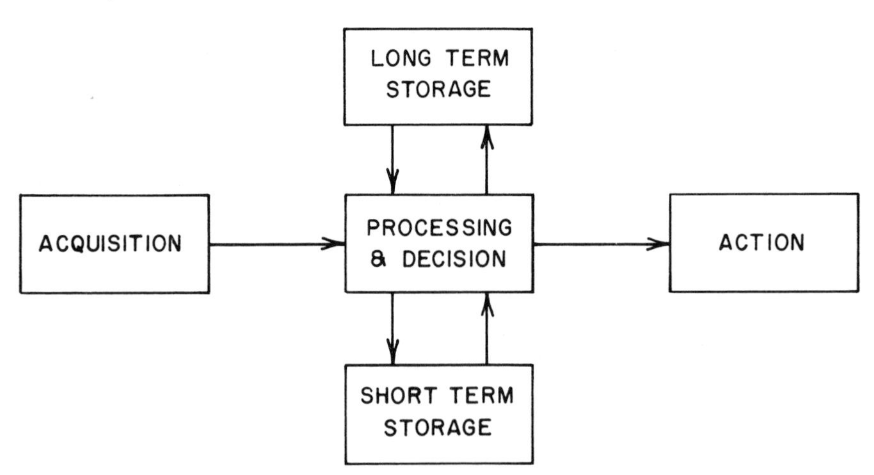

Fig. 3.1a Information processing: man or machine

Table 3.1b
CHARACTERISTIC COMPARISON: MAN VS. MACHINE

Man Excels in	*Machines Excel in*
Detection of certain forms of very low energy levels.	Monitoring (both men and machines)
Sensitivity to an extremely wide variety of stimuli	Performing routine, repetitive, or very precise operations
Perceiving patterns and making generalizations about them	Responding very quickly to control signals
Detecting signals in high noise levels	Exerting great force, smoothly and with precision
Ability to store large amounts of information for long periods—and recalling relevant facts at appropriate moments	Storing and recalling large amounts of information in short time-periods
Ability to exercise judgment where events cannot be completely defined	Performing complex and rapid computation with high accuracy
Improvising and adopting flexible procedures	Sensitivity to stimuli beyond the range of human sensitivity (infrared, radio waves and so forth)
Ability to react to unexpected low-probability events	Doing many different things at one time
Applying originality in solving problems, i.e., alternate solutions	Deductive processes
Ability to profit from experience and alter course of action	Insensitivity to extraneous factors
Ability to perform fine manipulation, especially where misalignment appears unexpectedly	Ability to repeat operations very rapidly, continuously, and precisely the same way over a long period
Ability to continue to perform even when overloaded	Operating in environments which are hostile to man or beyond human tolerance
Ability to reason inductively	

Information Theory

The relatively new discipline of information theory defines the quantitative unit of evaluation as a binary digit or "bit." For messages containing a number of equal possibilities (n), the amount of information carried (H) is defined as

$$H = \log_2 n, \qquad 3.1(1)$$

and the contact of a pressure switch (ps) carries one bit of information

$$H_{ps} = \log_2^2 \text{ (alternatives)} = 1 \qquad 3.1(2)$$

Most of the binary information in process control is such that probabilities of alternatives are equal. If this is not so, as with a pair of weighted dice, the total information carried is reduced. In an extreme case (if the switch were shorted), probability of a closed contact is 100%, probability of an open contact is 0% and total information is zero.

Characteristics of Man

Much has been learned about man as a system component since organized research in human factors was initiated during World War II. Anthropometric (body measurement) and psychometric (psychological) factors are now available to the designer of control systems as are many references to information on, for instance, ideal distances for perception of color, sound, touch and shape. The engineer ought to keep in mind that generally the source of data of this sort is a selected group of subjects (students, soldiers, physicians and so on) that may or may not be representative of the operator population available for a given project.

Body Dimensions

The amount of work space and the location and size of controls or indicators are significant parameters on operator comfort and output and require knowledge of body size, structure and motions (anthropometry). Static anthropometry deals with the dimensions of the body when it is motionless.

Dynamic anthropometry deals with dimensions of reach and movement of the body in motion. Of importance to the instrument engineer is the ability of the operator to see and reach panel locations while he is standing or seated.

Figure 3.1c and Table 3.1d illustrate dimensions normally associated with instrumentation (additional information in Reference 1).

Information Capability

Human information processing involves sensory media (hearing, sight, speech and touch) and memory. The sensory channels are the means by which man recognizes events (stimuli). The range of brightness carried by a visual stimulus, for example, may be considered in degrees and expressed in bits, and discrimination between stimuli levels is either absolute or relative. In relative discrimination, the individual compares two or more stimuli; in absolute discrimination, a single stimulus is evaluated.

Man performs much better in relative than in nonrelative discrimination. Typically, a subject can separate 100,000 colors when comparing them to each

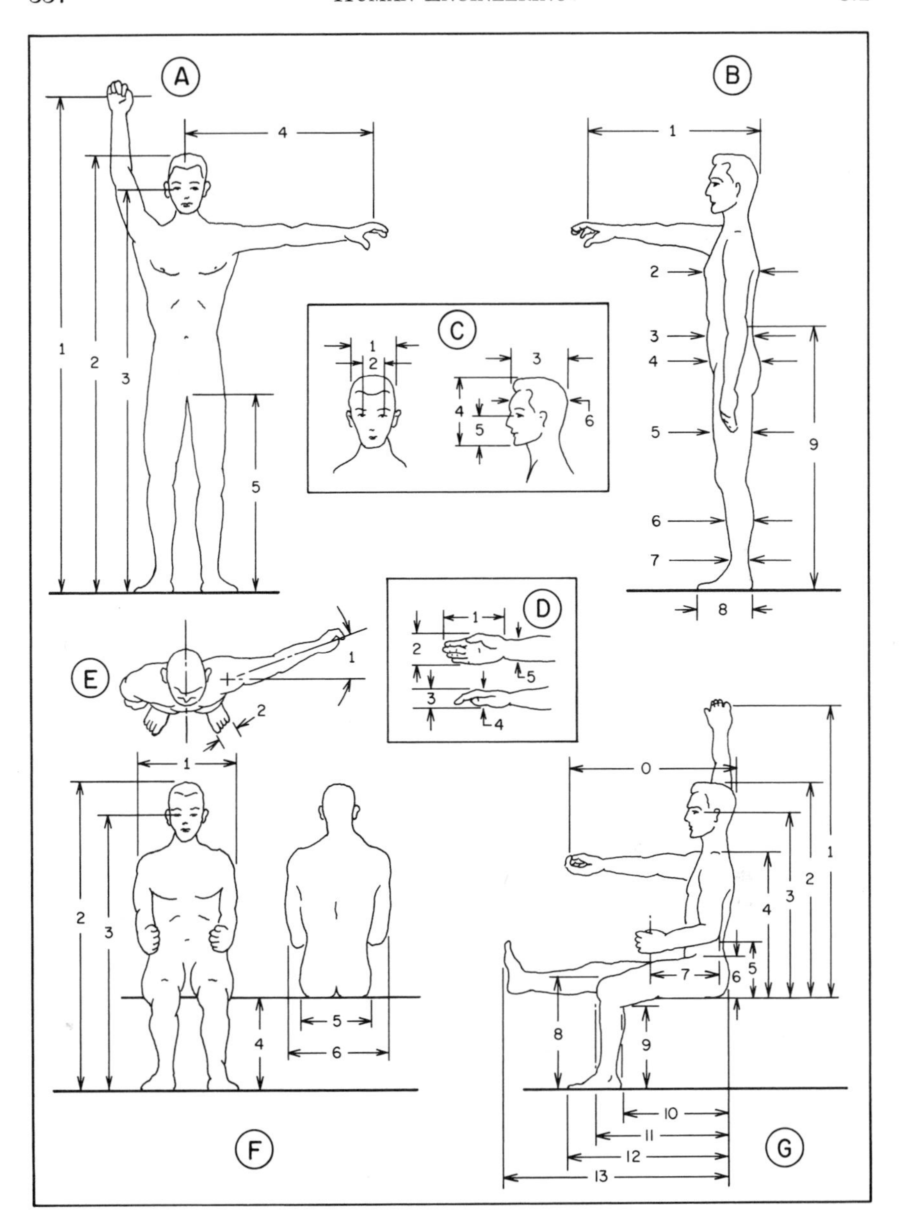

Fig. 3.1c Body dimension chart. Dimensions are summarized in Table 3.1d.

other; the number is reduced to 15 colors when each must be considered individually. Data in Table 3.1e deal with absolute discrimination encountered in levels of temperature, sound and brightness.

In addition to the levels of discrimination, sensory channels are limited

Table 3.1d
MALE HUMAN BODY DIMENSIONS

Selected dimensions of the human body (ages 18 to 45). Locations of dimensions correspond to those in Figure 3.1c.

		Dimensional Element	*Dimension (in inches except where noted)* 5th Percentile	95th Percentile
		Weight in pounds	132	201
A	1	Vertical reach	77.0	89.0
	2	Stature	65.0	73.0
	3	Eye to floor	61.0	69.0
	4	Side arm reach from center line of body	29.0	34.0
	5	Crotch to floor	30.0	36.0
B	1	Forward arm reach	28.0	33.0
	2	Chest circumference	35.0	43.0
	3	Waist circumference	28.0	38.0
	4	Hip circumference	34.0	42.0
	5	Thigh circumference	20.0	25.0
	6	Calf circumference	13.0	16.0
	7	Ankle circumference	8.0	10.0
	8	Foot length	9.8	11.3
	9	Elbow to floor	41.0	46.0
C	1	Head width	5.7	6.4
	2	Interpupillary distance	2.27	2.74
	3	Head length	7.3	8.2
	4	Head height	—	10.2
	5	Chin to eye	—	5.0
	6	Head circumference	21.5	23.5
D	1	Hand length	6.9	8.0
	2	Hand width	3.7	4.4
	3	Hand thickness	1.05	1.28
	4	Fist circumference	10.7	12.4
	5	Wrist circumference	6.3	7.5
E	1	Arm swing, aft	40 degrees	40 degrees
	2	Foot width	3.5	4.0
F	1	Shoulder width	17.0	19.0
	2	Sitting height to floor (std chair)	52.0	56.0
	3	Eye to floor (std chair)	47.4	51.5
	4	Standard chair	18.0	18.0
	5	Hip breadth	13.0	15.0
	6	Width between elbows	15.0	20.0
G	0	Arm reach (finger grasp)	30.0	35.0
	1	Vertical reach	45.0	53.0
	2	Head to seat	33.8	38.0
	3	Eye to seat	29.4	33.5
	4	Shoulder to seat	21.0	25.0
	5	Elbow rest	7.0	11.0
	6	Thigh clearance	4.8	6.5
	7	Forearm length	13.6	16.2
	8	Knee clearance to floor	20.0	23.0
	9	Lower leg height	15.7	18.2
	10	Seat length	14.8	21.5
	11	Buttock-knee length	21.9	36.7
	12	Buttock-toe clearance	32.0	37.0
	13	Buttock-foot length	39.0	46.0

NOTE: All except critical dimensions have been rounded off to the nearest inch.

Table 3.1e
ABSOLUTE DISCRIMINATION CAPABILITY
(Maximum Rates of Information Transfer)

Modality	*Dimension*	*Maximum Rate (Bits/Stimulus)*
Visual	Linear extent	3.25
	Area	2.7
	Direction of line	3.3
	Curvature of line	2.2
	Hue	3.1
	Brightness	3.3
Auditory	Loudness	2.3
	Pitch	2.5
Taste	Saltiness	1.9
Tactile	Intensity	2.0
	Duration	2.3
	Location on the chest	2.8
Smell	Intensity	1.53
	(Multi-Dimensional Measurements)	
Visual	Dot in a square	4.4
	Size, brightness, and hue (all correlated)	4.1
Auditory	Pitch and loudness	3.1
	Pitch, loudness, rate of interruption, on-time fraction, duration, spatial location	7.2
Taste	Saltiness and sweetness	2.3

in the acceptance rates of stimuli. The limits differ for the various channels and are affected by physiological and psychological factors. A study by Pierce and Karling suggests a maximum level of 43 bits per second.

Several methods of improving information transmission to man include coding (language), multiple channels (visual and auditory) and organization of information (location of lamps; sequence of events). Simultaneous stimuli, similar signals with different meanings, noise, excessive intensity, and the resulting fatigue of the sensors—stimuli that approach discrimination thresholds—are all detrimental to the human information process.

The limit of the operator's memory affects information processing and is considered here in terms of long and short span. Work by H. G. Shulmann[2] indicates that short-term memory uses a phonemic (word sounds) and long-term memory uses a semantic (linguistic) similarity. These approaches tend to make codes that are phonemically dissimilar less confusing for short-term recall, and codes that are semantically different less confusing for long-term recall. Training, for example, uses long-term memory.

Performance (Efficiency and Fatigue)

Motivation, annoyance, physical condition and habituation influence the performance of primarily non-physical tasks. These influences makes the subject of efficiency and of fatigue controversial. Both the number and frequency of stimuli to which the operator responds affect efficiency and fatigue.

Performance increases when the number of tasks increase from a level of low involvement to high involvement. When rates of information processing are raised beyond the limits of the various senses, performance rapidly deteriorates.

Application of Human Engineering

System Definition

The seemingly commonplace task of defining the man-machine system is more difficult to formalize than expected because 1.) operators themselves are too closely involved for objectivity; 2.) engineers tend to make assumptions based on their technical background; and 3.) equipment representatives focus their attention mainly on the hardware.

The definition can be more highly organized around the purpose components and functions of the system. The purpose of the system is its primary objective; it significantly affects design. In a system for continuous control of a process, operator fatigue and attention span are of first importance. Attention arousal and presentation of relevant data are important when the objective is the safety of personnel and equipment, and the operator must reach the right decision as quickly as possible.

Components of the system are divided into those that directly confront the operator and those that do not. The distinction is used to locate equipment. For example, a computer process console should be located for operator convenience, but maintenance consoles should be out of the way so as to reduce interference with operation.

System functions are primarily the operator's responsibility, and operator function should be maximized for tasks that involve judgment and experience and minimized for those that are repetitive and rapid. Table 3.1b is an efficient guideline for task assignment.

Statistics of Operator Population

As was mentioned in the introduction to this section, the statistical nature of human engineering data influences individual applications. A specific problem demands full evaluation of the similarity among operators. For example, if color blindness is not a criterion for rejecting potential operators, color coding must be backed by redundant means (shape) to assure recognition, because one out of four subjects has some degree of color blindness.

Variations and exceptions are to be expected and recognized, men being significantly different in physical size, native aptitude and technical acculturation. Control systems and instruments should be designed for the user. A frequent mistake made by the instrument engineer is to assume that the operator will have a background similar to his.

Some operators may have never ridden a bicycle, much less driven a car or taken a course in classic physics. In arctic regions, massive parkas and bulky gloves change all the figures given in body dimension tables. It is only in western culture that left-to-right motion of controls and indicators is synonymous with increasing values.

Setting Priorities

The diverse functions of the system confronting the operator vary in importance. The most accessible and visible areas should be assigned to important and frequently used items. Vertical placement is critical, the ideal elevations being from 3′11″ to 6′3″ (centered around the average eye level of the [5′1″] man). Above this area, visibility is good, but accessibility falls off (Figure 3.1f). At the lower segment even grocers know that objects are virtually hidden. Canting the panel helps to make subordinate areas more usable. (For more on control panel design, see Section 5.1 of Volume II.)

Instruments should be arranged from left to right (normal eye scan direction) in 1.) spatial order, to represent material movement (storage and distribution); 2.) functional order, to represent processing of material (distillation, reaction and so on); and 3.) sequential order, to modify the functional order approach to aid the operator in following a predetermined sequence even under stress.

Criteria for wrap-around consoles are different because the operator does not move along the panel, but sits in front of it. The primary objective, however, is still to locate instruments on the basis of function and importance (Figure 3.1g).

Locating similar functions in relatively the same place gives additional spatial cues to the operator by associating location with function. On this account it is advisable to divide alarms into groups related to meaning, including 1.) alarms requiring immediate operator action (for example, high bearing temperatures) which must be read when they register; 2.) alarms for

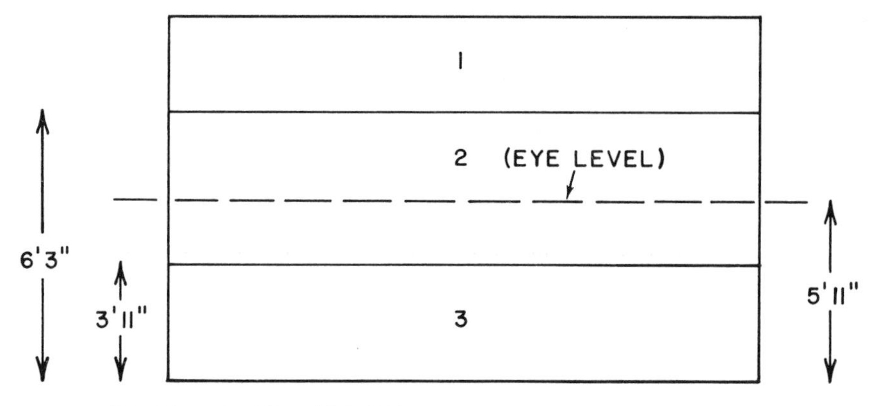

Fig. 3.1f Panel for standing operator. Variables are grouped by function, with indication and controls in area 2, seldom used controls or pushbuttons in area 3, indicators (analog or digital) only in area 1 and alarm and on-off indicators in area 1.

abnormal conditions which require no immediate action (standby pump started on failure of primary) which must be read when they register; and 3.) status annunciator for conditions which may or may not be abnormal. These are read on an as needed basis.

Alarms can be separated from other annunciators, have different physical appearance and give different audible cues. For example, group 1 alarms can have 2″ × 3″ flashing lamp windows with a pulsating tone. Group 2 alarms can have 1″ × 2″ flashing lamp windows with a continuous tone and group 3 can have 1″ × 1″ lamp windows without visual or auditory cues.

Alarms are meaningless if they occur during shutdown or as a natural result of a primary failure. Whether usable or not, they capture a portion of the operator's attention. Several alarms providing no useful information can be replaced by a single shutdown alarm. Later, the cause of the shutdown will be of keen interest to maintenance personnel; to provide detailed information about the cause, local or back-of-panel annunciators can be used.

Hardware Characteristics

Digital indicators are best when rapid reading or obtaining accurate values is of utmost concern. However, they give only a useless blur at a high rate of change. Analog indications give rapid evaluation of changing values. The reading can be related to zero and full-scale for interpretation of rate of change (estimated time to reach minimum or maximum levels). Corre-

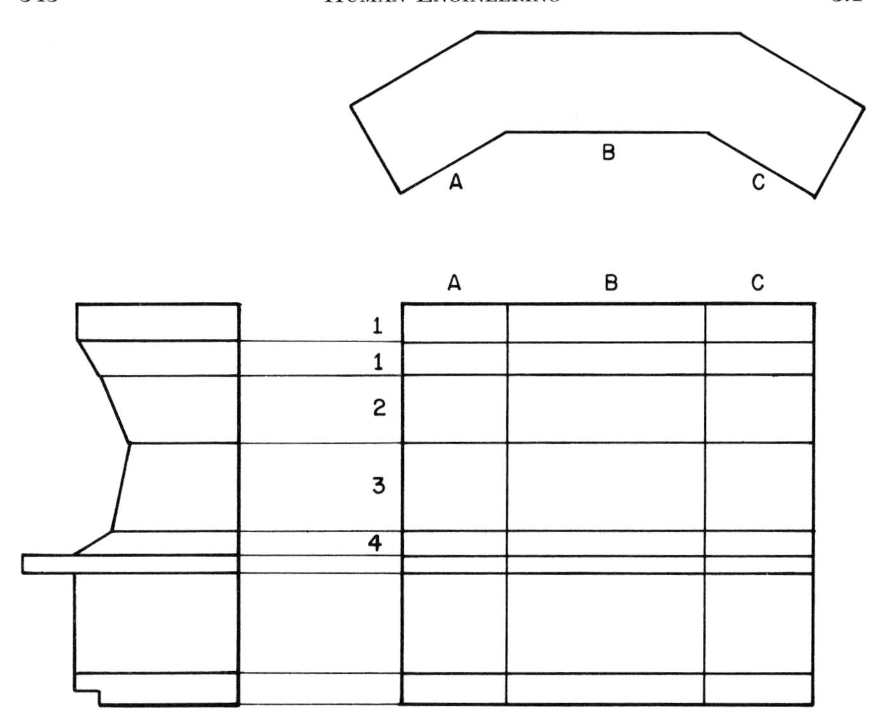

TYPE OF DEVICE	FREQUENCY OF USE	RECOMMENDED LOCATION
INDICATORS & CONTROLS	HIGH LOWER	3B 3A, 3C
CONTROLS ONLY	HIGH LOWER	4B 4A, 4C
INDICATORS ONLY	HIGH LOWER	2B 2A, 2C
STATUS INDICATION AND ALARMS	HIGH LOWER	1B 1A, 1C

Fig. 3.1g Wrap-around console for seated operator

spondingly, analog information is difficult to read when exact values are needed.

Display patterns should be natural to the eye movement—in most cases, left-to-right. An operator scanning only for abnormal conditions will be aided by this arrangement. The approach has been so successful that many of the major control instrument manufacturers produce a line of vertical scale deviation indicators (Figure 3.1h).

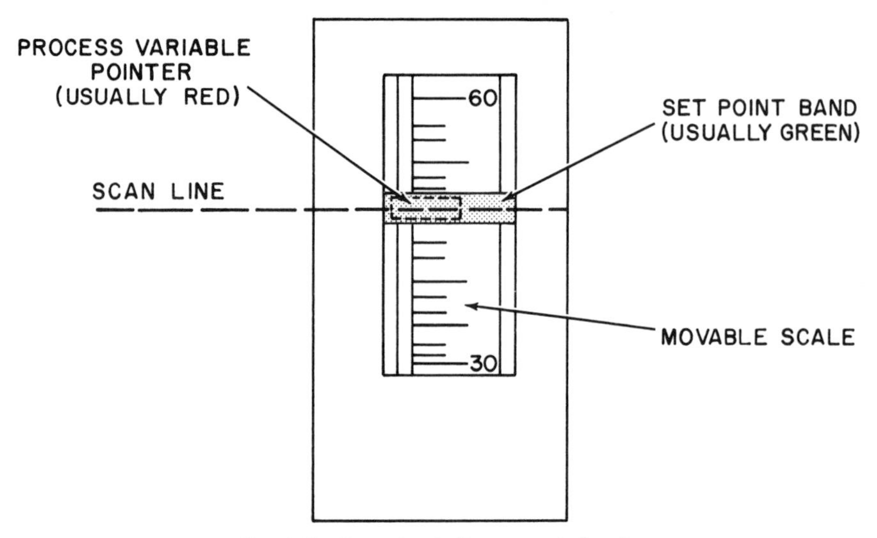

Fig. 3.1h Deviation indicator vertical scale

An alternate system is shown in Figure 3.1i, in which two rows are combined to give the effect of one line containing twice the number of readings. In our culture, horizontal scanning is eight times faster than vertical scanning.

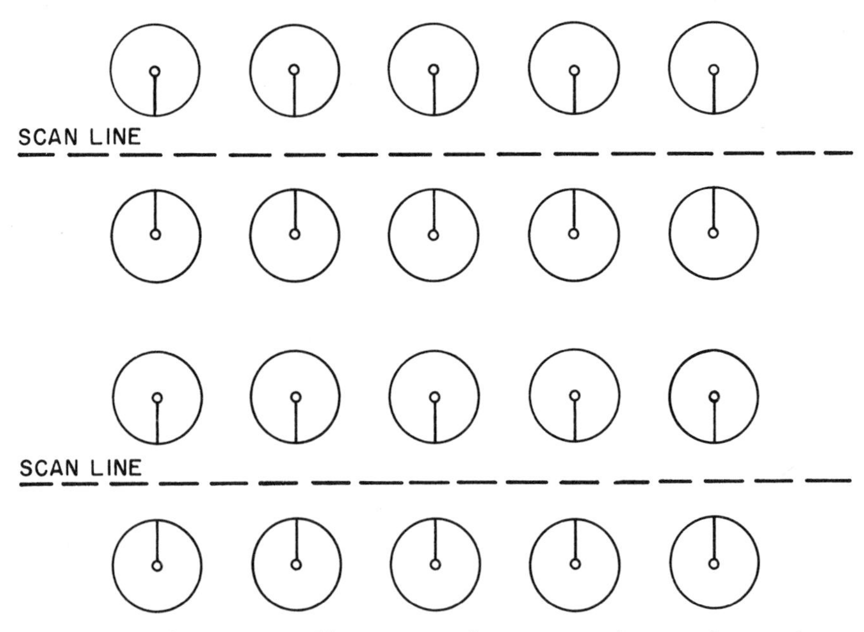

Fig. 3.1i Dial orientation enabling two rows of instruments to be scanned at one time

Viewing distances are important in displays with cathode ray tubes (CRT). Typical viewing distances are given in Table 3.1j:

Table 3.1j
VIEWING DISTANCES

Screen Size (inches)	*Distance from Screen (feet)*
9	1½–2½
15–17	2½–6
17–19	6–10
19–23	10–20
23–30	20–30

Black-and-white screens are visible at nearly any light level, provided that reflected light is eliminated by a hood or circularly polarized filter. Flicker due to slow refresh rate is a potential fatigue factor. Rates above 25 flickers per second are reasonably stable. Color CRTs require a lighting level above 0.01 lumen to insure that color vision (retinal rods) is activated and color discrimination is possible.

Information Coding

Several factors affect the ability of people to use coded information. The best method to predict difficulty is by the channel center load, which uses the basic information formula of equation 3.1(1). By this method the informational load is determined by adding the load of each character.

In a code using a letter and a number, the character load is

letter (alphabet) $L_h = \text{Log}_2\ 26 = 4.70$
number (0–9) $N_h = \text{Log}_2\ 10 = 3.32$
Code information $(C_h) = 4.70 + 3.32 = 8.02$

This method applies when the code has no secondary meaning, such as A 3 used for the first planet from the sun, Mercury. When a code is meaningful in terms of the subject, such as O 1 for Mercury (the first planet), the information load is less than that defined by the formula.

Typical laboratory data indicate that when a code carries an information load of more than 20, man has difficulty in handling it. Tips in code design include 1.) using no special symbols that have no meaning in relation to the data; 2.) grouping similar characters (HW5, not H5W); 3.) using symbols that have a link to the data (MF for male-female, not 1–2); and 4.) using dissimilar characters (avoiding zero and the letter O, and I and the numeral 1).

The operator environment can be improved by attending to details such as scale factor, which is the multiplier relating the meter reading to engineering units. The offset from mid-scale for the normal value of the variable, and the desired accuracy of the reading for the useful range are constraints. Within these constraints, the selected scale factor should be a one-digit number, if possible. If two significant digits are necessary, they should be divisible by (in order of preference) 5, 4 or 2. Thus, rating the numbers between 15 and 20 as to their desirability as scale factors would give 20, 15, 16, 18 (17, 19).

Operator Effectiveness

The operator is the most important system component. His functions are all ultimately related to those decisions that machines cannot make. He brings into the system many intangible parameters (common sense, intuition, judgment and experience)—all of which relate to the ability to extrapolate from stored data which sometimes is not even consciously available. This ability allows the operator to reach a decision which has a high probability of being correct even without all the necessary information for a quantitative decision.

The human engineer tries to create the best possible environment for decision making by selecting the best methods of information display and reducing the information through elimination of noise (information type), redundancy (even though some redundancy is necessary) and environmental control.

Operator Load

The operator's load is at best difficult to evaluate, since it is a subjective factor and because plant operation varies. During the past twenty years, automation and miniaturization have exposed the operator to a higher density of information, apparently increasing his efficiency in the process. Since increases in load increase attention and efficiency, peak loads will eventually be reached that are beyond the operator's ability to handle.

Techniques for reducing operator load include the simultaneous use of two or more sensory channels.

Vision is the most commonly used sensory channel in instrumentation. Sound is next in usage. By attracting the operator's attention only as required, it frees him from continuously having to face the display and its directional message. The best directionality is given by non-pure tones in the frequency range of 500 to 700 Hz.

Information quantity can be decreased at peak load. In a utility failure for example, more than half the alarms could be triggered when many of the control instruments require attention. It would be difficult for the operator to follow emergency procedures during this shutdown. A solution gaining acceptance in the industry involves selectively disabling nuisance alarms during various emergencies. Some modern high density control rooms have more than 600 alarms without distinction as to type or importance.

Environment

The responsibilities of human engineering are broad and include spatial relations—distances that the operator has to span in order to reach his equipment and his fellow operators. Controlling temperature and humidity to maintain operators' and equipment efficiency is also an important responsibility of the human engineer.

Light must be provided at the level necessary without glare or superfluous eyestrain. Daylight interference caused by the sun at a low angle can be an unforeseen problem. Last-minute changes in cabinet layout can create undesirable shadows and glare from glass and highly reflective surfaces. Flat, brushed or textured finishes reduce glare. Poor illumination can be reduced by light colors, low light fixtures and special fixtures for special situations. Fluorescent lighting, because of its 60 Hz flicker rate, should be supplemented by incandescent lamps to reduce eye fatigue, particularly important in the presence of rotating equipment, in order to eliminate strobing effect. Tables 3.1k and 3.1 l give detailed information on typical lighting applications.

Glare is the most harmful effect of illumination (Figure 3.1m). There is a direct glare zone which can be eliminated, or at least mitigated, by proper placement of luminaires and shielding, or, if luminaires are fixed, by rearrangement of desks, tables and chairs. Overhead illumination should be shielded

Table 3.1k
GENERAL ILLUMINATION LEVELS

Task Condition	*Level (foot candles)*	*Type of Illumination*
Small detail, low contrast, prolonged periods, high speed, extreme accuracy	100	Supplementary type of lighting; special fixture such as desk lamp
Small detail, fair contrast, close work, speed not essential	50–100	Supplementary type of lighting
Normal desk and office-type work	20–50	Local lighting; ceiling fixture directly overhead
Recreational tasks that are not prolonged	10–20	General lighting; random room light, either natural or artificial
Seeing not confined, contrast good, object fairly large	5–10	General lighting
Visibility for moving about, handling large objects	2–5	General or supplementary lighting

Table 3.1 l
SPECIFIC RECOMMENDATIONS, ILLUMINATION LEVELS

Location	*Level (foot candles)*	*Location*	*Level (foot candles)*
Home		School	
Reading	40	Blackboards	50
Writing	40	Desks	30
Sewing	75–100	Drawing (art)	50
Kitchen	50	Gyms	20
Mirror (shaving)	50	Auditorium	10
Laundry	40	Theatre	
Games	40	Lobby	20
Workbench	50	During intermission	5
General	10 or more	During movie	0.1
Office		Passenger Train	
Bookkeeping	50	Reading, writing	20–40
Typing	50	Dining	15
Transcribing	40	Steps, vestibules	10
General correspondence	30	Doctor's Office	
Filing	30	Examination room	100
Reception	20	Dental-surgical	200
		Operating table	1800

to about 45 degrees to prevent direct glare. Reflected glare from the work surface interferes with most efficient vision at a desk or table and requires special placement of luminaires.

Eyeglasses cause disturbing reflections unless the light source is 30 degrees or more above the line of sight, 40 degrees or more below or outside the two 15-degree zones as shown in Figure 3.1m.

Noise and vibration affect performance by producing annoyance, interfering with communication and causing permanent physical damage. Prolonged exposure to high sound levels causes hearing loss, both temporary and permanent. The noise exposure values in Table 3.1n are those stated in the Walsh Healy Public Contracts Act as being permissible. They are derived from the curves in Figure 3.1o and reflect the variation in sensitivity with frequency. (For more on this topic see Section 5.5.)

Annoyance and irritability levels are not easily determined, because they are subjective factors, and habituation significantly affects susceptibility. A quantitative tolerance limit has not yet been established. One aspect of background noise is deterioration of speech communication. Noise reduction

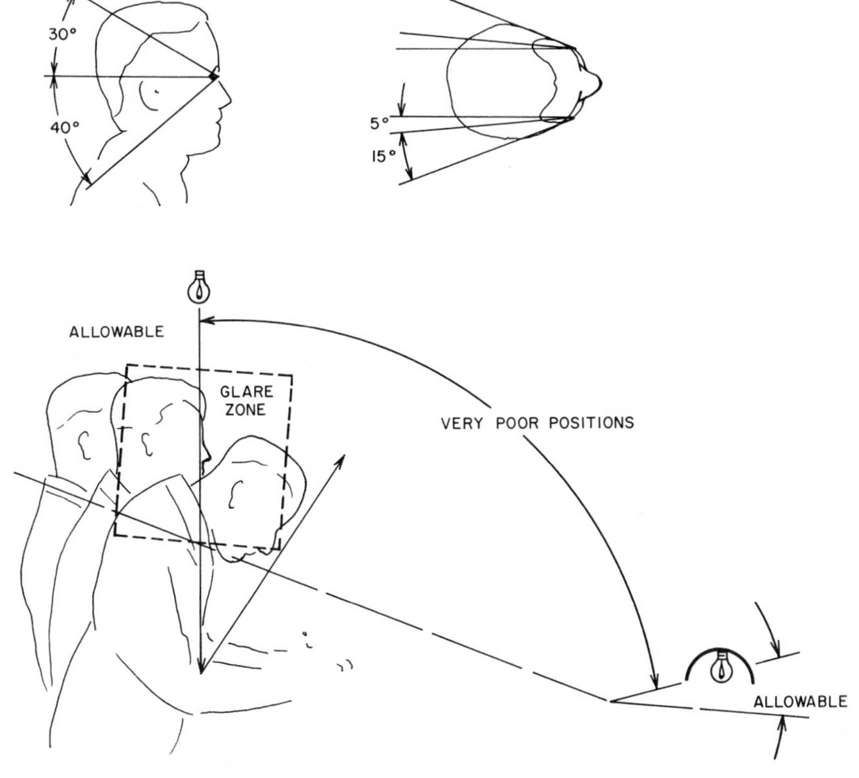

Fig. 3.1m Typical lighting chart

Table 3.1n
PERMISSIBLE NOISE EXPOSURES

Duration per Day (hours)	*Sound Level (dbA)*
8	90
6	92
4	95
3	97
2	100
1½	102
1	105
½	110
¼ or less	115

When the daily noise exposure is composed of two or more periods of noise exposure of different levels, their combined effect should be considered, rather than the individual effect of each. If the sum of the following fractions: C1/T1 + C2/T2. . . . Cn/Tn exceeds unity, then, the mixed exposure should be considered to exceed the limit value. Cn indicates the total time of exposure at a specified noise level, and Tn indicates the total time of exposure permitted at that level.

Exposure to impulsive or impact noise should not exceed 140 dbA peak sound pressure level.

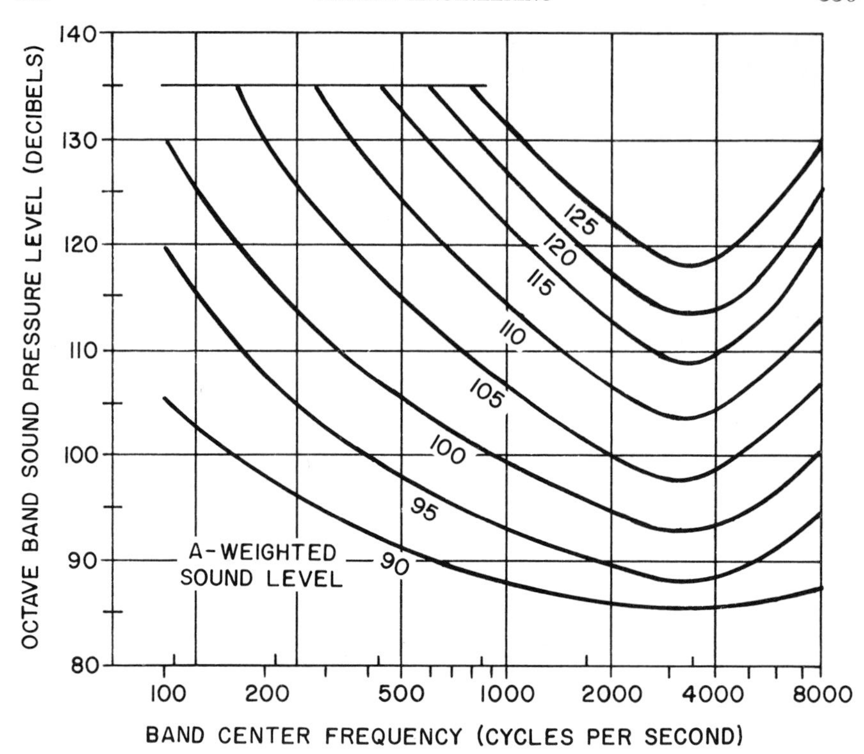

OCTAVE BAND SOUND PRESSURE LEVELS MAY BE CONVERTED TO THE EQUIVALENT A-WEIGHTED SOUND LEVEL BY PLOTTING THEM ON THIS GRAPH AND NOTING THE A-WEIGHTED SOUND LEVEL CORRESPONDING TO THE POINT OF HIGHEST PENETRATION INTO THE SOUND LEVEL CONTOURS. THIS EQUIVALENT A-WEIGHTED SOUND LEVEL, WHICH MAY DIFFER FROM THE ACTUAL A-WEIGHTED SOUND LEVEL OF THE NOISE, IS USED TO DETERMINE EXPOSURE LIMITS, FROM TABLE 3.1n.

Fig. 3.1o Equivalent sound level contours

may take the form of reducing noise emission at the source, adding absorbent material to the noise path or treating the noise receiver (having operators wear protective equipment like ear plugs and ear muffs). The last precaution reduces both speech and noise, but the ear is afforded a more nearly normal range of sound intensity and thus can better recognize speech. Figure 3.1p illustrates typical speech interference levels.

Summary

This section has described the tools needed, but the solutions that will fulfill the engineer's mandate from the industrial sector must come out of his

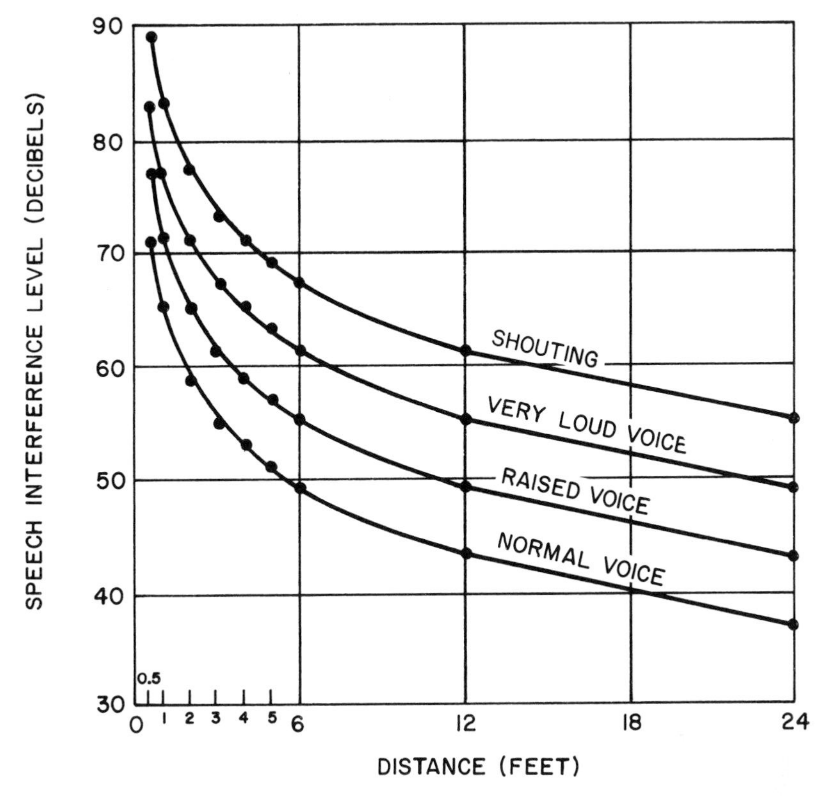

Fig. 3.1p Speech interference levels

own creativity. He must develop an original relationship with each new assignment and introduce creative innovations consistent with sound engineering practices. Decisions must not be based solely on data from current projects reflecting only what has not worked, and human engineering must not be left to the instrument manufacturer who, however well he fabricates, still does not know how the pieces must fit together in a specific application. Compilations of data on mental and physical characteristics must be approached cautiously, as they may reflect a group of subjects not necessarily like the operators in a specific plant.

It is the engineer's task to insure that the limitations of men and machines do not become liabilities, and to seek professional assistance when means of eliminating the liabilities are not evident.

REFERENCES

1. Woodson, V. J. and Conover, H. N., "Human Engineering Guide for Equipment Designers," University of California Press, Berkeley, California, Second edition, 1964.
2. Shulmann, H. C., Psychological Bulletin, Volume 75, 6, June, 1971.
3. Chapanis, A., "Man-machine Engineering," Wadsworth Publishing Company, Inc. Belmont, California, 1965.
4. McCormick, E. J., "Human Factors Engineering," (2d ed.). New York: McGraw-Hill, 1964.

3.2 INDICATOR LIGHTS

Types:	(a) Incandescent (b) Neon (c) Solid state *Note:* In the feature summary below the letters (a) to (c) refer to the listed indicator light types.
Operating Power Ranges:	(a) 1 to 120 VAC, VDC; 8ma to 10 amps; (b) 105 to 250 VAC, 90 to 135 VDC; 0.3 to 12ma; (c) 1 to 5 VDC; 10 to 100ma
Color of Unfiltered Light:	(a) White; (b) orange; (c) red
Relative Brightness:	(a) 1.0, 1.0; (b) 1.0, 0.5; (c) 2.0, 2.0
Average Useful Life:	(a) 10 to 50,000 hrs; (b) 5 to 25,000 hrs; (c) 50 to 100,000 hrs
Application Limitations:	(a) Shock and vibration can cause early failures and generate considerable heat; (b) require high voltages and current-limiting resistors; have relatively low light output; (c) are expensive; brightness is high but total light output is low.
Partial List of Suppliers:	Amp, Inc.; Clare-Pendar Co.; Dialight Corp.; Drake Manufacturing Co.; General Electric Co.; Hewlett-Packard; Industrial Devices, Inc.; Marco-Oak; Master Specialties Co.; Shelly Associates, Inc.

Lighted indicators convey several types of information to the operator including binary information, in which an on-off or open-closed condition can be displayed by a lighted (on) or unlighted (off) indicator; and status information, in which normal, abnormal or alarm conditions are expressed by legends, colors and flashing or non-flashing indicators.

The amount of information that can be displayed is directly proportional to equipment or system size and complexity. Redundant indicators should be omitted because the attention value of all indicators is reduced and confusion results if large numbers of them are used.

Characteristics of Light Sources

The commonest types of lighted indicators are the incandescent, neon and solid state lamps, and their spectral response curves are shown in Figure 3.2a. The relative response ordinate at 100 gives the peak sensitivity of the human eye and the peak wavelengths emitted by the lamps. Standard vision, for example, is most sensitive at about 560 millimicrons, which is in the yellow and yellow-green band. The peak output wavelength of the gallium arsenide phosphide (GaAsP) solid state lamp is 650 millimicrons, which is in the red and red-orange band. The curves reflect the relative efficiencies of the light sources; the more efficient ones will have an output that matches and falls within the standard vision curve. This output is approximated both by the neon and by the solid state lamps. They convert most of their input power into light and emit little heat.

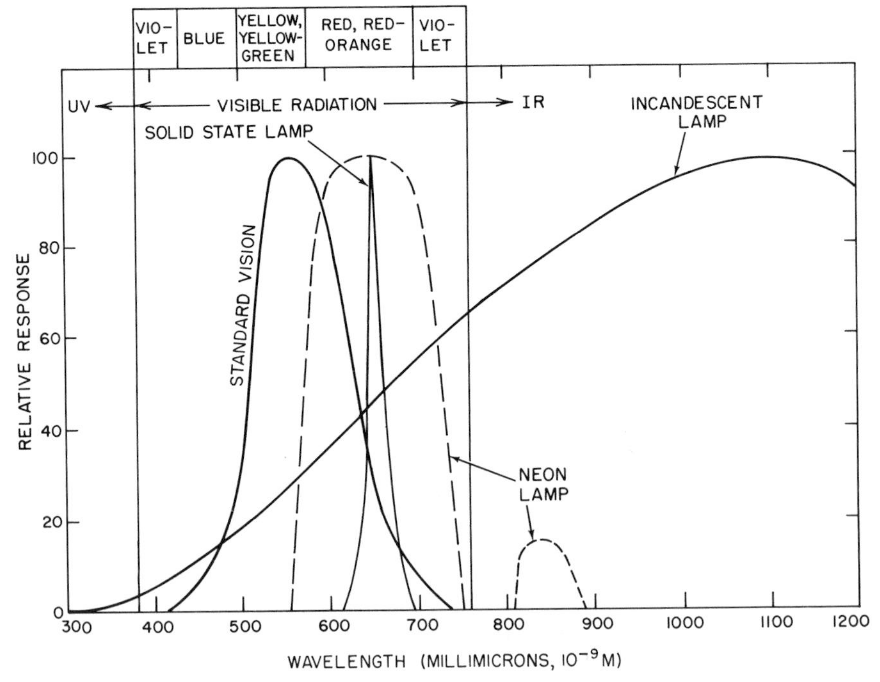

Fig. 3.2a Spectral response curves of the human eye and common lamps

Selection and Application

Important human factors in selecting and using lighted indicators are visibility and arrangement. To transmit information, the indicator must be visible to the operator. Variables that affect visibility are location, brightness, contrast, color, size and whether the indicator is flashing or non-flashing. Critical indicators should be located within 30° of the visual line of sight and should be at least twice as bright as the surface of the mounting panel. Dark panels are recommended because they furnish strong contrast with the indicator and reflect little light to the operator. For high ambient light levels, alarm legends should have dark characters imprinted on a light background; routine messages should use the reverse combination.

Colors can be a powerful tool when properly used for lighted indicators. To avoid confusion, however, only a few colors should be used to code different types of information. General information should be lighted in white; normal conditions may be green; and for cautions or abnormal conditions amber (yellow with a reddish tint) is excellent because it affords maximum visibility. Red should be used only for critical alarms that require immediate response by the operator. The use of blue or green lenses should be kept to a minimum. All lamps emit most strongly in the red and red-orange band. Consequently, much light is lost if it is filtered so as to appear blue or green. For important indicators, use the largest size that is compatible with the panel scale.

Flashing greatly improves visibility but its use should be limited to critical alarms. The rate should be 3 to 10 flashes per second with "on" time about equal to "off" time. Light indicators should be arranged according to a functional format. Indicators associated with a manual control (pushbutton or switch) should be placed closely above the control device. It is best to arrange related indicators into separate subpanel areas. Displays requiring sequential operator actions should be arranged in the normal reading pattern from left-to-right or from top to bottom. Critical indicators should have dual lamp assemblies for additional reliability. A lamp check switch should be supplied to test for and locate burned-out lamps. Another important design target is easy lamp replacement.

The selected lenses should be diffusive and eliminate glare and lamp hot spots. The lens should also provide a wide angle of view (120° minimum) and if side visibility is required, it should protrude over the mounting surface. The lens must be large enough to accommodate the required legends. Legends are commonly produced by hot stamping, engraving or photographically reproduced transparencies. Ordinarily, hot stamping is the most economical, whereas phototransparencies furnish the sharpest characters and are the most versatile.

Environmental parameters also affect the operation of the indicator lights. Special designs are available for shock, vibration or high temperature applications. Rapid dissipation of heat is important if the indicator generates heat.

Drip-proof or watertight designs should be selected if indicators are to operate in high humidity or corrosive atmospheres, or if panel washdown resistance is a requirement.

Components of Indicator Lights

Major components of most indicator light assemblies (Figure 3.2b) include the lampholder, lamp and lens. Panel light types are usually secured to a panel by a nut and lockwasher. Cartridge models can be held by a speednut friction

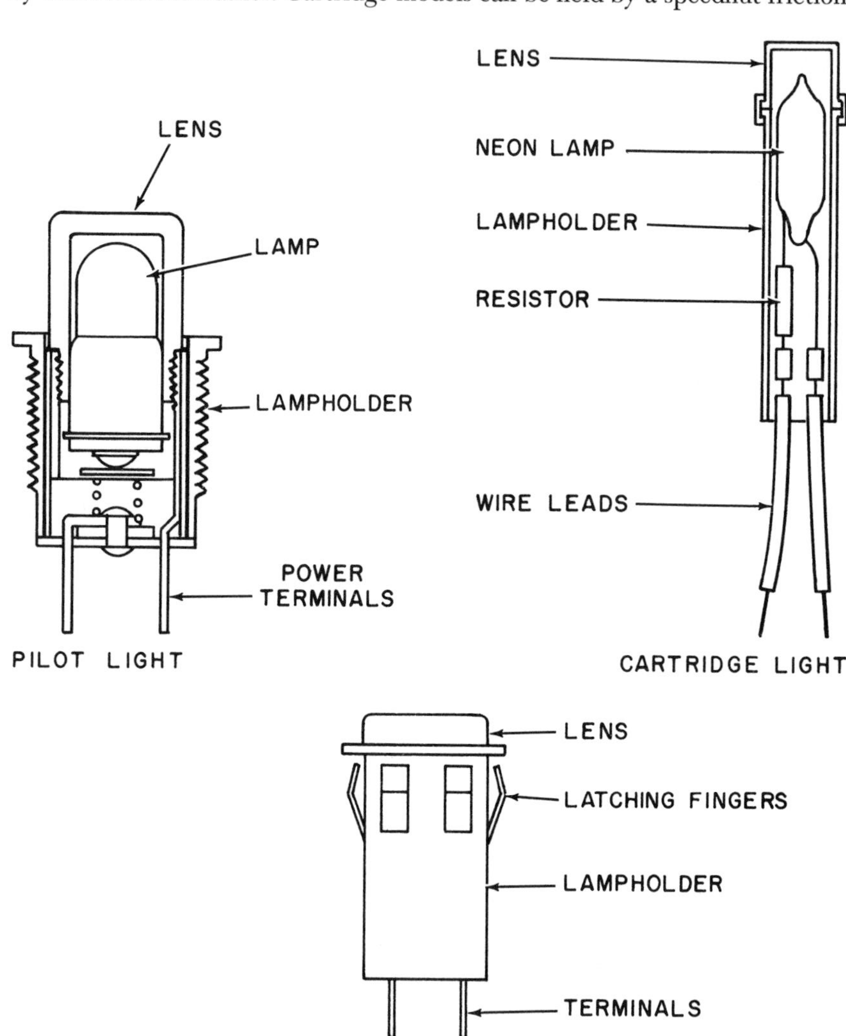

Fig. 3.2b Indicator light assemblies

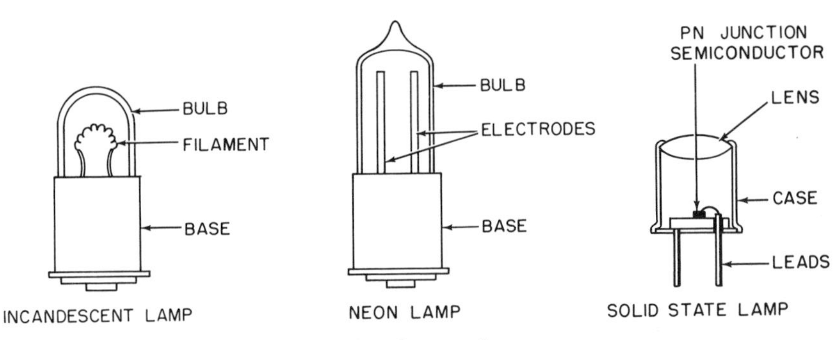

Fig. 3.2c Common lamps

clip. Snap-in lights are retained by expandable latching fingers. Power can be supplied through wire leads and solder, screw or quick-connect terminals.

The heart of all lighted indicators is the lamp or light source. The three types (Figure 3.2c) in common use are incandescent, neon and solid state. Major parts of a lamp are the bulb (containing the light emitter) and the base. Lamps are also classified according to bulb shape, size and type of base. Bulb shape and size are designated by a letter describing shape and a number that gives the nominal diameter in eighths of an inch. For example, a T-1 lamp has a tubular shaped bulb that is one-eighth of an inch in diameter. Common bases are the bayonet, screw, flanged, grooved and bi-pin. Some lamps have no base and are supplied with wire terminals.

Incandescent Lamps

The incandescent lamp (Figure 3.2c) consists of a coiled tungsten filament mounted on two support wires in an evacuated glass envelope. When current is passed through the filament, its resistance causes it to glow and emit both light and heat. Note from Figure 3.2a that only about one-third of the radiation emitted from this lamp is in the visible band (white light); the rest is in the infrared band (heat). This means that about two-thirds of the input power is emitted as heat. If large quantities of incandescent lamps are to be operated continuously and mounted closely together, special allowance for adequate heat dissipation is necessary.

If the lamp is to operate under shock and vibration, a low voltage, high current design should be used because it has strong filaments. Lamps of 6 volts or less usually have short, thick filaments, whereas lamps of more than 6 volts generally have longer and thinner filaments. In all cases, however, the lamp should be tested under simulated operating conditions.

Incandescent lamps can operate from 1 to 120 volts AC or DC. Current drain will be from 10 milliamperes to 10 amperes. Figure 3.2d shows the relationship between the applied voltage, lamp life, current and light output. Variations in applied voltage have a drastic effect on lamp life. It is common

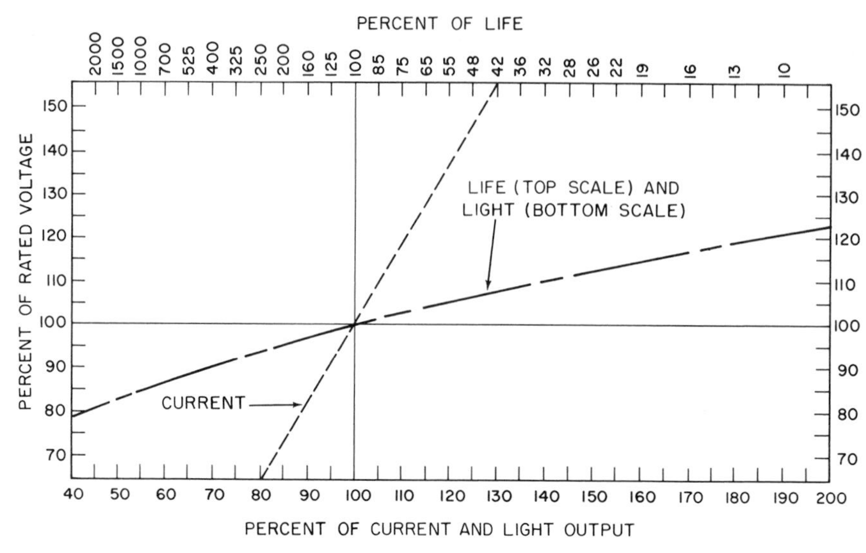

Fig. 3.2d Incandescent lamp characteristics

practice to improve life and sacrifice some light by operating the lamp at slightly below rated voltage. For example, a 6-volt lamp operated at 5 volts will have roughly 8 times the normal life and still provide 60 percent of normal light. Overvoltage results in nearly the opposite effect: a 6-volt lamp operated at 7 volts will have only about one-sixth of its normal life and provide one and one-half times the normal light. Therefore controlling applied voltage is very important.

Space permitting, a large lamp is preferred to a small one because its cost will be lower and its life and reliability higher. The large lamp will also emit less heat than the small one of equal light output, because the former has more surface area and its filament operates at a lower temperature.

Neon Lamps

The neon lamp (Figure 3.2c) consists of two closely spaced electrodes mounted in a glass envelope filled with neon gas. When sufficient voltage is applied across the electrodes, the gas ionizes, conducts a current and emits light and heat. All neon lamps require a current-limiting resistor in series with the lamp to guarantee the designed life and light characteristics, an example of which is the cartridge light in Figure 3.2b. The orange light emitted by this lamp is easily seen because a large portion of it falls within the standard vision curve (Figure 3.2a). Since most of its emitted radiation is in the visible band, little heat is emitted. An important consideration, however, is that although the neon lamp is an efficient light source, its total light output is low. A clear or lightly diffusing lens should be used with it so that only a small portion of the light is absorbed.

Neon lamps are very satisfactory for use under conditions of severe shock and vibration. The rugged mechanical construction avoids the use of the fragile filament of the incandescent lamp. Neon lamps will operate only on high voltages, and commonly run directly from standard 120 or 240 volt AC line voltages. Because of the high voltage they require little current and power.

Solid State Lamps

The solid state lamp (Figure 3.2c) is commonly called a light-emitting diode (LED) and it is a valuable by-product of semiconductor technology. (See Section 4.6 in Volume II.) It is basically a P-N junction diode mounted in a hermetically sealed case with a lens opening at one end. Light is produced at the junction of the P and N materials by two steps. A low voltage DC source increases the energy level of electrons on one side of the junction. In order to maintain equilibrium, the electrons must return to their original state; they cross the junction and give off their excess energy as light and heat. The light output of the popular gallium arsenide phosphide LED (Figure 3.2a) has a very narrow bandwidth centered in the red band. These lamps are very efficient and have exceptionally long life but are small and have low light output.

The electrical characteristics of the LED are similar to those of the silicon diode, are compatible with integrated circuits and operate directly from low level logic circuits. Solid state lamps, like neon lamps, perform satisfactorily under shock and vibration. Presently, they are rather expensive but the cost can be expected to fall as additional applications are found for it.

Checklist

1. Determine operating voltage.
2. Select lamp type and size.
3. Select lens for type, color, size and shape. The last two features should be large enough to hold the necessary legends.
4. Select a lamp holder that is compatible with both lamp and lens. Also consider the allowable panel space and the methods of mounting and providing electrical connections.
5. Test the indicator under simulated operating conditions.

Conclusions

The possible uses of indicator lights to display information are limited only by the designer's imagination; there is usually one particular combination of lamp, lens and lampholder that is best suited for an application. Incandescent lamps are preferred for most applications because they are available in the widest range of light output, sizes and voltages. The low light output of both the neon and solid state lamps limits their use to on-off indicators. Amber lenses should be widely used because they absorb little lamp light and they are the most visible of all colors. Snap-in lampholders should be used wherever possible because they require no mounting hardware and take little assembly time.

3.3 ALPHANUMERIC (DIGITAL) READ-OUTS

Types:	(a) Mechanical and electrical counters (b) Gaseous discharge tubes (c) Cathode ray tubes (d) Rear projection displays (e) Matrix displays *Note:* In the feature summary below the letters (a) to (e) refer to the listed alphanumeric readout types.
Display:	(a) numeric; (b,e) alphanumeric; (c,d) alphanumeric and symbols
Character Sizes:	(a) 0.12″ to 0.5″, (e) 0.1″ to 0.8″, (c) 0.38″ to 0.62″, (b) 0.3″ to 2.0″, (d) 0.12″ to 3.38″
Maximum Viewing Distance:	(a) 30 ft, (c) 40 ft, (e) 50 ft, (b) 100 ft, (d) 150 ft
Viewing Angle:	(a,c) 90°, (d) 100°, (b,e) 120°
Brightness in Foot-Lamberts:	(a) A function of ambient lighting, (d) 75, (c) 100, (b) 150, (e) 200
Life:	(d) 10,000 hrs, (c,e) 20,000 hrs, (a,b) 100,000 hrs
Operating Volts:	(e) 1.6–28 DC, (d) 5–28 DC, (a) 0–115 AC or DC, (b) 170–300 DC, (c) 2,500–3,000 DC
Relative Cost:	(a) low, (b) medium, (e) high, (c,d) highest
Partial List of Suppliers:	Amperex Electronics Corp. (b,c); Burroughs Corp. (b,e); Dialight Corp. (e); Industrial Electronic Engineers, Inc. (c,d,e); Master Specialties Co. (e); Monsanto Co. (e); Pinlites, Inc.

(e); RCA Electronic Components (e); Readouts, Inc. (e); Veeder-Root, Inc. (a)

Section 5.2 in Volume II and Section 3.2 in this volume contain information that is directly applicable to alphanumeric (hereafter called a/n) readouts, and should be read to supplement the information in this section on selection and use of a/n readouts.

As their name implies, a/n readouts furnish alphabetical (letters or legends) and numerical (digits or numbers) information. Common examples are the television cathode ray tube and the automobile odometer. These read-outs provide accurate, easily understood displays, and require no operator interpretation because they present clear, concise legends or exact numerical values.

Versatility is the primary advantage of a/n read-outs because they have the capability to display many different types of information. Some read-outs display only numbers, whereas others display numbers, letters and symbols. The contents of the display can be tailored to the type of variable being displayed. For example, the read-out of a digital voltmeter or electronic calculator may accommodate the maximum value and the decimal point location. In a process control display the read-out may include the loop name and tag number with the value of the variable being measured.

The common types of a/n read-outs are classified as mechanical and electrical counters, gaseous discharge and cathode ray tubes and rear projection and matrix displays.

Human Factors

To transmit information efficiently, read-outs must have sharp character resolution, which is a measure of the character image sharpness or clarity and relates directly to readability. Readability is the quality that allows an observer to perceive information with speed and accuracy and is a function of character style, proportions, height, contrast, color and refresh rate.

Characters that are pleasing to the eye are generated by simple continuous lines. To aid in character recognition, critical details should be prominent, and special features such as openings or breaks should be readily apparent. Many a/n read-outs use a matrix of dots or bar segments for character generation. Closely spaced dots are more legible and natural looking, whereas bar segments usually form boxlike characters containing noticeable intersegmental spaces. Character proportions should be predicated on a height-to-width ratio of roughly 3 to 2. Line width should be about one-seventh of character height. Minimum spacing between characters should be two line widths with about six line widths between words. Upright characters are preferable to slanted types.

Size and Contrast Considerations

A guideline for determining character height based on viewing distance is shown in Table 3.3a. The heights can be modified slightly for high ambient illumination or high brightness displays. A display mockup is recommended if a particular viewing distance is critical.

Table 3.3a
CHARACTER HEIGHTS
BASED ON VIEWING DISTANCE

Required Viewing Distance (feet)	*Nominal Character Height (inches)*
2.3 (28 ins.)	.12
5	.18
10	.25
15	.31
20	.38
30	.50
40	.68
50	.75
65	1.38
100	2.00

Contrast is the ratio between character image brightness and its background brightness when measured under normal ambient illumination. Acceptable ratios for a/n read-outs are about 5 or 10 to 1. Assuming the background to be a typical control console surface with a brightness of 20 to 50 foot-lamberts (ft-L), a nominal brightness range for read-outs should be 100 to 500 ft-L. Contrast depends on ambient illumination. Consequently, locations near an outside window or directly under a lighting fixture may require a filter to prevent image washout or to reduce objectionable reflections. An anti-reflective filter reduces reflections and improves contrast by passing proportionately more self-generated image light then reflected image background light.

The optimum read-out color is yellow, the eye being most sensitive to it (Figure 3.2a). Amber, red or orange are the next best choices. A filter can be used to obtain a desired color. A red filter, for example, will pass only the red light of an incandescent read-out while it absorbs all other colors. Several types of a/n read-outs generate character images by addressing time-displaced current pulses to dots or segments. The number of times (per second) that the image is generated is called the refresh rate. Low refresh rates can result in lowered brightness and occasionally cause flicker.

Two additional human factors involve viewing angle and change or update rate. If the read-out must be seen from several operator positions, it should have a viewing angle of about 120°. Recessed characters are satisfactory only for direct viewing. Maximum read-out update rate is twice per second if the operator is expected to read consecutive values. Numerical read-outs are not recommended for determing rate of change, direction or tracking.

Application Notes

For critical read-outs there should be a test function or switch to locate burned-out lamps, tubes or lost character portions. The loss of the horizontal center section of some matrix read-outs, for example, can lead to misreading an integer as zero when eight is really intended. The read-out bezel should be aesthetically pleasing and in contrast to the panel mounting surface.

Most a/n read-outs require an electrical power supply to provide memory and keep the read-out continuously lighted, and the read-outs are updated or changed by a decoder driver that accepts binary coded decimal inputs. Read-out modules containing standard decoder drivers with memory are preferred. Characteristics of common a/n read-outs are listed in Table 3.3b, and a description of each type follows.

Mechanical and Electrical Counters

Mechanical and electrical counters (Figures 3.3c and 3.3d) are relatively simple and inexpensive devices that provide a cumulative count of sequential events. Usual configurations consist of a series of numbered wheels mounted on a common input shaft rotated by mechanical or electrical pulses. The numerical read-out is visible through a window placed over the foremost row of digits. Digits should be as close as possible to the window so as to provide the maximum viewing angle and eliminate shadows from ambient lighting. Internal lamps should be supplied if these counters are to be used in darkened areas.

The driver output and the counter input should be compatible. Mechanical counters are driven by mechanically rotating or oscillating the input shaft. Rotation changes the read-out through internal gearing, and oscillations use ratchet-pawl mechanisms. Electrical counters work similarly except that a solenoid or electromagnet actuates the input shaft. Counters operating with mechanically rotating inputs usually provide one or ten counts per complete rotation of the input shaft. Inputs can be either clockwise or counterclockwise, consequently the count adds in one direction and subtracts in the other. Oscillating inputs provide one count per oscillation and counting is in one direction only.

Table 3.3b
ALPHANUMERIC READ-OUT CHARACTERISTICS

Type/Feature	*Method of Operation*	*Operating Characteristics*	*Typical Applications*
Mechanical and Electrical Counters	Numbered wheels are rotated behind a viewing window.	Contain inherent memory; power required only to change display; usually illuminated by ambient light.	Digital clock; water meter; gasoline pump
Gaseous Discharge Tubes	Shaped electrodes, in form of characters, ionize surrounding neon gas.	Require external memory and operate on high voltages (170–300 DC)	Electronic calculators; electrical meters
Cathode Ray Tubes	Shaped electron beam is projected on phosphor screen.	Require character generation and memory circuitry; operate on very high voltages (2,500–3,000 DC)	Computer-controlled displays
Rear Projection Displays	Miniature optical projectors containing incandescent lamps display images on viewing screen.	Require external memory; models are available for wide range of operating voltages.	Control-console displays
Matrix Displays	Characters are formed by selectively lighting dots or bar segments within a matrix.	Require external memory; models are available for wide range of operating voltages.	Electronic calculators; electrical meters

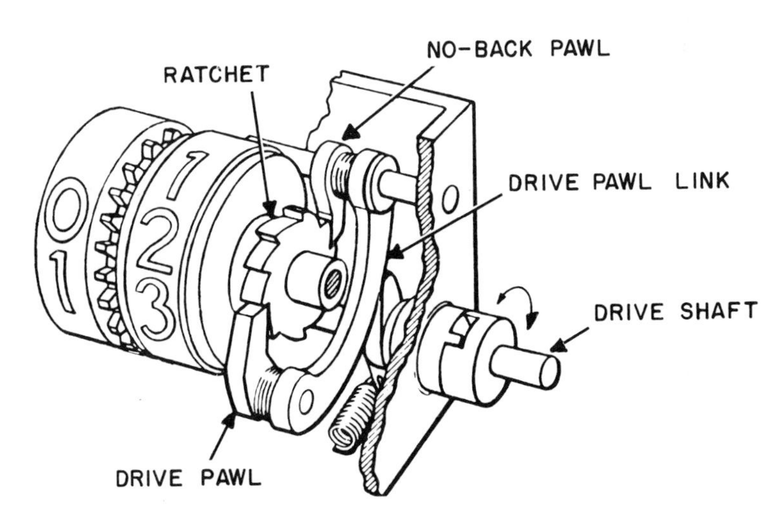

Fig. 3.3c Mechanical counter

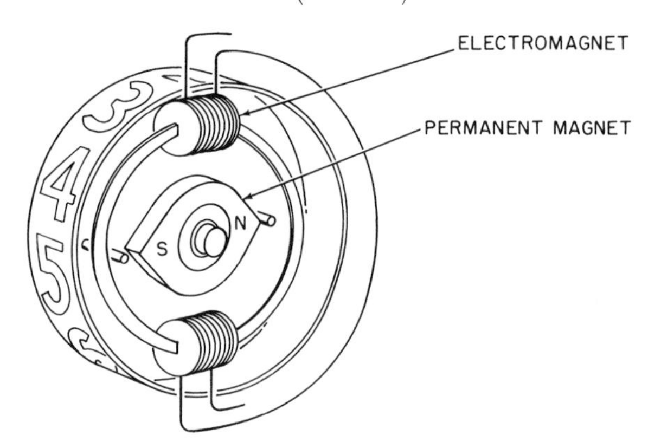

Fig. 3.3d Electrical counter

Electrical counters ordinarily supply one count per pulse and are available for AC or DC operation with a wide selection of voltage coils. Counters with two coils will add one count for a pulse through one coil and subtract one count for a pulse through the other coil. Zero reset counters should be used for applications requiring a cumulative count between random times or events. A single knob depression for mechanical counters or pulse for electrical counters will set all wheels to zero and a new count can be established.

Gaseous Discharge Tubes

Gaseous discharge tubes (Figure 3.3e) are widely used electronic a/n read-outs in which the mode of operation is similar to the neon lamp. Standard numerical display tubes contain a common anode and a stack of ten independent cathodes shaped like numbers. When a large negative voltage is applied to a selected cathode, the gas around it glows from being ionized and emits an orange light. If the inmost cathode is lighted, for instance, it is slightly recessed and the observer looks through the remaining unlit cathodes above it. A/n read-out tubes are similar to numerical tubes except that all the letters of the alphabet and numbers 0 to 9 can be generated by a single plane fourteen segment bar matrix. The ends of the segments are closely spaced and sharp character resolution is obtained without the noticeable gaps found in other types of segmented displays.

Gaseous discharge tubes require a high voltage DC power supply to ionize the neon gas. The positive side of the power supply is connected (in series) through a current-limiting resistor to the common anode. An integrated circuit decoder with memory accepts binary coded decimal inputs, connects the appropriate cathode to the negative side of the power supply and keeps the cathodes lighted. Standard supplies and decoder drivers should be used whenever they are available.

Fig. 3.3e Gaseous discharge tube

Both end and side view tubes are available. A filter should be used to improve character contrast and reduce reflections. Complete read-out assemblies containing tubes, operating circuitry, filter and bezel should be used because the savings in design and rework will more than offset the higher initial cost.

Cathode Ray Tubes

Cathode ray tubes, commonly called CRTs, are versatile a/n read-outs that operate like miniature television sets. Typical CRT components are shown in Figure 3.3f. A high voltage supply places the cathode at a low negative potential with respect to the anode. The heater causes the cathode or gun

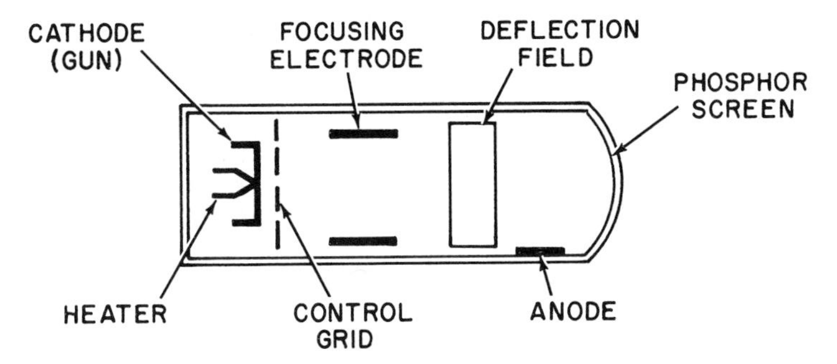

Fig. 3.3f Cathode ray tube components

to emit a steady stream of electrons that is either stopped (blanked) or accelerated (unblanked) by the control grid. The focusing electrode channels the electron stream into a narrow beam that is deflected by an electrostatic or electromagnetic field to excite the phosphor screen at the proper position.

Numerical display tubes contain ten electron guns (cathodes) mounted behind a control grid, and a metallic screen that contains openings etched in the shape of numbers. When the proper voltages are applied to a specific electron gun and the control grid, the desired number is projected on the phosphor screen. Available models of these tubes can display one, four or six numbers per tube.

More complicated than numerical types are a/n read-out CRTs since they can also display letters, numbers and symbols. They are usually electrostatic-deflecting, single-gun devices that generate characters by a line raster technique similar to that employed in television tubes. These small tubes employ a pattern of 20 horizontal lines, whereas television uses 525. Characters are made from a 5 wide × 7 high dot matrix caused by unblanking the electron beam sweep (exciting the phosphor) at the proper times.

Figure 3.3g shows the operation of a/n CRTs. Major components are the high voltage supply, character generator and decoder, of which the first-named is the origin of the electron beam. The character generator determines the display capabilities of the tube because it contains the information (memory) required to produce the characters. The character generator can be either self-contained or supplied from a computer memory. The decoder is connected to the character generator and accepts binary coded decimal

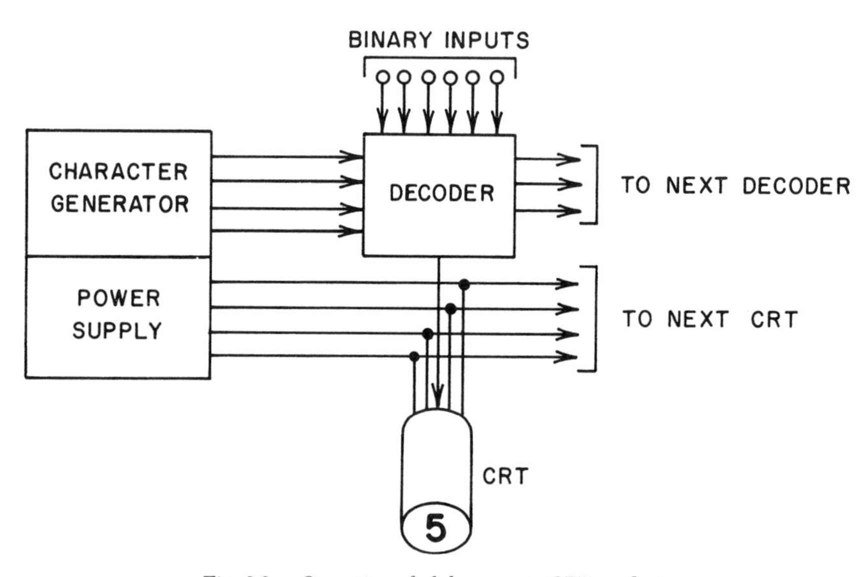

Fig. 3.3g Operation of alphanumeric CRT readouts

inputs that are converted to unblanking control grid signals that generate the desired characters. A common supply and character generator may be used for several tubes, but each tube requires its own decoder.

Rear Projection Displays

Rear projection displays (Figure 3.3h) are miniature optical projectors stacked in cordwood fashion. They can display anything that can be put on film, including symbols, words and colors. Twelve incandescent lamps at the rear of the display ordinarily illuminate the corresponding filmed messages, which are focused through a lens system and projected onto a single plane, non-glare viewing screen. Rear projection displays exhibit excellent character resolution and readability and produce a very natural a/n read-out appearance. They can be operated directly from electrical relays or from self-contained decoder drivers.

Matrix Displays

Matrix displays (Figure 3.3i) are a/n read-outs forming characters by selectively lighting dots or bar segments from within a matrix. Standard dot matrices are 5×7, 6×8 and 7×9, and can display all letters and numbers. The 7×9 matrix is preferred, affording as it does the best character resolution. Standard bar matrices utilize 7, 12, 14 and 16 bar segments. Only the last-

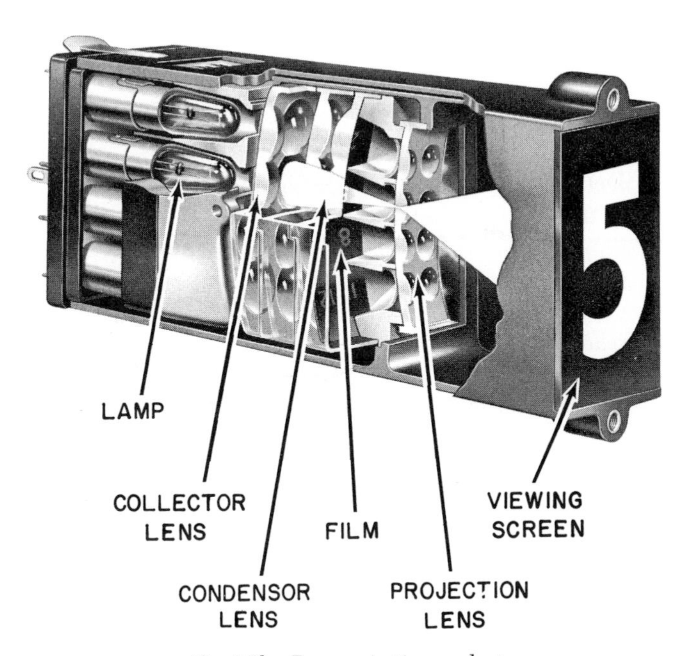

Fig. 3.3h Rear projection readout

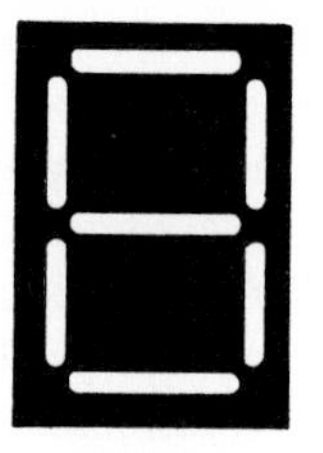

Fig. 3.3i Matrix displays

named can form all alphabetic and numeric characters without ambiguity. Matrix displays provide only fair to good character legibility. For example, the numbers 0, 1 and 2 can be misinterpreted as the letters O, I and Z, respectively.

Dots or bar segments can be lighted by incandescent, neon, fluorescent or solid state lamps. Incandescent lamps light dot matrices through fiber optic light guides and form high contrast characters. They can also be placed directly behind small acrylic plates acting as bar segments. Individual neon or fluorescent lamps can be used as the actual bar segments. Solid state lamps are usually arranged behind a filter as separate dots or bars. Decoders and driving circuitry for matrix displays are similar to those previously described for other read-outs.

Selection Guide

The following checklist should help to narrow the choice of read-outs to one or two acceptable types:

1. Determine the exact kind and content of information that must be displayed.
2. Determine the character size based on required viewing distance.
3. If the display must be seen from the side under high ambient illumination, confine the choice to types that have a viewing angle greater than 90° and a brightness of at least 100 ft-L.
4. Check the availability of the required power supply for the read-out.
5. Obtain actual cost and life figures from manufacturers' catalogues and determine their acceptability.

3.4 ANNUNCIATOR SYSTEMS

Types:

AUDIOVISUAL ANNUNCIATORS: integral, remote and semigraphic systems with audible and visual display and electromechanical (relay) or solid state (semiconductor) design.

RECORDING ANNUNCIATORS: integral, solid state, high speed systems with recorded print out.

VOCAL ANNUNCIATORS: integral, solid state system with audible command message.

Cost per Alarm Point:

integral cabinet, \$50; remote system, \$60; semigraphic system, \$75; recording annunciator, \$80; vocal annunciator, \$150. Equipment cost per point decreases as system size increases. The semigraphic system cost does not include the price of the graphic display. The vocal annunciator cost does not include the price of the communication system equipment.

Partial List of Suppliers:

AUDIOVISUAL ANNUNCIATORS: Beta Corp.; Compudyne Corp.; Electro-Devices; Rochester Instrument Systems, Inc.; Ronan Engineering Co.; SCAM Instrument Corp.; Swanson Engineering and Manufacturing Co. RECORDING ANNUNCIATORS: Hathaway Instruments, Inc.; Rochester Instrument Systems, Inc.; SCAM Instrument Corp. VOCAL ANNUNCIATORS: Transmation, Inc.

History and Development

The literature contains little to document the development of current industrial annunciator systems. The term "drop" was initially applied to individual annunciator points in process applications, from which we may infer that annunciator systems developed from paging systems of the type used in hospitals and from call systems used in business establishments to summon individuals when their services were needed. These systems consisted of solenoid operated nameplates which dropped due to gravity when de-energized. The drops were grouped at a central location and were energized by pressing an electrical pushbutton in the location requiring service. The system also included an audible signal to sound the alert condition.

Similar systems were used for fire and burglar alarms. The drops were operated either by manual switches or by trouble contacts which monitored thermal and security conditions in various building locations. the use of these systems in the chemical processing industry was a logical development, as the necessary monitoring switches became available.

This development, however, was preceded by explosion-proof, single station annunciators which were designed to operate in the petroleum and organic chemical process plants constructed immediately before, during and after World War II. They were usually installed on control panels located either outdoors at the process unit or in local control houses. These locations, being electrically hazardous, prevented the use of drop type systems.

By the late 1940s centralized control rooms were introduced from which the plant could be remotely operated. A drop type annunciator could be used in these general purpose central control rooms. However, more compact, reliable and flexible annunciators were subsequently introduced.

In the early 1950s the plug-in relay annunciator was developed. Instead of utilizing solenoid operated drops, it used electrical annunciator circuits with small telephone type relays to operate alarm lights and to sound a horn when abnormal conditions occurred. The alarm lights installed in the front of the annunciator cabinets were either the bull's-eye type or backlighted nameplates. The annunciators were compact and reliable and because of the hermetically sealed relay logic modules, they could be used in certain hazardous areas in addition to the general purpose control rooms. Miniaturization of instruments and the use of graphic control panels initiated the development of remote annunciator systems consisting of a remotely mounted relay cabinet connected to alarm lights installed at appropriate points in the graphic or semigraphic diagram.

Solid state annunciator systems with semiconductor logic modules were developed in the late 1950s permitting additional miniaturization and lowering both the operating power requirements and the heat generated. The semigraphic annunciator was introduced in the late 1960s and fully utilized the high-density capabilities of solid state logic. It has furnished very compact and flexible semigraphic control centers. The trend toward additional miniaturization is owing to the greater availability and reliability of integrated circuit logic components.

Principles of Operation

The basic annunciator system consists of multiple individual alarm points, each connected to a trouble contact (alarm switch), a logic module and a visual indicator (Figure 3.4a). The individual alarm points are operated from a common power supply and share a number of annunciator system components including an audible signal generator (horn), a flasher and acknowledge and test pushbuttons. In normal operation the annunciator system and individual alarm points are quiescent.

The trouble contact is an alarm switch which monitors a particular process variable and is actuated when the variable exceeds preset limits. In electrical annunciator systems it is normally a switch contact which closes (makes) or opens (breaks) the electrical circuit to the logic module and thereby initiates the alarm condition. In the alert stage the annunciator turns on the visual indicator for the particular alarm point and the audible signal and flasher

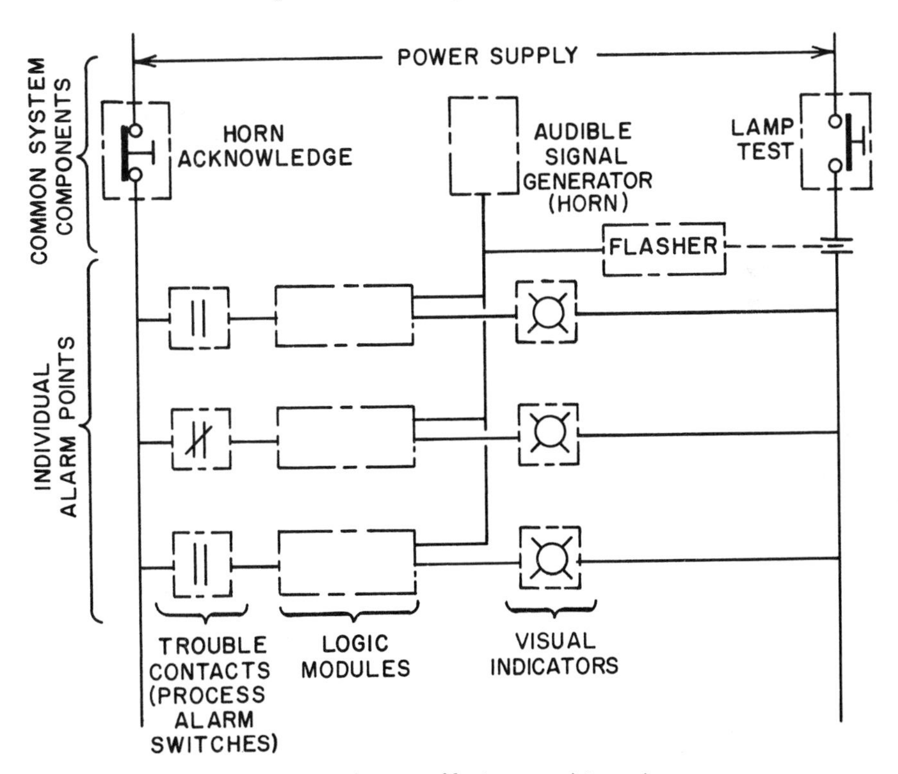

Fig. 3.4a Elements of basic annunciator system

for the system. The visual indicator is usually a backlighted nameplate engraved with an inscription to identify the variable and the abnormal condition, but it can also be a bull's-eye light with a nameplate. The audible signal can be a horn, buzzer or bell.

The flasher is common to all individual alarm points and interrupts the circuit to the visual indicator as that point goes into the alert condition. This causes the light to continue to flash intermittently until either the abnormal condition returns to normal or it is acknowledged by the operator. The horn acknowledgement pushbutton is provided with a momentary contact and when it is operated, it changes the logic module circuit to silence the audible signal, stop the flasher and turn the visual indicator on steady. When the abnormal condition is corrected, the trouble contact returns to normal and the visual indicator is automatically turned off. The lamp test pushbutton with its momentary contact tests for burned-out lamps in the visual indicators. When activated, the pushbutton closes a common circuit (bus) to each visual indicator in the annunciator system, turning on those lamps which are not already on due to an abnormal operating condition.

Operating Sequences

The operation of an individual alarm point in the normal, alert, acknowledged and return-to-normal stages is the annunciator sequence, a wide variety of which can be developed from commercially available logic components; many special sequences have been designed to suit the requirements of particular process applications. The five most commonly used annunciator sequences are shown in Table 3.4b, identified by the code designation of the Instrument Society of America (ISA). (For additional details on less frequently used sequences, see the ISA recommended practice RP-18.1.)

ISA Sequence 1B, also referred to as flashing sequence A, is the one most frequently used. The alert condition of an alarm point results in a flashing visual indication and an audible signal. The visual indication turns off automatically when the monitored process variable returns to normal. ISA Sequence 1D (often referred to as a dim sequence) is identical to sequence 1B except that ordinarily the visual indicator is on dim rather than off. A dimmer unit, common to the system, is required. Because all visual indicators are always turned on—either for dim (normal), flashing (alert) or steady (acknowledged)—the feature for detecting lamp failure is unnecessary. ISA Sequence 2A (commonly referred to as a ringback sequence) differs from sequence 1B in that following acknowledgment the return-to-normal condition produces a dim flashing and an audible signal. An additional momentary contact reset pushbutton is required for this sequence. Pushing the reset button after the monitored variable has returned to normal turns off the dim-flashing light and silences the audible signal. This sequence is applied when the operator must know if normal operating conditions have been restored.

Table 3.4b
MOST COMMONLY USED ANNUNCIATOR SEQUENCES

ISA Code for the Sequence	*Annunciator Condition*	*Process Variable Condition (trouble contact)*	*Visual Indicator*	*Audible Signal*	*Use Frequency*
1B	Normal	Normal	Off	Off	55%
	Alert	Abnormal	Flashing	On	
	Acknowledged	Abnormal	On	Off	
	Normal Again	Normal	Off	Off	
	Test	Normal	On	Off	
1D	Normal	Normal	Dim	Off	1%
	Alert	Abnormal	Flashing	On	
	Acknowledged	Abnormal	On	Off	
	Normal Again	Normal	Dim	Off	
2A	Normal	Normal	Off	Off	4%
	Alert	Abnormal	Flashing	On	
	Acknowledged	Abnormal	On	Off	
	Return to Normal	Normal	Dim Flashing	On	
	Reset	Normal	Off	Off	
	Test	Normal	On	Off	
2C	Normal	Normal	Off	Off	5%
	Alert	Abnormal	Flashing	On	
	Acknowledged	Abnormal	On	Off	
	Return to Normal	Normal	On	Off	
	Reset	Normal	Off	Off	
	Test	Normal	On	Off	
4A	Normal	Normal	Off	Off	28%
	Alert	Abnormal			
	—Initial		Flashing	On	
	—Subsequent		On	Off	
	Acknowledged	Abnormal			
	—Initial		On	Off	
	—Subsequent		On	Off	
	Normal Again	Normal	Off	Off	
	Test	Normal	On	Off	
All Others					7%

ISA Sequence 2C is like sequence 1B except that the system must be reset manually after operation has returned to normal in order to turn off the visual indicator. This sequence is also referred to as a manual reset sequence and, like sequence 2A, requires an additional momentary contact reset push-button. It is used when it is desirable to keep the visual indicator on (after the horn has been silenced by the acknowledgment pushbutton) even though the trouble contact has returned to normal.

ISA Sequence 4A, also known as the first out sequence, is designed to identify the first of a number of interrelated variables which have exceeded normal operating limits. An off-normal condition in any one of a group of process variables will cause some or all of the remaining conditions in the group to become abnormal. The first alarm causes flashing and all subsequent points in the group turn on the steady light only. This sequence monitors interrelated variables. The visual indication is turned off automatically when conditions return to normal after acknowledgement.

Optional Operating Features

Annunciator sequences may be initiated by alarm switch trouble contacts which are either open or closed during normal operations. These are referred to as normally open (NO) or normally closed (NC) sequences respectively, and the ability to use the same logic module for either type of trouble contact is called an *NO-NC option*. It is important because some alarm switches are available with either an NO or an NC contact but not with both, and therefore without the NO-NC option in the logic module two types of logic modules would be required. The logic module is converted for use with either form of contact by a switch or wire jumper connection.

The relationship between the NO and NC sequences required in the logic module to match the various trouble contacts and analog measurement signal actions is shown in Figure 3.4c. A high alarm in a normally closed annunciator system requires a normally closed trouble contact operated by a direct acting analog input. If an increase in the measured variable results in an increased output signal, the detector is direct acting; if the output signal is reduced, it is a reverse acting sensor. If the trouble contacts in all alarm switches in the plant are standardized such that normal operating conditions will cause all trouble contacts to be NC (or NO), the required annunciator sequence is also NC (or NO) and Figure 3.4c need not be referred to.

Annunciator systems are fail-safe or self-policing if they initiate an alarm when the logic module fails because of relay coil burnout. The feature is standard for most NO and NC annunciator sequences; annunciators using NC trouble contacts are also fail-safe against failures in the trouble contact circuit.

The lock-in option locks in the alert condition initiated by a momentary alarm until the horn acknowledgment button is pushed, preventing loss of a transient alarm condition until the operator can identify it. The logic module is usually changed from lock-in to non-lock-in operation by either adding a wire jumper or operating a switch. The lock-in feature is for monitoring unstable or fluctuating process variables or during plant start-up.

The test pushbutton in the standard annunciator serves only to test for burned-out lamps in the visual indicators. The operational test feature provides

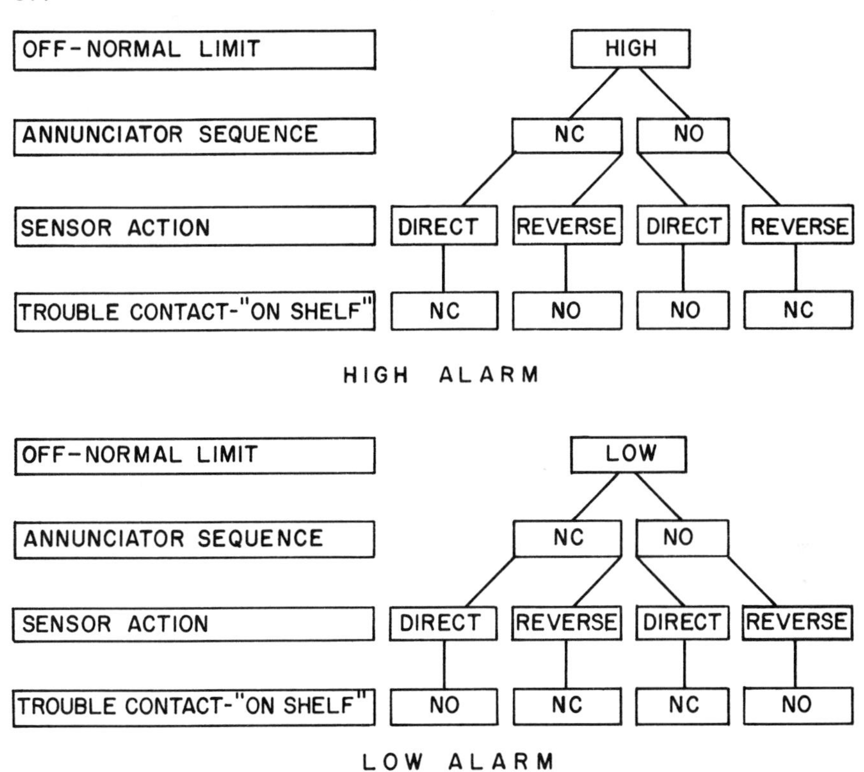

Fig. 3.4c Logic tree for NO and NC annunciator sequences

a test of the complete annunciator system including logic modules, lamps, flasher, audible signal and acknowledgment circuits. The operational test circuit usually requires an additional momentary contact pushbutton, which can replace the regular lamp test pushbutton. The logic module of relay type annunciators may have spare (electrically isolated) auxiliary contacts that can operate shut-down and interlock systems when alarm conditions occur. The auxiliary contacts are wired to terminal blocks in the annunciator cabinet for connection to external circuitry.

Repeater lights are located away from the common logic module and serve to alert operators in other areas. Annunciator cabinet terminals for connecting these repeater lights in parallel with the annunciator visual indicator are available.

It may also be desirable to actuate a horn in more than one location. The electrical load of multiple audible signals requires an interposing relay called a horn isolating relay operated by the logic modules. This relay has contacts of adequate capacity to operate multiple audible signals. Horn-isolating relays may be installed either in the annunciator cabinet or in a separate assembly. Annunciator systems can be used for several operational sequences without changing system wiring and many logic modules can supply more than one operational sequence. This multiple sequence capability is sometimes useful when the sequence has not yet been determined.

Audible-Visual Annunciators

The audible-visual annunicator may be packaged as an integral, remote or semigraphic annunciator.

Integral Annunciator

The integral annunciator, a cabinet containing a group of individual annunciator points wired to terminal blocks for connection to external trouble contacts, power supply, horn, acknowledge and test pushbuttons, in terms of cost per point is the most economical of the various packaging methods available. It is also the simplest and cheapest to install.

Two methods of packaging integral annunciators are illustrated in Figure 3.4d. In the non-modular type, plug-in logic modules are installed inside the cabinet and connected to alarm windows on the cabinet door through an interconnecting wiring harness; in the modular type, individual plug-in alarm point assemblies of logic module and visual indicator are grouped together. The non-modular and modular cabinet styles are both designed for flush panel mounting with the logic modules and visual indicators accessible from the front. Electrical terminals for the external circuitry are located in the rear of the cabinet and are accessible from the back.

Integral annunciators are used on non-graphic and on semigraphic control panels in which physical association of the visual indicators with a specific location in the graphic process flow diagram is not required. Integral annunciator cabinets occupy more front but less rear panel space than the equivalent remote designs. The electrical terminals are in a general purpose enclosure at the rear of the cabinet and trouble contacts can be wired directly to them, thus eliminating the need for and resultant costs of intermediate terminal blocks for trouble contact wiring.

The modular type cabinet has the advantage of expansion by enlarging the panel cutout and adding additional modular alarm point assemblies. Non-modular cabinets cannot be expanded, and new cabinets must be installed to house additional alarm points. Consequently, more spare points should be included when specifying the cabinet size for a non-modular system. The modular cabinet is also more compact, takes up less panel space and has a greater visual display area per point than has the non-modular design. Figure 3.4e illustrates various configurations of visual indicators which can be supplied with integral annunciator cabinets. Many of these groupings are also available in single unit assemblies for remote annunciator systems.

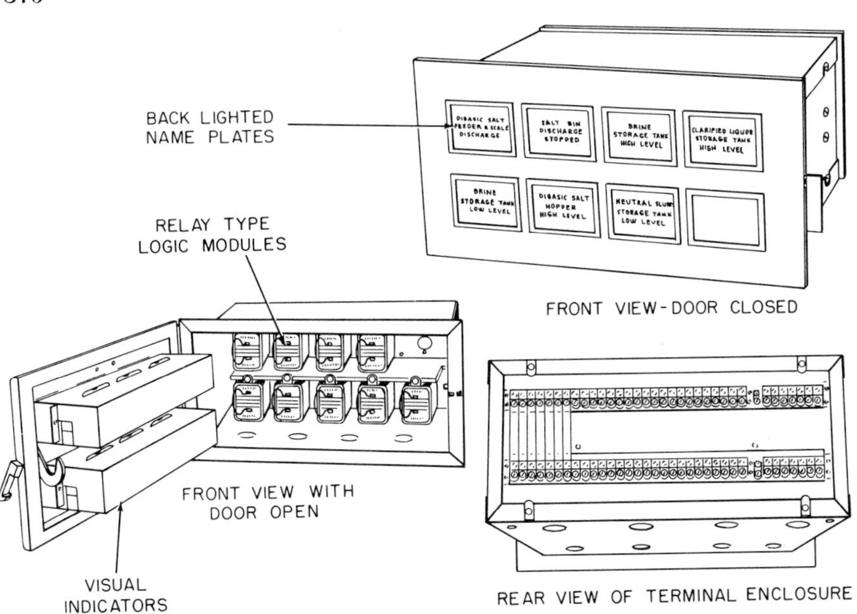

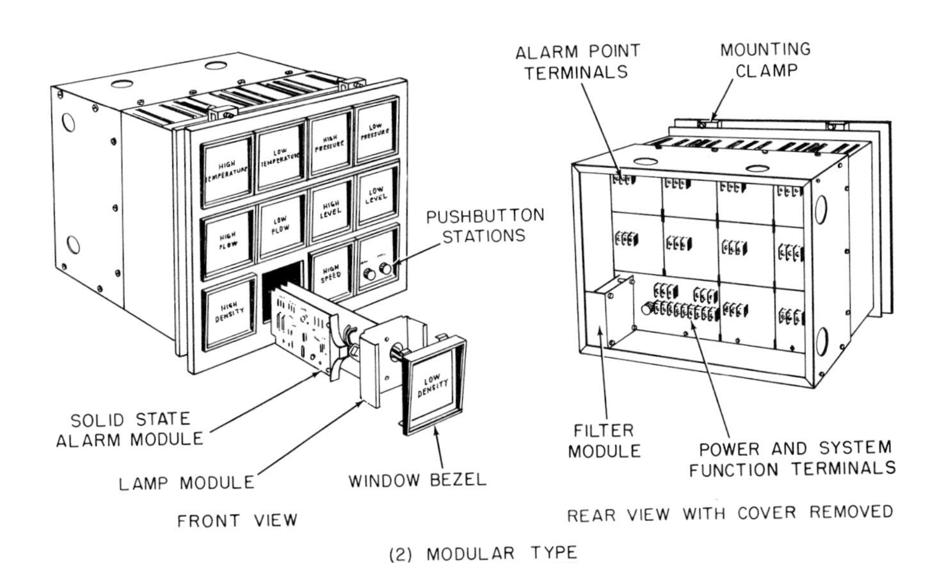

Fig. 3.4d Integral annunciator cabinets, modular and nonmodular

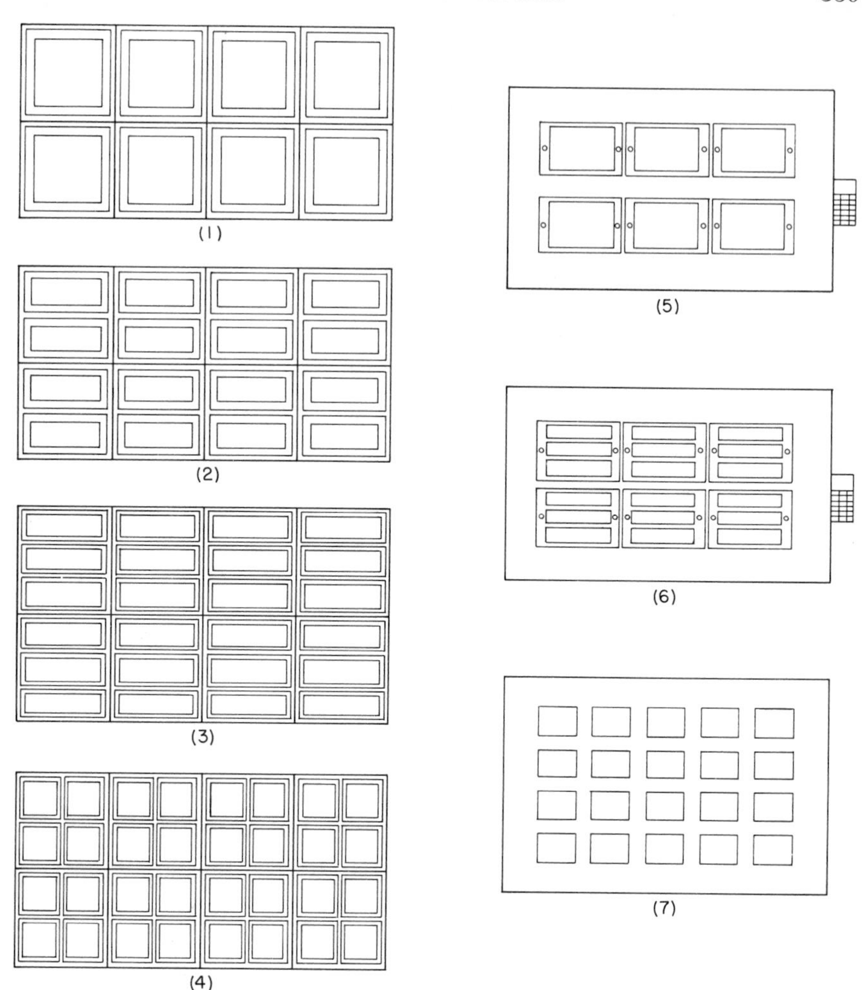

Fig. 3.4e Integral annunciator window configurations. 1. Modular single point annunciator; 2. Modular double point annunciator; 3. Modular triple point annunciator; 4. Modular quadruple point annunciator; 5. Nonmodular single point; 6. Nonmodular triple point; and 7. Nonmodular single point with small nameplate.

Remote Annunciator

The remote annunciator differs from the integral annunciator in that the visual indicators are remote from the cabinet or chassis containing the logic modules. Remote annunciators were developed to allow the visual indicators to be located in their actual process location in the graphic flow diagram. They are used with full and semigraphic control panels and in nongraphic applications in which an integral annunciator cabinet may require too much front panel space. Figure 3.4f illustrates a remote annunciator chassis with

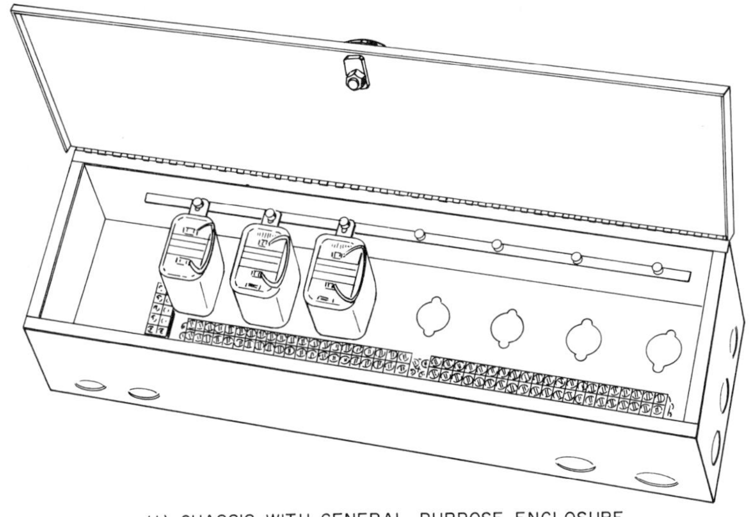

(1) CHASSIS WITH GENERAL PURPOSE ENCLOSURE

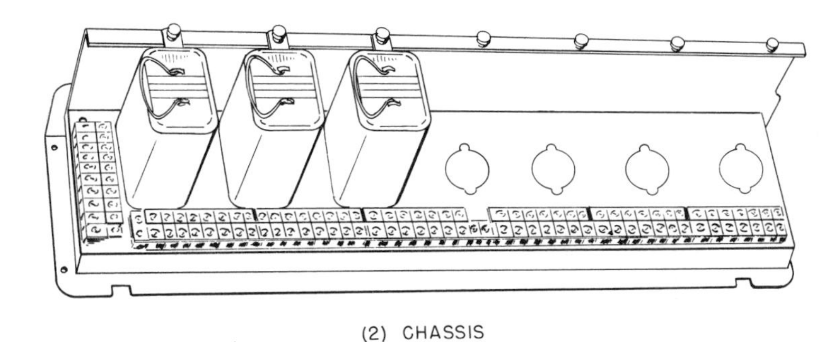

(2) CHASSIS

Fig. 3.4f Remote annunciator cabinets

optional cabinet enclosure. The chassis contains spare positions for plug-in logic modules and a system flasher. Auxiliary system modules such as horn isolating relays may also be plugged into the logic module chassis positions. The chassis and cabinet enclosure are designed for wall or surface mounting behind the control panel. Each chassis position has terminal points for connecting the visual indicator and trouble contact. In addition, the chassis has a system terminal block for connecting electrical power, horn, flasher and acknowledge and test pushbuttons.

Remote annunciator disadvantages include higher equipment and installation costs and increased back panel space. In addition, the wiring connections from field trouble contacts must be made to intermediate terminal blocks rather than directly to the cabinet terminals as with the integral annunciator.

These terminal blocks, the terminal enclosure and the required wiring results in high installation costs and extra space requirements. Finally, the remote annunciator is difficult to change, and modification costs of remote systems are substantially higher than those of the integral type, partially because spare visual indicators cannot be installed initially.

Semigraphic Annunciator

The semigraphic annunciator developed in the late 1960s combines some of the advantages of the integral annunciator with the flexibility to locate visual indicators at appropriate points in a graphic flow diagram. Figure 3.4g illustrates a semigraphic annunciator. It consists of a cabinet containing annunciator logic modules wired to visual indicators inserted in a $^3/_4$-inch lamp insertion matrix grid forming the cabinet front. The semigraphic display is placed between the lamp grid and a transparent protective cover plate, and the visual indicators are positioned to backlight alarm nameplates located in the graphic display. The protective cover and lamp grid are either hinged or removable so as to provide access to the logic module and lamp assemblies. The lamp assemblies are connected to intermediate terminals located behind the lamp grid, and the terminals in turn are connected to the logic modules. Terminal points for trouble contact wiring are in the back of the cabinet.

The semigraphic annunciator is flexible, and changes to the annunciator system, graphic display and related panel modifications can be made easily and cheaply. It is practical to prepare the graphic displays in the drafting room or model shop, thus protecting proprietary process information of a confidential nature. The graphic display has little or no effect on completing either the annunciator or the control panel because it can be installed on site or at any time. The semigraphic annunciator has a high density of 40 alarm points per linear foot and a solid state rather than relay type logic design. Power supplies are self-contained in the semigraphic annunciator cabinet.

Front panel layouts illustrating integral, remote and semigraphic annunciators are shown in Figure 3.4h. Integral systems similar to the one shown at the left in the figure are normally specified on nongraphic control panels. The graphic panel in the center utilizes a remote annunciator with backlighted nameplates (shaded rectangles) and pilot lights (shaded circles) for visual indication. The remote system may also be used with miniature lamps in a semigraphic display similar to the one shown at the right.

Recording Annunciators

In recent years, solid state, high speed recording annunciators have been developed, substituting a printed record of abnormal events for visual indication. These systems print out a record of the events and identify the variable, the time at which the alarm occurs and the time at which it returns to normal. They can also discriminate between a number of almost simultaneous events

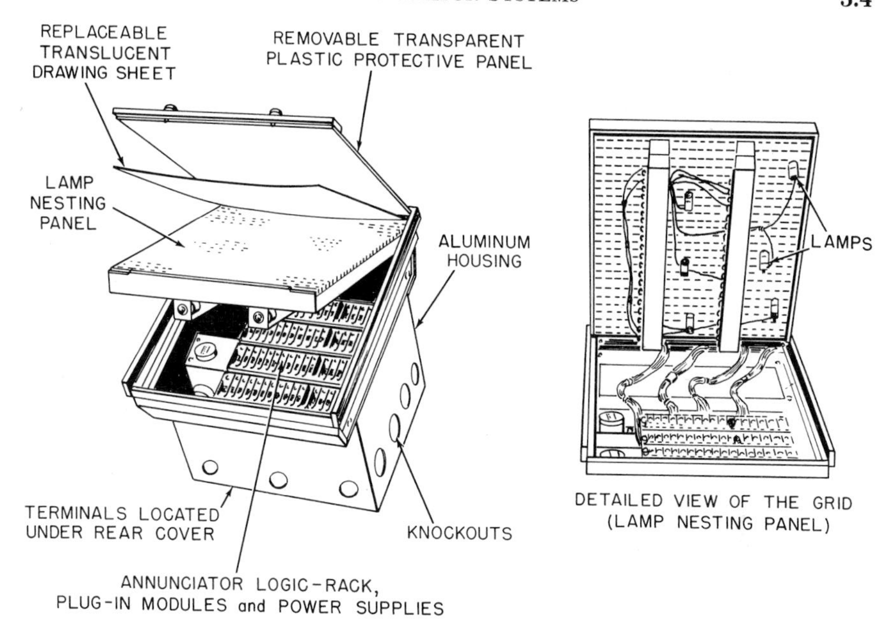

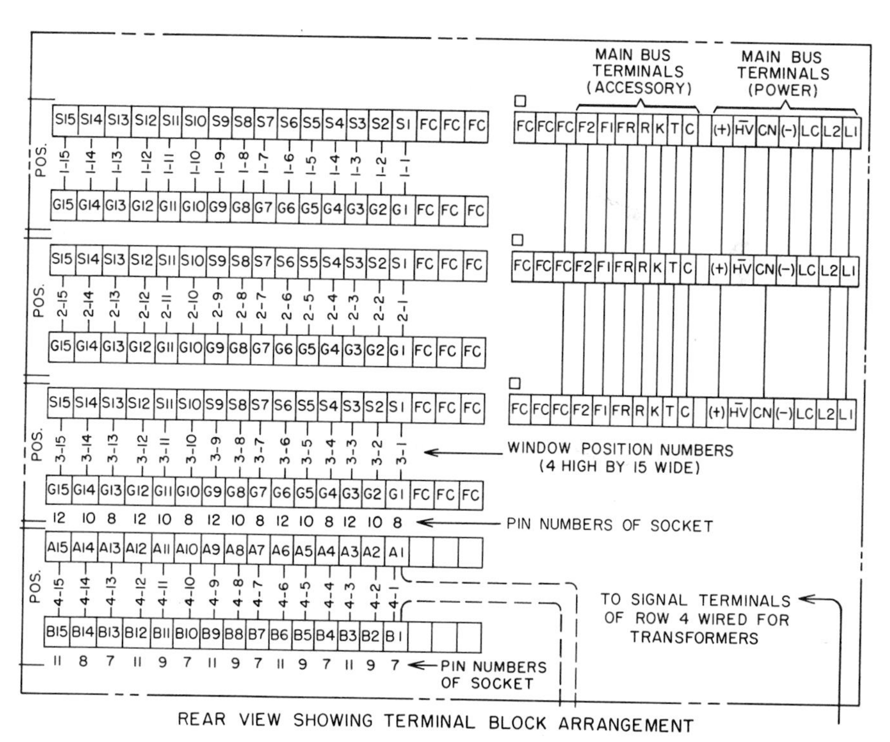

Fig. 3.4g Typical semigraphic annunciator

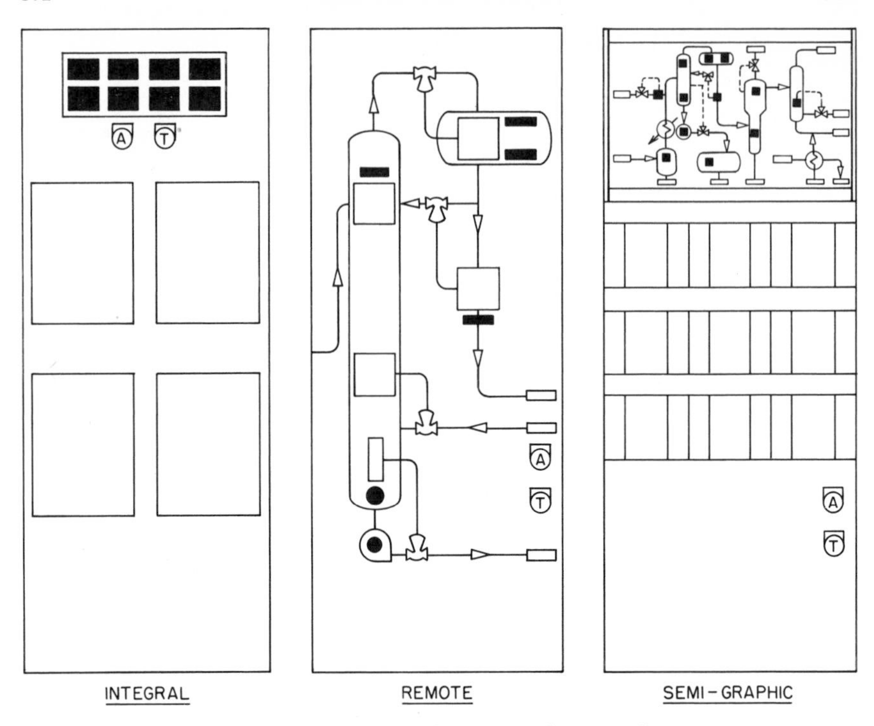

Fig. 3.4h Control panels with integral, remote and semigraphic annunciators

and print them out in the time sequence in which they occurred. A number of optional features including secondary printers at remote locations, supplementary visual indication and computer interfacing are also available. The typical unit consists of logic, control and printer sections.

The input status is continuously scanned and if a change in the trouble contact has occurred since the preceding scan cycle, the central control operates to place the exact time, the alarm point identification and new status of the trouble contact (normal or abnormal) into the memory and initiates the operation of the output control unit (Figure 3.4i). The output control unit accepts the stored information and transfers it to the printer, which logs the event, following which the memory is automatically cleared of the data and is ready to accept new information. In addition to or in place of the printer (if a permanent record is not required) a CRT display can also serve as the event read-out. Trouble contacts are connected to terminal points in the logic cabinet, and a cable connects the cabinet and the printer.

A recording annunciator can perform more sophisticated monitoring than an audible-visual annunciator and is correspondingly more expensive on a per point basis. System cost per point decreases as the system size increases. Higher equipment cost, however, is offset in part by savings in control panel space

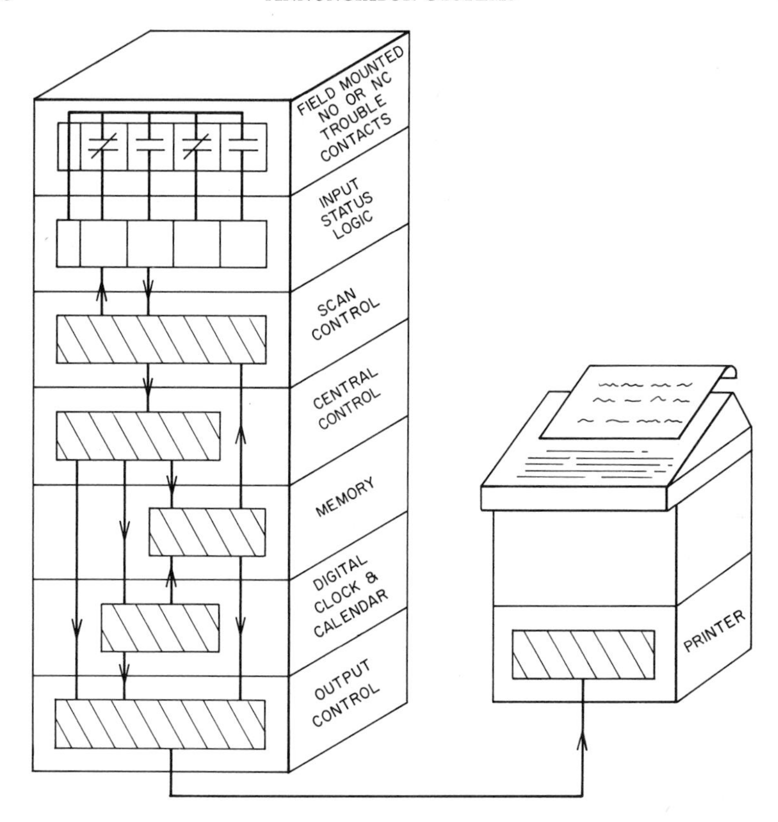

Fig. 3.4i Functional block diagram of recording annunciator

and in installation costs. Recording annunciators are frequently used by the electrical power generating industry, but may be applied to advantage in any industrial process which must monitor large numbers of operating variables and analyze abnormal events efficiently.

Vocal Annunciators

Vocal annunciators are unique in the type of abnormal audible message they produce. The audible output is a verbal message identifying and describing the abnormal condition when it occurs and repeating the message until the operator acknowledges the difficulty. The system continuously scans the trouble contacts and when an abnormality is found, it turns on a flashing visual indicator and selects the optional proper verbal message for broadcast. The visual indicator is turned off by the system when the point returns to normal. The control unit also arranges the messages to be broadcast in the order in which the difficulties occur. In the event of multiple alarms, the second message is played only after the first has been acknowledged. The flashing visual indicator for each point, however, turns on when the point becomes abnormal. The verbal message may be broadcast simultaneously in the control room and related operating areas, thus permitting operators at the operating unit to correct immediately.

Electromechanical Relay Type Annunciators

The basic element of this annunciator is an electrical relay wired to provide the logic functions required to operate a particular sequence (see Section 4.5 of Volume II). At least two relays are required per logic module for most sequences. The relays are installed and wired in a plug-in assembly which is the logic module for a single alarm point. The plug-in module assembly is usually hermetically sealed in an inert atmosphere to prolong the life of the relay contact. The sealing also makes the logic module acceptable in certain hazardous electrical areas.

Figure 3.4j is a semischematic electrical circuit for a remote system with sequence operation according to ISA Sequence 1B. Two logic modules are shown; one is in normally closed operation and the other is in normally open operation. The remote visual indicators for each alarm point, the horn, the flasher and the acknowledge and test pushbuttons common to the system are also shown. Each logic module has two relays, A and B, shown in their de-energized state according to normal electrical convention. The operation of these circuits during the various stages of sequence operation is as follows:

Normal. The trouble contact of the NC alarm point is wired in series with the A relay coil at point terminals H and NC. In the normal condition, the trouble contact is closed, relays A and B are energized and all A and B relay contacts are in the state opposite to that shown. Relay A is energized from power source H through the closed trouble contact, relay coil A, resistor R, terminal K and jumper to the neutral side of the line N. Relay B is also energized from H through the normally closed acknowledge pushbutton to terminal C, closed contact A2, relay coil B to N. Relay B is locked-in by its own contact B2 which closes when relay B is energized. The visual indicator is turned off by open contact A5. The audible signal and the flasher motor are turned off by open contacts A3 and B4 in the same circuit.

Alert. The trouble contact opens de-energizing relay A and returning all A relay contacts to the state shown in Figure 3.4j. The visual indicator is turned on, flashing through circuit H, lamp filament, terminal L, closed contacts A5 and B6, bus F, flasher contact F1 to N. The flasher motor is driven through circuit C, closed contacts A3 and B3, R bus, flasher motor to N and the audible signal is turned on by the same circuit.

Acknowledged. Relay B is de-energized by operating (opening) the momentary contact horn acknowledgment pushbutton and is locked out by open contact A2. All B relay contacts are returned to the state shown in Figure 3.4j. The visual indicator is turned on steady through closed contact B5 to

OPERATIONAL SEQUENCE

CONDITION	TROUBLE CONTACT	SIGNAL LAMPS	AUDIBLE SIGNAL	"A" RELAY	"B" RELAY
NORMAL	NORMAL	OFF	OFF	ENERGIZED	ENERGIZED
ALERT	ABNORMAL	FLASHING	ON	DEENERGIZED	ENERGIZED
AUDIBLE SILENCED (ACKNOWLEDGED)	ABNORMAL	STEADY-ON	OFF	DEENERGIZED	DEENERGIZED
NORMAL AGAIN	NORMAL	OFF	OFF	ENERGIZED	ENERGIZED
LAMP TEST	NORMAL	STEADY-ON	OFF	ENERGIZED	ENERGIZED

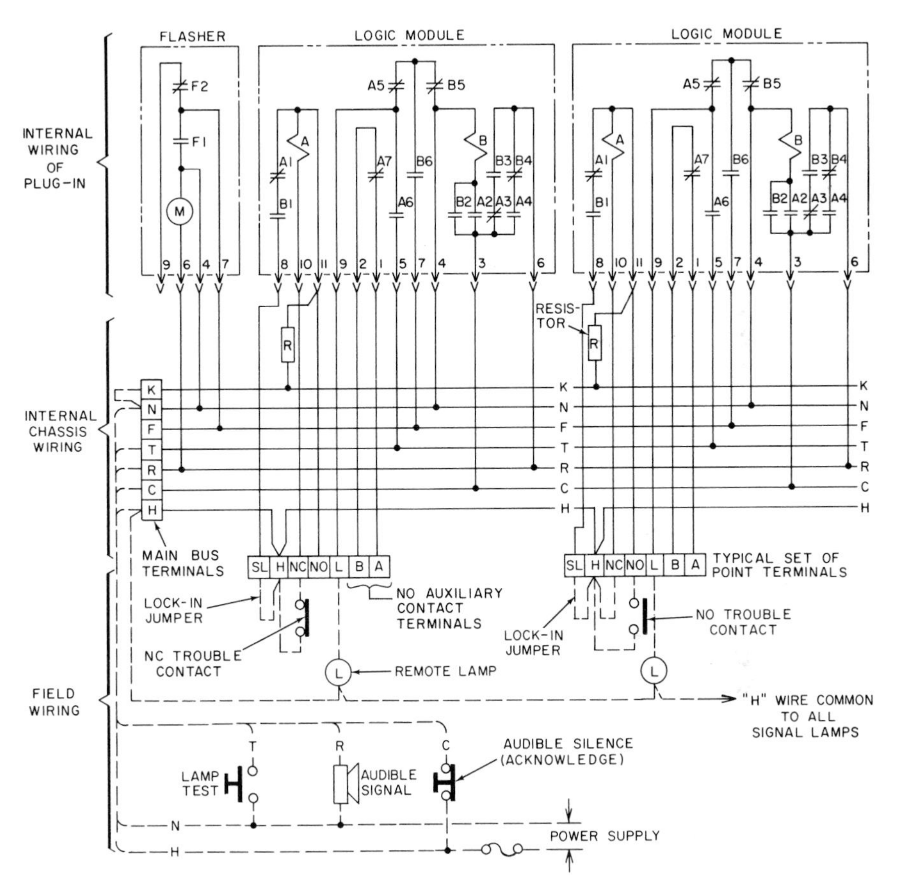

Fig. 3.4j Semischematic diagram of a relay type annunciator for ISA sequence 1B

N and is disconnected from flasher contact F1 by open contact B6. The flasher motor and audible signal are turned off by the horn acknowledgment push-button and remain off due to open contact B3.

Normal Again. When the variable condition returns to normal, the trouble contact closes to energize relay A, the visual indicator is turned off by open contact A5, relay B is energized by closed contact A2 and all circuits are again in the state described under normal.

Lamp Test. The lamp test circuit operates the visual indicators of only those alarm points which are in the normal condition. The circuit is completed through power source H, lamp filament, terminal L, closed contact A6, bus T, normally open momentary contact lamp test pushbutton to N. Closing the lamp test pushbutton to N completes the circuit and lights the visual indicators. Alarm points which are in the off-normal condition (either alert or acknowledged) do not operate because their A relays are de-energized and the A6 contact is open. The visual indicators of these abnormal alarm points are already turned on (either flashing or steady) through the operation of the alarm sequence.

Lock-in. The lock-in feature operates to prevent an annunciator alert condition (caused by a momentary alarm) from returning to normal until the horn acknowledgment button is pushed. Point terminals H and SL are jumpered to provide the lock-in feature (Figure 3.4j). When the trouble contact opens, the A relay is de-energized, and power source H is applied to the N side of the relay through closed contacts A1 and B1 and the power is dissipated through resistor R, terminal K and jumper to N. If the trouble contact returns to normal, relay A will remain de-energized because potential H is on both sides of the coil. If the acknowledgment button is pushed before the trouble contact closes again, relay B will be de-energized, opening contact B1 and the look-in circuit, thus permitting the system to return automatically to normal when the trouble contact closes. If the acknowledgment button is pushed after the trouble contact has reclosed, contact B1 opens momentarily, allowing the A relay to re-energize. Contact A1 opens and the circuits are re-established in their normal operating state.

Operational Test. Full operational test is incorporated in the annunciator sequence shown by replacing the jumper connection between main bus terminals K and N with a normally closed momentary contact pushbutton, which when pushed opens all annunciator circuits, thus initiating the alert condition of all alarm points in the normal condition.

Auxiliary Contacts. Normally closed contact A7 connected to point terminals A and B is available for auxiliary control functions.

Relay Fail-Safe Feature. Two parallel circuits, one consisting of closed contact A3 and open contact B3 and the other of open contact A4 and closed contact B4, operate an alert signal when there is a failure of either the A or B relay coil. A failure of the former initiates a normal alert in the same way as the trouble contact. A failure of the B relay turns on the audible signal through closed contacts A4 and B4.

Normally Open Trouble Contacts. The annunciator sequence and features described for NC trouble contacts operate in essentially the same way when NO contacts are used. In the NO system, however, the trouble contact is wired in parallel with the A relay coil to point terminals H and NO, and a wire jumper is installed between point terminals H and NC. Normally, the trouble contact is open and the A relay is energized from terminals H and jumper to terminal NC. In the alert condition, the trouble contact closes to de-energize the A relay by applying power source H to the N side of the relay.

Electromechanical relays are available for use with a variety of AC and DC voltages, but 120 AC, 50 to 60 Hz and 125 DC are the most popular. Power consumption of the logic modules is normally less than 10 volt-amperes (AC) and 10 watts (DC). Special low drain and no drain logic modules are available which consume no power during normal operation. Visual indicators consume different amounts of power, depending on the type. Small bull's-eye lights and backlighted nameplates use approximately 3 watts, whereas large units require 6 to 12 watts, depending on whether one or two lamps are used.

Electromechanical annunciator systems are reliable and may be used at normal atmospheric pressures and ambient temperatures in the 0 to 110°F range. They are not position sensitive. They will generate a substantial amount of heat during plant shutdown when a large number of points are askew, and therefore power should be disconnected during these periods. The principal disadvantages of the relay type annunciator are size, power consumption and heat generation.

Solid State Annunciators

A solid state logic module consists of transistors, diodes, resistors and capacitors soldered to the copper conductor network of a printed circuit board supplying the required annunciator logic functions. The modules terminate in a plug-in printed circuit connector for insertion into an annunciator chasis; it may also contain mechanical switching or patching devices to provide lock-in and NO-NC options.

Figure 3.4k is a semischematic electrical circuit for a remote system with ISA sequence 1B. The logic module shown is in normally closed operation. Remote lamps for two points and a flasher-audible module, speaker, acknowledge and test pushbuttons common to a system are also included. Switch S1 is the NO-NC option switch and is shown in the NC operating position. Switch S2 is the lock-in option switch and is shown in the lock-in position. The following description uses negative logic, i.e., a high = a negative voltage, whereas a low is approximately zero volts.

Normal. The trouble contact of the NC alarm point is connected to an input filter circuit consisting of resistors R13 and R50 and capacitor C1, and provides transient signal suppression as well as voltage dropping. The slide switch S1 connects the trouble contact and filter network to resistor R14. In this state transistor T1 is conducting, causing the full negative voltage to be dropped across resistor R17, resulting in a low voltage at the bottom end of resistor R20. Transistors T2 and T3 are the active elements of the input memory and are roughly equivalent to the A relay of an electromechanical module (Figure 3.4j). The base of T2 has four inputs including resistor R20, either directly from the trouble contact in NO operation or from the collector of T1 in NC operation; resistor R19 with a locking signal from the alarm memory transistor T5; resistor R28 and capacitor C2 which form a regenerative feedback from T3; and resistor R15 from the test circuit. The base of T3 has one input, resistor R29 from the collector of T2. In normal operation, all four inputs to the base of T2 are low, T2 is not conducting and T3 is conducting. Conversely, when a high signal is present at any one of the four inputs to the base of T2, T2 conducts and T3 turns off.

Transistors T4 and T5 are the active elements of the alarm memory and are approximately equivalent to the B relay of an electromechanical module. T4 and T5 together with bias resistors R30, R33 and cross coupling resistors R31 and R32 form a bistable (flip-flop). In normal operation T4 is off and T5 conducts. The upper end of capacitor C4 is connected to the collector T2. When T2 is off, its collector is at a high and capacitor C4 will change from top to bottom, minus to plus. Transistor T7 is a high capacity lamp amplifier and T6 is its preamplifier. In normal operation the base of T6 is high and T6 is on, T7 is off and the visual indicator is off.

Before completing the description of the normal condition, the operation of the flasher-audible module in Figure 3.4k will be described. The module has two oscillators. The first is a 3 Hz unit generating a signal which is amplified and supplied to the logic modules through the F1 bus. The second is a 700 Hz oscillator generating a signal which is amplified and supplied to the audible signal through the R bus.

The audible signal is a permanent magnet type transducer (speaker) which converts the electrical energy into sound. Initiated by an audio oscillator the active elements of which are transistors T1 and T2, these transistors together with passive components (capacitors C1 and C2 and diodes D3 and D4) form an unstable multivibrator when an input is present on the FR bus. In the normal conditions there is no FR signal, the voltage necessary to turn on transistor T1 is missing and the oscillator will not operate. This is the normal or quiescent condition.

Alert. The trouble contact opens and the base of T1 becomes low, turning off T1. This action produces a high on the base of T2 through R20, which turns on T2 and turns off T3. The negative end of C4 is clamped to common through T2, causing a positive pulse at the base of T5 through diode D12, turning T5 off and T4 on. With T2 conducting, the base of T6 becomes low through resistor R6 and with T5 off the clamp on the flasher signal is removed through diode D4 at the junction of R1 and R2. The flasher source provides an alternating high and low voltage at the F1 bus, which turns T6 on and off, which in turn turns T7 and the light off and on. The flasher signal is generated by transistors T8 and T9 which are the active elements of an unstable multivibrator used as an on-off signal to the output driver stage. Resistor R24 and capacitor C5 decouple the oscillator from the power lines

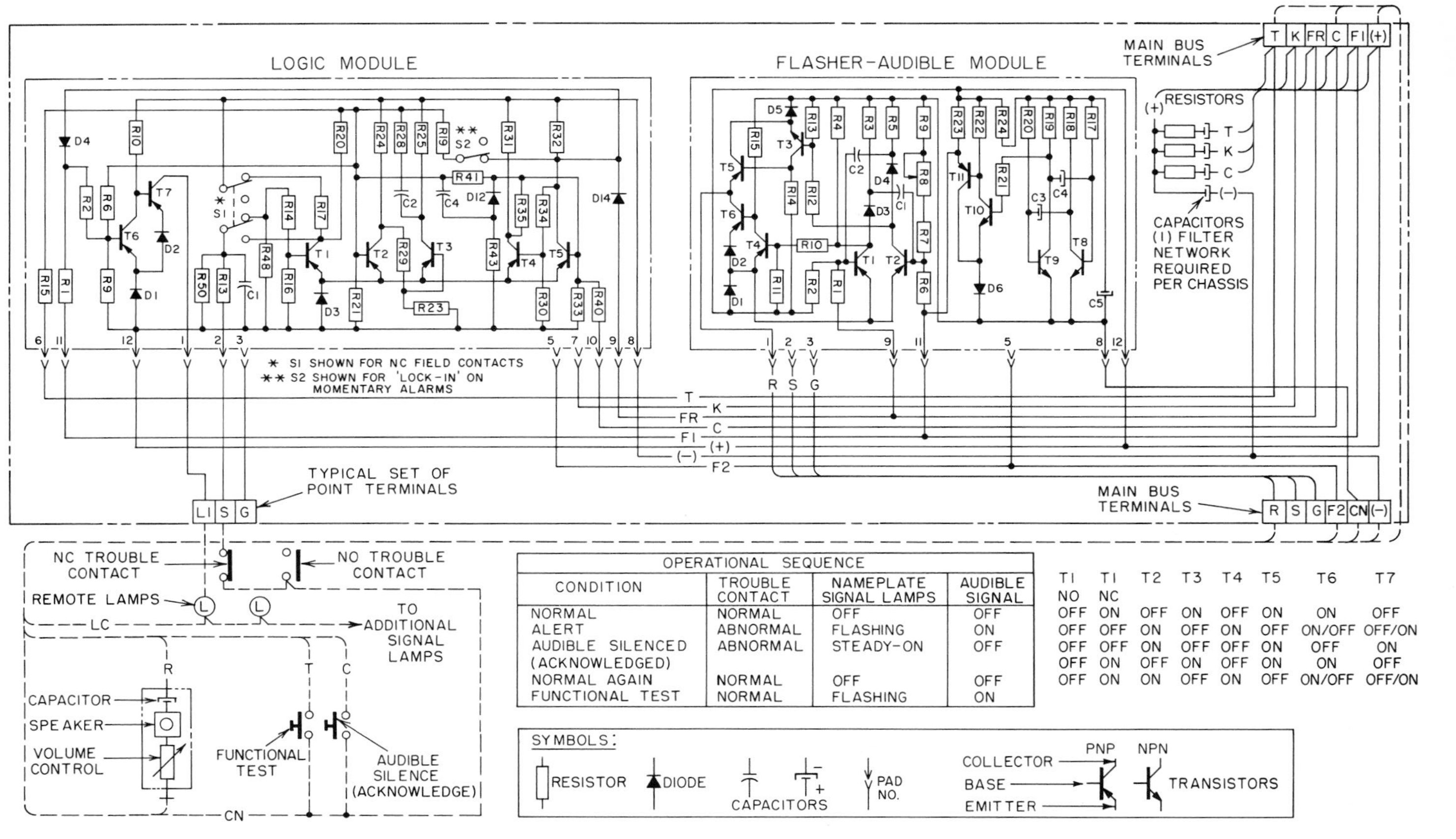

OPERATIONAL SEQUENCE			
CONDITION	TROUBLE CONTACT	NAMEPLATE SIGNAL LAMPS	AUDIBLE SIGNAL
NORMAL	NORMAL	OFF	OFF
ALERT	ABNORMAL	FLASHING	ON
AUDIBLE SILENCED (ACKNOWLEDGED)	ABNORMAL	STEADY-ON	OFF
NORMAL AGAIN	NORMAL	OFF	OFF
FUNCTIONAL TEST	NORMAL	FLASHING	ON

T1 NO	T1 NC	T2	T3	T4	T5	T6	T7
OFF	ON	OFF	ON	OFF	ON	ON	OFF
OFF	OFF	ON	OFF	ON	OFF	ON/OFF	OFF/ON
OFF	OFF	ON	OFF	OFF	ON	OFF	ON
OFF	ON	OFF	ON	OFF	ON	ON	OFF
OFF	ON	ON	OFF	ON	OFF	ON/OFF	OFF/ON

Fig. 3.4k Semischematic diagram of a solid state annunciator for ISA sequence 1B

so that its frequency is not affected by that of the other oscillators. The output driver stage consists of transistors T10 and T11, a switching inverter and an emitter follower stage respectively, which produces an alternating high and low voltage at F1 bus through R23. Transistor T11 is a high current transistor capable of driving a multiple lamp load.

The audible signal is initiated by a high on the FR bus, which turns on an audio oscillator the elements of which are transistors T1 and T2. The audio oscillator output is amplified in an audio amplifier stage composed of transistors T3, T4, T5 and T6 connected in two pairs—one T3 and T5; the other T4 and T6. The input components to the stage from the audio oscillator are opposite each other, i.e., whenever one is high (negative voltage) the other is low (near zero), causing only one pair of transistors to conduct at a time. When T5 is off, T6 is on and the capacitor is discharged. This alternating action causes an alternating current to flow in the speaker coil, giving an audible signal.

Acknowledged. A negative voltage is applied to the base of T5 through resistor R40 by closing the acknowledge pushbutton. This turns T5 on and T4 off and the FR bus becomes low, thus silencing the audible signal. When the point is acknowledged, T5 conducts; this restores the clamp at R1 and R2 which removes the flash source voltage. T6 is off all the time and T7 is on all the time and the light is on steady.

Normal Again. When the variable condition returns to normal, the trouble contact closes and the base of T1 becomes high, turning T1 on. This produces a low on the base of T2, which turns T2 off and T3 on. All circuits are again in the state described under the normal condition.

Lamp Test. No separate lamp test is normally provided. A full operational test is initiated by pushing the test button, which applies an alternate input signal through resistor R15 to the base of transistor T2. This turns T2 on, initiating a full operational test of the system as already described.

Lock-in. The lock-in feature is provided by a switch S2. If the switch is in the lock-in position (Figure 3.4k), when T5 turns off (on the alarm condition) a high at the collector of T5 is coupled to the base of T2 through R19, which keeps T2 turned on even if the trouble contact returns to normal. Transistor T5 will remain off and keep T2 turned on until the acknowledgment button is pushed.

If the switch is in the non-lock-in position, the circuit between the collector of T5 and the base of T2 is open; therefore, T2 will turn off if the trouble contact returns to normal before acknowledgment and return all circuits to normal.

Operational Test. See description under lamp test.

Auxiliary Contacts. Auxiliary contacts are not supplied as part of the logic module. Adapter assemblies consisting of relays operated by the semiconductor logic, however, are available.

Relay Fail-Safe Feature. Not available in solid state circuits.

Normally Open Trouble Contacts. The annunciator sequence and features already explained for NC trouble contacts operate in essentially the same way for NO contacts. The NO-NC option switch by-passes inverter stage transistor T1. When the contact closes on an abnormal condition, it turns on T2 through R20 and the sequence operation proceeds exactly as described for the NC operation. Solid state annunciators are for use with DC voltages ranging from 12 to 125 DC. Power consumption of the logic modules ordinarily is less than 5 watts. Visual indicators consume different amounts of power, depending on the type. Bull's-eye lights use approximately 1 watt, whereas backlighted nameplates use from 1 to 6 watts, depending on the number and wattage of lamps.

Solid state annunciators are very reliable and are not position sensitive. They offer the advantages of compactness, low power consumption and little heat generation, factors that make them particularly useful in large integral annunciators. The per point cost of solid state systems is slightly higher than their relay type equivalent, owing to the cost of power supplies and interfacing accessories which may be required with solid state systems. The cost of the logic modules, visual indicators and cabinets themselves is not excessive. Integrated circuit components using recently developed microcircuits will most likely reduce size, power consumption and heat dissipation of annunciator systems.

Annunciator Cabinets

Annunciator systems are installed in areas ranging from general purpose to hazardous. (For more on area classifications, see Section 10.10 of volume I.) Annunciator cabinets are installed indoors and outdoors in a variety of dusty, moisture-laden and other adverse environments. Industrial annunciator cabinets are usually designed for general purpose, dry indoor use. Special cabinets and enclosures are used in hazardous, moist and outdoor locations.

The requirements of class 1, division 2 hazardous locations as defined in Article 500 of the National Electric Code (NEC) are satisfied by the visual indicators and logic modules (either relay or solid state) of most annunciator systems. A manually operated or door interlocked power disconnect switch is used with annunciator cabinets in those locations to turn off power when relamping or changing logic modules.

Annunciator equipment for class 1, division 1 areas is installed in cast steel or aluminum housings approved for the hazardous environment. They are bulky and expensive to purchase and install. Integral annunciators are limited in size from one to three point units. Remote annunciators installed in explosion-proof enclosures and wired to explosion-proof bull's-eye lights are used for larger systems. Annunciator power must be disconnected either manually or automatically before opening the enclosures to prevent an accidental arc or spark when relamping or changing logic modules.

Annunciator cabinets installed in either general purpose or hazardous areas (class 1, division 2) are made weatherproof either by housing them in a suitable enclosure or by covering the exposed cabinet front with a weatherproof door. Housings which comply with class 1, division 1 requirements are also weatherproof. Figure 3.4 l illustrates several weatherproof and hazardous area enclosures.

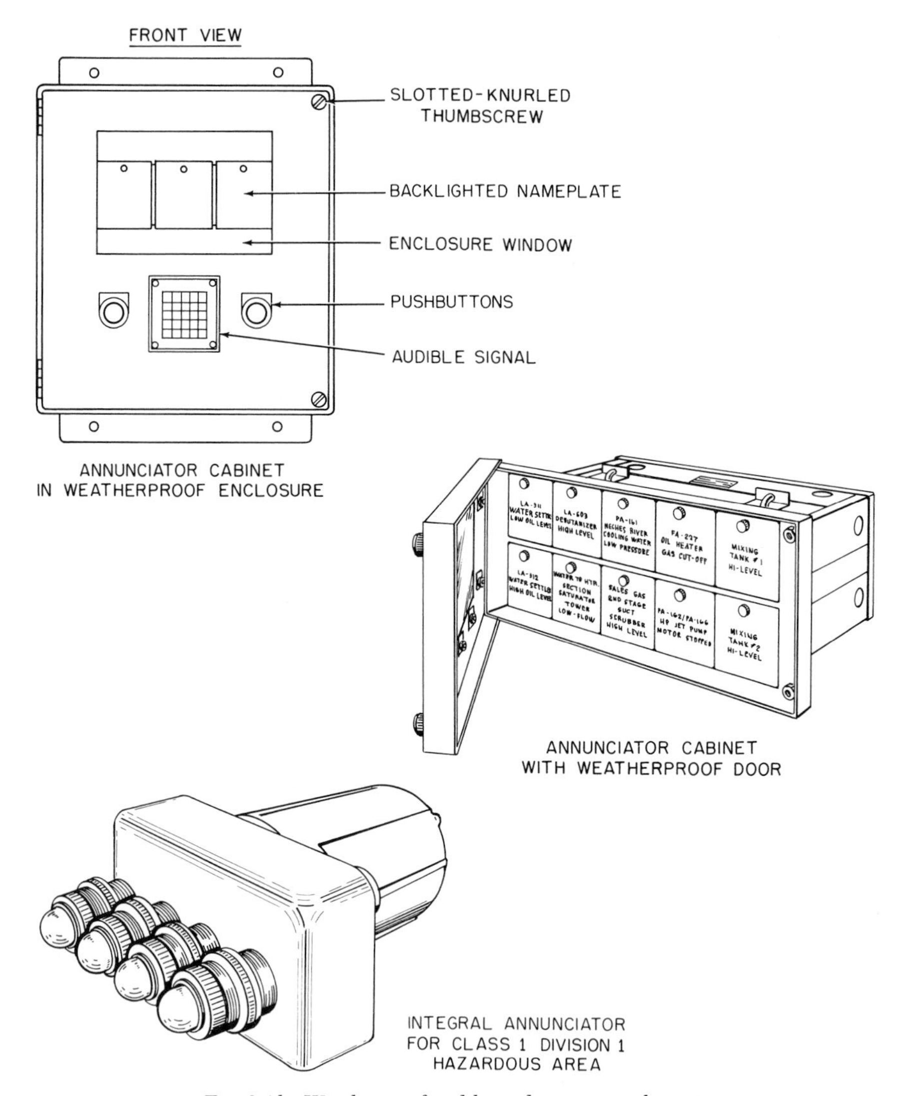

Fig. 3.4 l Weatherproof and hazardous area enclosures

Annunciators are classified intrinsically safe if they are designed to keep the energy level at the trouble contact below that necessary to generate a hot arc or spark. Care must also be taken in installing the system to install wiring so as to preclude a high energy arc or spark at the trouble contact due to accidental short circuit or mechanical damage. Thus, general purpose trouble contacts may be used with intrinsically safe annunciator systems even though they are installed in a hazardous area. The annunciator logic modules and visual indicators, however, must conform to the electrical classification of the area in which they are installed. (For more on intrinsically safe designs, see Section 10.11 of volume I.)

Pneumatic Annunciators

Pneumatic annunciators consist of air operated equivalents of the trouble contact, logic module and visual indicator stages of an electrical annunciator system. A single point system furnishing high tank level monitoring is shown in Figure 3.4m. Power supply to the system is instrument air at 80 to 100 PSIG, which is reduced to the required operating pressure by pressure

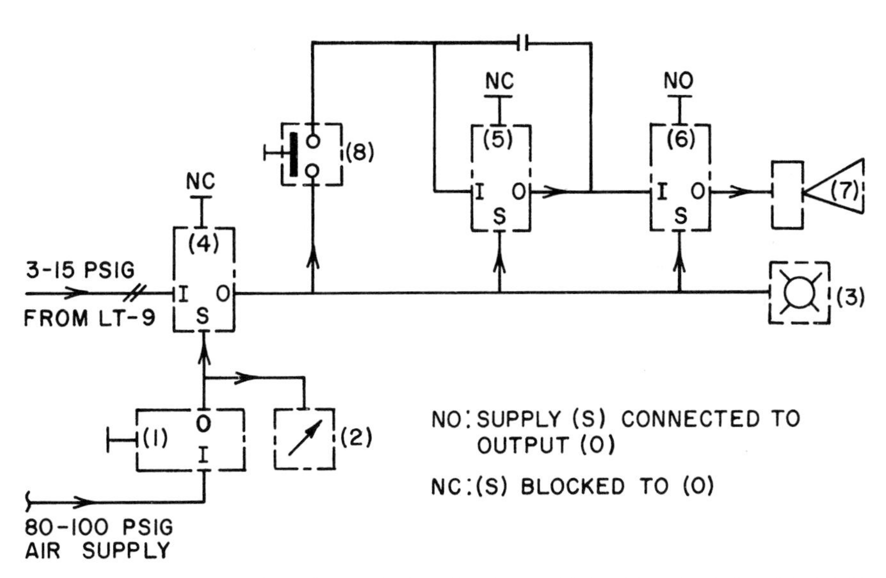

CONDITION	RELAY (4)	RELAY (5)	RELAY (6)	HORN (7)	INDICATOR (3)
NORMAL	CLOSED	CLOSED	OPEN	OFF	OFF
ALERT	OPEN	CLOSED	OPEN	ON	ON
ACKNOWLEDGE	OPEN	OPEN	CLOSED	OFF	ON
NORMAL AGAIN	CLOSED	CLOSED	OPEN	OFF	OFF

Fig. 3.4m Pneumatic annunciator circuit

regulator (1). The operating pressure is indicated on pressure gauge (2). A 3 to 15 PSIG analog input signal from a direct acting level transmitter (LT-9) enters high pressure limit relay (4) which is normally closed and set to open when the high level limit is exceeded. When this happens (alert condition), an input at supply pressure from (4) turns on a pneumatic visual indicator (3) and a normally open high pressure limit relay (6) allows supply air flow to air horn (7), turning it on. Simultaneously, the air output from (4) enters normally closed high pressure limit relay (5) and momentary contact push-button (8) which is a normally open acknowledgment pushbutton for the system. In the alert condition, the pneumatic indicator and horn are both on.

The alert condition is acknowledged by pushing button (2), closing it and thereby opening high pressure limit relay (5). Supply air pressure from (5) closes high pressure limit relay (6), which cuts off the operating air to the horn, thereby turning it off. Simultaneously, operating air pressure from (5) is fed back to the inlet of (5). The feedback pressure locks up (5) so that it will not close when the acknowledgment pushbutton (8) is released. In the acknowledged condition, the pneumatic indicator is on and the horn is off. The system returns to normal when the 3 to 15 PSIG analog input falls below the set point. This closes high pressure limit relay (4) which turns off the pneumatic indicator (3). It also closes high pressure limit relay (5) by venting the lock-in circuit through relay (4).

Pneumatic annunciators are used when one or two alarm points are needed but electrical power is not readily available and in hazardous electrical areas where an electrical annunciator might not be practical. Pneumatic annunciators require a substantial amount of installation space and are expensive to manufacture.

3.5 CRT DISPLAYS IN PROCESS CONTROL

Types:	Storage tube, refreshed raster (T.V.) scan, refreshed X-Y positioned.
Screen Size:	From 6 × 8 inches to 23 × 30 inches
Refresh Rate:	40 to 60 Hz
Character Capability:	500 to 4,800 characters
Characters per Line:	64 to 128 characters
Number of Character Lines:	12 to 74 lines
Character Set:	64 to 96 characters
Vector Modes:	Relative or absolute, or both
Vector Capability:	500 to 5,000 per frame
Cost:	$6,000 to $175,000
Partial List of Suppliers:	Adage; Computek; Control Data; Data Disc; Digital Equipment; Foxboro; Hendrix; IBM; Impac; Information Displays; Metra Instruments*; Monitor Displays; Sanders Associates; SEL; Tektronix; Univac.

As larger and more complex plants are built with central control rooms containing greater numbers of instruments, the man-process communication problems grow. When digital computer control was first applied in the process industry, alphanumeric control panels were utilized in addition to the conventional analog instrumentation displays. With these systems the user was able to display and manipulate the control loop parameters usually, however, only singly.

Plant Communications

More recently, the man-process communication requirement has been expanded to include not only the needs of the process operator but all com-

* This unit displays as many as 40 analog inputs in a bargraph format.

munication between the process and plant personnel. Table 3.5a defines seven levels of communication in a process plant involving process operating, engineering, programming and management personnel. A process operator, for example, would utilize information from levels 1 and 3; an instrument engineer would require information from level 2 and a system engineer would require information from level 4. A manager needs information from level 7.

The Total System

Although there are multi-channel bargraph instruments utilizing cathode ray tubes (CRTs) for displaying analog inputs, we will assume that the CRTs

Table 3.5a
COMMUNICATION LEVELS IN A PROCESS PLANT

Level 1:	*Emergency Indicators and Alarms* Includes both indicators and safety alarms that warn of impending difficulty; assist operator either in moving process to a safe operating point or in shuting it down.
Level 2:	*Component Diagnosis and Maintenance* Includes information required to diagnose plant and computer system component failure; make maintenance checks and assist maintenance man in correcting faults.
Level 3:	*Process Operation Information* Includes all information required by operating personnel to keep plant running safely and as close to economic optimum as possible.
Level 4:	*Process Evaluation and Diagnosis* Includes information needed to determine how well process is operating; to investigate potential technical or efficiency problems and to diagnose rapidly complex process failures when they occur.
Level 5:	*Process Supervision Information* Includes plant parameters which affect over-all economy and efficiency: information, e.g., on current schedules, feed stock availability and quality, utility usage and product qualities and costs required to make day-by-day or minute-by-minute adjustments of operating conditions to achieve optimum plant operation.
Level 6:	*System Maintenance and Improvement Information* Includes program information needed to derive the most from computer system and to make online system modifications as better operating methods are developed by plant personnel.
Level 7:	*Process Accounting and Scheduling* Includes information on quantities of production, feed stock supplies, shipping and labor to assist in establishing production schedules.

are employed in a computer controlled plant and are operated by digital logic devices. A block diagram of a typical digital computer control system is shown in Figure 3.5b. The CRT hardware is contained in the two consoles and consists of a CRT display, a keyboard (containing alphanumeric, functional and cursor

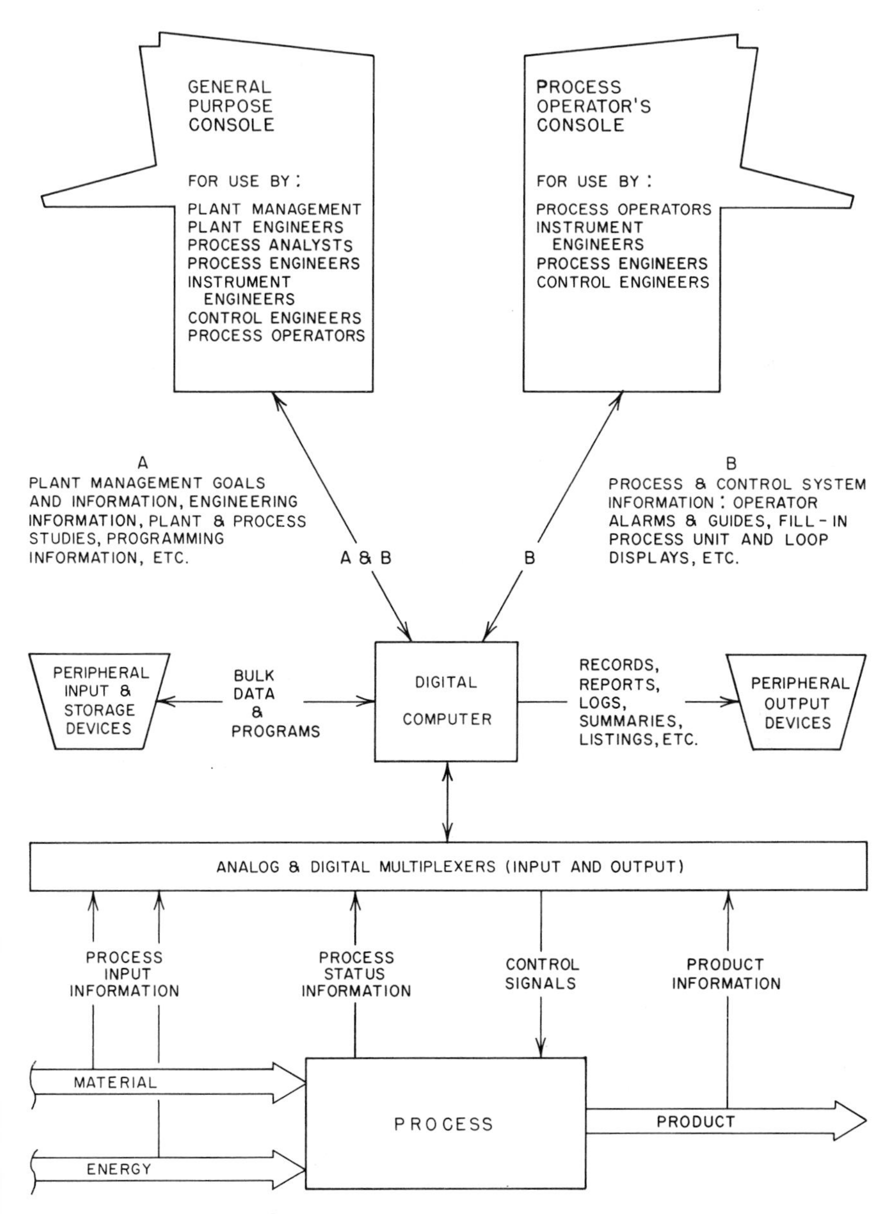

Fig. 3.5b Computer control system with CRT display

control keys), alarm light switches, a refresh memory, an alphanumeric character generator and format control and associated control logic. A vector generator can be supplied (optional) for graphic displays. Figure 3.5c illustrates the CRT hardware.

The digital computer memory stores the operating system and data lists. An auxiliary bulk storage device (drum or disk) is sometimes used to store additional programs and data files, and the computer utilizes a priority interrupt scheme and two bi-directional information channels for communication with other devices. One of these channels, commonly referred to as the programmed input-output (PIO) channel, transfers control information and single data words to and from a specified register (usually an accumulator) in the computer. The second channel transfers multiple words or blocks of data to and from the computer memory and is usually referred to as the direct memory access (DMA) channel, or simply the channel input-output (CIO).

The process control program requires a list of control tasks to be performed at specified intervals. These tasks include acquiring process data through analog and digital inputs and computation of appropriate control or alarm actions, or both, based on the input data. The task list contains the necessary data for the process control programs to carry out the desired control operations. These data include input and output addresses, constants, point names, digital status-information and current input-output values. This task

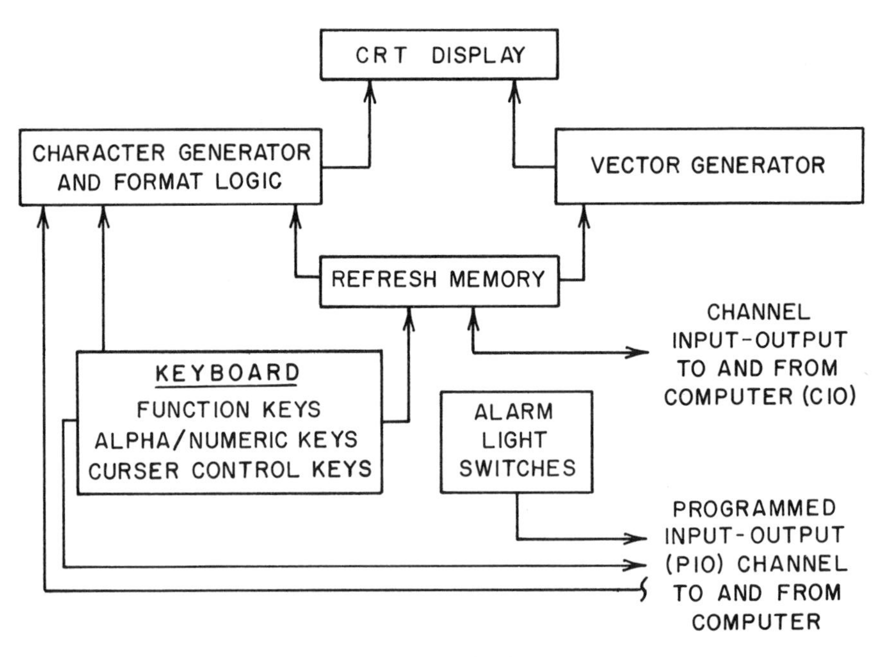

Fig. 3.5c CRT hardware

list is called the *process data base* and is the prime source of process data for display.

Servicing console data transmission requests from programs operating in the central processing unit and handling console keyboard interrupt requests are functions of the real time executive system. Programs performing tasks requiring data to or from a process display console pass their requests for service to the real time executive function programs. The calling programs receive acknowledgment of successful completion of the requested operation or indicators describing an aberrant condition.

Communication

Data Display

Cathode ray tubes display large amounts of information and are selective in displaying only relevant parameters or complex relationships between parameters. By fully exploiting the alphanumeric and graphic capabilities, the CRT is more efficient and economical than other methods of data display. Several choices of CRT implementation include a storage tube display, a raster (T.V.) scan display and random X-Y positioned, refreshed display. What follows is a description of the random X-Y positioned, refreshed display.

The size of the usable display area, and hence the size of the CRT, is determined primarily by the size of the character, the number of characters per line and the number of lines required. A secondary consideration is the required amount of graphic display. A typical display of a process plant unit is shown in Figure 3.5d. In order that this display be legible from a distance

PROCESS DISPLAY 2 FEED SPLITTER DATE 3-1-71 TIME 1515

										Alarms	
Loop	Block	Input	Meas	Units	Set Point	Output	Scan	CNT	Mode	ABS	Dev
FSM100 →	FSC100	FSP100	340.5	PSIG	300.0	....	ON	ON	Auto	HI	
			....		....	....					
	FSC101	FSF101	5.8	TCFT/H	5.6	....	ON	ON	Comp		
			....		....	....					
FSM300 →	FSC300	FSL300	12.8	FT	13.0	....	ON	ON	Auto		
			....		....	....					
	FSC301	FSF301	120	TGPH	124	....	ON	ON	Bkup		
			....		....	....					
FSM400 →	FSC400	FST400	550.7	DEGF.	555.0	....	ON	ON	Auto		
			....		....	....					
	FSC401	FSF401	128.5	TP/H	125.0	....	ON	ON	Comp		
			....		....	....					
FSM450	FSC450	FSF450	132.7	TGPH	134.0	....	ON	ON	Comp		
	→↑: Cursors		....		....	....					

Fig. 3.5d Typical process display on a CRT

of five to ten feet, the character height should be between $\frac{3}{8}$ and $\frac{1}{4}$ inch (Table 3.3a). To display the information shown in Figure 3.5d, a character format of 80 to 96 characters per line and 20 to 30 lines are required. These characteristics dictate a diagonal measurement on the CRT of at least 19 or 21 inches. The mixed display of alphanumeric characters and graphics shown in Figure 3.5e would also fit comfortably on a 21-inch CRT.

In a block diagram of a typical CRT display unit (Figure 3.5f), electromagnetic deflection and low voltage electrostatic focus maintains display quality at all locations on the CRT screen. P-31 phosphor (green) is usually preferred over P-1 phosphor (white) because it is more durable. The block of input data to the display unit shown in Figure 3.5f (X and Y position data and blanking) is supplied by a character generator, format control or vector generator. The data are digital in nature. The body of X and Y position data is loaded into output registers connected to high speed, digital-to-analog (D-A) converters, the output of which drives a linearity corrector and deflection amplifier. The linearity corrector compensates for geometrical distortion in the CRT, and the deflection amplifier must be capable of furnishing as much as 5 amperes of current to the deflection yoke. The blanking amplifier provides a signal to turn the electron beam in the CRT either on or off.

The information supplied to the CRT display unit must be continually repeated or refreshed. So that flickering or a "swimming" effect does not occur on the display screen, the refresh rate should be synchronous with the power line frequency—ordinarly 60 (or 50) Hz.

Cursor Control and Data Protection

When a CRT display is used in process control, a pointer or cursor is required to indicate the parameter to be acted upon. Cursors (Figure 3.5d) are manipulated from a keyboard so that they are beneath the line, value or character to be selected for the next operation. Cursor control keys (Figure 3.5g) include the four arrow keys ←, →, ↑, ↓ for movement in one of the four primary directions, a FAST key to increase the rate of movement, a HOME

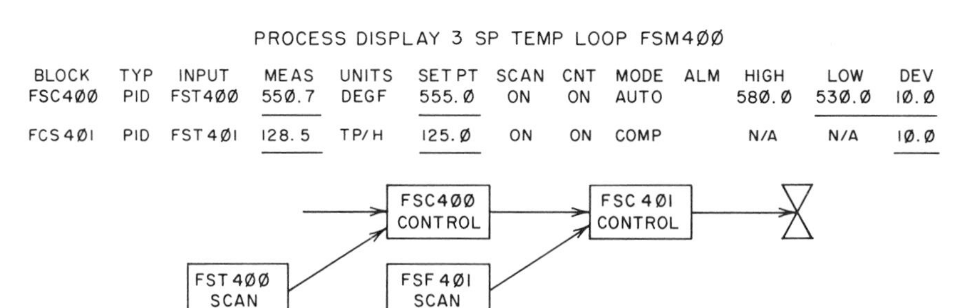

Fig. 3.5e Mixed alphanumeric and graphic display on a CRT

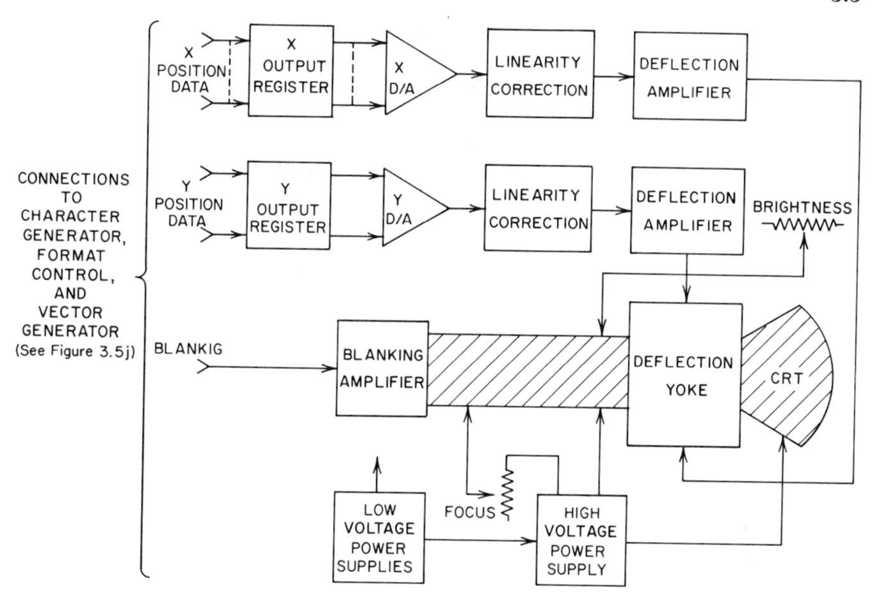

Fig. 3.5f CRT display components

key to return the cursor to the upper left corner of the CRT screen and a JUMP key which will be subsequently explained together with the protect feature. The position of the cursor can also be controlled or interrogated from the computer.

It is often undesirable to enable a user to modify values or characters on the CRT screen. A "protect" feature protects characters specified by a program in the digital computer from being modified. This feature might be implemented by a bit associated with each character, such that when it is set to a "1" state, the character can be modified only by the computer, not directly by the user. This feature also enables the computer to read selectively only unprotected information in the refresh memory.

The cursor control JUMP key allows the cursor to move from a current position to the next unprotected character following a protected one, thus by-passing (protected) characters that cannot be changed by the user—a very useful feature in a fill-in-the-blanks operation.

A "blink" feature permits individual characters displayed on the CRT to be blinked on and off several times per minute; this is useful for special conditions, such as alarm indication. This too is controlled by a bit associated with each character in the refresh memory. Supplying solid, dashed and dotted lines is useful for graphic displays. For example, a solid and dashed line might differentiate between a measurement and a set point when displaying trend information.

Alphanumeric Keyboard

Alphanumeric keys (Figure 3.5g) modify or make additions to the display on the CRT screen. Entries can be made only into unprotected locations, and are themselves unprotected. The operations are performed by the hardware associated with the CRT and require no response from the computer. The keys resemble those commonly found on a typewriter and when depressed, a code (usually USASCII-8) corresponding to the key legend is entered into a refresh memory location corresponding to one directly above the cursor on the CRT screen, and the cursor is incremented by one location. With the key code entered into the refresh memory, the character is displayed at the corresponding location on the CRT screen.

A depression of the Space bar (key) causes a space (blank) character to be entered into the refresh memory and the cursor to be incremented by one location. A Back Space key, when depressed, causes a space character to be entered into the refresh memory, and the cursor to be decreased by one location. By depressing the repeat key and a character key, the normal operation of the character key is repeated at a predetermined rate.

Function Keyboard

The function keys (Figure 3.5g) request a specific action of the digital computer. When a function key is depressed, a priority interrupt signal is sent to the computer, and the computer reads a code on the PIO channel corresponding to the depressed key and executes the request, which might be to place all the control loops displayed on the CRT on manual control; or the request might be to display a directory of the display library on the CRT. In other words, each key requests a unique function, which is programmed in the computer.

Alarm Light Switches

Alarm light switches operate very much like function keys, with one notable exception—the former are lighted pushbutton switches whose light is controlled either from the computer or by an external (field) contact closure. When depressed, the buttons primarily request new displays; when lighted, they indicate alarm conditions associated with the corresponding display. Depressing an alarm light switch causes a hardware action identical to a function key depression. When the computer program detects conditions which should turn an alarm light on or off, the computer addresses the appropriate light on the PIO channel, and by setting a unique bit to 1 or 0, the light is turned on or off, respectively.

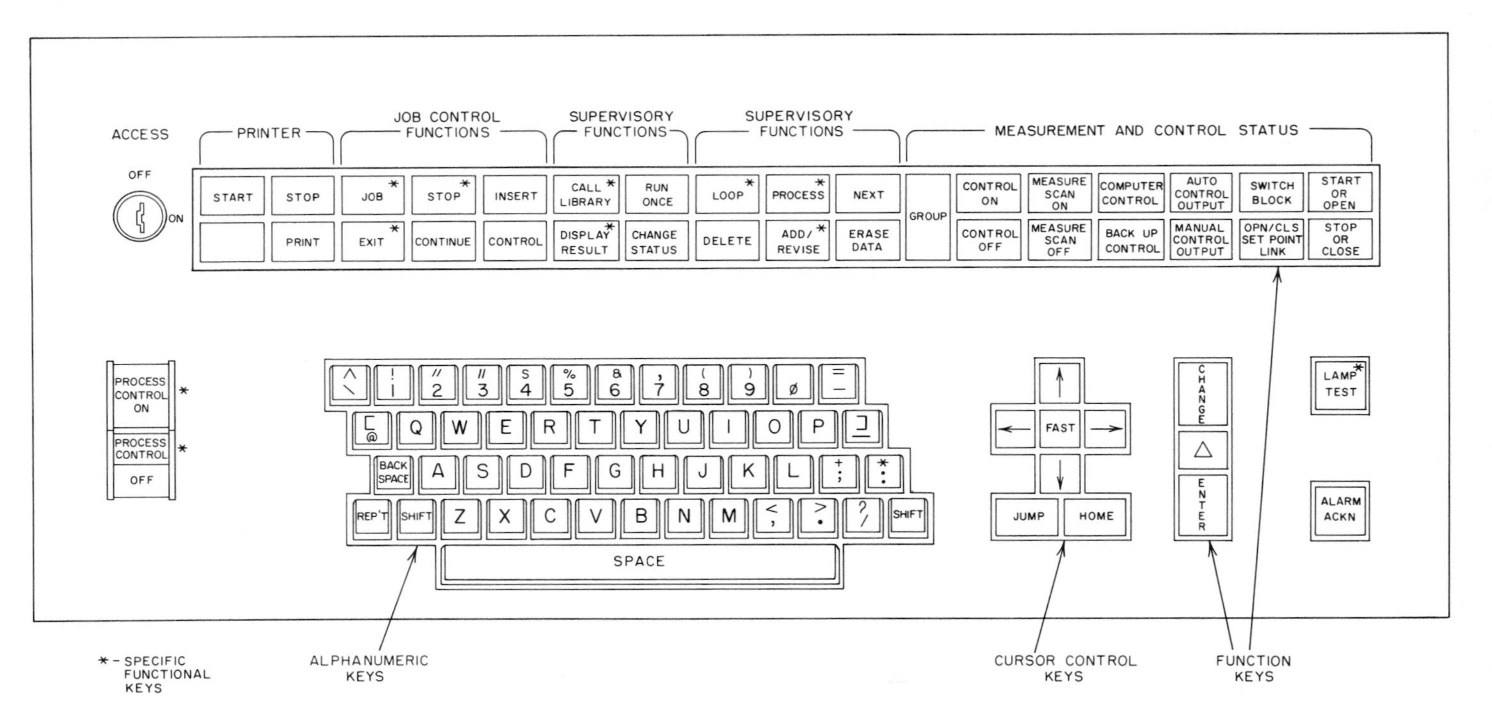

Fig. 3.5g General purpose keyboard

Data Display Methods

Refresh Memory

The refresh memory stores information (in coded form) displayed on the CRT screen. Since the duration of the CRT phosphor is several hundred microseconds, the displayed information must be regenerated and displayed at a nominal rate of sixty times per second. The refresh memory may consist of magnetic or acoustic delay lines, semiconductor shift registers, magnetic cores, magnetic disk or drum or semiconductor memory cells. The particular size, organization and bit coding can vary (Figure 3.5c). It can furnish information to a computer or to a character and vector generator. It can also accept information from a computer and a keyboard.

For example, a refresh memory associated with a 2,000 alphanumeric character display (80 characters per line, 25 lines) or with a display having 3,000 inches of vectors (straight line segments for graphic displays) may consist of semiconductor memory cells organized into 2,000 words, each word twelve bits of information. For display generation, each word is sequentially accessed and sent either to a character generator or to a vector generator. For a memory word that stores a character code, the bit structure shown in Figure 3.5h might be utilized.

The mode bits differentiate between characters and several types of vectors. For example, when the mode bits are logical ØØ, the word is defined as containing alphanumeric character information. The protect bit determines whether or not the character code can be changed from the keyboard or selectively accessed from the computer. The blink bit determines whether or not the character will blink. The character code defines the alphanumeric character (usually in USASCII code) which will be accessed from this memory location and displayed by the character generator.

If the mode bits of a memory word are logical 1Ø or 11, the current word and the word in the next memory location are defined as containing either relative or absolute vector information, respectively. A relative vector is a straight line, its origin the current beam position on the CRT screen and its end point defined as a change in X and Y position with respect to this origin. An absolute vector is a straight line the origin of which is the current

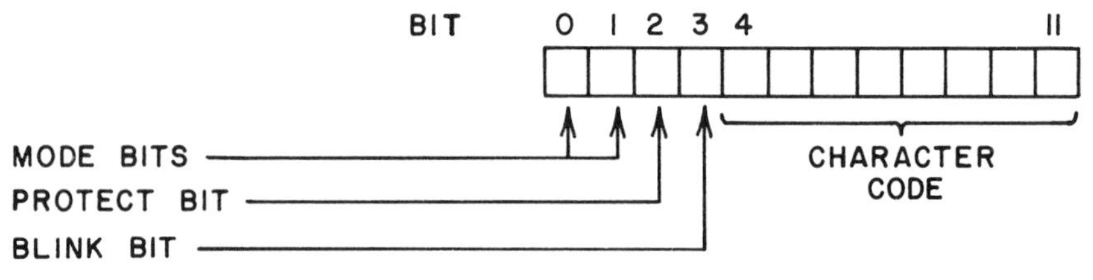

Fig. 3.5h Bit structure for character code

beam position on the CRT screen and the end point of which is defined as an X and Y position in a fixed grid with the grid origin of X = Ø and Y = Ø at the lower left corner of the CRT screen. The bit structure in Figure 3.5i might be utilized to define a vector.

If the mode bits (in word 1) are logical 1Ø, the two words define a relative vector, and therefore the X displacement contains a ΔX value and the Y displacement contains a ΔY value. If the mode bits are logical 11, the two words define an absolute vector, and therefore the X displacement contains an X value and the Y displacement contains a Y value. The line type determines whether the vector to be generated will be a solid, dashed, dotted or invisible line (blanked movement).

Character Generation and Format Control

Alphanumeric characters may be generated utilizing several techniques. Analog stroke, a "race-track," character mask scanning and read-only-memory character generation are a few examples. The following example is based on read-only-memories.

Figure 3.5j illustrates a character generator and format control logic. Since there are 2,000 memory locations containing character codes for each of 2,000 character positions on the CRT screen, the value contained in the refresh memory address register (Figure 3.5j) must be unique for each character position. The refresh memory is accessed at the location specified by the contents of the address register, and data are loaded into the data register from this location. The format control accepts the contents of the address register as an input and generates absolute values of X and Y data which positions the CRT beam to the starting position of the appropriate character location on the CRT screen.

The contents of the data register are then used as an address for the read-only-memory, and the body of X and Y data is accessed and loaded into the appropriate shift registers (Figure 3.5j). As the information is shifted out

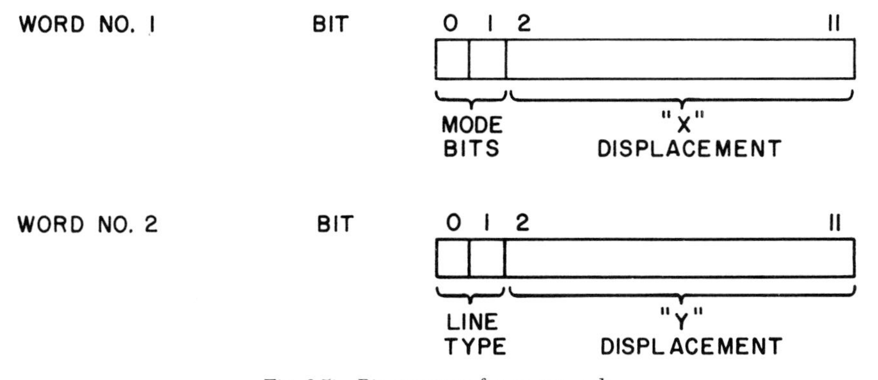

Fig. 3.5i Bit structure for vector code

of these registers bit by bit, it is decoded and sent to the X and Y output registers (Figure 3.5f). This mass of decoded data causes the output registers to increment (or count) up or down, which in turn causes the appropriate character (specified by the contents of the data register) to be written on the CRT screen.

Vector Generator

Vectors are straight line segments used to construct graphic displays. Typical methods of vector generation utilize analog ramp generators, binary rate multipliers or digitial arithmetic units. The example to be described uses the digitial arithmetic units.

The vector generator receives data from the refresh memory (Figures 3.5i and 3.5j) and based on this body of data provides incremental data to the X and Y output registers (Figure 3.5f). When the vector generator receives the X and Y displacement information from the refresh memory, if the vector

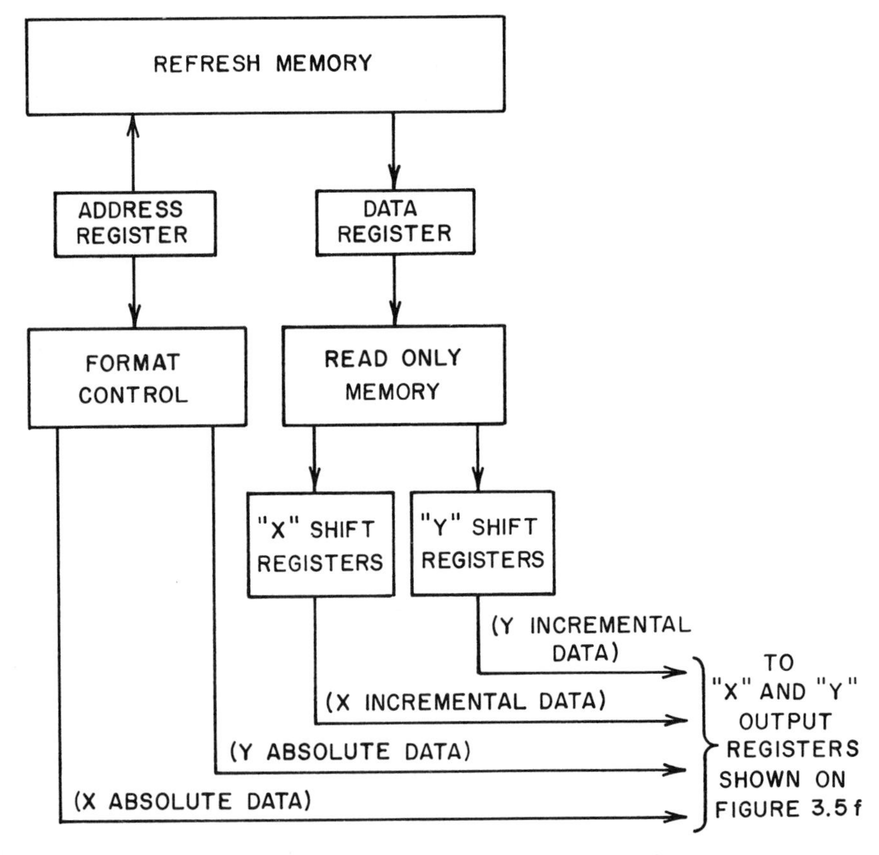

Fig. 3.5j Character generator and format control logic

was specified as a relative vector, the information will consist of a ΔX and a ΔY value. The vector generator operates directly on this body of data. If the information received from the refresh memory has been specified as an absolute vector, an auxiliary operation of computing ΔX and ΔY will occur, by subtracting the current beam coordinates (X and Y values) from those obtained from the refresh memory.

The operation (algorithm) of the vector generator is such that an incremental step (or movement) is made to minimize the value of ΔX and ΔY. A new value of ΔX and ΔY is then computed, and another incremental step is made to minimize the value of ΔX and ΔY. The process is repeated until the computed values of ΔX and ΔY are both zero. The result is a best fit to the desired straight line segment displayed on the CRT screen. The solid, dashed or dotted lines are generated by turning the CRT beam on or off (blanking) at desired intervals.

Display Initiation

Process display initiation comprises a chain of events that begins with an operator key action and ends when the selected display has been transmitted to the console refresh memory and real time update has commenced. From the operator's point of view, one or more key actions are required to fetch a display. From the point of view of program or software, these key actions identify what the operator wants to see. The operator must have a method of observing both the process variables and the response to actions taken by the control programs in the computer. The operator also needs to be notified of alarm conditions and requires a method of communicating directives to the process control program.

To accomplish these objectives the process display programs must allow the operator to 1.) initiation process data display requests which will be updated to reflect process variable changes; 2.) manipulate reference values and states (on-off or automatic-manual) of block or loop records in the process data base; 3.) terminate a process display; 4.) request other relevant programs such as directories or plant efficiency calculations; and 5.) respond directly to an alarm.

Keys and Key Sequencing

Inherent in each of the operations just mentioned is the process display console keyboard. Considerations of key sequence include 1.) how much data must be entered from memory; 2.) how many key strokes are required to achieve a display response; 3.) how many key strokes are required to recover from a data entry error; and 4.) how many operator decisions (choices) are required to proceed through a desired sequence.

The function keys (Figure 3.5g) may be divided and arranged in groups as shown. From the point of view of software, the keys are also arranged by purpose. Keys supplying a constant response can be grouped by key code. All other keys are conditional response keys. It is useful to think of these two groups as specific (constant response) keys and conditional (sequence dependent) keys. Alarm key lights are specific functional keys. Conditional keys manipulate process reference values, control states and select data from a recipe. Specific functional keys are indicated by an asterisk in Figure 3.5g; all other keys are conditional. Although in general it is desirable to minimize key operations which serve to initiate process displays, it does not always follow that every action that an operator might take should have minimum key activity. On the one hand, operations such as modification of numerical values may require visual verification before entry into the data base is attempted. On the other hand, state changes of process data blocks or loops should require minimum key actions. Thus, the design of operator key activity and key sequencing must be related to the display tasks and to the keyboard design.

The key sequence for any operation may be constructed as follows: first a specific key is used. This produces a fixed (by-key-code) visual response. Operator data are entered by the alphanumeric keys followed by a conditional key. The alphanumeric keys do not transmit data to the central processing unit (CPU), but only to the display memory. The cursor is manipulated by the operator to enter data at appropriate display locations. Key sequence diagrams are useful in planning process display-process operator interaction. Figures 3.5k and 3.5 l show two typical sequences. The alarm key sequence (Figure 3.5k) is used when an alarm key light comes on owing to a process upset. The operator presses the key, initiating a process unit display (Figure 3.5d). The process loop display sequence (Figure 3.5 l) requires entry of data before the loop to be displayed can be selected. The sequence begins with a specific key operation (loop key) and continues through entry of data (Figure 3.5m) and initiation of loop display (Figure 3.5e). These examples are initiating sequences. Operator interaction with a live loop or process display is a continuation of the techniques described, using conditional keys.

Format Construction

Display formats consist of the fixed or static information (column titles, headings, operator instructions and recipes) and the address for each piece of data to be retrieved from the process data base, displayed and appropriately updated on the display. Static format data may be conveniently separated from the data base-related information, allowing independent modification of titles and headings. The references to process data base information should be symbolic. Usually, the process data base is referenced by block or loop name which points to a complete set of measurement and control data about one process control input or output, or both. Within this set of data the

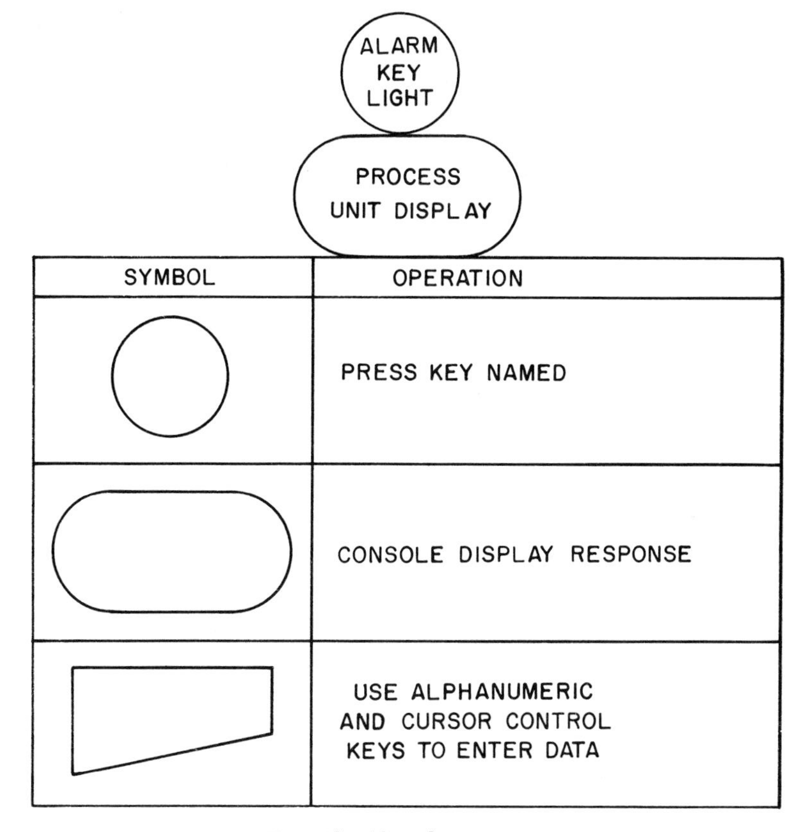

Fig. 3.5k Alarm key sequence

references to particular information such as a set point or measurement should be symbolic.

Symbolic references to data items in each block record simplify the display initiating program. The references are passed as arguments to a subroutine set which locates the appropriate item and performs internal to external format conversions.

Process Data Retrieval and Display

The CRT based process display allows considerable flexibility about what process data are to be shown. Various special purpose displays may be constructed to suit individual processes and operating policies. Individual blocks or a single piece of information for several loops (Figure 3.5n), groups of connected blocks (a control loop, Figure 3.5e) or sets of data related to process unit performance (Figure 3.5d) may be displayed. If each of these displays

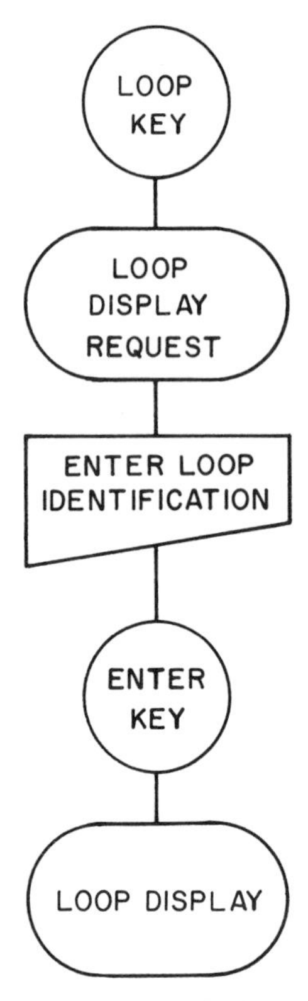

Fig. 3.51 Process loop display sequence

is considered a standard type of display, one may then construct displays using different sets of block or loop names for each process unit display, the format of which will remain constant—the block or set of loop data, or both, being changed to reflect the set of names chosen. Each set of blocks or loops is referred to by an identity code.

Process unit displays may be automatically displayed in response to an alarm key action by constructing a list of alarm key codes and associating a "process unit display" identity code with each. The identity code retrieves the list of block or loop names to be displayed. At the same time the name list is retrieved, the identity of the format (both the fixed part and the data base related part) is also retrieved. Thus, *a process unit display identity is used to point to a predefined display format and a set of process data.*

The body of data necessary to initiate a display of the type shown in Figure 3.5d includes 1.) the names of the blocks or loops, or both, to be displayed; 2.) the format of the display; and 3.) for each block in the display, a.) the symbolic references to values in the block record, e.g., MEAS (meas-

```
LOOP OR BLOCK DISPLAY REQUEST

LOOP OR BLOCK ID................        DISPLAY TYPE................
                                        (LEAVE BLANK FOR LOOP DISPLAYS OR BASIC FORMAT
COMPLETE FOR TYPE 5 DISPLAY:            BLOCK DISPLAYS)
TREND PEN NO................
MEASUREMENT SCALE: MIN................ PCT MAX................ PCT (LEAVE BLANK FOR
                                                                  0 TO 100 PCT)

        TYPE NO.     DISPLAY DESCRIPTION
           1              BASIC FORMAT
           2              MEASUREMENT FORMAT
           3              BAM PROCESS OPERATORS DISPLAY
           4              BAM INITIATING DISPLAY
           5              TREND RECORDER
           6              TREND DISPLAY (CRT)
           7
           8
           9
          10
```

Fig. 3.5m Loop display request

urement), ABS (absolute); and b.) for each symbolic block reference value, its relative or absolute display address; and c.) whether or not the value is to be updated in real time, whether or not the operator is to be allowed to modify the value and how many characters are to be displayed.

Display Propagation and Termination

The display initiating program retrieves the appropriate data for building the display and supplies real time display control and functional key service data through files or lists to the respective programs. The updating program is responsible for maintaining the displayed measurements and other values in a current or real time state. The functional key service program is responsible for all operator-requested modifications of the data display. Typically, these changes are of two types: data entry or value manipulation, and state changes, e.g., on-off or automatic-manual.

In Figures 3.5d, 3.5e and 3.5m, unprotected underscores define the appropriate areas for data entry for the operator. All other data displayed are protected and cannot be modified or changed by the operator at the console. The display initiating program also must set a flag or bit in each block record requiring update, and this bit or flag is referred to as the display capture bit.

PROCESS DISPLAY: 8 PLANT FLOW MONITOR DATE: 3-1-71 TIME: 1515

Point ID	Description	Status	Value	Alarm
FSF101[10]	Plant 2 feed flow	On	5.8 TCFT/H	
FSF301	Splitter flow to heater	On	120.8 TGPH	
FSF401	Splitter steam flow	On	128.5 TP/H	
FSF450	Acc flow to splitter	On	132.7 TGPH	
HRF502	Heater fuel flow	On	2500 CFT/H	
PFF600	Product A flow	On	60.7 TCFT/H	
PFF550	Rct. feed flow to fractionator	On	106 TGPH	
PFF701	Fractionator steam flow	On	107.4 TP/H	
PFF750	Product B flow	On	75.5 TGPH	

Fig. 3.5n Plant flow monitoring display

Data Capture, Conversion and Routing

In display propagation the values displayed are changed to reflect the variations in the controlled process. Measurements, alarm states, internally modified set points or reference values are examples of data requiring continuous updating. Update frequency may be other than the normal processing interval, which is inconvenient and requires additional program logic. If the update frequency is the same as the processing (scan) interval, the process control program may be constructed to examine the data capture bit in each block record. When the bit is on, the block record data are set aside in a temporary display file or list. When the process control program has completed its tasks for the current interval, it calls on the display update program for execution. It should be noted that display update is called on only when data have been captured for update.

The display update program finds the block record in the display file or list and with the display control information assembled by the display initiator converts the appropriate mass of data in the captured block record to external format and transmits it to the display. Typically, for each block record* the display update program includes 1.) block name; 2.) display console number if more than one; 3.) symbolic data names for all items to be updated; and 4.) display memory address (where the data is to be displayed).

Operator Interaction, Data Entry and State Changes

Operator interaction with a process display consists of manipulating the blocks displayed, e.g., changing a block from ON to OFF, and of entering new numerical data, e.g., changing a set point.

Figure 3.5d shows a process display. In the column labeled set point there

* Note that the use of symbolic block data item reference is carried in the display update function as well as in the display initiator.

is (for each block) a numerical value, and directly under the value a line of underscores which shows the operator where a new value for the set point is to be entered. The cursor is moved to the underscore field and the new value is entered, using the numerical keys. The underscores are unprotected characters and can be overwritten from the keyboard. The operator may then either move the cursor to other underscores and enter more data or press the Enter key, which causes an interrupt to occur in the CPU, and in response to the interrupt a program is called for execution which will service the operator's console. The program will in this case read the unprotected characters in the display memory and attempt to modify the appropriate values in the process data base block records.

The functional key service program uses data supplied by the display initiator to determine the set of data that goes with the blocks on display. Appropriate visual feedback to the operator is obtained by overwriting the existing value (the one above the entered value) with the new value and restoring the data entry area to unprotected underscores. Should the new value be unacceptable, the underscores are not restored and the offending value may be set to blink. An error diagnostic message may also be displayed.

Block state changes may be accomplished by pointing the cursor to the block name and pressing one of the measurement and control status function keys. These keys have been appropriately labeled control on, control off and so forth (Figure 3.5g). The same interrupt response takes place as already described. The program responding to operator requests for service must be able to activate and deactivate function keys. Also, the interrupt-causing keys must be identified. Activating and identifying keys is performed by a key mask table for each console keyboard. It contains the code for each key that is currently active. An active key is one that has been put in the table by the program servicing the console. Thus, the servicing program can at any time determine the function keys that the operator may legitimately use. The key mask table is used by the console-interrupt-handler segment of the real time executive to determine if the servicing program has to be called.

If many consoles are operating concurrently, the servicing program attends to each as requests occur. There need be only one servicing program for each console that has a currently active process display, and the information that it requires includes 1.) the total number of unprotected characters on display; 2.) a sequential list of the block name and symbolic value name for each data entry field; 3.) the length of each data entry field and the display address of its related protected value; and 4.) access to the same data as the display update program. This set of data is needed to insure that visual feedback for every requested state change is available. If visual feedback is not possible, the requested state change is erroneous, and an appropriate diagnostic measure is displayed.

Terminating the Display

After observation and manipulation, the operator indicates that the operation is complete by requesting another display by a specific function key or alarm key light, or both. The console-interrupt-handler segment of the real time executive determines if process display termination is necessary by keeping track of real time update operations on a console by console basis. If the current display on a console is not being updated in real time, termination is unnecessary; the requested program is responsible for clearing the display.

The process display termination program determines which blocks in the process data base were being "captured" for display on the console and resets or stops their ensuing capture and display by resetting the display capture bit in the block record. The program also purges the data files supplied by the display initiation to the update program. When termination is complete the operator-requested function is allowed to proceed. If many process displays on many consoles are being updated in real time, the termination must take care not to terminate capture of blocks that are being displayed on other consoles.

3.6 ELECTRICAL SWITCHES AND PUSHBUTTONS

Types:

(a) Pushbutton
(b) Toggle
(c) Rotary
(d) Thumbwheel

Note: In the feature summary below the letters (a) to (d) refer to the listed electrical switch types.

Features:

(a) Keyboard, panel, industrial oil-tight,
(b) Lever, rocker, thumbwheel selectors,
(c) Selector, adjustor,
(d) Selector

Actuation:

(a) Push in,
(b) Pivot,
(c) Twist,
(d) Push up and down

Sizes:

(a) Standard, miniature,
(b) Standard, miniature, subminiature,
(c) Standard, miniature,
(d) Standard

Costs:

(a) $5 to $40,
(b) $2 to $15,
(c) $10 to $100,
(d) $4 to $8.

Partial List of Suppliers:

Arrow-Hart, Inc., C & K Components, Clare-Pendar, Controls Research Corp., Cutler-Hammer, Inc., General Electric Co., Licon, Marco-Oak, Microswitch Div. of Honeywell, Inc., Oak Manufacturing Co., Switchcraft.

Proper selection and use of manual controls are complex tasks. Usually a best choice exists for each application among several which would do the job.

Principles of Operation

An electric switch type manual control (Figure 3.6a) consists of the switching contacts, an actuator to bring them into proximity, terminals to connect the contacts to the conductor of an electric circuit, the insulating mounting provisions and the enclosure, either separate or integral with the mounting for contacts and terminals. The gap between contacts is filled by air or an inert gas. The control is operated by moving the contacts together until the applied voltage causes an electrostatic failure of the gap. The ensuing electron flow heats the cathode and cools the anode contact until molecular welding occurs, creating a continuous metallic path and completing the circuit.

Contact resistance is a measure of the degree to which the insulating gas remains between the contacts, preventing welding over their entire bearing area. It is directly proportional to contact pressure.

Switching Action

Switching action refers to the positions assumed by the contact in response to actuator motion. The two basic types (Figure 3.6b) include momentary action, which provides one contact closure and subsequent reopening with one actuation; and maintained action, which requires separate actuations to transfer the contacts from one extreme position to another, and back.

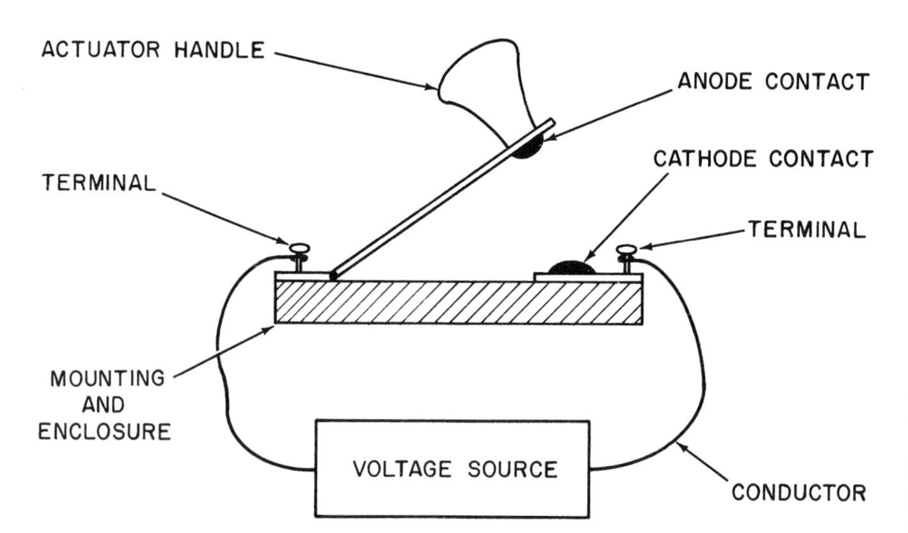

Fig. 3.6a Operating elements of a manual control

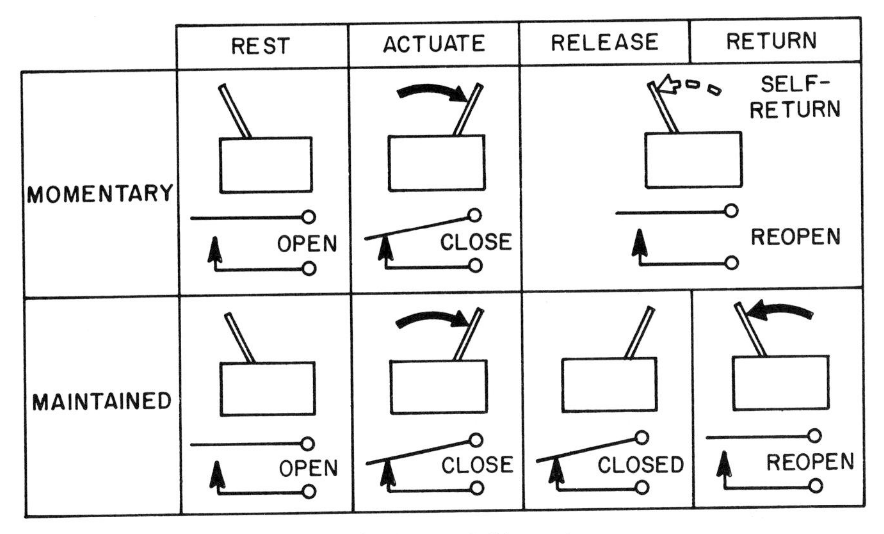

Fig. 3.6b Basic switching actions

Other special types include 1.) mechanical bail, which allows operation of only one control at a time, physically inhibiting depression of others. Automobile radio selectors are a typical example; 2.) electrical bail, which allows delayed or remote actuation including sequencing of several controls. Solenoid operators frequently sequence rotary selectors or ganged pushbuttons; and 3.) sequential action, which opens one contact set in a specified order in relation to other contacts in the same control. This type may involve simultaneous operation of several contact sets and repeated opening and closing of one contact set during one actuation.

Contact Arrangements

Switching capacity refers to the number, type and arrangement of contacts and circuits within the control. Table 3.6c contains the related terminology. (For more on contact configuration standards, see Figure 4.5b in Volume II.)

Normally open contacts (NO). The circuit is open (current does not flow) when the contacts are not actuated. Moving them to the other extreme position completes (closes) the circuit.

Normally closed contacts (NC). The circuit is complete when the switch is not operated. Moving the contacts opens the circuit, interrupting current flow.

Number of poles (P). The number of conductors in which current will be simultaneously interrupted by one operation of the switch.

Number of throws (T). The number of positions at which the contacts

	DIAGRAM	OPERATION	FEATURES
SPST NO	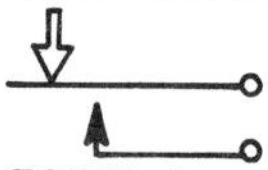 FORM A	MAKE (PULSE)	PULSES WHEN ACTUATOR IS DEPRESSED, BUT NOT WHEN RELEASED.
SPST NC	 FORM B	BREAK (PULSE)	SAME AS FORM A EXCEPT OPENS CIRCUIT INSTEAD OF CLOSING IT
SPDT NO/NC	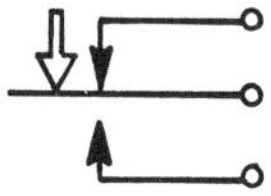 FORM C	BREAK-MAKE (TRANSFER)	PROVIDES COMPLETE CIRCUITS AT BOTH CONTACT POSITIONS
SPDT NO/NC	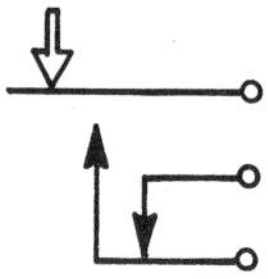 FORM D	MAKE-BEFORE-BREAK (CONTINUITY TRANSFER)	MOMENTARY OVERLAP OF CONTACTS PROVIDES CIRCUIT CONTINUITY
SPDT NO	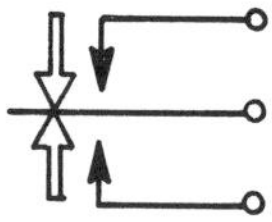 FORM K	CENTER OFF (PULSE)	PROVIDES PULSES IN TWO CONDUCTORS FROM ONE SWITCH. USEFUL FOR "UP, DOWN" JOGGING.
DPST NO	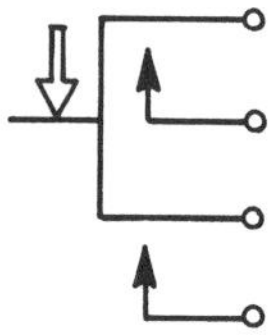 FORM AA	DOUBLE MAKE (DOUBLE PULSE)	INTERRUPTS CURRENT IN TWO INDEPENDENT CONDUCTORS OF A CIRCUIT. REPLACES TWO FORM A SWITCHES.
DPDT NO/NC	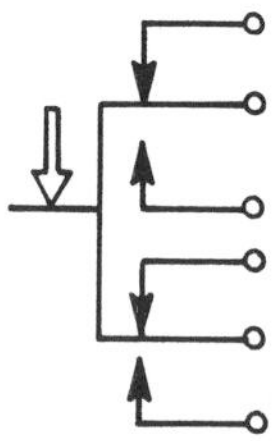 FORM CC	DOUBLE BREAK-DOUBLE MAKE (DOUBLE TRANSFER)	SIMULTANEOUSLY OPERATES TWO CIRCUITS. SEVERAL WIRING POSSIBILITIES. REPLACES TWO FORM C SWITCHES.

Table 3.6c Switching contact configurations

will provide complete circuits. Since contacts usually have only two extreme positions, switches are either single-throw or double-throw.

Switching Elements and Circuits

Switching circuits are classified as low-energy or high-energy. Low-energy circuits do not develop sufficient volt-amperes to break down insulating contaminant film (that may have built up on the contacts) by contact heating. Consequently, infrequent use degrades the controls in these circuits. Accordingly, reliable low-energy switching elements must maintain low contact resistance. Solutions include 1.) wiping action under high contact pressures; 2.) contact materials with high oxidation resistance; and 3.) hermetically sealed enclosures, either evacuated or filled with an inert atmosphere.

High-energy circuits generate sufficient energy to sustain an arc between contacts. Therefore, a reliable high-energy switching element must perform under very high contact temperatures. Unlike low-energy circuits, frequent operation degrades reliability. Large contact areas, thick plating and thermally conductive alloys combat high temperatures. Special features divert, vent or extinguish the arc.

Switching elements are summarized in Table 3.6d. Mechanical contactors either wipe the contacts over each other or strike them together. Wiping contacts move in a series of jerks, during which the contacts alternately weld and break loose. Striking contacts are either elastically or plastically deformed on impact.

Magnetic reed contacts are overlapping ferromagnetic beams, cantilevered from a glass tube filled with a dry inert gas. A small gap separates the overlapping ends, and magnetic induction in the gap eventually overcomes their stiffness, bending them until they touch. The circuit is reopened by removing the magnetic influence. Reed contacts provide 1.) high contact pressure with low actuator force, since the actuator does not directly move the contacts; 2.) very fast response; and 3.) long life. Since the reeds are cantilevered springs, a major problem is contact bounce.

Snap-action switches are very rapidly struck together by an overcenter spring mechanism. They provide 1.) irreversible switching (since the operator can do no more than start the switching action); 2.) precise timing, including virtually simultaneous switching of many circuits with one actuator; 3.) precise, consistently repeatable travel due to only one moving part—the spring; and 4.) high load capacity and long life, since arcing is limited by the very fast contact transfer time.

Wiping blade contacts employ a stiff, blade-shaped contactor which passes over spring mounted stationary contacts under high pressure. Alternately, the moving blade is the spring and the fixed contacts are stiff. The contacts are sometimes knurled to assist the wiping action. Hard silver plating

	NAME	DIAGRAM	OPERATION	OPERATING CHARACTERISTICS	ADVANTAGES AND DISADVANTAGES	APPLICATIONS
MECHANICAL CONTACTORS	MAGNETIC REED	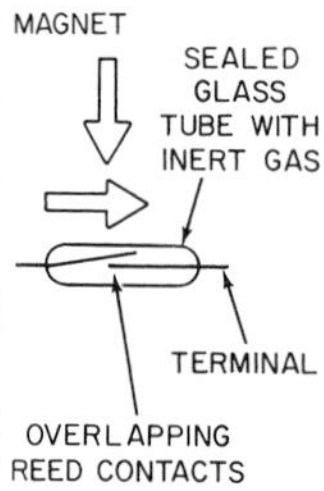 MAGNET; SEALED GLASS TUBE WITH INERT GAS; TERMINAL; OVERLAPPING REED CONTACTS	MAGNETIC INDUCTION BENDS REEDS. STRIKING CONTACTS.	ACTUATION FORCE INDEPENDENT OF CONTACT PRESSURE. HIGH CONTACT PRESSURE. FAST, PRECISE TIMING. EXCELLENT FOR LOW-ENERGY CIRCUITS.	HERMETICALLY SEALED CONTACTS. BOUNCE IS PROBLEM, REQUIRES EXTERNAL COMPENSATING CIRCUITRY. CAN BE ACTUATED BY VIBRATION. REQUIRES SHIELDING FROM EXTERNAL MAGNETS. LONG LIFE, INEXPENSIVE.	KEYBOARD PUSHBUTTON, PANEL PUSHBUTTON, (INTEGRAL).
MECHANICAL CONTACTORS	SNAP-ACTION	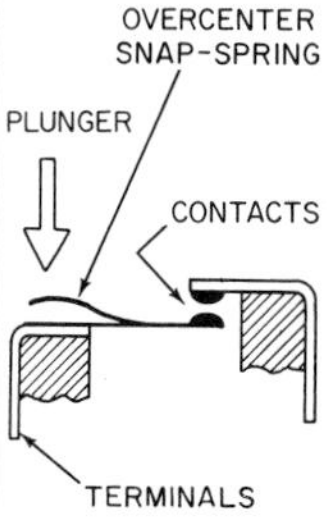 OVERCENTER SNAP-SPRING; PLUNGER; CONTACTS; TERMINALS	LEAF SPRING SNAPPED. OVER-CENTER. STRIKING CONTACTS.	IRREVERSIBLE ACTION NO "TEASING". FAST, PRECISE, SHORT, REPEATABLE TIMING AND TRAVEL. BEST FOR SIMULTANEOUS SWITCHING OF MANY CIRCUITS WITH ONE ACTUATOR. HIGH CONTACT PRESSURE, HIGH ACTUATION FORCE.	ENCLOSED CONTACTS. SUITABLE FOR LOW OR HIGH-ENERGY CIRCUITS. RELATIVELY INEXPENSIVE. VERY SMALL SIZES AVAILABLE.	TOGGLE, PANEL PUSHBUTTON (BUILT-UP), INDUSTRIAL PUSHBUTTON (MEDIUM AND HEAVY DUTY).
MECHANICAL CONTACTORS	WIPING BLADE	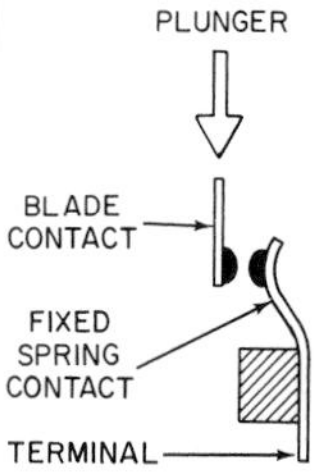 PLUNGER; BLADE CONTACT; FIXED SPRING CONTACT; TERMINAL	CANTILEVER BLADE CONTACT WIPES PAST FIXED SPRING CONTACT.	LOW CONTACT PRESSURE. WIPING CLEANS CONTACTS, BUT WEARS AWAY PLATING. TIMING AND TRAVEL NOT PRECISELY REPEATABLE.	HARD SILVER PLATING USED TO RESIST ABRASION, THUS NOT GENERALLY SUITED TO LOW ENERGY USE. ROTARY IS EXCEPTION SINCE INFREQUENT USE ALLOWS GOLD PLATING. RELATIVELY INEXPENSIVE.	INTEGRAL PANEL PUSHBUTTON, ROTARY, INDUSTRIAL PUSHBUTTON (LIGHT DUTY).
CONTACTLESS	INTEGRATED CIRCUIT	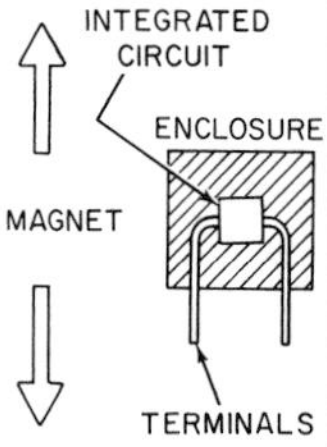 INTEGRATED CIRCUIT; ENCLOSURE; MAGNET; TERMINALS	(HALL EFFECT) VOLTAGE IS DEVELOPED ACROSS EDGES OF CURRENT-CARRYING CONDUCTOR BY MAGNETIC INDUCTION.	ACTUATOR FORCE INDEPENDENT OF SWITCHING ACTION. VERY FAST SWITCHING, PRECISE TIMING.	NO CONTACT BOUNCE. ALWAYS-ON BIAS VOLTAGE SOMETIMES REQUIRED. EXCELLENT FOR REPETITIVE, LONG-LIFE USE IN LOW-ENERGY CIRCUITS. EXPENSIVE. SEALED.	KEYBOARD PUSHBUTTON.

Table 3.6d Basic switching elements

alloys are required to resist wear, which generally restricts their use to high-energy circuits.

Contactless switches are useful in low-signal-level electronic circuits. The absence of contact bounce eliminates the need for external filtering circuits. Other advantages are virtually unlimited life, very fast switching and freedom from the effects of contamination. Most contactless switches employ semiconductors, which require an always-on bias voltage. This can be a disadvantage when compared to a mechanical switch.

Types and Grades

Table 3.6e summarizes the features of four types of manual control devices including pushbutton, toggle, rotary and thumbwheel, all of which are available in several grades and types of construction.

Appliance grade switches are not suitable for process control; they are low-cost devices designed for light, non-abusive environments in which precise timing and current control are not required, performance degradation is not critical during operating life and exacting operator feedback is not required.

Commercial grade switches are suitable for the control-room. They provide consistent actuation and switching performance at specified reliability in average environments. Some degree of contact sealing is present, and corrosion-resistant contact materials are used. Actuation and switching mechanisms afford precise control of light to medium electrical loads.

Industrial grade switches are specialized for local installation in abusive environments. Construction is rugged in order to accommodate abusive actuation and unusually heavy electrical loads. Actuators and contacts are sealed against liquid and solid contaminants, especially oil.

Military grade switches are suitable for process control use, but are usually over-specified in some areas, underspecified in others and carry a price premium.

Built-up construction utilizes separate housings for actuator and switching element. Circuit flexibility is the principal advantage, since switching modules can be ganged at will, although standard capacities are available. Disadvantages are unsealed interface between actuator and switch modules, back-of-panel volume and price.

Integral construction combines actuator and switching element in a common housing. Advantages include satisfactory sealing and smaller over-all dimensions. The major disadvantage is fixed switching capacity.

Pushbuttons

Pushbuttons provide the simplest, most naturally comfortable actuation motion. They alone accommodate extensive labeling without the use of separate nameplates. Disadvantages include switching capacity lower than found in other types due to only two contact positions, and lack of inherent indication of switching status (such as by handle position) other than lighting. Two methods of producing colored light are by transmission and projection. Color selection has been discussed in Section 3.2 in this volume.

Transmitted colors are produced by transmitting white light through colored button-lenses; they should be used whenever possible. Colors are

Table 3.6e
CHARACTERISTICS OF ELECTRIC SWITCHES

	Pushbutton	*Toggle*	*Rotary*	*Thumbwheel*
Typical Applications	Keyboard, motor, machine tool, instrument; best for pulsing; action dynamics match digital display.	Computer data register; power on-off, test; "center off" type suitable for jog, adjustment about reference.	Computer control panels, ammeter adjustment (potentiometer); action dynamics match fine analog adjustment.	Encoded data entry to digital computing circuits; instrument settings; exact value setting.
Advantages and Disadvantages	Easiest, most comfortable operation; self-contained label flexibility.	Least expensive 2,3 position selector; simple non-precision operation, reflex action; best for actuation of multiple switches simultaneously.	Most capacity in one control; selector available for 4+ positions in high-energy circuits; self-contained encoding optional; also mounting for components.	Most capacity in least panel area; self-contained encoding, dial read-out of value setting; single alphanumeric character-position or complete messages.
Handles and Feedback Features	Integral handle, lighting very useful; no feedback from handle position.	Integral, special "decorator" style available; handle position positive feedback (indicator) of status; no light or label except rocker (marginal).	Separate, many shapes and color options; label on skirts; no lighting.	Integral; small tabs adjacent to indicator dial; same number as number of positions; lighting optional.
Environmental Protection	Sealed switch elements standard; sealed actuators optional; standard industrial grade.	Enclosed switch elements standard; sealed actuator handle optional.	Enclosed switch elements optional; sealed actuator shaft optional.	Enclosed switch elements and actuator handle optional.
Number of Positions	1,2	2,3	2–24 (potentiometer)	8–16
Comparable Cost, Position	2.5	1.0	5.0	—
Ratings	0.5–60A @ 125VAC or 30VDC; 1A @ 600VAC; 0.5A @ 600VDC	20A @ 125VAC 12A @ 30VDC	0.3A @125VAC 1A @ 30VDC	0.13A @ 125VAC, or @ 125VDC
Life	10,000,000 operations	1,000,000 operations	100,000 360° cycles	100,000 360° cycles
Mounting	Front or back of panel; printed wiring board.	Back of panel; printed wiring board.	Back of panel; printed wiring board.	Front or back of panel; printed wiring board.

intense, saturated, uniformly distributed over the button, do not degrade from lamp heat and can be read both from great distances and through wide viewing angles.

Projected colors are produced by projecting colored light on a white button-screen. The colored light is produced by transmitting white light through colored filters which slip over the lamps. Projected colors are weak, dilute, easily overpowered by normal room ambient light, have restricted viewing distances and angles and degrade as the colored filter is progressively destroyed by lamp heat.

PANEL PUSHBUTTONS

A variety of integral and built-up configurations is available (Figure 3.6f), incorporating molded or metal housings, various force and travel options and

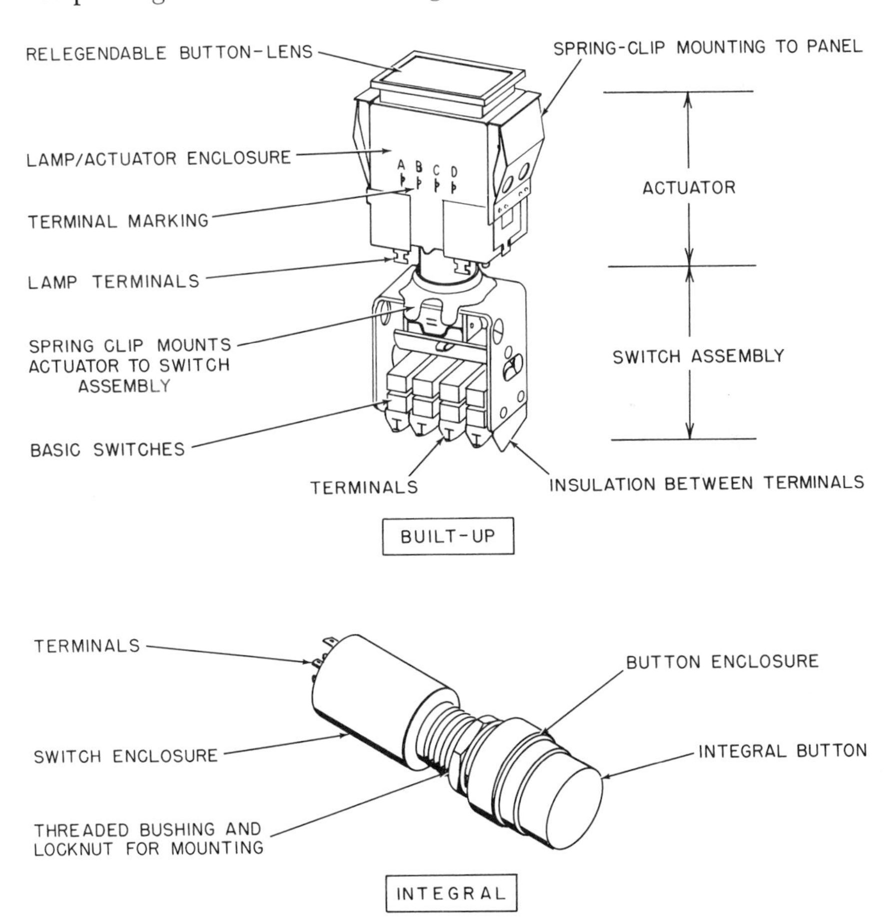

Fig. 3.6f Panel pushbuttons

a wide range of lighted and unlighted buttons. Square and rectangular buttons are most easily mounted and provide the most labeling area. Labeling is accomplished by hot-stamping, engraving, two-shot molding, silk-screening or photographic film inserts. Panel pushbuttons are usually mounted in individual openings; built-up configurations mount from front of panel, using spring clips or barriers. Integral configurations mount from back of panel, using threaded bushings and locknuts. Buttons and lamps in each configuration are accessible from the front of panel.

Each mounting method has definite panel thickness limits, which can, however, be circumvented by subassembly to a bracket which in turn mounts on the back of the panel. The buttons are available for both low-energy and high-energy circuits.

INDUSTRIAL PUSHBUTTONS

These buttons are ruggedly made to withstand abusive environments and to resist abusive operation (Figure 3.6g). Compact and standard sizes are available, and both kinds use a built-up construction to accommodate custom switching for specific application. Actuators and contacts are both liquid-tight and dust-tight, and contacts are available for three grades of duty including 1.) electronic (wiping action, gold plating); 2.) standard (wiping action, silver plating); and 3.) heavy duty (striking action, silver, silver alloy or cadmium alloy plating).

KEYBOARD PUSHBUTTONS

These buttons are integral pushbuttons used in low-energy electronic circuits (Figure 3.6h). Force-travel characteristics are carefully designed for precise tactile feedback during repetitive, high-speed touch typing. Simplified, durable, low-cost actuator mechanisms and switching elements furnish a high degree of reliability for many operations. Construction is as functional as possible for mass-production, group mounting, and group connecting. Techniques include molded housings, two-shot molded buttons and labels and printed wiring board mounting and connecting. Both assembled keyboards and a variety of standard and custom arrangements, encoding, operating forces and buttons are available from most manufacturers of keyboard pushbuttons.

Toggle Switches

Toggle switches are the cheapest two-position or three-position selectors for a large number of circuits of any level of complexity or energy, and integral and built-up configurations are available. A major advantage is positive indication of switching status from handle position. The three basic handle configurations include 1.) lever, 2.) rocker and 3.) thumbwheel.

The lever handle configuration (Figure 3.6i) is the most common and gives the most positive handle position indication. Careful consideration of length,

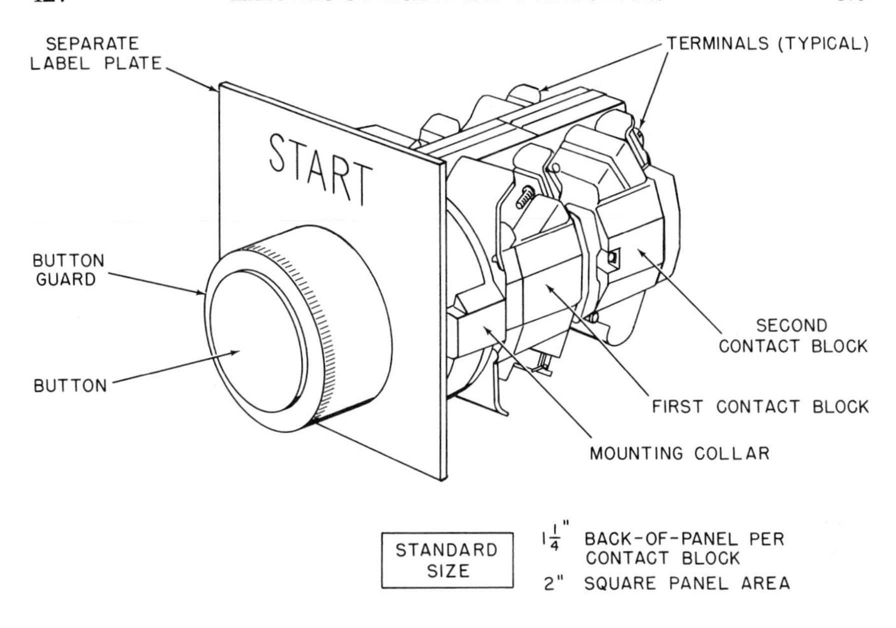

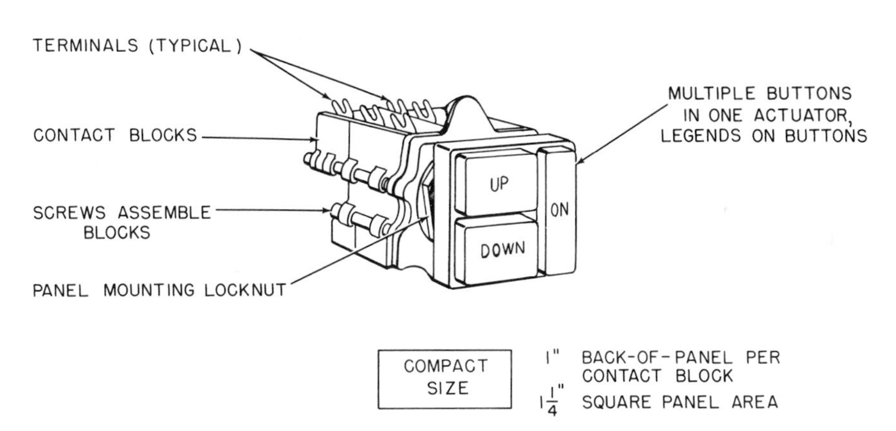

Fig. 3.6g Industrial pushbuttons

shape, protrusion above the panel and panel graphics design is required to capitalize on this advantage. A wide selection of shapes, sizes, colors and trim hardware is available to create various front-of-panel appearances, accentuate handle positions and safeguard against accidental actuation.

The rocker handle (Figure 3.6i) combines positive handle indication with the operating simplicity of pushbuttons. Limited labeling and lighting are also available, but both features are marginal owing to space limitations.

The thumbwheel handle (Figure 3.6i) is occasionally used to reduce front-of-panel protrusion and to accommodate special actuation motion.

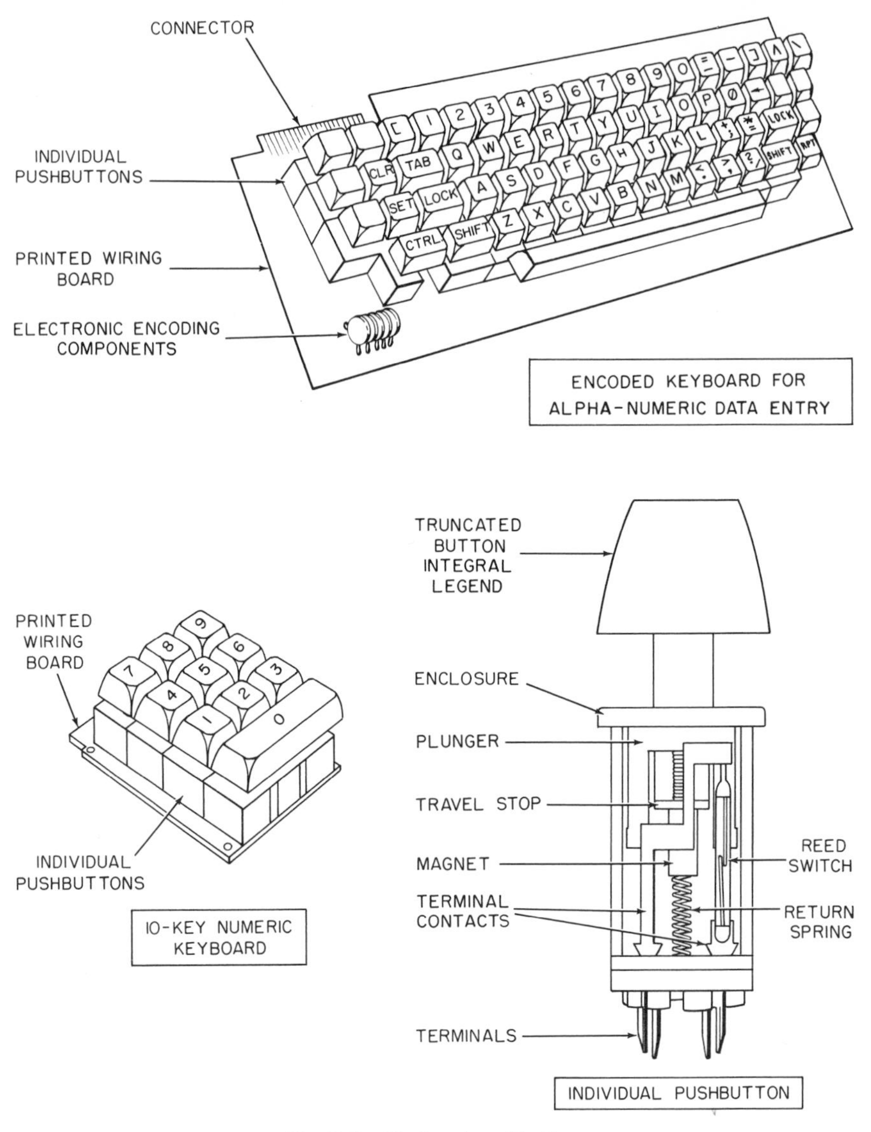

Fig. 3.6h Keyboard pushbuttons

Disadvantages include 1.) lack of positive indication of switch position by handle position, and 2.) control of travel and force are difficult since the finger moves in and out at the same time that it moves side to side. The latter characteristic makes the handle especially awkward for use as an adjustment control about a spring-loaded "center off" position.

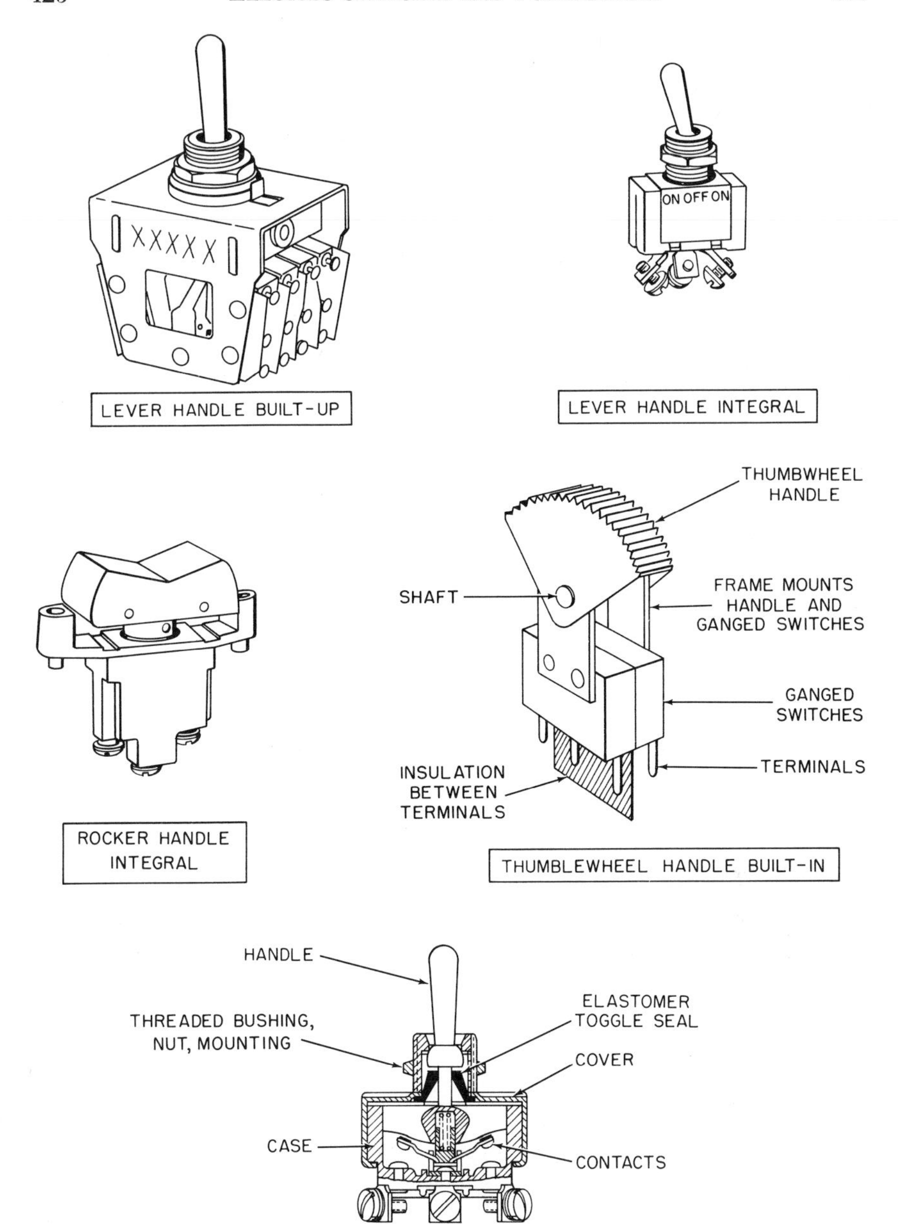

Fig. 3.6i Toggle switches

Rotary Switches

Rotary controls (Figure 3.6j) are selectors or adjustors for low-energy circuits and afford greater switching capacity in less volume than any other control. Switching capacity is expanded by adding switching modules, and a major advantage is that entire operational sequences can be contained on one control. Disadvantages include 1.) increased operating force with increased switching capacity, and 2.) no direct access to individual positions without actuating intermediate positions.

Builtup construction is almost universal, either enclosed or openframe. Mounting is either to printed wiring boards or back of panel, using standoffs or threaded bushings and locknuts. Switch modules consist of fixed contacts mounted on round printed wiring board decks. Each deck has a movable contactor attached to a common shaft. The printed wiring affords flexibility in contact arrangement, contact spacing, total rotational travel, switching action and mounting of electronic components for self-contained encoding.

The inherent flexibility of these controls is commonly abused by packing too much switching capacity into one control, based on the rationale of saving the cost of another rotary switch. Angular spacing of contacts becomes so small, and operating force so high, that actuation is awkward and imprecise. Very high contact pressures result and can mechanically damage contacts and

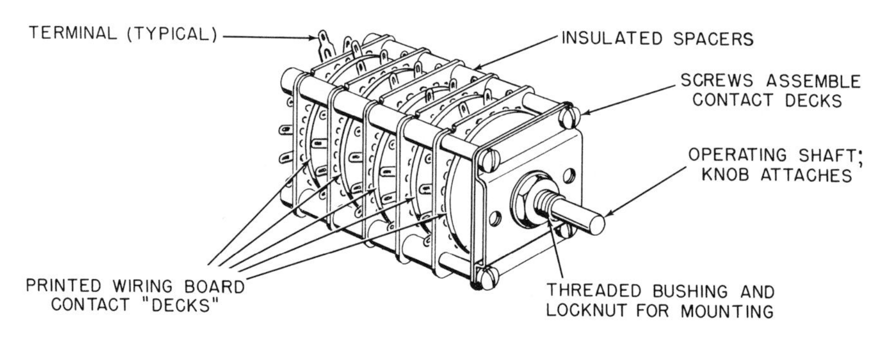

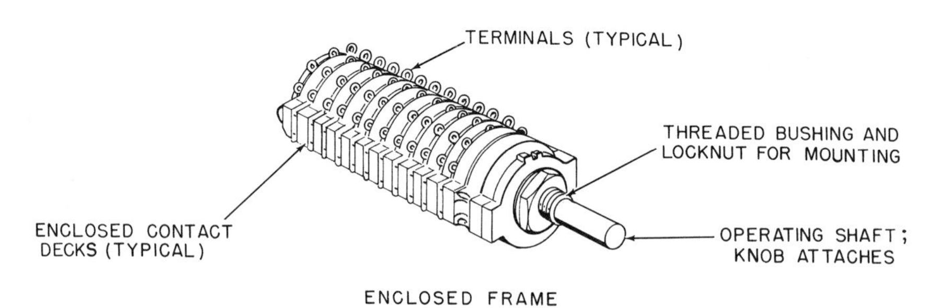

Fig. 3.6j Rotary selectors

actuator mechanism. Considerable panel area is required to label a large number of positions. Large diameter handles although they decrease force, also consume panel area. The other extreme of excessively wide contact spacing causes awkward wrist and arm movements.

The guidelines generally accepted include

1.) Minimum angular spacing = 15°
2.) Maximum angular spacing = 90°
3.) Minimum positions = 4 in 360°
4.) Better minimum = 4 in 240°
5.) Maximum positions = 24 in 360°
6.) Better maximum = 20 in 300°

Thumbwheel Switches

Thumbwheel switches (Figure 3.6k) are a form of rotary selector for encoded data entry to low-energy electronic circuits. Encoding is accomplished by internally mounted electronic components. Thumbwheels take up very little front of panel area owing to parallel arrangement of contact disk, moving contactor and handle; and self-contained indicator dials. Contact disks can be stacked horizontally to create controls of any size. Thus, thumbwheel switches can do the job of several rotaries, pushbuttons or toggles and use considerably less panel space. A major disadvantage of their use is that considerable concentration and dexterity are required for efficient operation. Also, 1.) actuator motion is usually opposite to dial motion; 2.) handles are very small and must be poked at with finger tips; 3.) incremental spacing on dials is small; and 4.) horizontal spacing between handles is small.

Application and Selection Considerations

The goal of the selection (Figure 3.6 l) is to provide the most direct link between operator and process by considering 1.) what is being controlled—the process; 2.) how it is being controlled—the control system; 3.) how the control is used—human factors; and 4.) the service conditions—electrical, mechanical and environmental. Process response to control action determines the operator's subsequent action. Important variables include time response, amplitude response, linearity and damping.

Control system translation separates the process from the operator's control action, and the operator from the process response. Important factors affecting selection of manual controls involve 1.) type of control—analog, digital; 2.) resolution—proportional, derivative, integral and combinations; 3.) accuracy—physical losses through transmitters; 4.) data processing errors—chopoff and roundoff; 5.) visual feedback—display resolution, speed, accuracy and 6.) safety provisions.

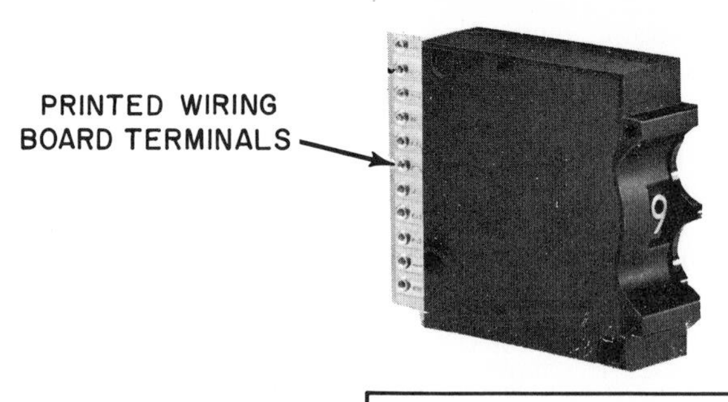

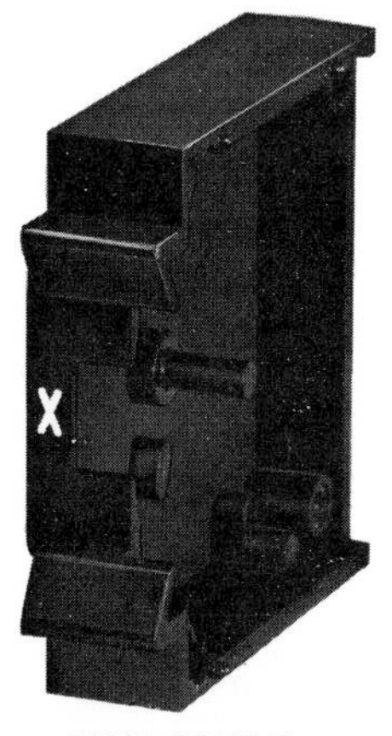

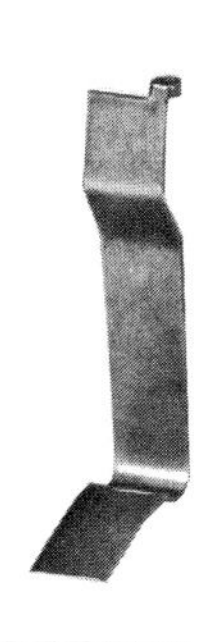

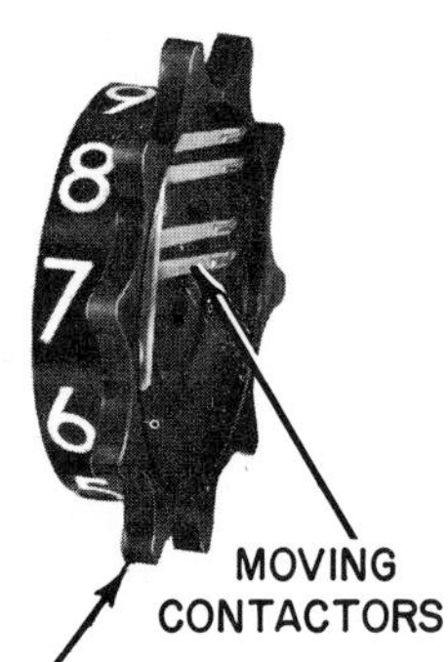

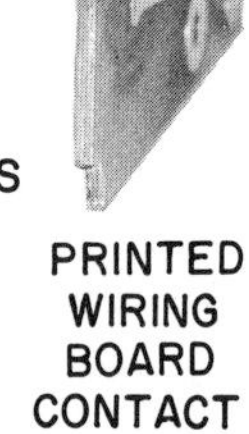

Fig. 3.6k Thumbwheel selector

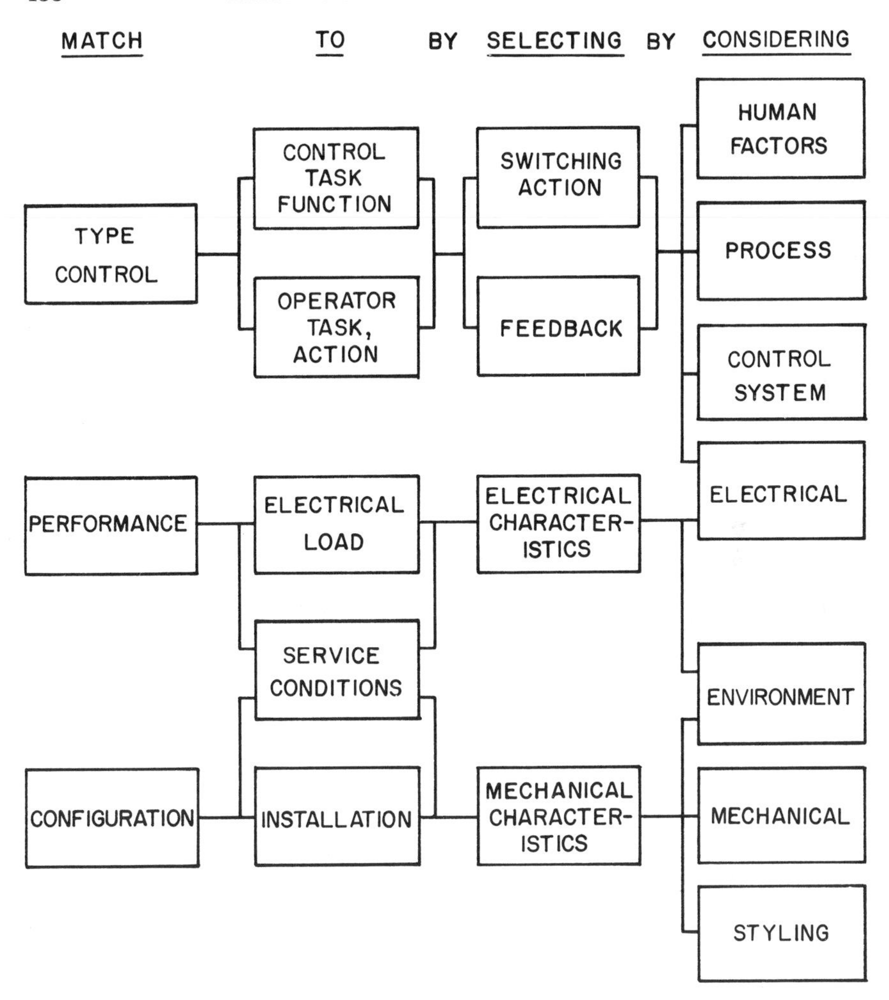

Fig. 3.6 l Selection procedure

Human Factors

Basic to the human factors is mind-eye-hand coordination. Figure 3.6m depicts the role of this neural aptitude in manipulating controls, and its efficiency depends on how well the controls are matched to 1.) the control task and actions required to perform it, 2.) the work station design and 3.) the displays associated with the controls.

Tasks (Figure 3.6n) are classified according to 1.) selection of alternates: start, stop, automatic and manual; 2.) adjustment: gross, fine, fast and slow; 3.) tracking: adjustment about a moving reference; and 4.) data entry: setting values, loop address and loading computer registers.

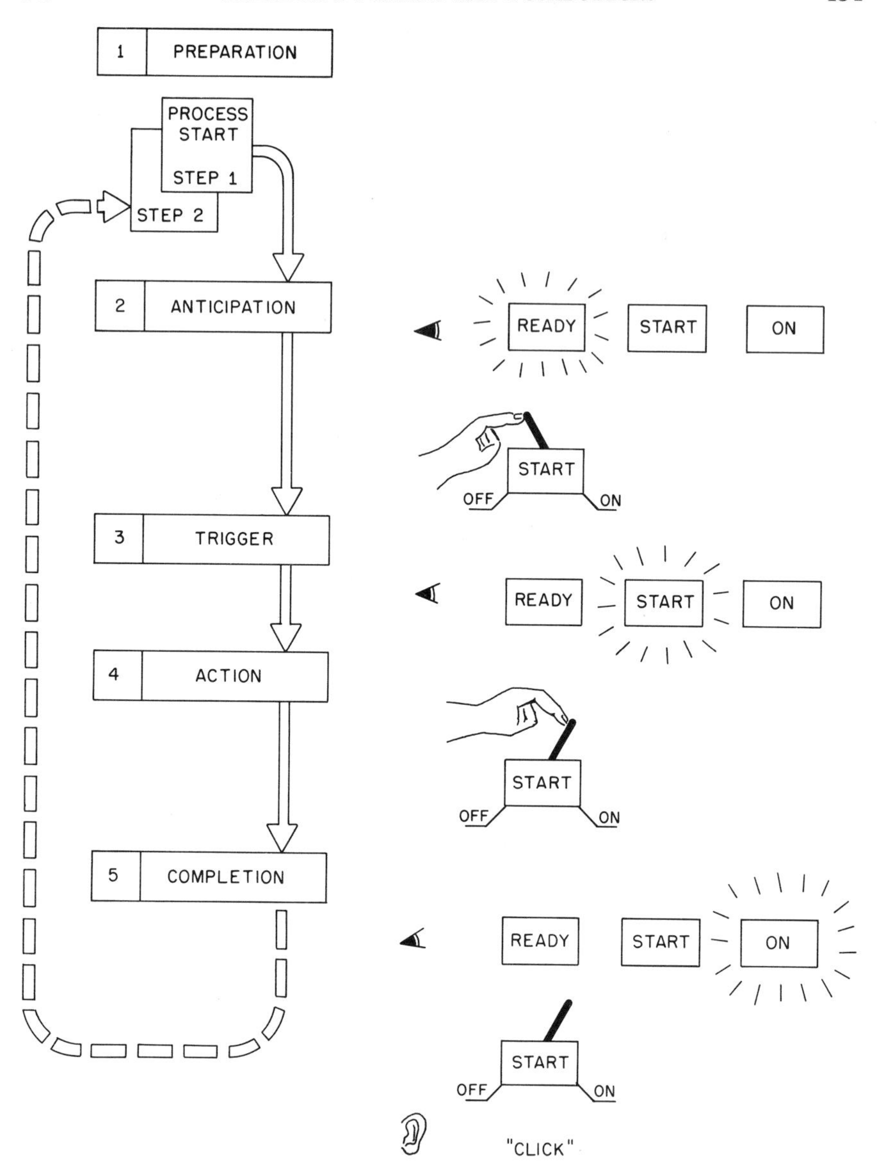

Fig. 3.6m Mind-eye-hand coordination in control manipulation

Actions (Figure 3.6n) required to perform these tasks are 1.) sequential or simultaneous; 2.) continuous or intermittent; 3.) exact or approximate; 4.) repetitive or different each time; 5.) frequent or infrequent; 6.) single pulse or hold-down; and 7.) normal or emergency.

Work station design objective is to match physical characteristics of

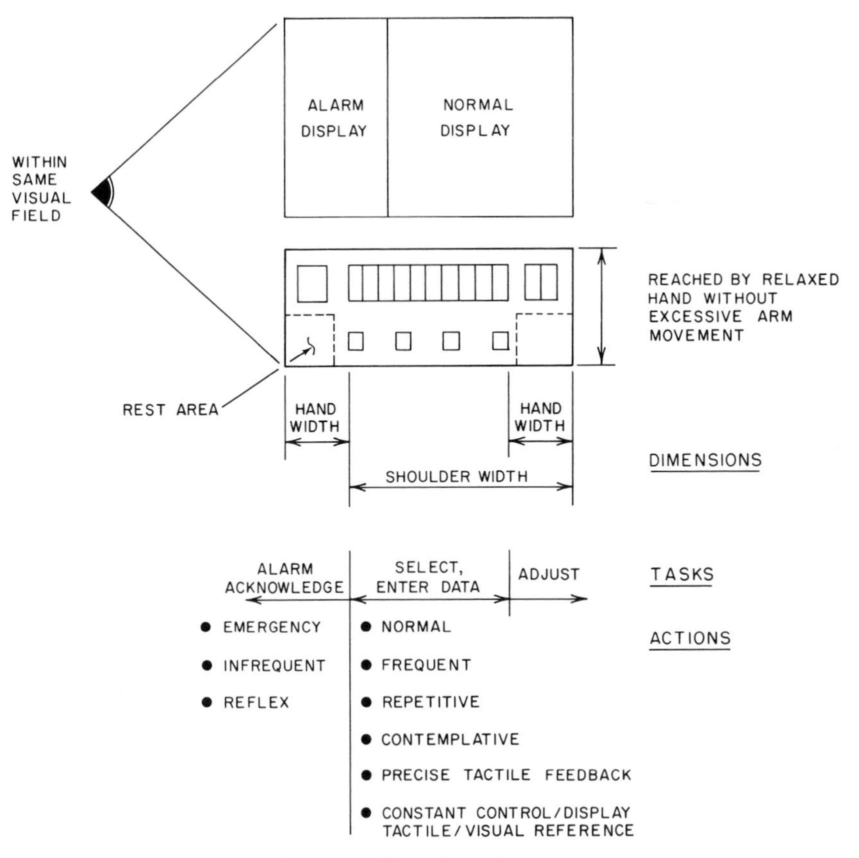

Fig. 3.6n Example work station

controls and displays to the nature of the control task, the actions required to perform it and the operator's capabilities.

Anthropometry and general arrangement are different depending on the operator's position (sitting, standing, stationary or mobile). Figures 3.6o and 3.6p show proper physical relationships. Angles between hand, fingers and controls affect the "feel" of the control. The relaxed hand assumes natural angles which determine the most comfortable and strongest line of force (Figure 3.6q). Controls should be oriented and arranged so as to achieve these angles over the entire control panel. Compromises are necessary to reach extremities of the panel without inordinate body movement. Detailed arrangement should include the following considerations:

1.) The same type of control should be placed differently for sitting and standing positions.
2.) A sequence of time or order of use should be clearly depicted.

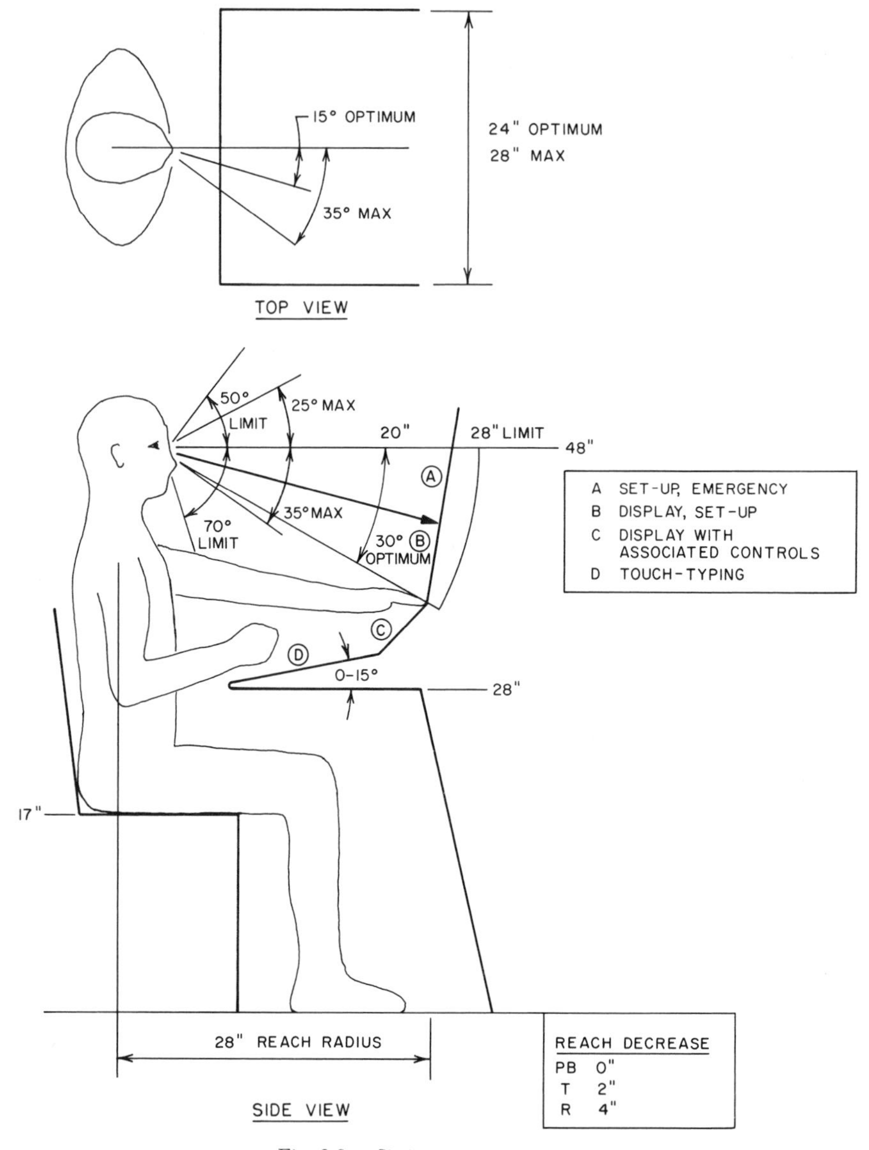

Fig. 3.6o Sitting operating position

The sequence should be maintained regardless of operator or panel orientation.

3.) Workload should be distributed fairly between left and right hands, most people being nimbler with their right than with their left hand.

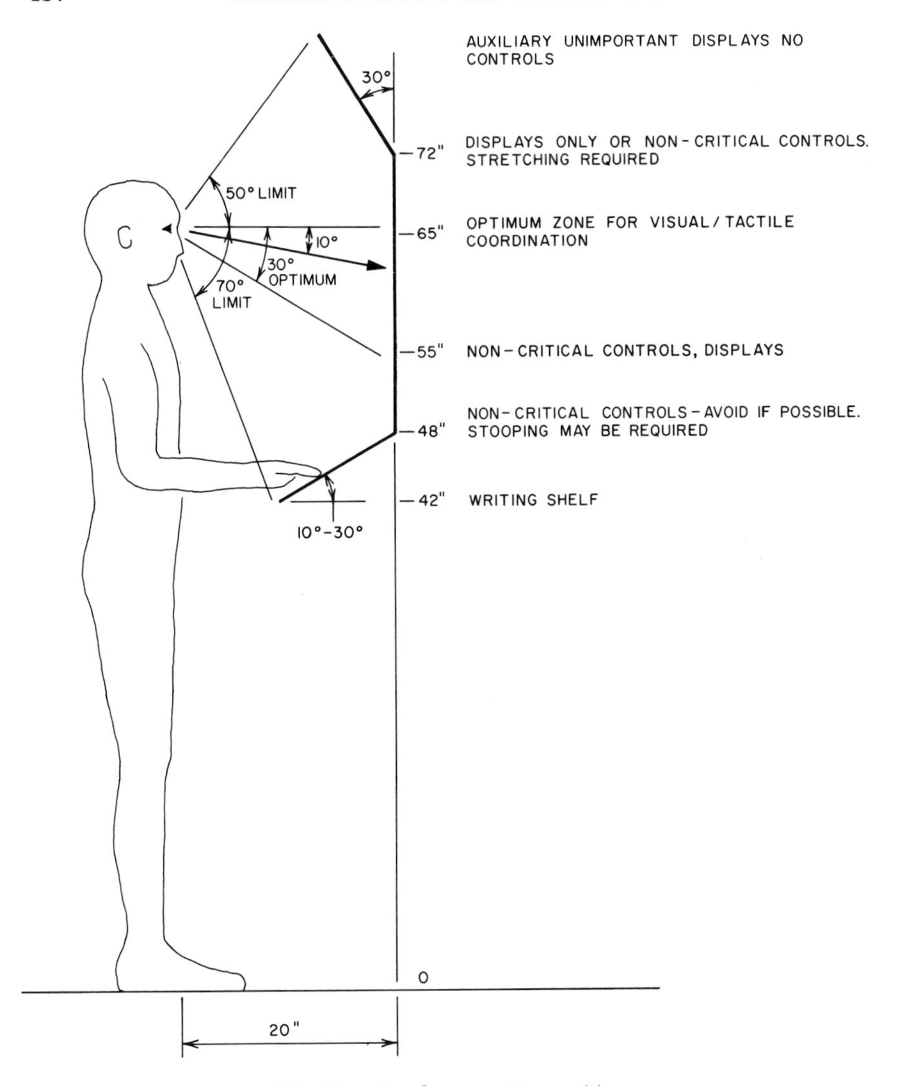

Fig. 3.6p Standing operating position

4.) Related controls and displays should be clearly associated and located within the same visual field to minimize head and eye movement. Actuation of controls should not obscure related displays, labeling or lighting.

5.) Controls should be located below or to the left of the display for left-hand operation, and below or to the right for right-hand operation.

6.) Similar controls performing similar functions should operate in

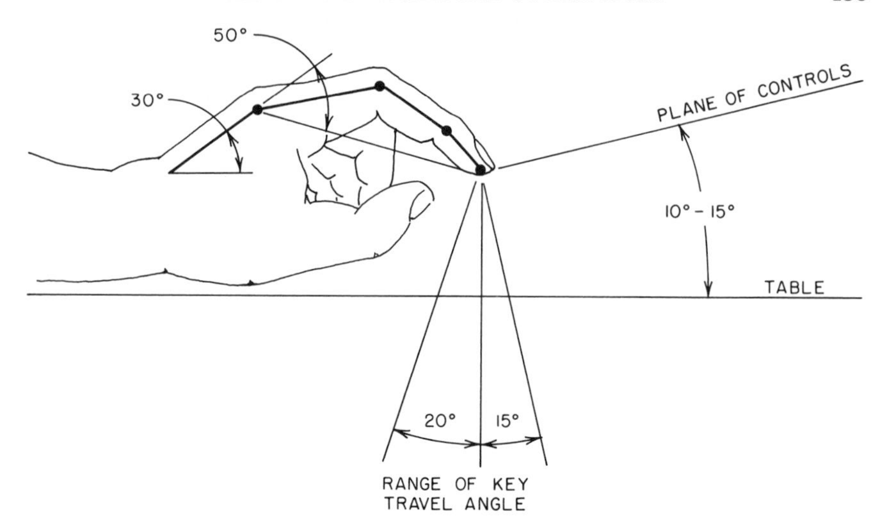

Fig. 3.6q Natural hand and finger angles

similar directions; similar controls performing dissimilar functions should operate in distinctly different directions.

7.) Critical controls may require safeguards against accidental actuation by clothing, general body contact and falling objects.

Display Movement

Control and display movement should be coordinated for direction, rate, accuracy, resolution, ratio and linearity. Certain natural relationships are required by human habit patterns and reflexes (Figure 3.6r). Small, precise display adjustments require long-travel, low-force control motion. Short travel and low force should be used for fast displays. Pulse controls are well-suited to incremental displays. Analog displays are tracked or adjusted most efficiently by a rotary potentiometer, or by a control with maintained "on" positions on either side of a spring-return center "off." Efficient identification of the control, its function and its use is a keystone of efficient operation. Considerable care is required to achieve this seemingly commonplace goal. It is important to recognize that most identification techniques involve double duty, and that redundancy of controls can both confuse the operator and lend a "busy" appearance to the panel (Figure 3.6s). Control design should blend with the over-all styling theme of the work station. Handles and mounting hardware are available in a variety of shapes, sizes, textures, materials, labeling and lighting (Figure 3.6t).

Error Prevention

Error prevention techniques are valuable in order to forestall simultaneous and missed actuations. Both activities are especially prevalent in repeti-

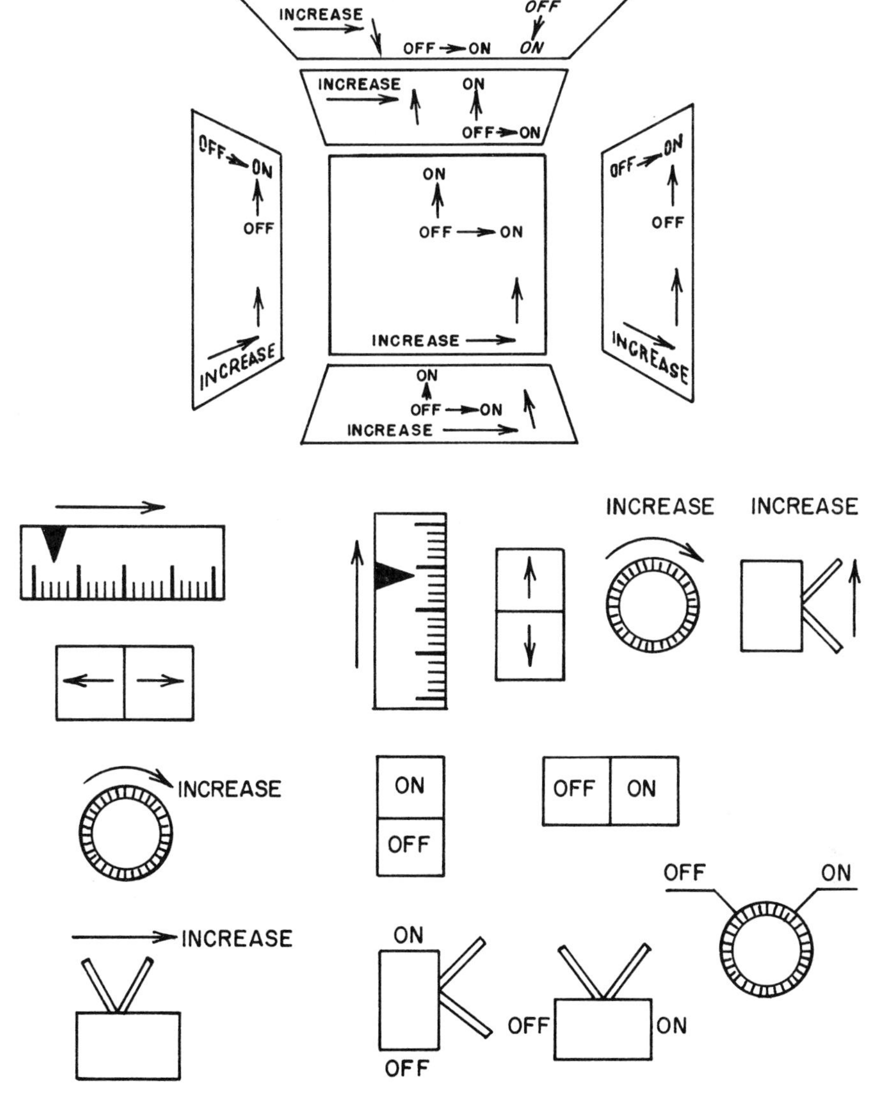

Fig. 3.6r Natural direction of motion relationships

tive, high-speed keyboard operation, and can be traced to speed limits imposed by switching mechanisms, actuator mechanisms and scanning speed of external monitoring equipment.

Mechanical interlock inhibits two types of simultaneous actuations so that 1.) two keys cannot be simultaneously depressed to the switching point, and 2.) a depressed key must be released before another can be depressed.

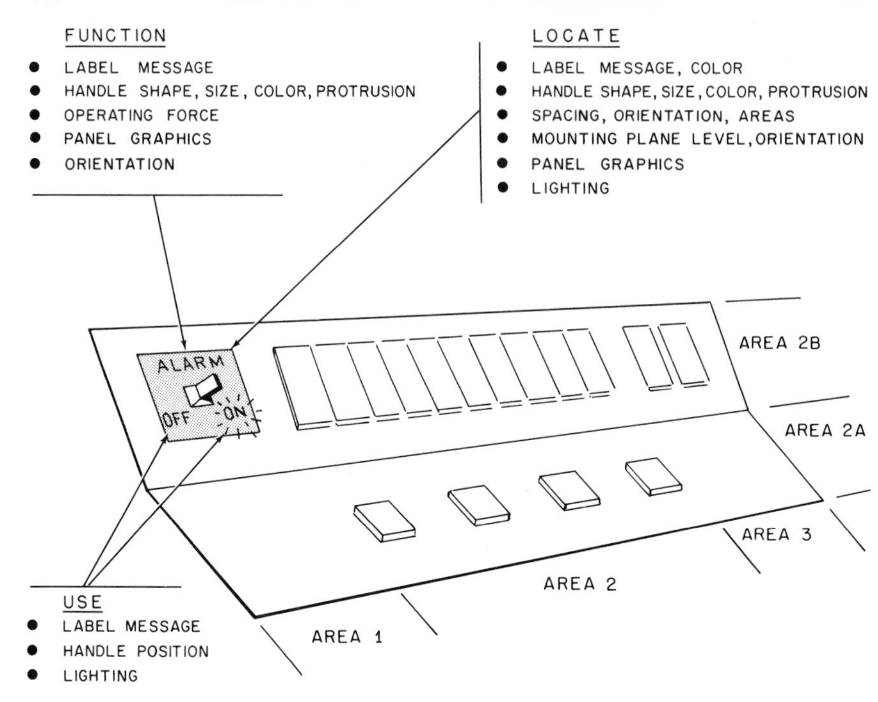

LABELING GUIDELINES

- TOO MUCH INFORMATION IS AS USELESS AS TOO LITTLE
- BE SPECIFIC, CONCISE, CLEAR, BUT NOT ELABORATE
- AVOID "ENGINEERING LANGUAGE," IDENTIFY WHAT IS BEING CONTROLLED NOT THE NAME OF THE CONTROL
- LABELS SHOULD BE ON CONTROLS WHENEVER POSSIBLE. WHEN ON PANEL, ASSOCIATION WITH IDENTIFIED CONTROL SHOULD BE OBVIOUS.
- ADJACENT LABELS SHOULD NOT CREATE ADDITIONAL "FALSE" LABEL
- PLACE LABELS CONSISTENTLY ABOVE OR BELOW IDENTIFIED CONTROL
- ORIENT LABELS HORIZONTAL TO READING LINE-OF-SIGHT, AVOID VERTICAL, SLANTED, OR CURVED ORIENTATIONS
- USE CAPITAL LETTERS, CORRELATE SIZE WITH VIEWING DISTANCE (REFER TO SECTION 3.3)
- COLORS OF LABELS SHOULD PROVIDE MAXIMUM CONTRAST WITH LABEL BACKGROUND

Fig. 3.6s Identification

Electronic interlocks artificially increase the actuation speeds by delaying scanning of consecutive key outputs. "Two-key roll over" generates codes during depression of one key and release of another by blocking the second until the first is released. "One-character memory" stores a character code until another is generated, erasing the first.

Tactile feedback in the form of a properly designed force-travel relationship is the best guarantee against missed actuations. Skilled operators develop a rhythm of depressing keys. They keep several steps ahead of their actual manipulation, counting on a change in their rhythm as a signal to stop. This change is usually the absence of a completion cue for an intended

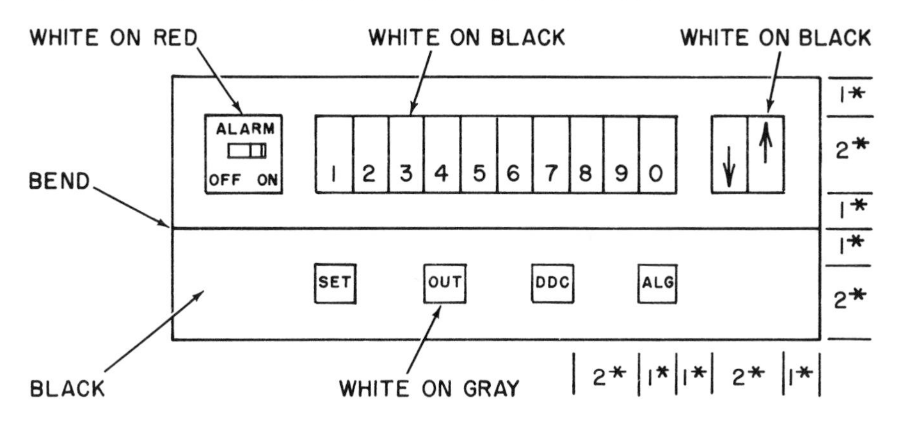

- NEAT, ORDERLY ARRANGEMENT
- RELATED SHAPES, SIZES
- MODULAR SPACING*
- RELATED COLORS - BLACK, GRAY
- KEY GROUPING BY SPACING, COLOR, SHAPE
- LABELS CONCISE, UNAMBIGUOUS, CONSISTENT
- LABEL COLORS UNIFORM, HIGH CONTRAST
- SPECIAL COLOR CODING FOR HABIT-PATTERN REFLEX - RED = ALARM

Fig. 3.6t Styling

actuation. Completion is signified by a sudden increase in operating force near or at the end of travel. Supplementary audible or visual cues also occur at this point. The force-travel relationship shown in Figure 3.6u is generally regarded as desirable, is available in solenoid-assisted electric keyboards and is closely approximated by some manual keyboard controls. However, the curve shown in Figure 3.6v is the most economical (a simple linear spring) and therefore the most available. Tests have shown that it is a primary source of missed keystrokes in high-speed data entry.

Environment variables on manipulation of controls include temperature, relative humidity, ambient illumination and noise. Sweating may limit the amount of force the operator can apply, degrade dexterity and cause fingers to slip off handles. Excessive light can mask lighted buttons, causing errors and loss of time and reducing safety. Noise detracts from concentration and may preclude audible feedback.

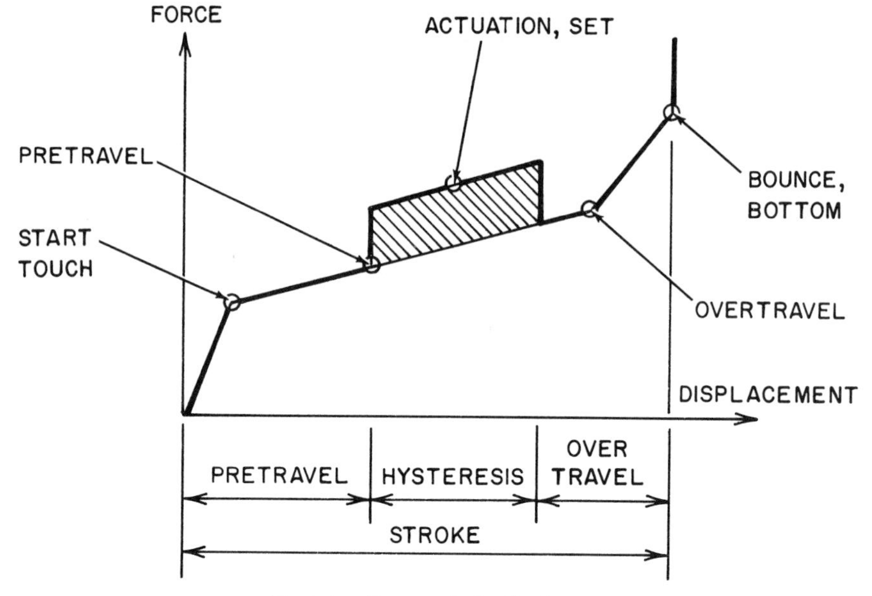

Fig. 3.6u Best tactile feedback

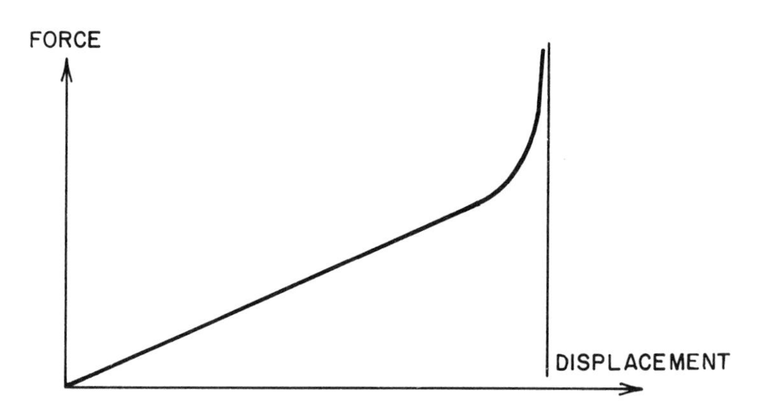

Fig. 3.6v Force-travel relationships with linear spring feedback

Electrical Rating and Performance

Electrical performance of the control is determined by current and voltage ratings, contact bounce, switching sequence and electrical interference from other equipment. Current rating measures resistance to welding of contacts. Underrated contacts fail either to open, or to open the required number of times during a specified interval. Current ratings are determined by contact configuration, plating materials and transfer time. Selection should account for potential overloads, which can exceed normal currents by as much as 30 times. Common sources of overload are 1.) inrush starting currents of motors and tungsten-filament lamps; 2.) relays and solenoid; 3.) equipment malfunctions such as short circuits and unstable power supplies; and 4.) lightning.

Voltage rating reflects resistance to arcing between contacts or terminals owing to voltage surges. Ratings are determined by air gap spacing between contacts and between terminals, and the insulation resistance of their mounting materials. Terminal spacing can cause a problem on very small subminiature switches, requiring supplementary insulating strips between terminals. Low insulation resistance can cause voltage leakage to ground or excessive dissipation within the dielectric itself. High-impedance analog computing circuits are especially sensitive to these phenomena. Certain insulating materials support formation of conductive surface films by repeated arcing, leading to eventual flashover between contacts or terminals.

Alternating current resistive loads are easily interrupted because the cyclical current reversals prevent arcing caused by continuous current build-up across the contacts. However, high AC frequencies (400 Hz) can approach DC in this respect. Direct current ratings are usually lower than AC ratings, and DC arcs tend to sustain themselves, since the applied voltage does not reverse as it does with AC. Special arc suppression techniques are sometimes required and are incorporated either within 1.) the switch using surge cavities, vents or magnets the field of which opposes that generated by the arcing current, or within 2.) external circuitry using resistors, capacitors and rectifiers to reduce or surmount the arcing current.

Contact bounce occurs partially in all mechanical contactors. Reed contacts are especially vulnerable owing to their cantilevered spring construction. Snap-action contacts have little bounce since the overcenter spring provides damping. Contactless switches, of course, are free from bounce. Bounce is equivalent to a sequence of contact openings and closings, the consequences of which are 1.) reduced contact life; 2.) generation of extraneous signals, which requires external filter circuitry; and 3.) radio frequency interference.

The switching sequence should be chosen so as to minimize the circuit complexities that can occur during contact transfer. Make-before-break sequences furnish momentary contact overlaps, useful maneuvers when attempting simultaneous switching of several fast, low-inductance circuits with a multi-pole switch.

Mechanical Features

Mechanical features of electrical switches and of parent equipment that affect work station design include 1.) type of mounting (front of panel, back of panel, printed wiring board, individual, in groups and vibration isolation); 2.) front of panel dimensions (spacing and size of openings); 3.) back of panel dimensions (spacing and access); 4.) access requirements (repair, replacement, maintenance and rearrangement of lamps, handles and wiring); and 5.) wiring (terminal configuration and location). Special precautions are occasionally required to preclude undesirable interaction between controls and parent equipment. Communications equipment operating at radio frequencies requires shielding from electrical noise generated by contact bounce. Reed switches must be shielded from heavy magnetic ferric materials (steel) and from magnetic fields stronger than their actuating magnet.

Environmental Considerations

Environmental factors pertinent to process control application include 1.) temperature; 2.) relative humidity; 3.) contamination from gases, liquids and solids; 4.) barometric pressure; and 5.) vibration and shock. Most commercial and industrial electrical switches function within specification in environments that the operator can also withstand. However, infrequently attended controls may be exposed to much more severe operating conditions. Electrical codes sometimes require low-energy circuits in hazardous locations, a circumstance that may necessitate intermediary isolation circuitry between control and controlled equipment, or sufficient air gap spacing to ensure potential differences below spark-generation levels.

The contact temperature is the sum of the room temperature, the rise within the parent equipment and the rise within the switch. High temperatures accelerate contact corrosion. Unequal expansion of components can cause cracking and binding. Forced cooling may be required for densely spaced lighted pushbuttons that operate continuously.

Relative humidity and temperature cycling produce condensation. Moisture accelerates contact corrosion (decreasing current rating) and decreases insulation resistance (decreasing voltage rating). Contamination from moisture, salts, oils, corrosive gases and solid matter degrades contacts and actuator mechanisms. Contacts are protected by glass-to-metal or plastic seals and inert atmospheres. Sealing can be done at the control panel, between actuator and contact mechanism or within the contact enclosure. Controls should remain in their original packing materials until just before installation. Special materials such as low-sulfur-content papers retard contact corrosion during shipping and storage.

Vibration and shock can cause contact chatter, arcing and outright structural failure of the contact mechanism, and can also affect the choice of operating force, handle shape, weight and mounting method. Vibration damping is afforded by snap-in spring mounting and the parent equipment structure. Orientation on the control panel should place handle travel at right angles to the vibration force. Barometric pressure affects the ability of air to extinguish arcs and dissipate heat. Current and voltage ratings must be reduced for use at high altitudes; hermetically sealed enclosures are occasionally needed for such atmospheric applications.

Chapter IV

PROCESS CONTROL SYSTEMS

T. J. Myron, Jr.
G. A. Pettit
R. E. Wendt, Jr.

CONTENTS OF CHAPTER IV

4.1 CONTROL OF EXTRUDERS

Extruder Types

Single screw extruders commonly convert granular resin feeds into sheets, films and shapes such as pipe, and are described by screw diameter in inches and L/D ratio—L being the screw length and D the screw diameter. They are supplied in sizes of 1, 1½, 2½, 3½, 4½, 6, 8 and 12 inches. L/D ratios from 20 to 30 are common. The single screw extruder is by far the type most frequently manufactured.

Machines using twin screws are generally large volume production units for pelletizing resins in petrochemical plants and are equipped with various combinations of intermeshing and non-meshing screws that corotate or contrarotate. Screw design features allow compounding, devolatilizing, melting, blending and other processing in a single machine. Twin screw machines are often melt-fed directly from polymerization reactors and perform multiple functions on the polymer prior to pelletizing and packaging it as a finished product.

Extruder Dies

The shape and ultimate use of the output are defined by the die shape. Dies are broadly classified as 1.) sheet dies extruding flat sheets as much as 120 inches wide and ½ inch thick; 2.) shape dies for making pipe, gasketing, tubular products and many other designs; 3.) blown film dies, using an annular orifice to form a thin-walled envelope. The diameter of the envelope is expanded with low pressure air to roughly three times the annular orifice diameter and forms a thin film. The process is for films 5 mils thick (0.005″) at the upper limit; 4.) spinnerette dies for extrusion of single or multiple strands of polymer for textile products, rope, tire cord or webbing; 5.) pelletizing dies for granular products in resin production, synthetic rubbers and scrap reclaiming. These dies form multiple strands roughly ⅛ inch in diameter. Rotating knives continuously cut the strands in short lengths after which the pellets drop into water for cooling; and 6.) cross-head dies for wire coating in which the bare wire or cable enters the die and emerges coated with semimolten polymer. The wire enters and leaves the die at an angle of 90° to the extruder axis—hence the term cross-head.

Many special arrangements of extruder dies are used for composite films; two polymers enter the die from two extruders and exit as a sheet, the top and bottom polymers being of different chemical composition so as to obtain a film or sheet of two colors or so as to have desirable characteristics of both materials. Dies for rigid foam production are similar to blown film dies; special dies with moving parts extrude netting continuously. Most dies require a short connecting pipelike piece, commonly called the adapter, to connect to the extruder head.

Polymer Types and Characteristics

Any list of commercial polymers is incomplete owing to the rapid progress of polymer science and production. The common resins such as polyethylene and styrene extrude rather simply, and are relatively stable and reasonably predictable. Polymers in the vinyl chloride group are not stable at extrusion temperatures. If not promptly removed from the process they decompose, emitting hydrogen chloride gas which is very corrosive to metal. Consequently, dies for vinyl chloride are commonly chrome plated and designed so as to eliminate pockets or crevices where the material may be entrained and decompose.

An important characteristic of synthetic polymers which complicates the extrusion is sensitivity to shear rate, a phenomenon known as non-Newtonian behavior (see Chapter VI in Volume I). The relationship between apparent viscosity and shear rate is also dependent on temperature and must be considered in designing control equipment for extruders[1]. Certain polymers such as polyvinyl chloride (PVC) and acrylonitrile-butadiene-styrene (ABS) are much more sensitive to shear rate than are more crystalline materials such as nylon and acrylics.

The basic formula for die output is

$$Q = k\Delta P/\mu \qquad 4.1(1)$$

where Q = output capacity in cu in. per second,
k = die constant in cu in.
ΔP is pressure difference between inlet and outlet (lb per sq in.)
μ is viscosity in lb sec per sq in.

Die pressure is frequently assumed to be a linear function of screw speed. As previously stated, the apparent viscosity of a given material is a function of temperature as well as of shear rate, and melt temperature measured at the adapter is also an important operational guide to extruder operation.

Extruder Barrel Temperature Control

The barrel is usually divided into temperature control zones roughly 12 to 18 inches long (Figure 4.1a). A $4\frac{1}{2}$ inch extruder, for example, may have

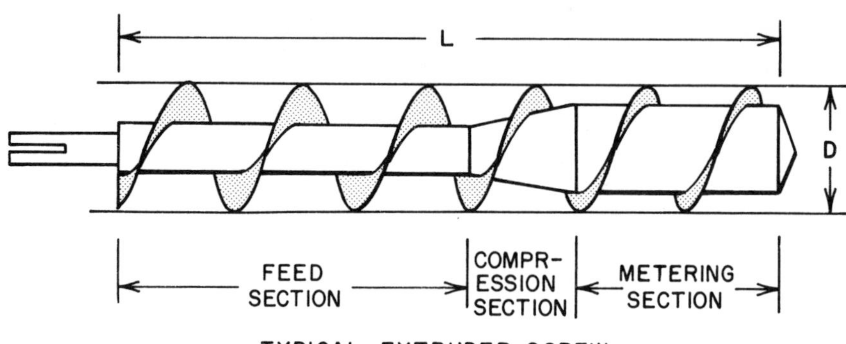

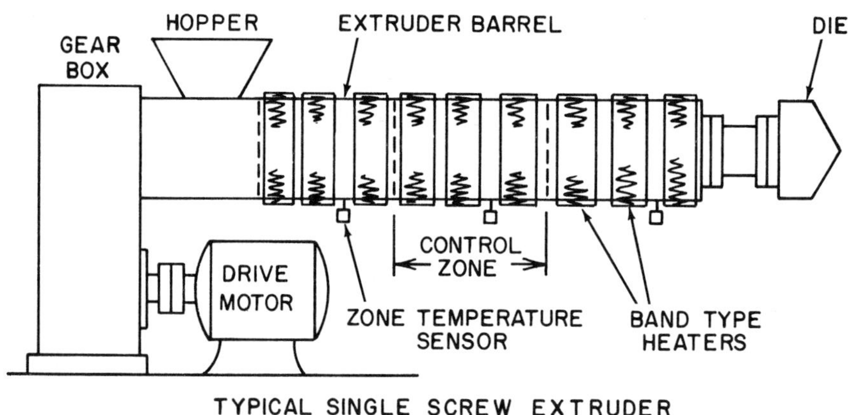

Fig. 4.1a Typical single screw extruder

4 to 6 barrel zones. Conventional control systems use a separate temperature control loop including power control device, temperature sensor and heater for each zone. Band type resistance heaters are of two piece construction to facilitate removal.

Extruders require large heater ratings to decrease heating time. After extrusion has begun, heat is internally generated from the friction and shear of the rotating screw. This heat is a function of the screw speed squared. At high production speeds the internal heat equals the external losses, resulting in an adiabatic operation requiring no conducted heat. At high speeds the melt temperature rises above the accepted maximum, and barrel cooling is required. The amount of internal heat generated is also influenced by screw design, head pressure and resin viscosity.

Controllers for the extrusion industry have developed very specialized forms. Rarely are recorder-controllers used. Instead the instruments are almost always of the electronic type, of small size and with combination outputs for extruder barrel heating, cooling, alarms and startup interlocks. (For more on the frequently used SCR controllers, see pp. 300 and 415 in Volume II.)

The available choices of control instrumentation features are summarized in Table 4.1b.

Electric heating directly at the barrel can also be done by induction, utilizing a coil of copper wire wound about the barrel to induce an electro-

Table 4.1b
CONTROLLER FEATURE SELECTION CHOICES

Sensor Type	*Temperature Display*	*Set Point Display*	*Control Mode*	*Power Device*
Thermistor	Hi-lo lamps	Arbitrary scale	On-off	Contactors
Thermocouple	Deviation meter	Calibrated dial	Time proportioning	SCR
Resistance	Calibrated scale	Calibrated scale	Two mode	Saturable reactor
	Digital display	Digital display	Three mode	Valves

magnetic field. The barrel steel is heated by internal circulating currents, a much more responsive technique than resistance type heating. The coil is energized with 60-cycle current and is usually controlled by contactors.

Fan Cooling

Many extruders use aluminum heating shells with cast-in heaters of encased nichrome coiled throughout the aluminum jacket. The shells are in two segments clamped together to surround the extruder barrel. Shells for fan cooling have fins cast at the outside surface to increase surface area. Motor driven fans or blowers are positioned directly below the zone and when switched on (manually or automatically) remove barrel heat by forced convection. A zone fan on a $2\frac{1}{2}$-inch extruder can typically remove the equivalent of 5 Kw per hour (Figure 4.1c).

The control panel is usually equipped with a three-position selector switch. In the automatic position the temperature controller turns the fan on at approximately 25% heater power level (in the proportioning band), which then stays on throughout the balance of the range to zero heat. The ON position allows continuous fan operation for shutdown and is operated by an auxiliary circuit in the control instrument.

Water Cooling

Extruders using water cooling have the aluminum shells with cast-in tubing as well as heaters. Treated water is continuously circulated through the coils by a common pump and heat exchanger. Solenoid valves, controlled by the zone temperature controller, allow circulation to each zone in order

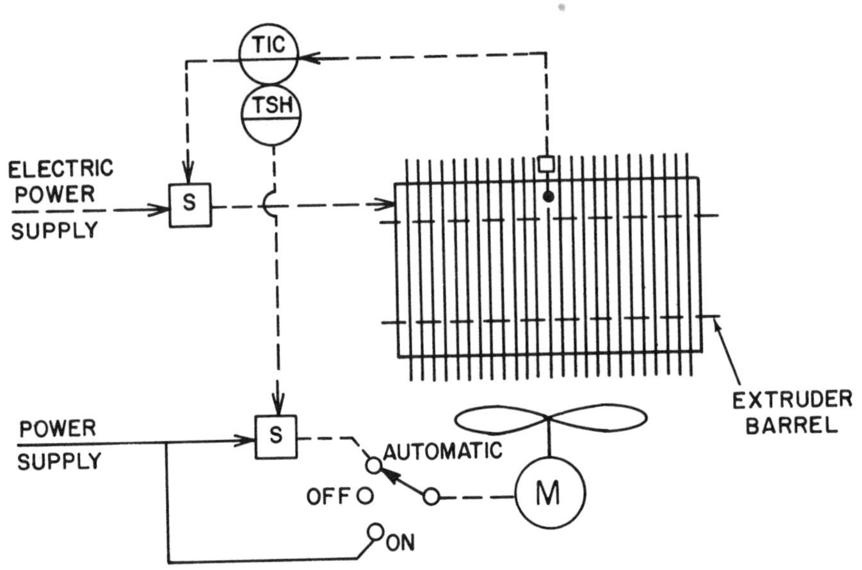

Fig. 4.1c Heater with fins for electric heating and fan cooling

to maintain the proper barrel zone temperature. The valves are operated from an auxiliary switch in the zone temperature control instrument in a time proportioning manner with extra slow cycle rate and capability for very short pulses.

Water cooling is frequently too effective compared to fan cooling. Cooling water is usually flashed into steam at most barrel temperatures used for processing thermoplastic materials. Some machine builders use compressed air to clear the water from the passages and eliminate trapped fluid and erratic cooling. By this method, a very short pulse of cooling water charging is followed by immediate removal to obtain less severe cooling. Running the heat exchanger sump at higher temperatures also reduces the severity of water cooling. Some extruder barrels have grooves machined in the outside surface, with cooling coils imbedded in the grooves and band heaters clamped directly over the barrel and tubing.

Hot Oil Systems

Extruders 6 inches and larger, both single and twin screw designs, frequently use circulated oil heating around the barrel through an outside jacket. The upper temperature limit is roughly 400°F. Such design is primarily for rubber, PVC, styrenes, polyethylene and acrylics. The extruder is zoned and pumped continuously from a common sump which contains both electric heating elements and cooling coils. Sump temperature is maintained by simple, self-contained controllers. The extruder barrel is heated or cooled by the oil,

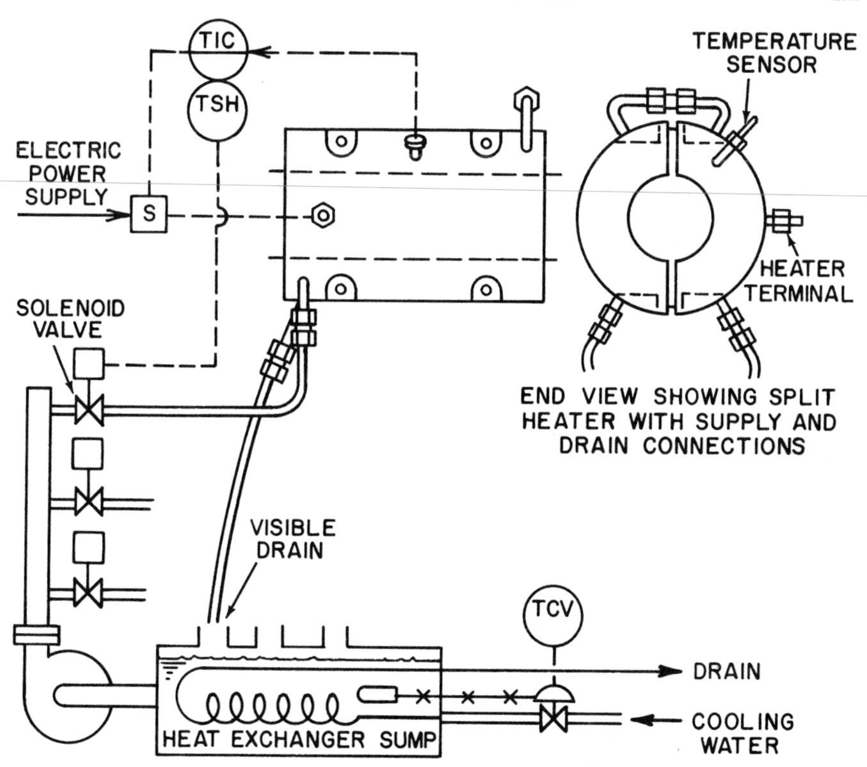

Fig. 4.1d Heater coils for water cooling

depending on internal heat generation. Circulation is at constant flow rate, and zone temperature controllers are not used.

Water is sometimes substituted for hot oil when the operation, although primarily cooling, requires initial heating. Water is capable of considerably greater heat transfer than is hot oil because of high specific heat, efficient thermal conductivity and low viscosity (which promotes turbulent flow). In these water circulating systems, the extruder zones are encased, each zone with its own pump and sump tank. The control system circulates water at constant temperature, which is returned through a shell and tube cooler to the sump.

Melt Temperature Cascade System[2]

The output melt temperature is a function of internal shear energy (converted to heat energy) plus conducted heat or minus the barrel cooling, depending on the operation. The temperature of the melt is as important as the output rate for quality extrusion.

The cascade control to reset the zone controller set points achieves con-

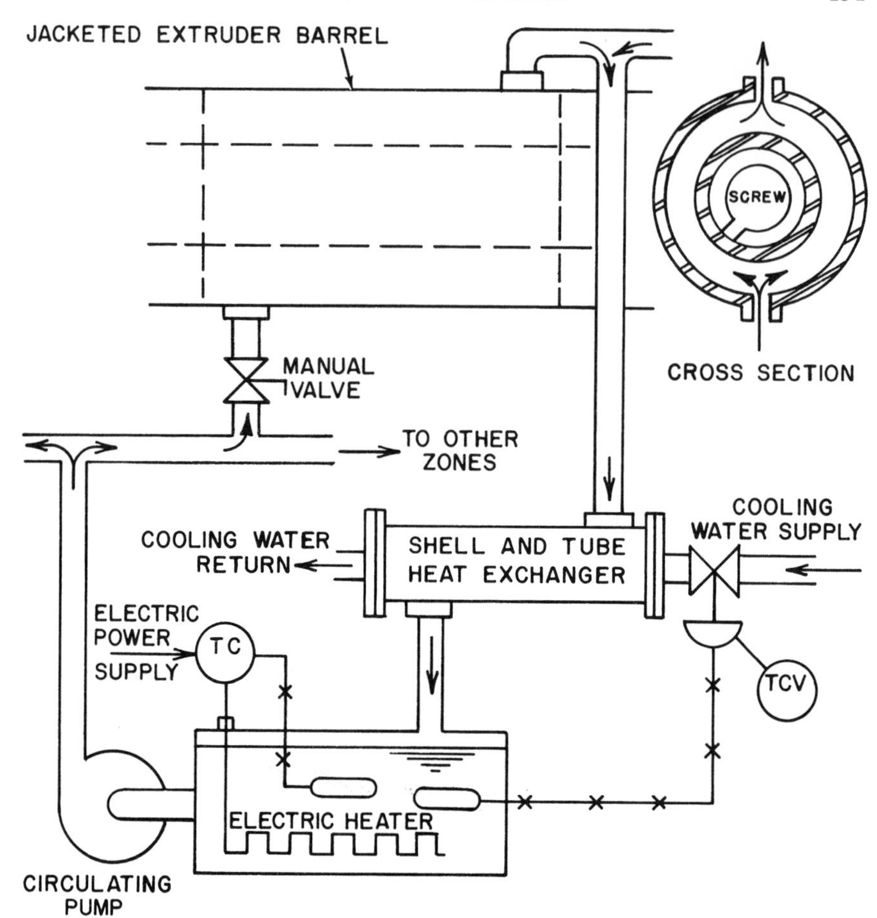

Fig. 4.1e Oil heated extruder zone

tinuous temperature control of the melt and approaches the ultimate in one-knob operation. Figure 4.1f shows the arrangement for cascade feedback. One special feature is that the system allows only depression of the zone temperature controller set points—a safety consideration because of heat degradation and pressure buildup in polymer systems. A safety interlock with the extruder screw drive is usually incorporated in order to prevent both polymer freezing during shutdowns and drive damage at startup.

Zone controllers provide both heating and cooling and their set points are regulated by the melt controller. Each zone can be adjusted individually to follow a certain percentage of the feedback cascade signal so that a preset program of zone depression will follow a definite barrel temperature zone profile curve. The extruder metering zones are capable of the greatest heat transfer and therefore receive the greatest percentage of feedback and most

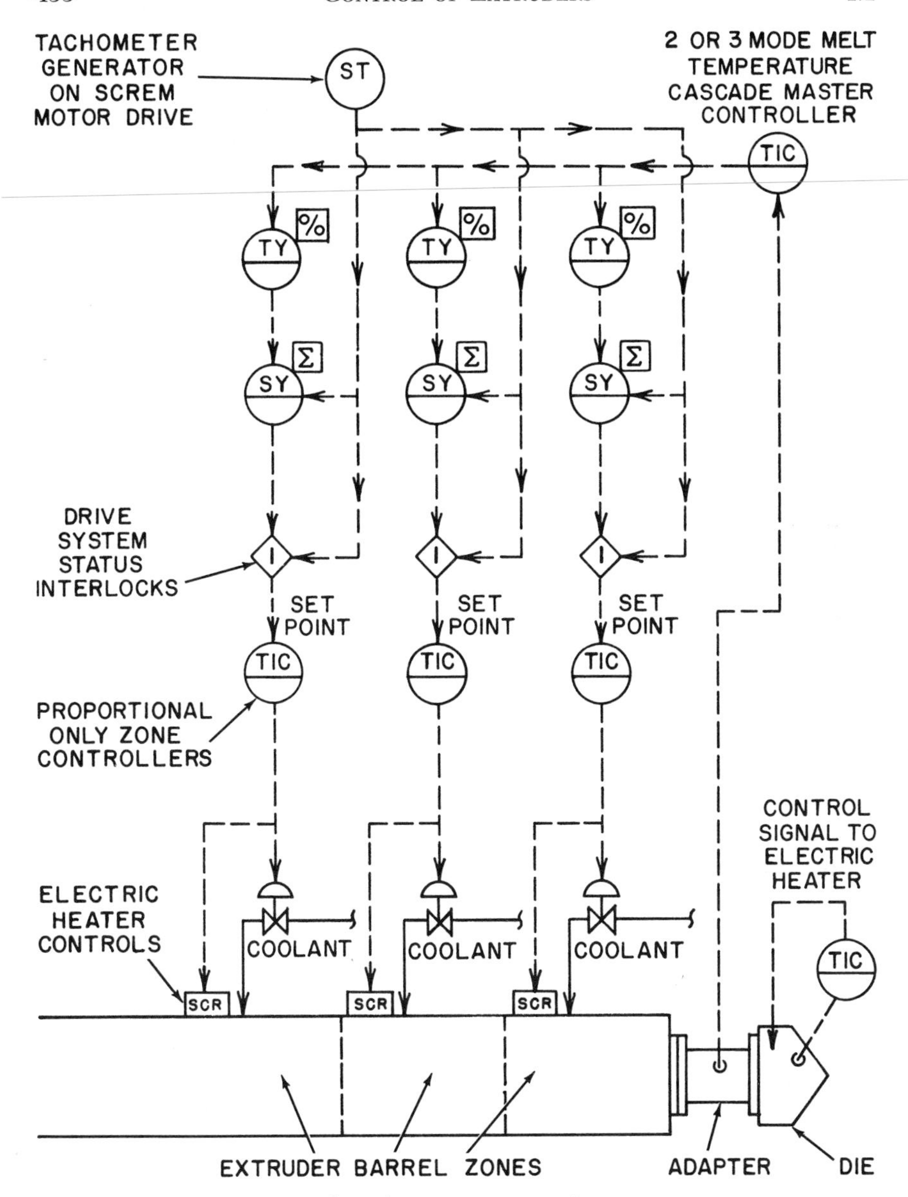

Fig. 4.1f Melt temperature cascade system

set point depression. Feeding and compression zones are most effective at relatively constant temperature to provide optimum wall friction for most efficient feedings; therefore they usually receive a small percentage of the feedback signal.

The melt controller must have proportional and integral control; according to Meissner[3], rate action is also beneficial. The zone controllers should be proportional for simplicity only, because reset action at the primary melt controller achieves the desired melt temperature without droop. Some cascade systems incorporated a tachometer feedforward signal to supply set point depression proportional to screw speed, a function that reduces temperature departures from the control point due to screw speed drift or intentional changes.

Extruder Pressure Measurement

Most extruders run at pressures from 500 to 10,000 PSIG, and the die output flow is directly proportional to extruder output pressure, which in turn depends on screw speed, die restriction and resin viscosity. Knowledge of extruder pressure is very important for efficient extrusion. Some common extruder pressure sensors include:

Grease sealed gauges. These are direct reading bourdon tube gauges with a tube tip capillary for complete filling with high temperature grease, usually silicone. The grease remains viscous at high temperatures and prevents the molten polymer from entering and solidifying in the gauge or piping. Disadvantages include the need for periodic greasing, and occasional contamination of the product. (For other designs of volumetric pressure seal elements see Figure 2.12f in Volume I. These may require temperature compensation.)

Pneumatic force balance transmitters. These provide linear output to low pressure indicators, recorders and controllers. Their force balance principle makes them less sensitive to process temperature variations (see Section 6.1 in Volume II for more details).

Strain gauge transducers. These are high quality pressure sensors, widely used in extruder applications as discussed in Section 2.8 in Volume I.

Pressure Control of Extruders

Synthetic fiber processes frequently utilize an extruder to melt and transport nylon and polyesters to a bank of gear pumps feeding individual spinning die heads. The shear characteristics of these polymers are near-Newtonian and thus display relatively constant viscosity at wide variations in shear rate. This characteristic is responsible for pressure being nearly proportional to screw speed, which is necessary for pressure control by regulation of the screw speed.

In contrast, PVC can be pumped through a restricted orifice at increasing rates without an increase in back pressure, due to its apparent viscosity dropping with increased pumping shear rate. The pressure of such a process is nearly impossible to control by manipulating screw speed.

Figure 4.1g illustrates the control scheme in fiber processes. Because some of the fiber spinning pumps frequently fail or are intentionally stopped, the

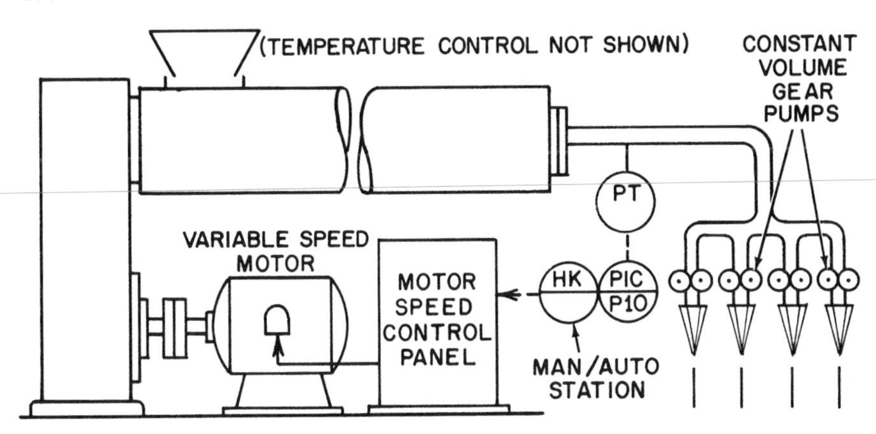

Fig. 4.1g Extruder pressure control in synthetic fiber production

pressure controller must immediately slow the extruder screw to a new output rate determined by the constant volume pumps remaining operable.

The screw drive motors are generally DC drives or eddy current clutches (Section 2.2 in Volume II) capable of accepting 4 to 20 madc control signals. For stable control the response of the entire control loop must be virtually instantaneous. Hysteresis in mechanical drive systems can be a source of control problems. Pressure control of extruders feeding extrusion dies directly has been used experimentally with variable results, depending on the shear to viscosity relationships of the polymers.

Use of control valves between extruder and die heads is popular with extrusion theorists but has not been commercialized because of the light demand and high cost of rapidly responding, high pressure and high viscosity control valves.

Produce Dimension Control

Blown film lines producing quality tubular and flat film to 5 mils thickness commonly use thickness measurement, and control. Thickness can be adjusted by the speed of the takeup rolls if the thickness variations (in machine direction) are consistent across the film. Increasing the takeup speed reduces the film gauge; decreasing the speed increases its thickness (Figure 4.1h).

Thickness variations in die direction (across the width) can be measured, but most attempts to automate sheet die lip adjustments have been unsuccessful. Sheet dies extrude materials up to ½ inch thick and 120 inches wide.

Measurement of film thickness to 100 mils is made with radiation instruments, using beta rays (Section 3.4 in Volume I). Among design variations are scanning heads that measure and record thickness over the entire width of materials. Infrared and mechanical devices have also been used. The viscos-

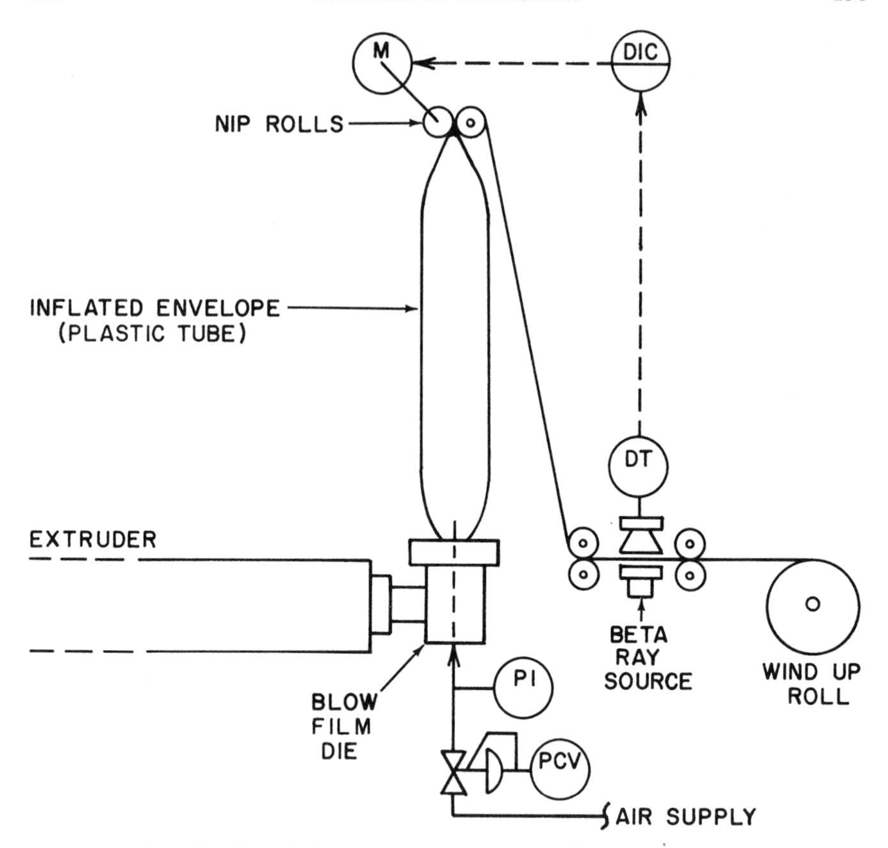

Fig. 4.1h Film thickness control through nip roll speed manipulation

ity of the polymer in the extruder and die head constitutes the most difficult aspect of extruder design, simulation and control. The art is steadily advancing and certain viscosity analyzers are available for extruder applications. (For more information see Section 6.14 in Volume I.) In most applications the viscosity measurement controls the polymerization, not the rate or dimensional quality of the pelletizing extruder output.

Computer Control of Extrusion

The digital computer with supervisory control has been successfully applied to composite film systems[4] and assumes functions such as speed, temperature, gauge, profile, and web tension. The role of the computer will be important in manipulating extruder pressure and temperature according to the correct mathematical model and will correct itself or adapt to variations in the process material. Product gauging, which is not readily applicable to shapes, such as synthetic fiber and blow molding applications, appears to be a much more difficult process to model and therefore is less likely to be controlled by computers.

REFERENCES

1. Bernhart, E. C., "Processing of Thermoplastic," New York: Reinhold, 1959.
2. Pettit and Ahlers, SPE Journal, Vol. 24, November 1968.
3. Meissner, S. C. and Meneges, E. P., Technical Papers, Volume XVII, Society Plastic Engineering.
4. Plastic World, June 1971, P. 84.

4.2 CONTROL OF EVAPORATORS

Evaporation is one of the oldest unit operations, dating back to the Middle Ages when any available energy source was used to concentrate thin brine solutions in open tanks. As archaic as this technique is, it is still widely used during the early spring in northern New England when maple sugar sap is tapped and concentrated in open-fired pans. Solar evaporation and bubbling hot gases through a solution are other examples of concentrating solutions. For our purposes, evaporation will be limited to concentrating aqueous solutions in a closed vessel or group of vessels in which the concentrated solution is the desired product, and indirect heating (usually steam) is the energy source. Occasionally, the water vapor generated in the evaporator is the product of interest such as in desalinization or in the production of boiler feed water. In other cases neither vapor nor concentrated discharge has any market value, as in nuclear wastes.

Evaporator Terminology

Single effect evaporation. Single effect evaporation occurs when a dilute solution is contacted only once with a heat source to produce a concentrated solution and an essentially pure water vapor discharge. Schematically the operation is shown in Figure 4.2a.

Multiple effect evaporation. Multiple effect evaporations use the vapor generated in one effect as the energy source to an adjacent effect (Figure 4.2b). Double and triple effect evaporators are the most common; however, six effect evaporation can be found in the paper industry where kraft liquor is concentrated, and as many as twenty effects can be found in desalinization plants.

Boiling point rise. This term expresses the difference (usually in °F) between the boiling point of a constant composition solution and the boiling point of pure water at the same pressure. For example, pure water boils at 212°F at 1 atmosphere and a 35% sodium hydroxide solution boils at about 250°F at 1 atmosphere. The boiling point rise is therefore 38°F. Figure 4.2c illustrates the features of a Dühring plot in which the boiling point of a given composition solution is plotted as a function of the boiling point of pure water.

Economy. This term is a measure of steam utilization and is expressed in pounds of vapor produced per pound of steam supplied to the evaporator

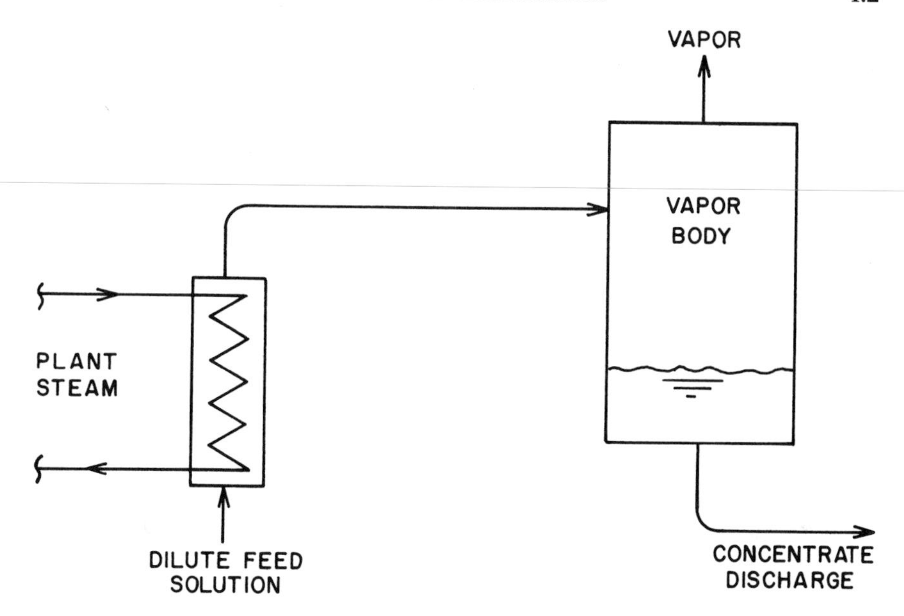

Fig. 4.2a Single effect evaporator

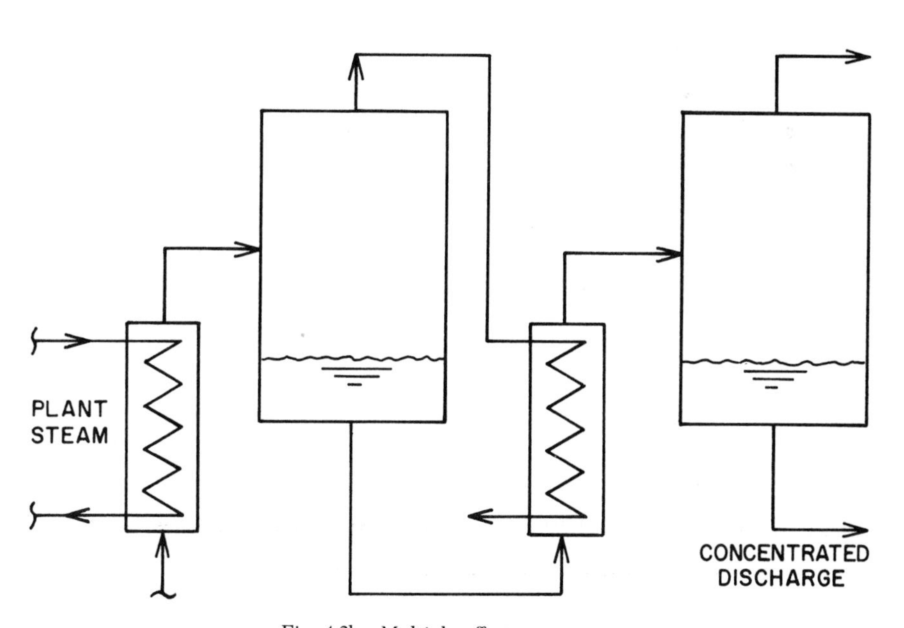

Fig. 4.2b Multiple effect evaporator

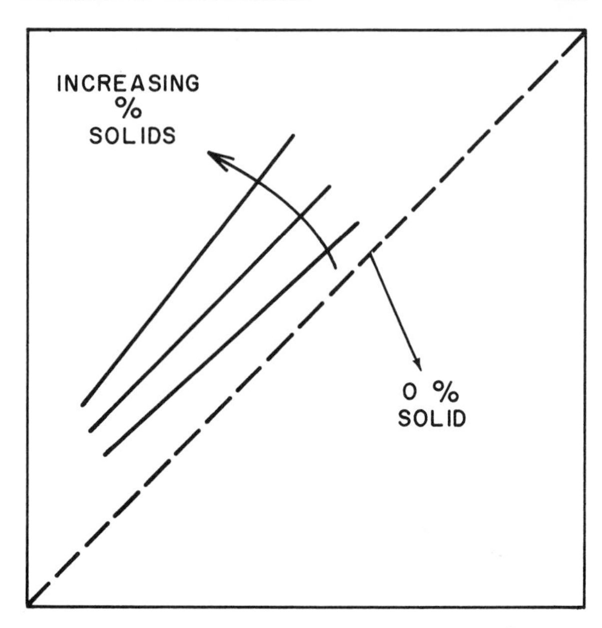

Fig. 4.2c Dühring plot of boiling point rise

train. For a well designed evaporator system the economy will be about 10% less than the number of effects; thus, for a triple effect evaporator the economy will be roughly 2.7.

Capacity. The capacity for an evaporator is measured in terms of its evaporating capability, viz., pounds of vapor produced per unit time. The steam requirements for an evaporating train may be determined by dividing the capacity by the economy.

Co-current operation. The feed and steam follow parallel paths through the evaporator train.

Countercurrent operation. The feed and steam enter the evaporator train at opposite ends.

Types of Evaporators

Six types of evaporators are used for most applications, and for the most part the length and orientation of the heating surfaces determine the name of the evaporator.

Horizontal tube evaporators (Figure 4.2d) were among the earliest types. Today, they are limited to preparation of boiler feed water, and in special construction (at high cost), for small-volume evaporation of severely scaling liquids such as hard water. In their standard form they are not suited to scaling or salting liquids and are best used in applications requiring low throughputs.

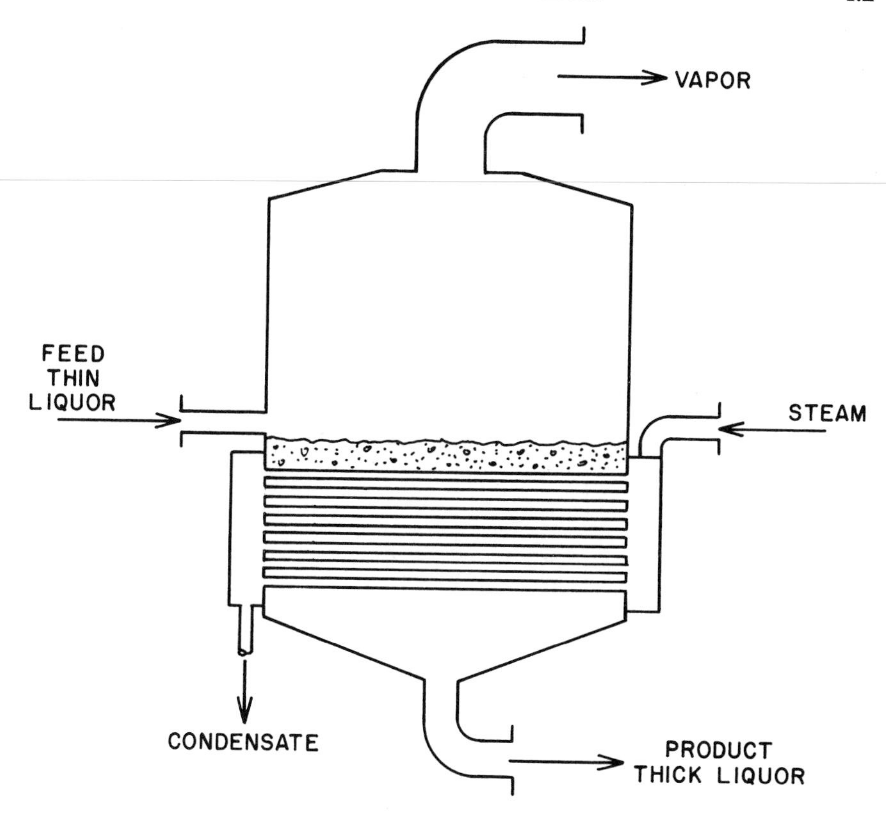

Fig. 4.2d Horizontal-tube evaporator

Forced circulation evaporators (Figure 4.2e) have the widest applicability. Circulation of the liquor past the heating surfaces is assured by a pump and consequently these evaporators are frequently external to the flash chamber so that actual boiling does not occur in the tubes, thus preventing salting and erosion. The external tube bundle also lends itself to easier cleaning and repair than the integral heater shown in the figure. Disadvantages include high cost, high residence time and high operating costs due to the power requirements of the pump.

Short-tube vertical evaporator (Figure 4.2f) is common in the sugar industry for concentrating cane sugar juice. Liquor circulation through the heating element (tube bundle) is by natural circulation (thermal convection). Since the mother liquor flows through the tubes, they are much easier to clean than those shown in Figure 4.2d in which the liquor is outside the tubes. Thus, this evaporator is suitable for mildly scaling applications in which low cost is important and cleaning or descaling must be conveniently handled. Level control is important—if the level drops below the tube ends, excessive scaling

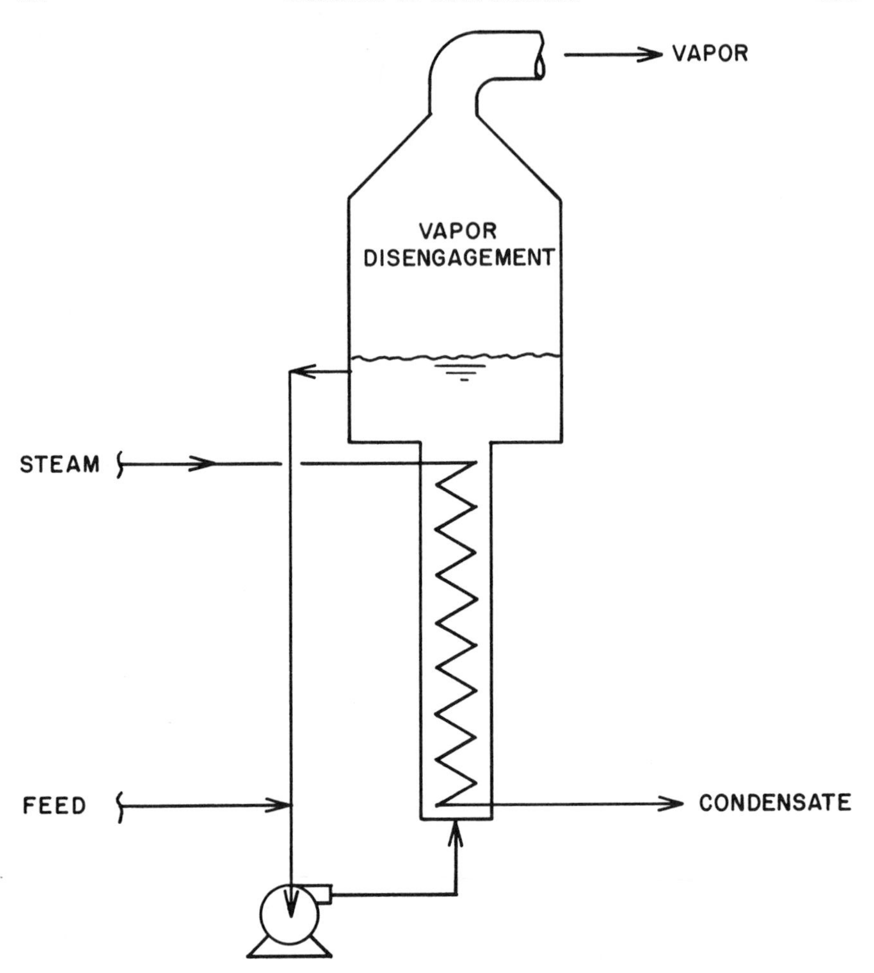

Fig. 4.2e Forced circulation evaporator

results. Ordinarily, the feed rate is controlled by evaporator level to keep the tubes full. The disadvantage of high residence time in the evaporator is compensated for by the low cost of the unit for a given evaporator load.

Long-tube vertical evaporator or rising film concentrator (RFC) shown in Figure 4.2g is in common use today, because its cost per unit of capacity is low. Typical applications include concentrating black liquor in the pulp and paper industry and corn syrup in the food industry. Most of these evaporators are of the single pass variety, with little or no internal recirculation. Thus, residence time is minimized. Level control is important in maintaining the liquid seal in the flash tank. The units are sensitive to changes in operating conditions, which is why many of them are difficult to control. They offer

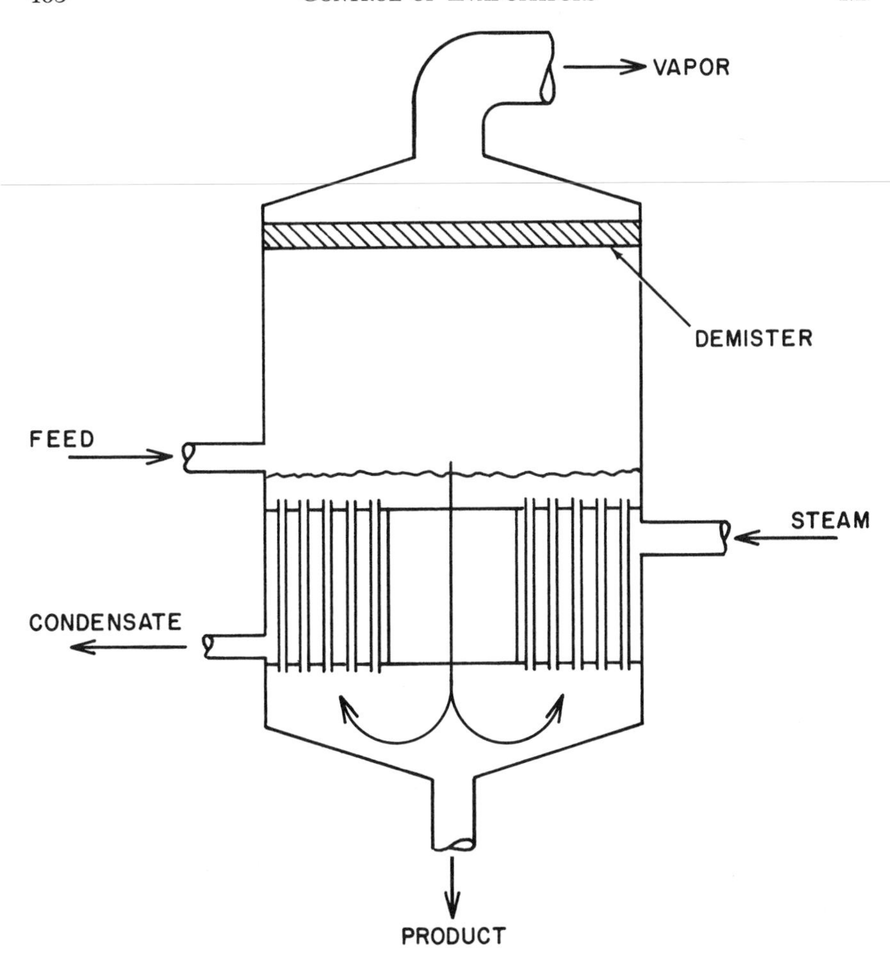

Fig. 4.2f Short-tube vertical evaporator

low cost per pound of water evaporated and have low holdup times but tend to be tall (20 to 50 feet), requiring more head room than other types.

Falling film evaporator (Figure 4.2h) is commonly used with heat sensitive materials. Physically, the evaporator looks like a long-tube vertical evaporator except that the feed material descends by gravity along the inside of the heated tubes, which have large inside diameters (2 to 10 inches).

Agitated film evaporator, like the falling film evaporator, is commonly used for heat sensitive and highly viscous materials. It consists of a single large diameter tube with the material to be concentrated falling in a film down the inside where a mechanical wiper spreads the film over the inside surface of the tube. Thus, large heat transfer coefficients can be obtained, particularly with highly viscous materials.

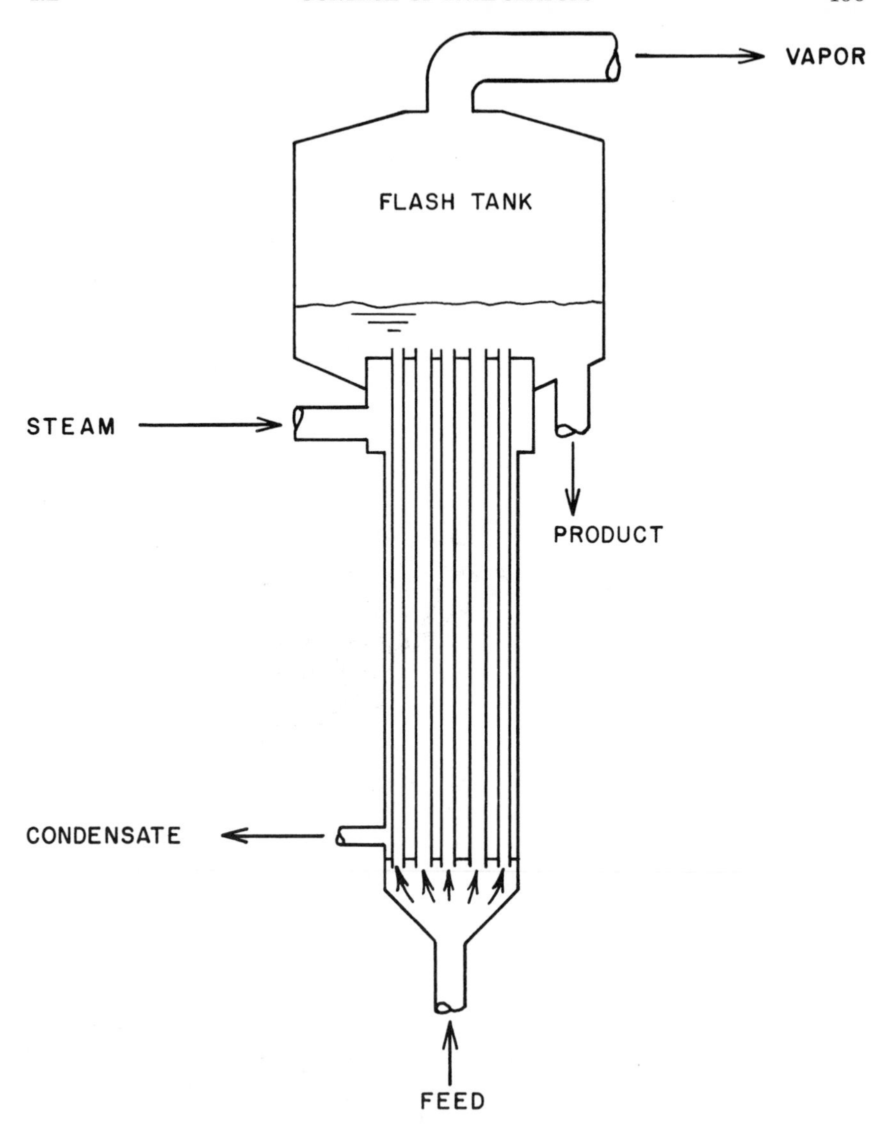

Fig. 4.2g Long-tube vertical evaporator

Control Systems for Evaporators

The control systems to be considered in achieving final product density include 1.) feedback, 2.) cascade and 3.) feedforward. For ease of illustration, a double effect, co-current flow evaporator will be used. Extension to more or less effects will not change the control system configuration.

The choice of system should be based on the needs and characteristics

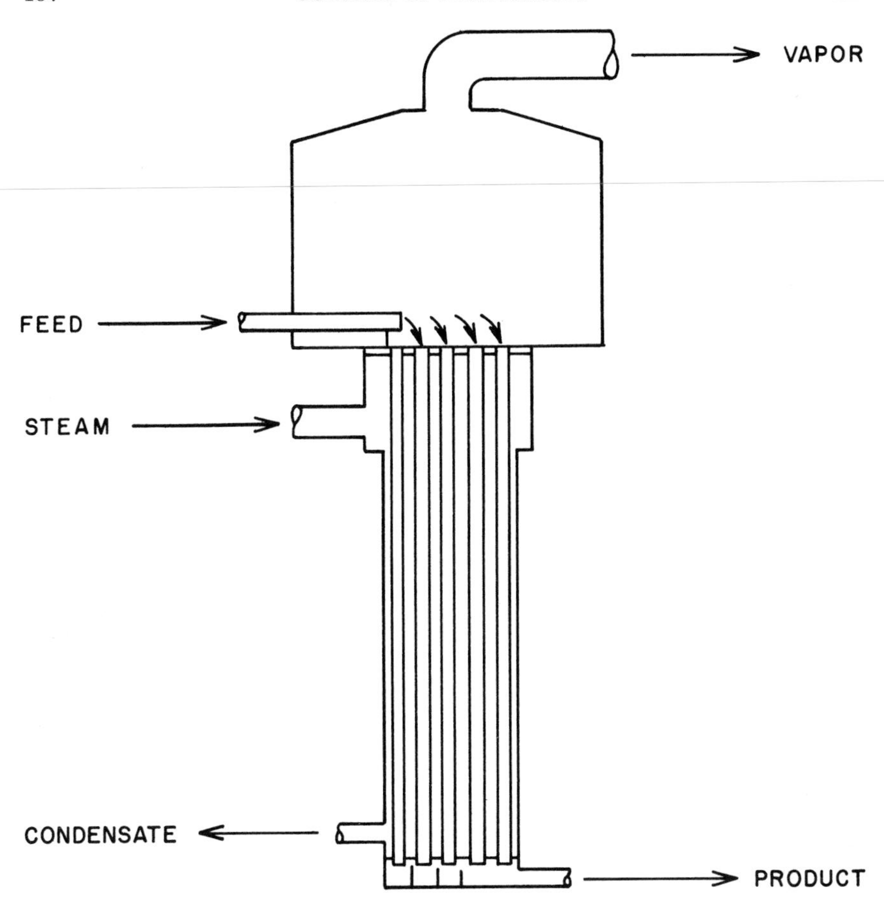

Fig. 4.2h Falling film vertical evaporator

of the process. Evaporators as a process class tend to be capacious (mass and energy storage capability) and have significant dead time (30 seconds or greater). If the major process loads (feed rate and feed density) are reasonably constant and the only corrections required are for variations in heat losses or tube fouling, feedback control will suffice. If steam flow varies because of demands elsewhere in the plant, a cascade configuration will probably be the proper choice. If, however, the major load variables change rapidly and frequently, it is strongly suggested that feedforward in conjunction with feedback be considered.

Feedback Control

A typical feedback control system (Figure 4.2i), consists of measuring the product density with a density sensor and controlling the amount of steam

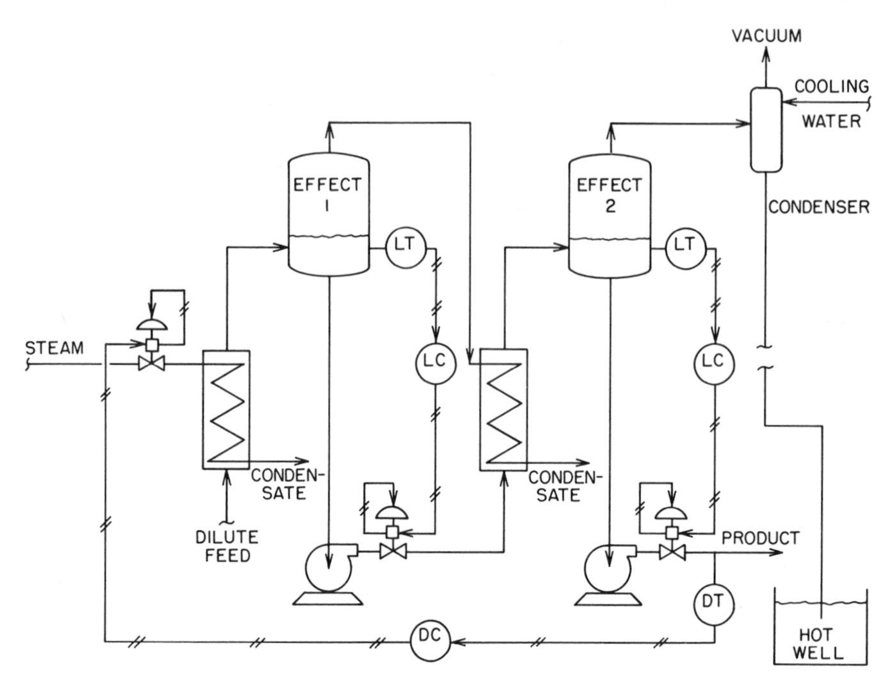

Fig. 4.2i Feedback control system

to the first effect by a 3-mode controller. The internal material balance is maintained by level control on each effect. (A brief description of the various methods of measuring product density will be found at the end of this section; additional discussion regarding density measurement may be found in Chapter 3 in Volume I.) Ritter et al. (1) have modeled the control system configuration shown in Figure 4.2i and have found it to be very stable. They investigated other combinations of controlled and manipulated variables, which however provided poor evaporation.

Cascade Control

A typical system is illustrated in Figure 4.2j. This control system, like the feedback loop in Figure 4.2i, measures the product density and adjusts the heat input. The adjustment in this instance, however, is through a flow loop which is being set in cascade from the final density controller, an arrangement that is particularly effective when steam flow variations (outside of the evaporator) are frequent. It should be noted that with this arrangement the valve positioner is not required. A valve positioner will actually degrade the performance of the flow control loop. (For more information see Section 7.8 in Volume II.)

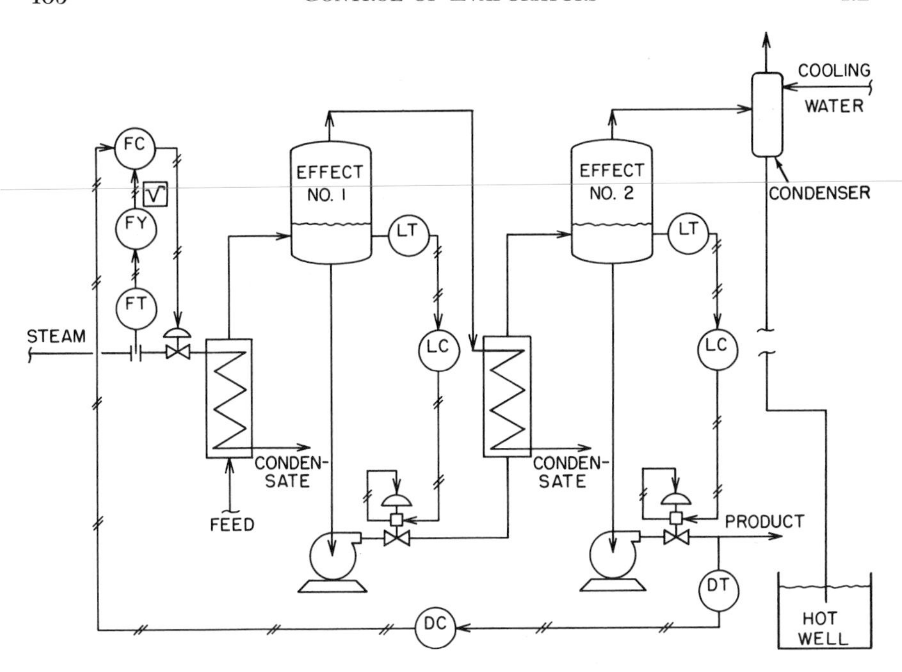

Fig. 4.2j Cascade control system

Feedforward Control

In most evaporator applications the control of product density is constantly affected by variations in feed rate and feed density to the evaporator. In order to counter these load variations, the manipulated variable (steam flow) must attain a new operating level. In the pure feedback or cascade arrangements this new level was achieved by trial-and-error as performed by the feedback (final density) controller.

A control system able to react to these load variations when they occur (feed rate and feed density) rather than wait for them to pass through the process before initiating a corrective action would be ideal. The technique is termed *feedforward control.* (For more on this technique see Section 7.6 in Volume II.) Figure 4.2k illustrates in block diagram form the features of a feedforward system. There are two types or classes of load variations—measured and unmeasured. The measured load signals are inputs to the feedforward control system where they compute the set point of the manipulated variable control loop as a function of the measured load variables. The unmeasured load variables pass through the process, undetected by the feedforward system, and cause an upset in the controlled variable. The output of the feedback loop then trims the calculated value of set point to the correct

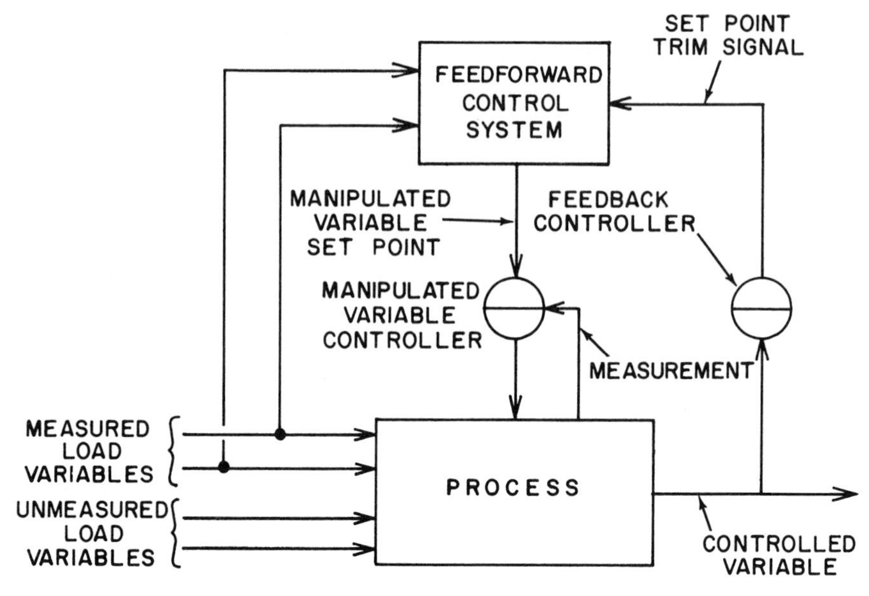

Fig. 4.2k Feedforward system

operating level. In the limit, feedforward control would be capable of perfect control if all load variables could be defined, measured and incorporated in the forward set point computation. Practically speaking, the expense of accomplishing this goal is usually not justified.

At a practical level then, the load variables are classified as either major or minor, and the effort is directed at developing a relationship which incorporates the major load variables, the manipulated variables and the controlled variable. Such a relationship is termed the steady state model of the process. Minor load variables are usually very slow to materialize and are hard to measure. In terms of evaporators, minor load variations might be heat losses and tube fouling. Load variables such as these are easily handled by a feedback loop. The purpose of the feedback loop is to trim the forward calculation to compensate for the minor or unmeasured load variations. Without this feature the controlled variable could be off the set point. Thus far we have discussed two of the three ingredients of a feedforward control system, viz., the steady state model and feedback trim. The third ingredient of a feedforward system is dynamic compensation. A change in one of the major loads to the process also modifies the operating level of the manipulated variable. If these two inputs to the process enter at different locations, there usually exists an imbalance or inequality between the effect of the load variable and the effect of the manipulated variable on the controlled variable, i.e.,

$$\frac{\Delta\text{controlled variable}}{\Delta\text{load variable}} \neq \frac{\Delta\text{controlled variable}}{\Delta\text{manipulated variable}}$$

This imbalance manifests itself as a *transient* excursion of the controlled variable from set point. If the forward calculation is accurate, the controlled variable returns to set point once the new steady state operating level is reached. In terms of a co-current flow evaporator, an increase in feed rate will call for an increase in steam flow. Assuming that the level controls on each effect are properly tuned, the increased feed rate will rapidly appear at the other end of the train while the increased steam flow is still overcoming the thermal inertia of the process. This sequence results in a transient decrease of the controlled variable (density), and the load variable passes through the process faster than the manipulated variable. This behavior is shown in Figure 4.2 l. The same arrangement is seen in Figure 4.2m except that the manipulated variable passes through the process faster than the load variable. Such behavior may occur in a countercurrent evaporator operation. This dynamic imbalance is normally corrected by inserting a dynamic element (lag, lead-lag or a combination thereof) in at least one of the load measurements to the feedforward control system. Usually, dynamic compensation of that major load variable, which can change in the severest manner (usually a step change), is all that is required. For evaporators this is usually the feed flow rate to the evaporator. Feed density changes, although frequent, are usually more gradual and the inclusion of a dynamic element for this variable is not war-

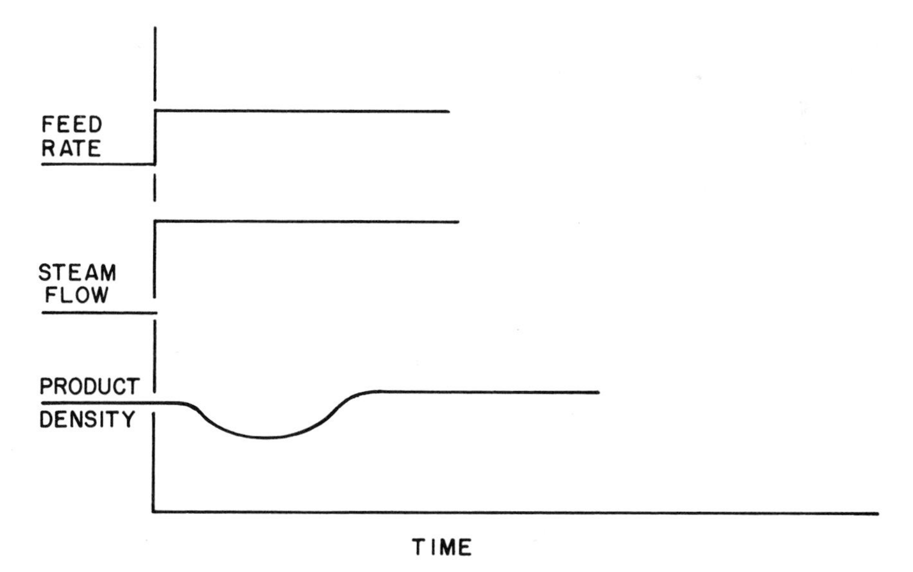

Fig. 4.2 l Load variable faster than manipulated variable

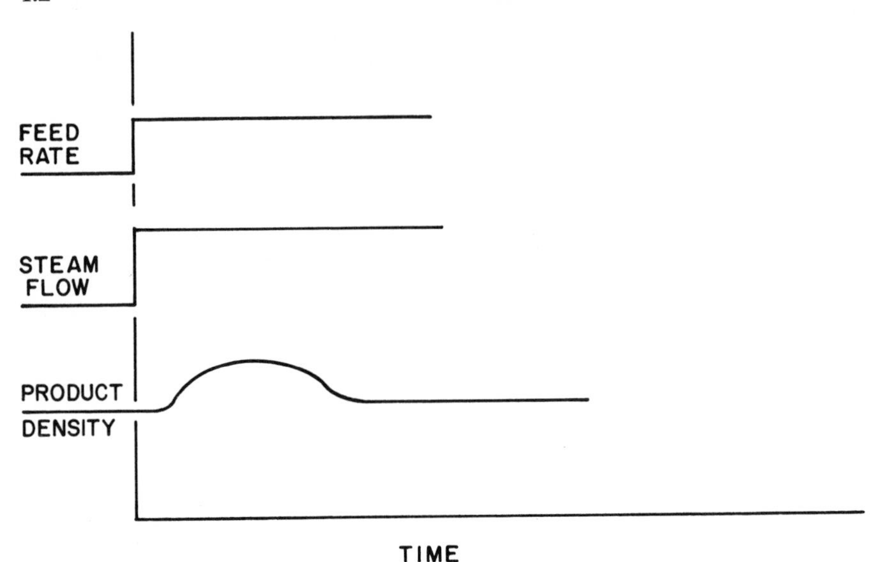

Fig. 4.2m Manipulated variable faster than load variable

ranted. In summary, the three ingredients of a feedforward system are 1.) the steady state model, 2.) process dynamics and 3.) addition of feedback trim.

STEADY STATE MODEL

Development of the steady state model for an evaporator involves material and energy balance. A relationship between the feed density and percent solids is also required and is specific for a given process, whereas the material and energy balances are applicable to all evaporator processes. Figure 4.2n

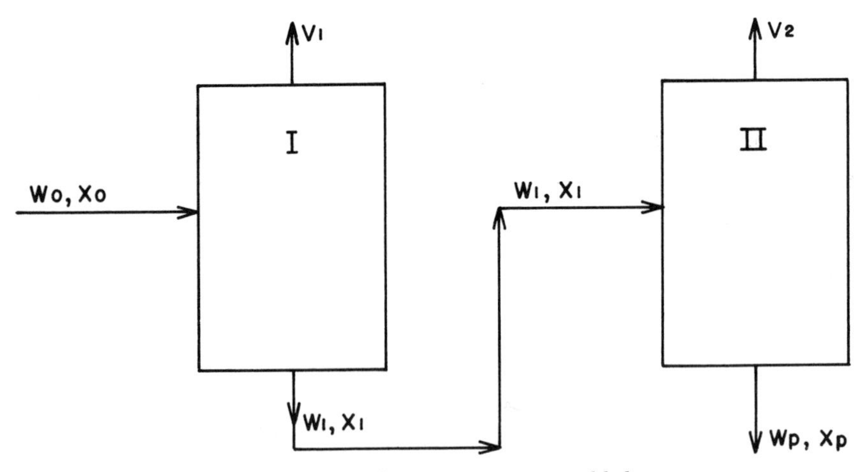

Fig. 4.2n Double-effect evaporator material balance

illustrates the double-effect evaporator from the standpoint of a material balance.

where W_o = feed rate in lbs per unit time
V_1 = vapor flow from Effect I in lbs per unit time
X_o = weight fraction solids in feed
W_1 = liquid flow rate leaving Effect I in lbs per unit time
X_1 = weight fraction solids in W_1
V_2 = vapor flow from Effect II in lbs per unit time
W_p = liquid product flow in lbs per unit time
X_p = weight fraction solids in product (the controlled variable)

Over-all balance in Effect I:

$$W_o = V_1 + W_1 \tag{4.2(1)}$$

Over-all balance in Effect II:

$$W_1 = V_2 + W_p \tag{4.2(2)}$$

Solid balance in Effect I:

$$W_oX_o = W_1X_1 \tag{4.2(3)}$$

Solid balance in Effect II:

$$W_1X_1 = W_pX_p \tag{4.2(4)}$$

Substituting 4.2(2) in 4.2(1):

$$W_o = V_1 + V_2 + W_p \tag{4.2(5)}$$

Combining 4.2(3) and 4.2(4):

$$W_oX_o = W_pX_p \tag{4.2(6)}$$

Solving 4.2(6) for W_p and substituting in 4.2(5) gives

$$W_o = V_1 + V_2 + \frac{W_oX_o}{X_p} \tag{4.2(7)}$$

The W_o term of 4.2(7) can be written in terms of volumetric flow (gallons per unit time), the usual method of measuring this variable:

$$W_o = V_oD_o = V_oD_WS_o \tag{4.2(8)}$$

where V_o = volumetric feed rate in gallons per unit time
D_o = feed density in lbs per gallon
D_W = nominal density of water, 8.33 lbs per gallon
S_o = specific gravity of feed

Substituting for W_o from 4.2(8) in 4.2(7) and combining terms,

$$V_oD_WS_o\left(1-\frac{X_o}{X_p}\right)=V_1+V_2=V_t \qquad 4.2(9)$$

where V_t = total vapor flow in lbs per unit time.

The total vapor flow (V_t) is proportional to the energy supplied to the train (plant steam) and the proportionality constant is the economy (E) of the system, i.e.,

$$V_t = W_sE \qquad 4.2(10)$$

where W_s = steam flow in lbs steam per unit time
E = economy in lbs vapor per lb steam

Substituting for V_t in 4.2(9)

$$V_oD_WS_o\left(1-\frac{X_o}{X_p}\right)=W_sE \qquad 4.2(11)$$

Equation 4.2(11) is the steady state model of the process and includes all of the load variables (V_o and X_o), the manipulated variable (W_s) and the controlled variable X_p. At this point the W_sE portion of equation 4.2(11) may be modified to include heat losses from the system and to include the fact that the feed may be subcooled; this is effectively a heat loss since in either case a portion of the steam supplied is for purposes other than producing vapor. Typical values of effective heat losses vary from 3 to 5%. If, for example, a 5% heat loss is assumed, equation 4.2(11) becomes

$$V_oD_WS_o\left(1-\frac{X_o}{X_p}\right)=0.95\ W_sE \qquad 4.2(12)$$

For an in-depth discussion relating to methods of computing the economy (E) of a particular evaporator system, see references (2) and (3).

The S_o $(1 - X_o/X_p)$ portion of equation 4.2(11) is a function of the feed density, $f(D_o)$, i.e.,

$$f(D_o) = S_o\left(1-\frac{X_o}{X_p}\right) \qquad 4.2(13)$$

For each feed material a relationship between the density of the feed material and its solids weight fraction has usually been empirically determined by the plant or is available in the literature. See reference (4) where the density = percent solids relationship of 70 inorganic compound is available.

Assume, for example, that a feed material (to be concentrated) has the solids-specific gravity relationship shown in Table 4.2n.

If this feed material was to be concentrated so as to produce a product having

Table 4.2n
WEIGHT FRACTION AND
SPECIFIC GRAVITY RELATIONSHIP

Solids Weight Fraction (X_o)	*Specific Gravity* (S_o)
0.08	1.0297
0.16	1.0633
0.24	1.0982

a weight fraction of 50% ($X_p = 0.50$), the $f(D_o)$ relationship of 4.2(13) can be generated as shown in Table 4.2o.

Table 4.2o
DENSITY TO WEIGHT FRACTION
RELATIONSHIP

Solids Weight Fraction (X_o)	*Specific Gravity* (S_o)	$\left(1 - \frac{X_o}{X_p}\right)$	$S_o\left(1 - \frac{X_o}{X_p}\right) = f(D_o)$
0	1.0000	1.0	1.0000
0.08	1.0297	0.840	0.865
0.16	1.0633	0.680	0.723
0.24	1.0982	0.520	0.571

This body of data is plotted in Figure 4.2p. In all the cases investigated, the $f(D_o) = S_o$ relationship is a straight line having an intercept of 1.0, 1.0. The $f(D_o)$ relationship can then be written in terms of the equation of a straight line: $y = mx + b$, i.e.,

$$f(D_o) = 1.0 + m\,(S_o - 1.0) \qquad 4.2(14)$$

where m = slope of line

Using the data of Table 4.2o the value of m is determined as

$$m = \frac{1.0000 - 0.571}{1.0000 - 1.0982} = -4.37 \qquad 4.2(15)$$

Therefore,

$$f(D_o) = 1.0 - 4.37\,(S_o - 1.0) \qquad 4.2(16)$$

Substituting 4.2(16) into 4.2(11) and solving for the manipulated variable, steam flow, W_s:

$$W_s = \frac{V_o D_w f(D)}{E} \qquad 4.2(17)$$

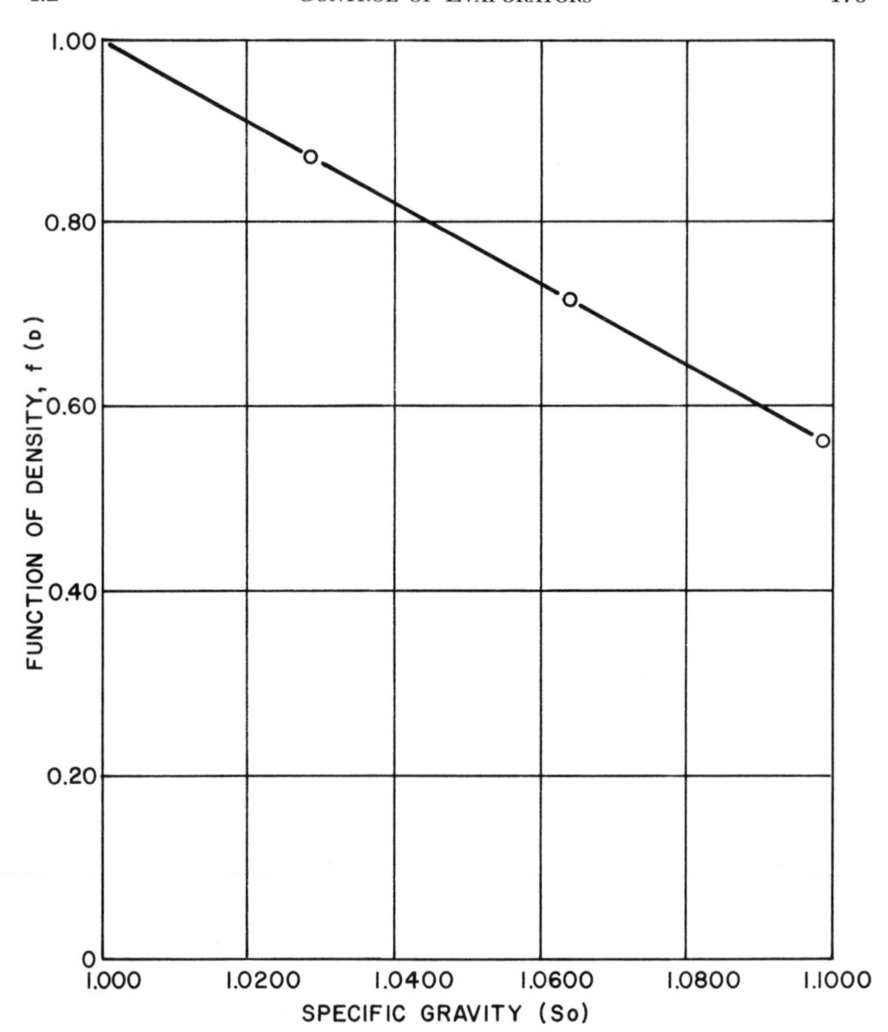

Fig. 4.2p Density to specific gravity relationship

SCALING AND NORMALIZING

With the equation for the steady state model defined, it can now be scaled and the analog instrumentation specified. (For more information see Section 9.2 in Volume II.) Scaling of analog computing instruments is necessary to insure compatibility with input and output signals and is accomplished most effectively by normalizing, i.e., assigning values from 0 to 1.0 to all inputs and outputs. The procedure involves 1.) writing the engineering equation to be solved, 2.) writing a normalized equation for each term variable of the engineering equation and 3.) substituting the normalized equivalent of each term in 2.) into 1.).

The first step of the procedure has already been done—equation 4.2(17) is written. To illustrate, let $V_o = 0$ to 600 GPH, $W_s = 0$ to 2,500 lbs per hr, $S_o = 1.000$ to 1.1000, E = 1.8 lbs vapor per lb of steam and $D_w = 8.33$ lbs per gallon. The scaled equations for each input are

$$V_o = 600\ V'_o \qquad 4.2(18)$$

$$W_s = 2500\ W'_s \qquad 4.2(19)$$

$$S_o = 1.0000 + 0.10000\ S'_o \qquad 4.2(20)$$

where V'_o = volumetric flow transmitter output 0 to 1.0 or 0 to 100%
W'_s = steam flow transmitter output 0 to 1.0 or 0 to 100%
S'_o = specific gravity transmitter output 0 to 1.0 or 0 to 100%

The values of D_w and E need not be scaled since they are constants. Since $f(D_o)$ is already on a 0 to 1.0 basis, the $f(D_o)$ term for the sake of completeness can be written

$$f(D_o) = 1.0\ f(D_o)' \qquad 4.2(21)$$

Operating on the $f(D_o)$ equation first, equations 4.2(21) and 4.2(20) are substituted into equation 4.2(16)

$$(D_o)' = 1.0 - 4.37\ (1.0000 + 0.1000\ S'_o - 1.0) \qquad 4.2(22)$$

$$'(D_o) = 1.0 - 0.437\ S'_o \qquad 4.2(22)$$

Substituting 4.2(18), 4.2(19) and 4.2(22) into 4.2(17) as well as the values of E and D_w

$$2500\ W'_s = \frac{600(V'_o)\ 8.33\ (1.0 - 0.437\ S'_o)}{1.8} \qquad 4.2(23)$$

$$W'_s = 1.11\ V'_o\ (1.0 - 0.437\ S'_o) = 1.11\ V'_o\ f(D_o)' \qquad 4.2(23)$$

PROCESS DYNAMICS

The dynamics of the co-current evaporator, in which steam is the manipulated variable, requires that a lead-lag dynamic element be incorporated in the system to compensate for the dynamic imbalance between feed rate and steam flow. In the example, it was arbitrarily assumed that steam flow is the manipulated variable resulting in equation 4.2(17). In some applications evaporators are run on waste steam, in which case the feed rate is proportionally adjusted to the available steam, which makes feed the manipulated variable and steam the load variable. Solving 4.2(17) for feed rate

$$V_o = \frac{W_s E}{D_w(D)} \qquad 4.2(24)$$

In this arrangement the dynamics do not change, but the manipulated variable advances through the process faster than the load variable, which requires a dynamic element having first order lag characteristics. The instrument arrangement for each case is shown in Figures 4.2q and 4.2r.

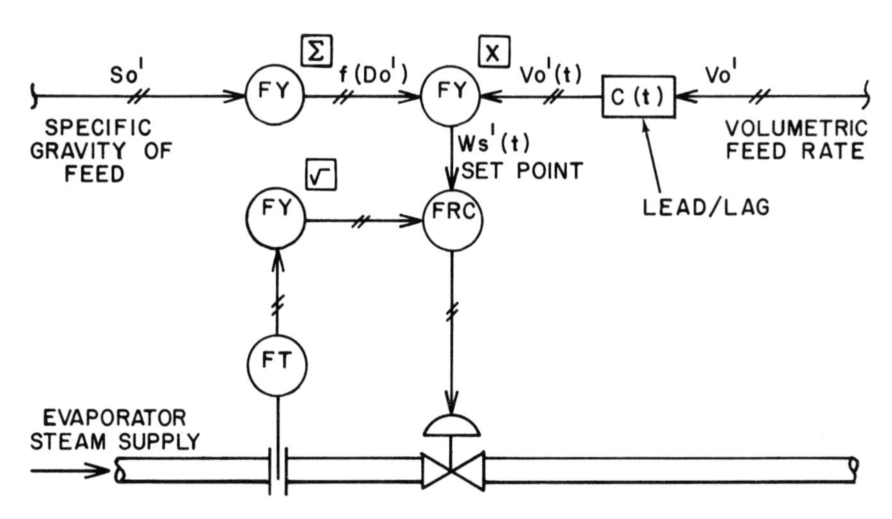

Fig. 4.2q Feedforward with dynamics (Equation 4.2[23])

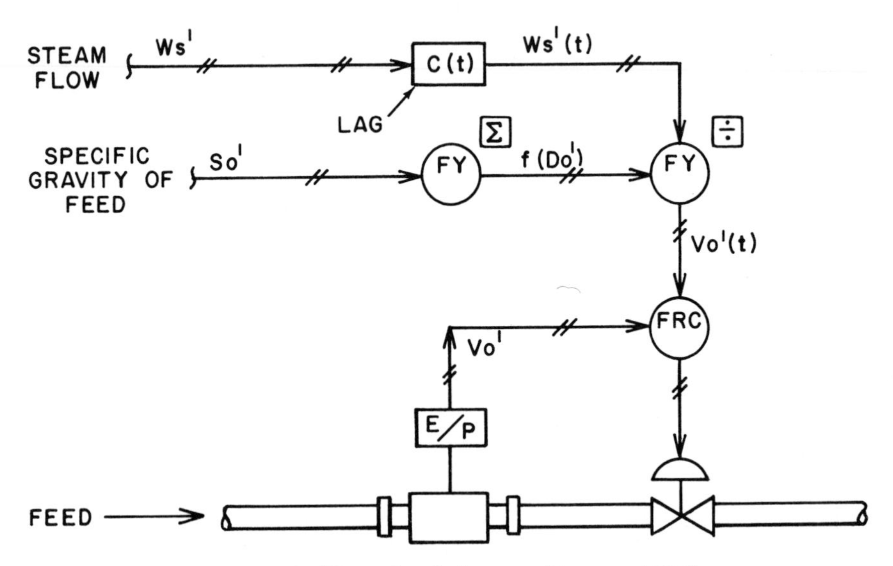

Fig. 4.2r Feedforward with dynamics (Equation 4.2[24])

FEEDBACK TRIM

The final consideration is feedback trim. As a general rule feedback trim is incorporated into the control system *at the point at which the set point of the controlled variable appears*. For the evaporator the set point is the slope of the $f(D_o)$ relationship (Figure 4.2p). If the value of X_p changes, the slope of the line changes too. To this point the value of m was assumed to be a constant (0.437) which is incorporated into the summing relay or amplifier (Figures 4.2q and 4.2r). The instrumentation was scaled to make one grade of product, i.e., 50% solids for the example. If a more or less concentrated product were desired, the gain term would have to be changed manually. In order to increase the flexibility of the control system, a multiplier and a final product density control loop are added. The controller output is now variable not only to permit changing the concentration of the product (slope adjust) but also to adjust the steam flow set point to compensate for the minor load variations which up to this point were not considered. Each arrangement (comparable to Figures 4.2q and 4.2r) is shown in Figures 4.2s and 4.2t.

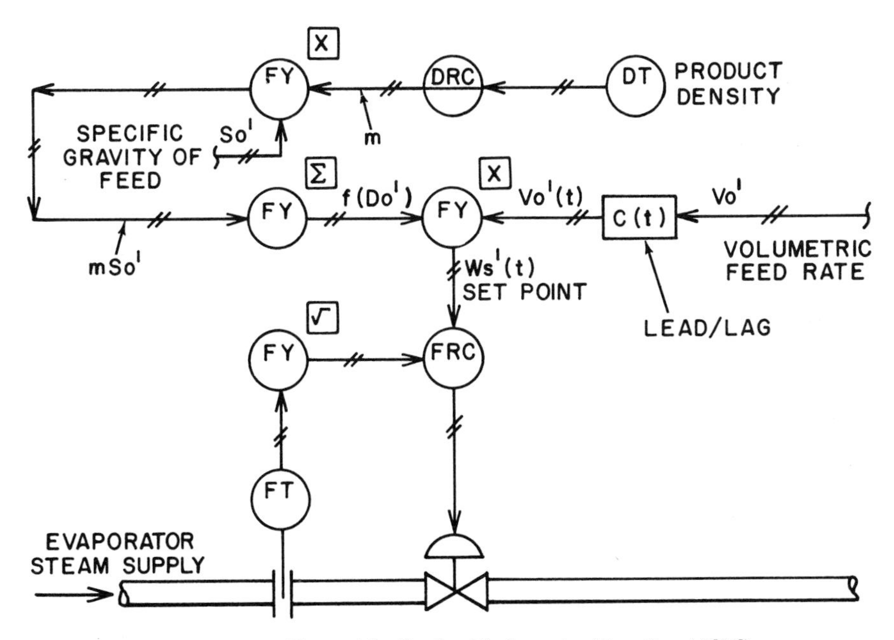

Fig. 4.2s Feedforward-feedback with dynamics (Equation 4.2[23])

Other Control Loops

STEAM ENTHALPY

So far we have only discussed load changes related to feed density and feed flow rate and have specified the instruments to accomplish the control

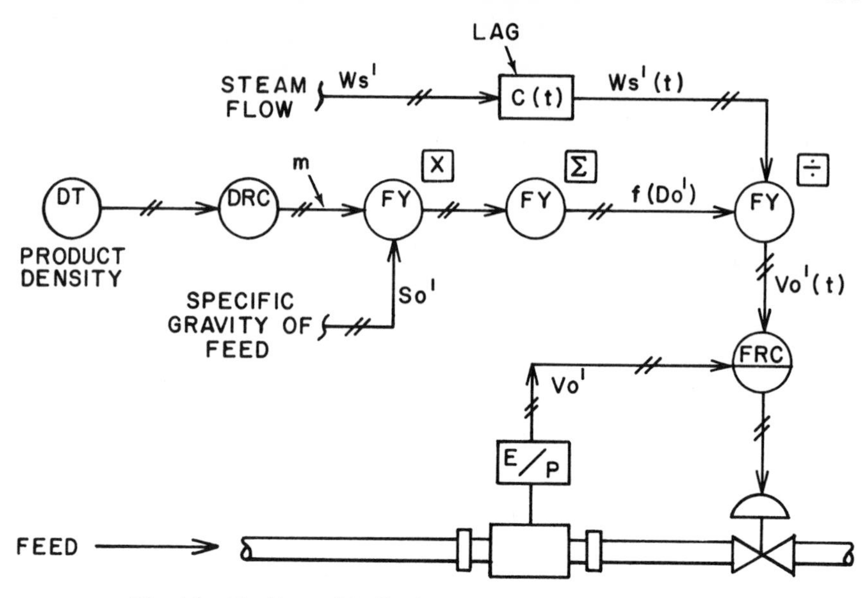

Fig. 4.2t Feedforward-feedback with dynamics (Equation 4.2[24])

objective. One additional load variable which if allowed to pass through the process would upset the controlled variable, i.e., product density, is steam enthalpy. In some applications the steam supply may be carefully controlled so that its energy content is uniform; in other applications substantial variations in steam enthalpy may be experienced. The objective is to consider the factors that influence the energy level of the steam supplied and to design a system that will protect the process from these load variables.

For saturated steam the energy content per unit weight is a function of the absolute pressure of the steam. If the flow of steam to the process is measured with an orifice meter, the mass flow of steam is

$$W_s = K_1 \sqrt{\frac{h}{v}} \qquad 4.2(25)$$

where W_s = steam flow in lbs per hour
h = differential head measurement in feet
v = specific volume in cu ft per lbm
K_1 = orifice coefficient dependent on the physical characteristics of the orifice.

The total energy to the system is

$$Q = W_s H_s \qquad 4.2(26)$$

where Q = energy to system per BTU per hour
W_s = steam flow in lbs per hour
H_s = heat of condensation in BTU per lb (enthalpy of saturated vapor, minus enthalpy of saturated liquid.)

Substituting Equation 4.2(25) into Equation 4.2(26) gives

$$Q = K_1 \sqrt{\frac{(H_s)^2}{v} h} \qquad 4.2(27)$$

For any particular application the steam pressure will vary within limits around a normal operating pressure. To demonstrate the design of a control system to compensate for variations in energy input to the process, assume that the steam pressure varies between 18 and 22 PSIA with a normal operating value of 19.7 PSIA. The pressure transmitter has a range of 0 to 25 PSIA. These values of pressure variation and operating pressure are typical for evaporator operation. The value of the $(H_s)^2/v$ term appearing in equation 4.2(27) will vary, depending on the steam pressure. Over a reasonably narrow range of pressures, the value of $(H_s)^2/v$ can be approximated by a straight line with the general form of

$$(H_s)^2/v = bP + a \qquad 4.2(28)$$

where P = absolute pressure in PSIA
a and b = constants

Table 4.2u shows the typical values of H_s, v, $((H_s)^2/v)'$ and p and p' selected from the specified range of pressures and from where the steam is condensed at 1.5 PSIG.

Table 4.2u
SPECIFIC VOLUME-ENTHALPY DATA

(p) *Steam Supply Pressure (PSIA)*	(H_s) *Heat of Condensation (BTU/lbm)*	(v) *Specific Volume* (ft^3/lbm)	$((H_s)^2/v)' = \frac{(H_s)^2/v}{((H_s)^2/v)_{normal}}$	$p' = \frac{p}{p_{max}} = \frac{p}{25}$
18	969.1	22.2	0.916	0.720
19	970.2	22.1	0.965	0.760
19.7*	970.7	20.4	1.00	0.788
20	971.2	20.1	1.02	0.800
21	972.1	19.2	1.07	0.840
22	973.0	18.4	1.11	0.880

* Normal operation

$$(H_s^2/v) \text{ normal} = \frac{(970.7)^2}{20.4} = 46{,}208$$

Rewriting equation 4.2(28) in scaled form

$$\left(\frac{(H_s)^2}{v}\right)' = bp' + a \qquad 4.2(29)$$

The designer can either linearize the data—using any two points from Table 4.2u—or utilize a least squares computation to find the best straight line.

In a least square computation the following values of a and b constants of equation 4.2(29) were obtained

$$\left(\frac{(H_s)^2}{v}\right)' = 0.085 + 1.161\ p' \qquad 4.2(30)$$

where $a = 0.085$
$b = 1.161$

Squaring equation 4.2(27) and substituting in equation 4.2(30)

$$(Q^2)' = (1.161\ p' + 0.085)\ hk_1 \qquad 4.2(31)$$

The parenthetical portion of equation 4.2(31) can be rewritten so as to make the sum of the two coefficients equal 1.0, which simplifies its implementation using conventional analog hardware. This is done by multiplying and dividing each term in the parenthesis by the sum of the two coefficients.

$$(Q^2)' = 1.246\left(\frac{1.161\ p'}{1.246} + \frac{0.085}{1.246}\right) hk_1 \qquad 4.2(32)$$

$$(Q^2)' = 1.246\ (0.932\ p' + 0.068)\ hk_1 \qquad 4.2(32)$$

The instrumentation to implement equation 4.2(32) is shown in Figure 4.2v.

INTERNAL MATERIAL BALANCE

The feedforward system described earlier imposes an external material balance as well as an internal material balance on the process. The internal balance is maintained by liquid level control on the discharge of each effect.

Analysis of a level loop indicates that a narrow proportional band ($<10\%$) can achieve stable control. However, because of the resonant nature of the level loop which causes the process to oscillate at its natural frequency, a much lower controller gain must be used (proportional bands 50 to 100%).[5] With the wider proportional bands, reset is required to help maintain set point. A valve positioner is also recommended to overcome the usual limit cycle characteristics of an integrating process and the non-linear nature of valve hysteresis. (See Section 7.8 in Volume II for more information.)

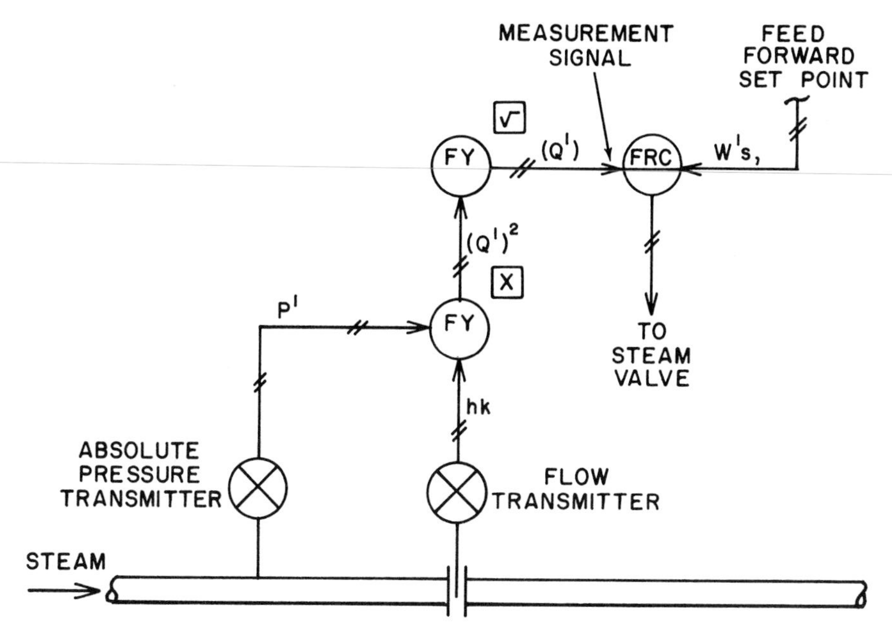

Fig. 4.2v Enthalpy compensation system

ABSOLUTE PRESSURE

The heat to evaporate water from the feed material is directly related to the boiling pressure of the material. In most multiple effect evaporations each effect is held at a pressure less than atmospheric in order to keep boiling points below 212°F. The lowest pressure is in the effect closest to the condenser, with pressures increasing slightly in each effect away from the condenser. Three possible methods of controlling the absolute pressure include 1.) controlling the flow of water to the condenser, 2.) bleeding air into the system with the water valve wide open and 3.) locating the control valve in the vapor drawoff line, manually setting the water flow rate and air bleeding as necessary.

Method 3 requires an extraordinarily large valve, since the vapor line may be 24 or 30 inches in diameter. Method 2 is uneconomical because the expense of pumping the water offsets the savings realized by using a smaller valve on the air line. Method 1 represents the best compromise between cost and controllability and is preferred (Figure 4.2w). (For more on condenser pressure control see Section 10.11 in Volume II.)

Product Density Measurement

Perhaps one of the most controversial issues in any evaporator control scheme is the method to measure the product density. Common methods

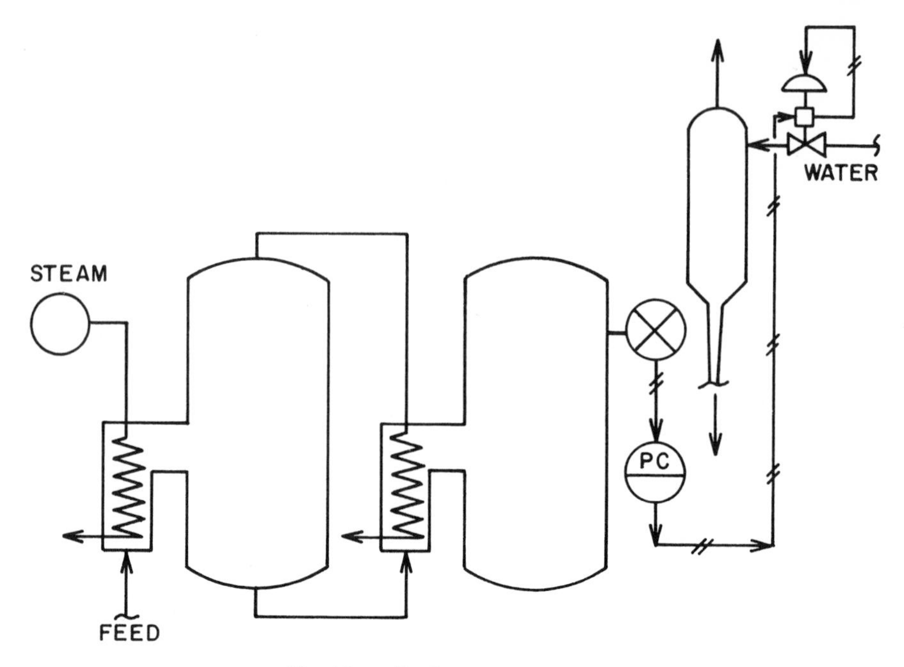

Fig. 4.2w Absolute pressure control

include 1.) temperature difference, boiling point rise (Section 1.9 in this volume); 2.) conductivity (Section 8.15 in Volume I); 3.) differential pressure (Section 3.3 in Volume I); 4.) gamma gauge (Section 3.4 in Volume I); 5.) U-tube densitometer (Section 3.5 in Volume I); 6.) buoyancy float (Section 3.2 in Volume I); and 7.) refractive index (Section 8.6 in Volume I).

Each method has its strengths and weaknesses. In all cases, however, care must be taken to select a representative measurement location to eliminate entrained air bubbles or excessive vibration, and the instrument must be mounted in an accessible location for cleaning and calibration. The location of the product density transmitter with respect to the final effect should also be considered. Long runs of process piping for transporting the product from the last effect to the density transmitter increase deadtime, which in turn reduces the effectiveness of the control loop.

BOILING POINT RISE

Perhaps the most difficult and controversial method of product density measurement is by temperature difference or boiling point rise. Dühring's rule states that a linear relationship exists between the boiling point of a solution and the boiling point of pure water at the same pressure. Thus, the temperature difference between the boiling point of the solution in an evaporator and the boiling point of water at the same pressure is a direct measurement of the concentration of the solution. Two problems in making this measurement are location of the temperature bulbs and control of absolute pressure.

The temperature bulbs must be located so that the measured values are truly representative of the actual conditions. Ideally, the bulb measuring liquor temperature should be just at the surface of the boiling liquid. This location can change, unfortunately, if the operator decides to use more or less liquor in a particular effect. Many operators install the liquor bulb near the bottom of the pan where it will always be covered, thus creating an *error due to head effects* which must be compensated for in the calibration.

The vapor temperature bulb is installed in a condensing chamber in the vapor line. Hot condensate flashes over the bulb at an equilibrium temperature dictated by the pressure in the system. This temperature minus the liquid boiling temperature (compensated for head effects) is the temperature difference reflecting product concentration.

Changes in absolute pressure of the system alter not only the boiling point of the liquor, but the flashing temperature of the condensate in the condensing chamber as well. Unfortunately, the latter effect occurs much more rapidly than the former, resulting in transient errors in the system which may take a long time to resolve. Therefore, it is imperative that absolute pressure be controlled closely if temperature difference is to be a successful measure of product density. These systems were more effective installations where control of water rate to the condenser rather than an air-bleed system was employed.

CONDUCTIVITY

Electrolytic conductivity is a convenient measurement to use in relationships between specific conductance and product quality (concentration), such as in a caustic evaporator. Problem areas include location of the conductivity cell so that product is not stagnant but is flowing past the electrodes; temperature limitations on the cell; cell plugging; and temperature compensation for variations in product temperature.

DIFFERENTIAL PRESSURE

Measuring density by differential pressure is a frequently used technique. The flanged differential pressure transmitter is preferred for direct connection to the process; otherwise lead lines to the transmitter could become plugged by process material solidifying in the lines. Differential pressure transmitters are more frequently used on feed density than on final density measurements.

GAMMA GAUGE

This measurement is popular in the food industry because the measuring and sensing elements are not in contact with the process. It is very sensitive and not subject to plugging. Periodic calibration may be required due to the half-life of source materials. Occasionally, air is entrained especially in ex-

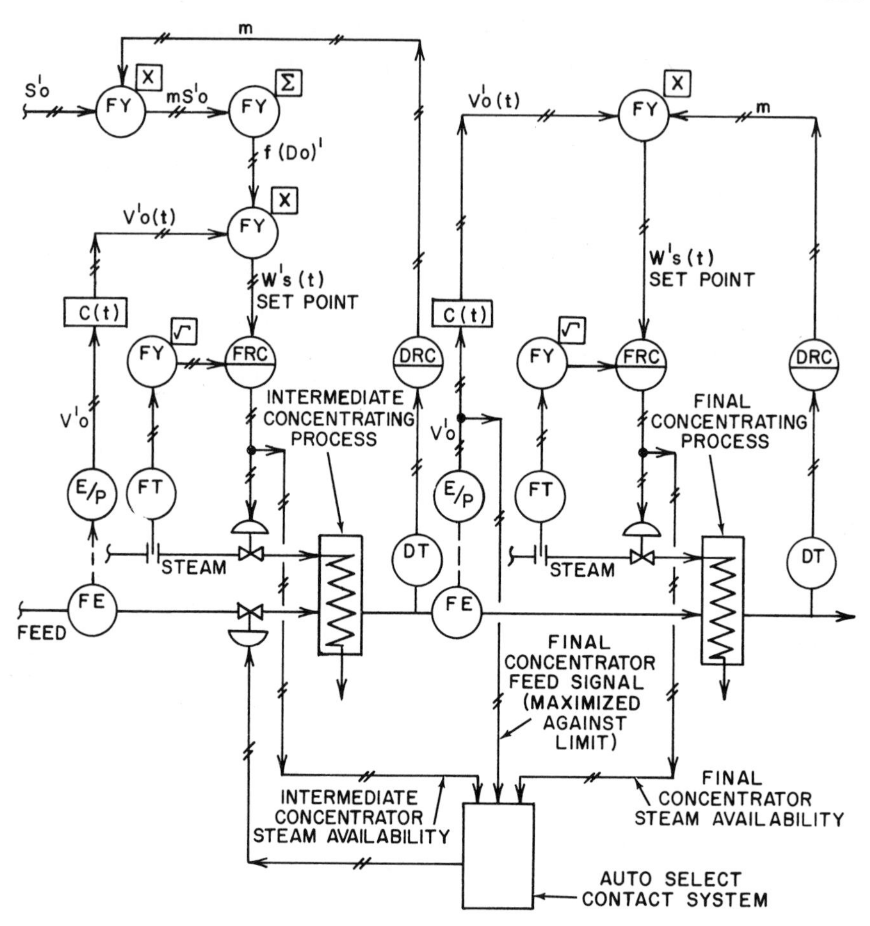

Fig. 4.2x Evaporator auto-select system

tremely viscous solutions. Therefore, the best sensor location is a flooded low point in the process piping.

U-TUBE DENSITOMETER

The U-tube densitometer, a beam-balance device, is also a final product density sensor. Solids can settle out in the measuring tube, causing calibration shifts or plugging.

BUOYANCY FLOAT

Primarily used for feed density detection, the buoyancy float can also be applied to product density if a suitable mounting location near the evaporator can be found. Because flow will affect the measurement, the float must be located where the fluid is almost stagnant or where flow can be controlled (by recycle) and its effects zeroed out. A teflon-coated float helps reduce drag effects.

Auto-select Control System

In many processes the final product is the result of a two-step operation. The first step produces an intermediate product which serves as the feed to a final concentrator. The aim is to insure that the process is run at the maximum throughput consistent with the process limitations, an example of which is shown in Figure 4.2x. In this two-step evaporation there are three limitations to the process including 1.) the steam supply to the intermediate concentrator can be reduced due to demands in other parts of the plant; 2.) the steam supply to the final concentrator can be reduced due to demands in other parts of the plant; and 3.) the final concentrator can accept feed only at or below a certain rate, and it is desired to run this part of the process at its limit.

Each part of the process has its own feedforward-feedback system. The intermediate concentrating process has the feedforward system shown in Figure 4.2s. The final concentrating process does not require feed density compensation so that it is only necessary to establish the ratio of steam to feed and adjust the ratio by feedback. Ordinarily, the system is paced by the final concentrator feed signal which is fed to the auto-select system. The final concentrator controller output manipulates the feed to the intermediate process so as to maintain the final concentrator feed at the set point limit.

Should either steam supply become deficient, its output would be selected to adjust the feed to the intermediate process. Thus, the process is always run at the maximum permissible rate consistent with the process limitation(s). (For more on auto-select control systems see Section 7.10 in Volume II.)

REFERENCES

1. Ritter, R. A. and Andre, H., "Evaporator control system design." *Canad. Journal of Chem. Engineer.* 48: 696, Dec. 1970.
2. McCabe, W. L. and Smith, J. C., *Unit Operations of Chemical Engineering.* New York: McGraw-Hill, 1956.
3. Brown, G. G., et al., *Unit Operations.* New York: Wiley, 1950.
4. Perry, J. H., et al., *Chemical Engineers' Handbook,* (4th ed.). New York: McGraw-Hill, Pp. 3-72 to 3-80.
5. Shinskey, F. G., *Process Control System.* New York: McGraw-Hill, P. 74.

4.3 FLUIDICS APPLICATIONS IN THE PROCESS INDUSTRIES

Fluidics is a practical art. Other technologies may be faster, more compact, more sparing of power or quieter, but fluidics is the safest—and for man and his equipment—the most reliable and the most tolerant of environmental adversity. This practicality emerges in examining applications in the process industries, in which very few all-fluidic systems are found, possibly because of the paucity of fluidic final control elements. Few processes can be manipulated by the small control range of fluidic final control elements presently available and the promise of interface-free control is yet to be realized.

The nature of existing fluidic applications is somewhat fragmented, as will be noted from the following discussion.

Final Control Elements

Fluidic final control elements have been successful as diverting valves ranging from a few inches to more than 6 feet in diameter. A fluid signal less than 1% of the controlled flow will divert a process stream from one conduit to another (Figure 4.3a). This valve can only divert the flow, it cannot stop it, because the flow must always have a place to go under the direction

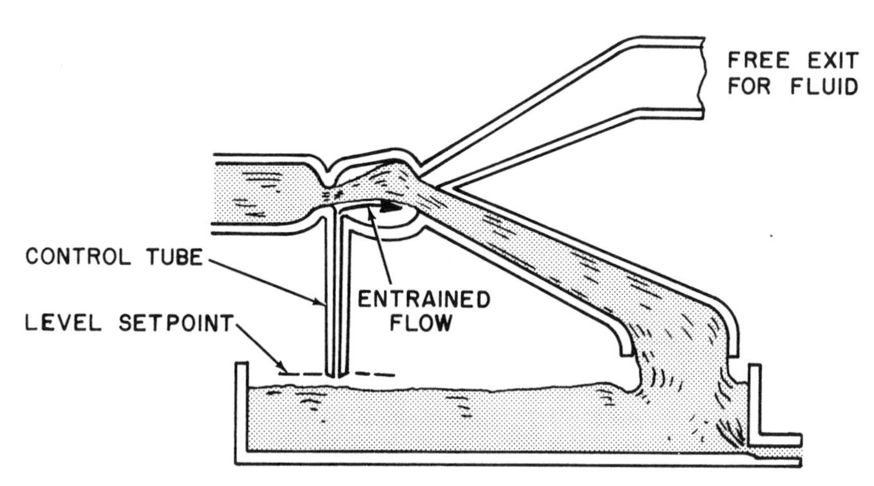

Fig. 4.3a Self-actuated level control using flow diverting valve

of its inertia as established in the interaction section. A disadvantage of the device is the large amount of pumping power required to keep the fluid in continuous motion. (For more details see Section 1.18 in Volume II.)

Analog Controllers

Pure fluidic operational amplifiers are available from several sources, but the well-developed techniques of the electronic operational amplifiers cannot be easily copied. For one thing, there is no pure fluid analog of the series capacitor. For another, the losses are much greater in fluid than in electronic systems. Various circuit schemes have been adapted from electronics or were newly developed to circumvent these shortcomings. Consider the operational amplifier circuit of Figure 4.3b in which the filtered positive feedback boosts the low frequencies. Mathematically, the transfer function is

$$\frac{Y(s)}{X(s)} = \frac{R3}{R2} \cdot \frac{R7 + R6 + R5 + R7(R6 + R5)sC}{R7 + R6 + R5 - (R2 + R3)R5/R2 + R7(R6 + R5)sC} \quad 4.3(1)$$

If $R7 + R6 + R5 = (R2 + R3)R5/R2$, then

$$\frac{Y(s)}{X(s)} = \frac{R3}{R2} \cdot \frac{R7 + R6 + R5}{R7(R6 + R5)sC} + 1 \quad 4.3(2)$$

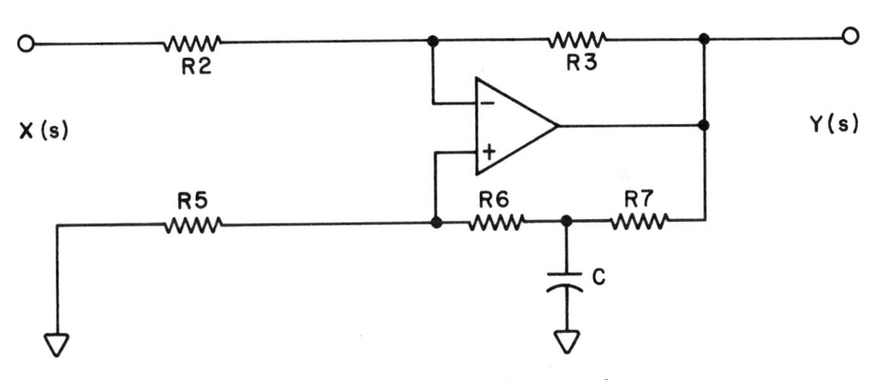

Fig. 4.3b Proportional-plus-integral controller element without a series capacitor

This is the transfer function of a proportional-plus-integral controller.

The same technique applied in the negative feedback path accentuates high frequencies (Figure 4.3c), where the transfer function is

$$\frac{Y(s)}{X(s)} = -\frac{R10 + R9(1 + R10sC)}{R8} \quad 4.3(3)$$

This function exhibits proportional-plus-derivative action, completing the repertoire of linear controller responses required.

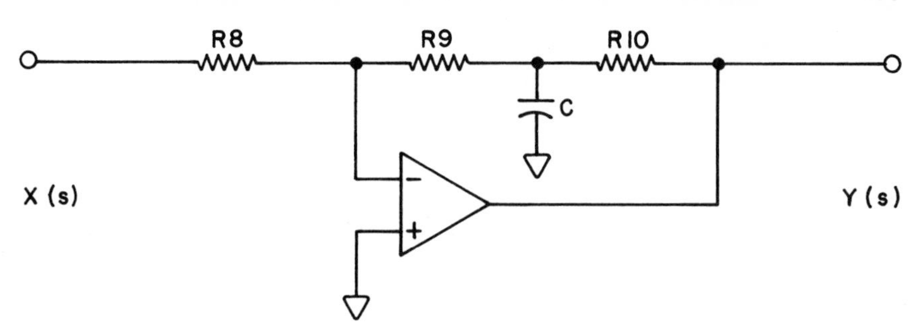

Fig. 4.3c Proportional-plus-derivative action without a series capacitor

Fluidic Controller

This completely different approach starts from the recognized non-linearity of many controlled processes and exploits analog control with a non-linear control, which oscillates at a self-determined frequency and amplitude. In Figure 4.3d, the comparator block has the amplitude characteristic shown in Figure 4.3e. For an error signal outside the hysteretic range of the comparator, the control has on-off characteristics. For small errors, the comparator's relatively rapid non-linear oscillations provide time-proportion for the driving force, so that the effect on the process is a variable gain between the values of maximum and minimum saturation of the output.

Web Thickness Detection

Web handling offers a range of opportunities for fluidic sensing and control. Thickness of relatively smooth materials passing over a roll can be measured with a back pressure sensor (Figure 4.3f). A stiff and accurate mounting bracket is necessary to hold the sensor at a definite distance from the roll, an unnecessary measure if the web is impervious as well as smooth,

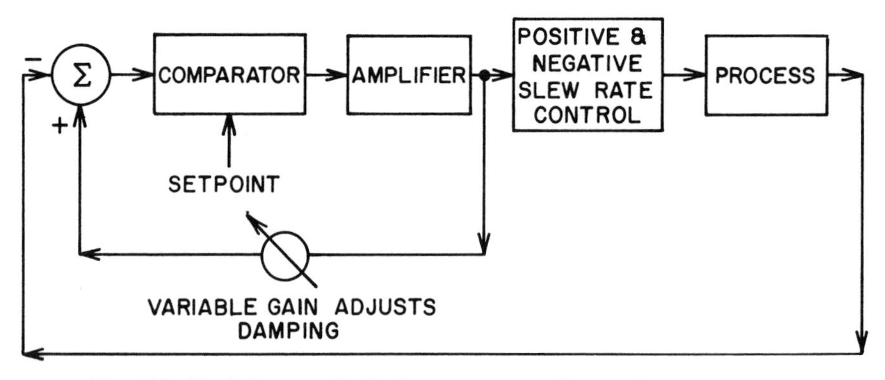

Fig. 4.3d Block diagram of a fluidic process controller (Applied Fluidics Inc.)

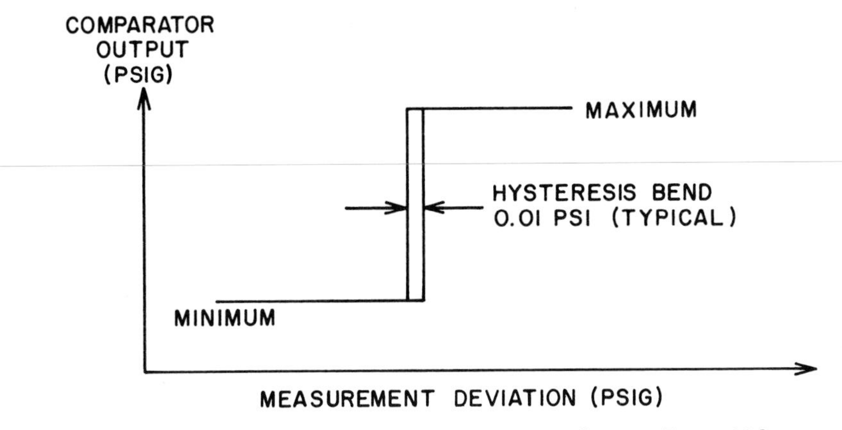

Fig. 4.3e Amplitude characteristic of comparator shown in Figure 4.3d

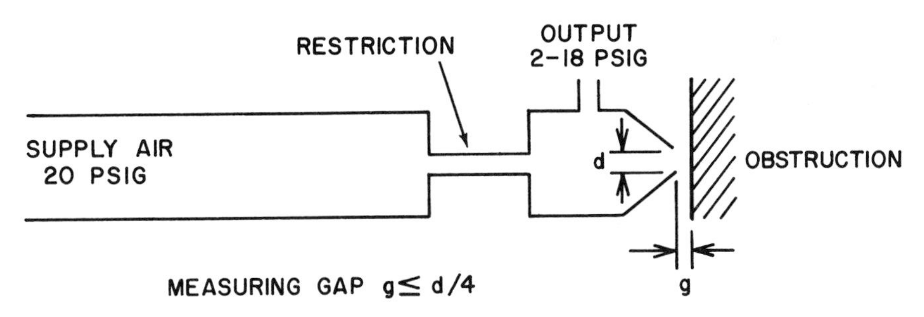

Fig. 4.3f Back pressure sensor for measuring proximity of obstructions

because a differential arrangement (Figure 4.3g) will measure the thickness of an obstruction anywhere in the gap.

To avoid undesirable effects, the web speed in both schemes should be no more than about 25% of the escaping air velocity. If the variations in thickness at nearly constant web speed alone are of interest, the web speed may approach 50% of the air speed in the measuring gap. Splice detection is best handled as a low-resolution variation of thickness measurement. Schmidt triggers and other comparators distinguish input variations from set point values. Edge detection and edge following employ a variation of an interrupted gap sensor, either acoustic or jet. An analog signal proportional to edge position can be generated within a limited range characterized by proportional receptor behavior. This interval of proportionality is a function of receptor area.

Broken web detection is the simultaneous extreme of the thickness and edge measurements. It is usually seen as a missing edge, sensed with an interruptable jet. Pinhole and tear detection in the body of the web is an unsatisfactory application for fluidics if the web speed is high, because the

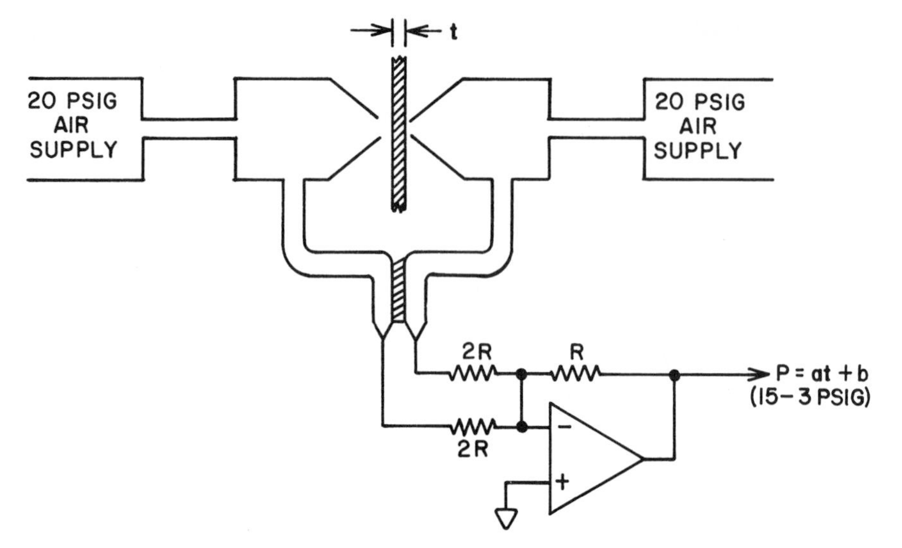

Fig. 4.3g Double back pressure sensor with summed pressure signals for differential thickness measurement

total quantity of air passing through the defect is insufficient to be distinguished from the normal leakage past the edge of the detector array.

Level Detection

Liquid level measurement is a variation on proximity sensing and the same kinds of sensors can be applied, if the nature of the nonrigid surface is properly considered. For continuous level measurement the classic dip tube (Figure 4.3h) supplies a gauge pressure output equal to the static head at the bottom tip of the immersed tube. (For more on air bubblers and performance, see Section 1.8 in Volume I.)

The self-actuated level control diverting valve (Figure 4.3a) utilizes a sensor operated at *negative* pressure. Since liquid rises within the aspirator control tube during flow diversion, a primary requirement is to make the tube wide enough to allow air bubbles to pass freely and drain promptly when the liquid seal is broken at the bottom.

Flow Detection

Fluid flow is the process variable most akin to fluidics. One of the more sophisticated analog fluidic extensions of head-producing sensors is the *square root extractor,* in which a fluid resistor with square law flow characteristics is the input element to the summing point of a fluidic operational amplifier, whose feedback element is a linear resistor. The square law characteristic of the input resistor cancels the square law characteristic of the head meter, and the output is scaled and linearized in one step.

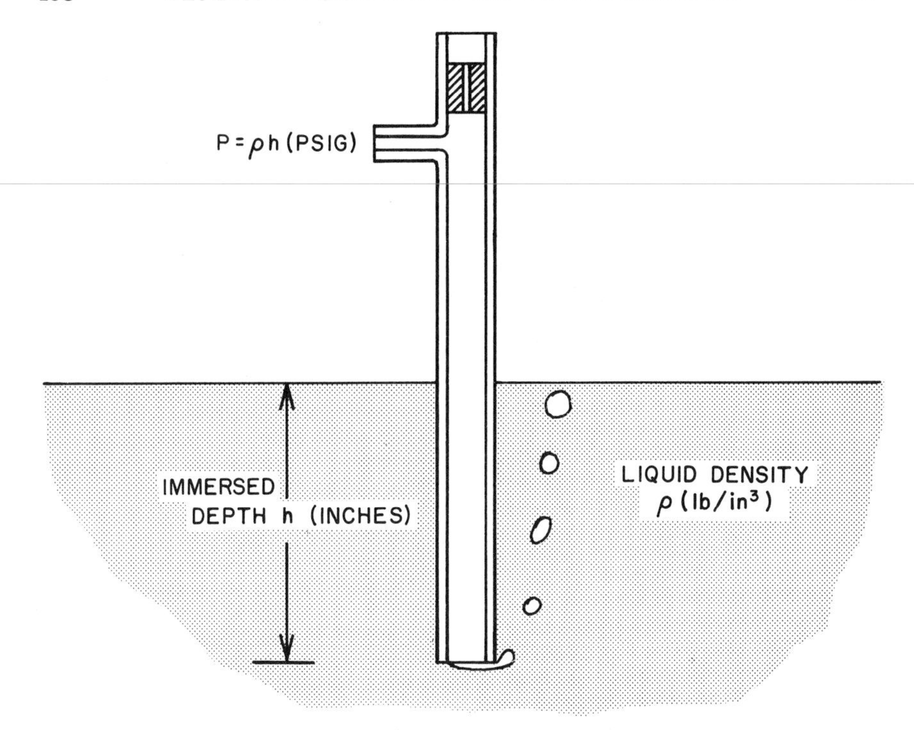

Fig. 4.3h Dip tube measures static head at bottom of tube

A sensing jet can be trained across a process stream obliquely or parallel to it to obtain information about the flow. The cross-flow direction can be implemented without protrusions into the conduit. Axial flow appears to be superior according to the published performance data on wind velocity sensors.

Hydrodynamic Oscillation

A promising phenomenon for flow measurement is hydrodynamic oscillation, which is well suited for operation with fluidic sensors. It is a small signal process, which exploits relatively stable fluid properties to arrive at a frequency of oscillation output.

One of the hydrodynamic oscillators for flow measurement is the *swirlmeter,* in which the fluid stream is eddied around the flow axis and is expanded through a diverging section, after which the eddy is no longer stable but is processed about the flow axis at a rate linearly proportional to the flow rate. (For more information, see Section 5.10 in Volume I.)

Another hydrodynamic phenomenon, recognized for decades (the Kármán vortex trails) but only recently applied to flow measurement, is the *shredding of vortices* from a symmetrical, relatively unstreamlined body obstructing the flow. The mechanics of the process guarantee that there will be a point on

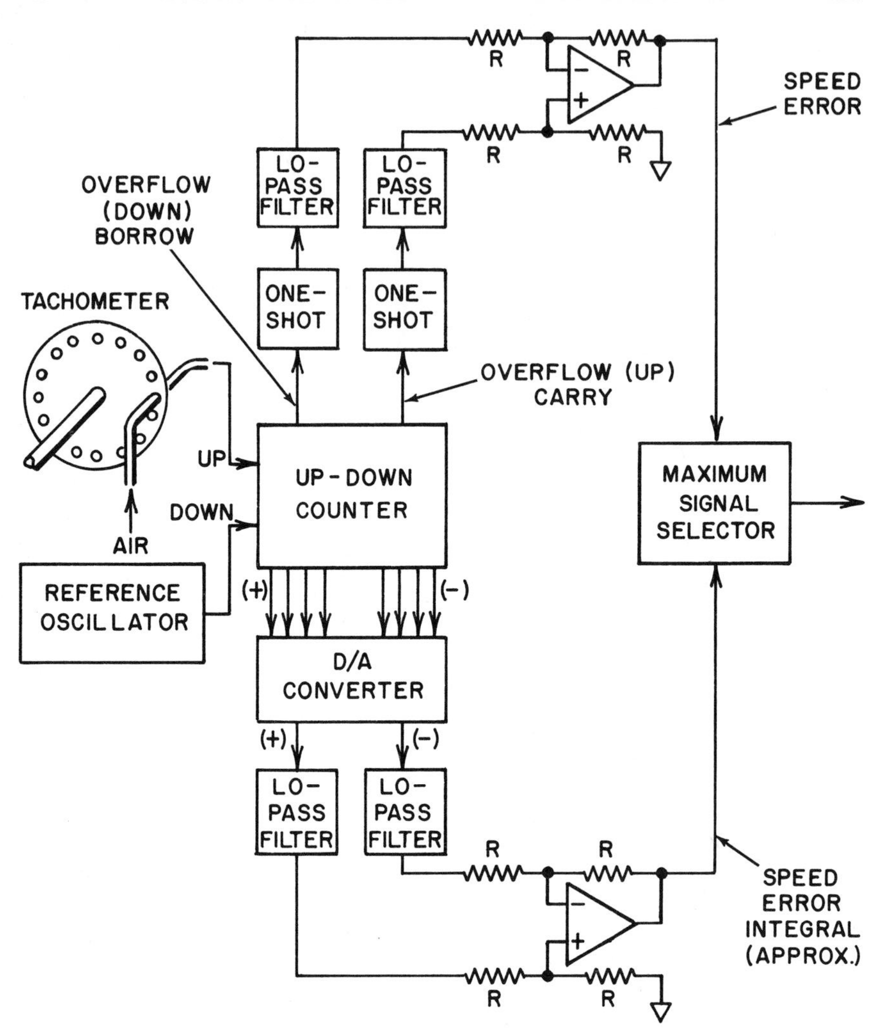

Fig. 4.3i Digital speed control

the body where flow will reverse in synchronism with the vortex shedding, the frequency of which is linear with flow rate. Devices based on this principle have been commercially introduced with self-heated thermistor sensors for the reversing flow, but a fluidic output signal is also feasible. (For more about this device see Section 1.5 in this volume.)

A wall attachment oscillator, acting through intentional feedback of the mass flow wave from its complementary outputs, measures the flow rate of the power jet by the frequency with which the jet alternates between the two output legs. Linearity is satisfactory, and an output signal in fluidic form is directly available. A principal drawback is the large pressure drop required to develop the power jet.

All the flow sensitive hydrodynamic oscillators share the common fault of converting the variable frequency fluidic signal within the sensor to a medium (such as air) usable by a fluidic signal processor outside the process stream. One approach is to install a tap on the process pipe and use a purging fluid which is compatible with the process fluids. On higher pressure processes a pair of bellows, statically balanced but dynamically unbalanced to the alternating fluidic signal, conveys the frequency impulses to an ambient pressure environment.

Temperature Measurement

Temperature measurement by fluidics can be carried out by a wall attachment oscillator with an acoustic wave feedback, rather than a mass flow front as in the flow measuring oscillator. A fluidic temperature detector is described on P. 406 in Volume I. Since acoustic velocity varies as the square root of absolute temperature, the output frequency of the temperature sensing oscillator is nonlinear to varying degrees, depending on the range required.

Speed Sensors

Speed control is compatible with either a digital or an analog fluidic approach. A digital system is best for fixed running speeds because it simplifies the implementation of integral control. A pulse-generating tachometer furnishes a variable frequency signal for comparison with an internally generated reference frequency (Figure 4.3i). As an example of fluidic control potential, the carburetor of an internal combustion engine can be replaced with a fluidic version which controls both the manifold pressure and the fuel flow rate.

An analog system is more compatible with variable speed operation. The analog fluidic tachometer is a straight vane centrifugal blower through which a small, constant reverse flow of air is forced. The flow prevents loss in viscosity from heating the contained air, changing its density or upsetting the tachometer calibration.

Chapter V

POLLUTION MEASUREMENT AND CONTROL

C. P. Blakeley
H. C. Roberts
C. J. Santhanam

CONTENTS OF CHAPTER V

5.1 WATER QUALITY MONITORING SYSTEMS

Specifications:	The "Specifications for an Integrated Water Quality Data Acquisition System," 8th edition, by the Federal Water Quality Administration (now part of the Environmental Protection Agency) applies to all purchases by that agency. Only some of the listed manufacturers meet the specification.
Measured Water Quality Parameters:	Conductivity, dissolved ozygen, ORP, pH, selective ions (Table 5.1e), temperature and turbidity.
Physical Variables:	Air temperature, rainfall, river flow or stage level, sunlight, wind direction and speed.
Cost:	A system to measure and record pH only will need equipment costing approximately $1,800. An integrated monitoring system for pH, conductivity, temperature and dissolved oxygen, consisting of sample system, signal conditioners and multi-point recorder, completely factory assembled, will cost about $7,000. Installation and a suitable shelter for the monitor are not included.
Partial List of Suppliers:	Automated Environmental Systems, Inc.; Beckman Instruments; Honeywell, Inc.; K.D.I. Poly-Technic Inc.; Robertshaw Controls Inc.; Schneider Instruments Co.; Union Carbide Corp.

Continuous water quality monitors range from the simplest one-parameter system (submerged sensor and recorder with integral signal conditioner) to sophisticated multi-parameter systems. In addition to water analysis meteorological parameters may also be determined. Data may be recorded at the site or remote from it, continuously or at fixed intervals and can be stored on tape. Any combination of these features is available. (For more on water pollution control systems see Section 10.14 in Volume II.)

The Purpose of Water Quality Monitors

From tributary to ocean, the quality of water changes continuously owing to diurnal influences, the seasons, natural contaminants and pollutants introduced by man. The quality of water remains an unknown factor until various parameters are measured and compared with standards, and it is of concern to virtually anyone using, regulating or researching water anywhere in the world. Water quality monitors are usually divided into five categories, depending on how the body of data is to be used.

1.) *Data collection.* Primarily fixed permanent installations owned by government agencies collect data for publication and general use. The instrumentation is characterized by a high degree of equipment standardization but with enough built-in flexibility to enable the use of various types of data recording and transmission.

2.) *Surveillance or enforcement monitoring.* Both permanent and portable installations monitor the quality of water in a river or stream in order to enforce compliance with established water quality standards.

3.) *Area studies.* Both permanent and portable installations are used to study a river basin or definite segment of a water course for a fixed interval, usually to determine the level of pollution and the major offenders thereto.

4.) *Industry.* Plants that use large quantities of water and discharge water directly into a natural water course commonly use fixed, permanent installations to monitor effluents in order to comply with standards.

5.) *Short-term surveys.* Short-term or instantaneous measurements along a water course are occasionally needed. Frequently, the surveys are made prior to the installation of permanent monitoring equipment by data collection or enforcement agencies, or prior to construction of an industrial plant.

A Water Quality Monitoring System

The instruments and components for continuous water quality measurements must always be considered as a system regardless of complexity; the system always consists of at least four components items (Figure 5.1a).

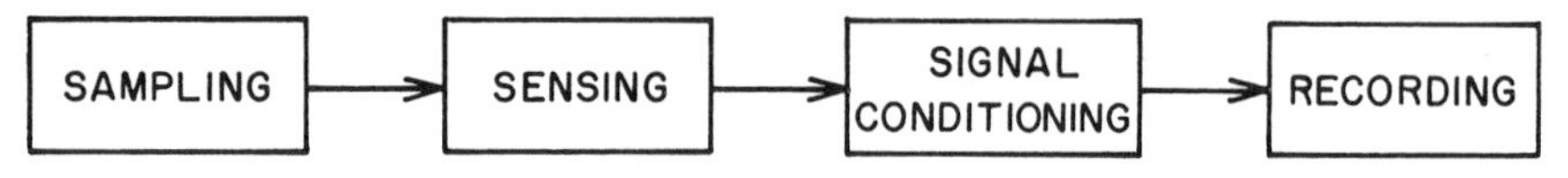

Fig. 5.1a Components of water quality monitoring system

As part of SAMPLING, water is pumped to a sampling tank, or sensor(s) are immersed in the water. SENSING includes the quantitative analysis of a constituent by devices exposed to the water (or air or sun) selectively sensitive to the parameter being measured. SIGNAL CONDITIONING is the conversion of the electrical signal from the sensor to a form acceptable to the recording device (or telemetering device). A signal conditioner may be an amplifier or—in its simplest form—a measuring circuit in a recorder. RECORDING is the act of making a record representative of the sensed parameter and can be an analog record, typewriter logged data or punched paper tape.

The basic system can be variously altered and expanded to meet individual requirements. The most frequent addition is a telemetering system for recording at a central location remote from the sensing station (Figure 5.1b). A water quality system can be as simple as a sensor placed in the water and connected to a conveniently located recorder, or as complex as a data collection system with many sensors placed at several locations and recording at a central location (Figure 5.1c).

Commonly Measured Water Quality Parameters

To measure all the parameters furnishing data on water composition is both costly and impractical. The components and characteristics that should be measured continuously are

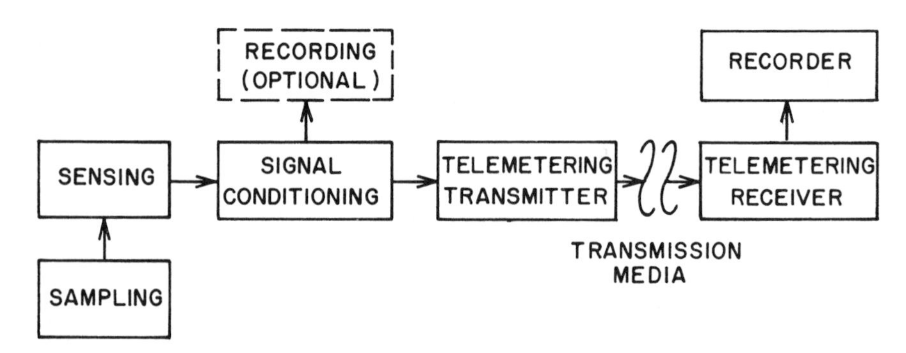

Fig. 5.1b Telemetering feature added to water quality monitoring system

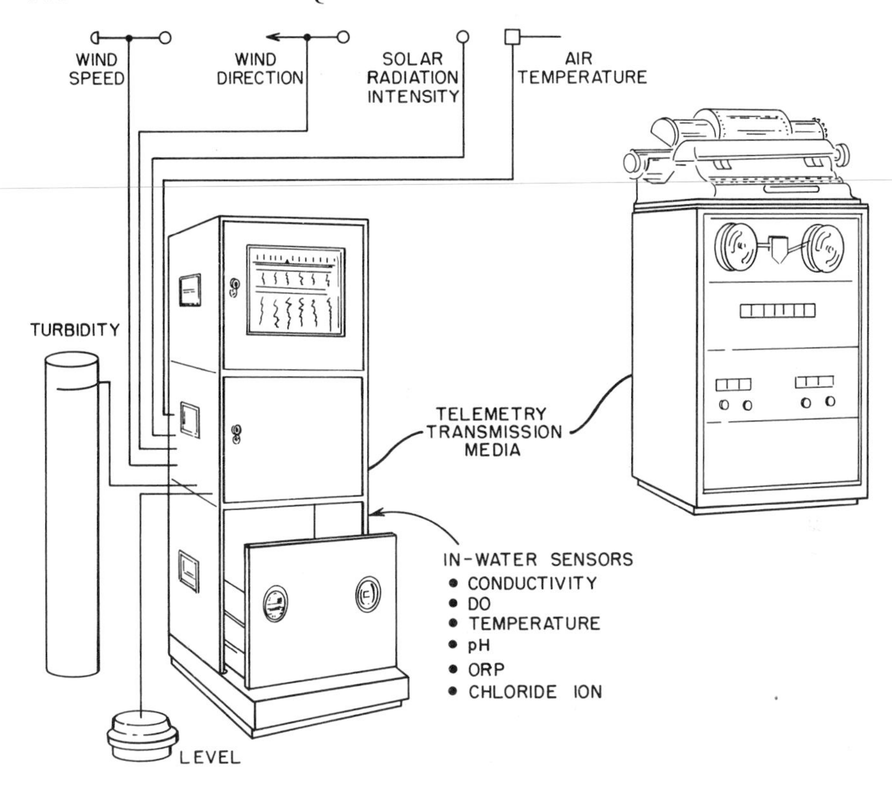

Fig. 5.1c A complex water quality data collection system

Dissolved oxygen (DO). Reported in mg per liter (ppm), dissolved oxygen is a measure of the water's ability to sustain the aquatic life on which fish and other organisms feed. The solubility of oxygen in water decreases with increase in water temperature. The chloride content of water also affects the solubility of oxygen (Table 5.1d). Almost every type of waste consumes oxygen from the receiving stream and from the surrounding atmosphere.

Specific conductance (conductivity). Conductivity is reported in micromhos per centimeter, one micromho being one-millionth of a mho, and mho is the reciprocal of an ohm. Conductivity is the measure of the dissolved ionized solids in water. Non-specific and measuring to some degree all ions present in the water, it is temperature dependent and, in water, has a tremendous range. Ultra-pure water, for example, has a conductivity of 0.055 micromhos per cm at 25°C, whereas a typical city potable water supply may reflect between 150 and 250 micromhos per cm at 25°C. Snow and ice control of highways using salt may increase conductivity of river water immediately after run-off. A sudden increase in conductivity is an indication of pollution by strong acids, bases or other highly ionized substances.

pH (hydrogen ion activity). This parameter is reported in pH units from 0 to 14, zero denoting maximum acidity, 7 denoting neutrality and near 14 denoting maximum alkalinity. pH is the logarithm of the reciprocal of the active (uncomplexed) hydrogen ion concentration in moles per liter. Natural waters exhibit only slight fluctuations in pH. Rapid changes in pH are almost always caused by pollution, such as mine acid drainage or dumps of acids or alkalies, or both.

Oxidation-reduction potential (ORP). An electrochemical measurement of the potential (emf) developed by oxidizing or reducing materials in water, natural waters contain both oxidants and reductants. By reaction with each other they tend to balance out so that a water supply at true equilibrium shows zero ORP. Municipal or industrial plant wastes may contain strong oxidizing or reducing agents (highly chlorinated water, for example) thus producing a plus or minus emf in ORP measurement. ORP measurement is dependent on temperature.

Turbidity. Reported in Jackson Turbidity Units (J.T.U.), this is a parameter measure of the undissolved solids in water including silt, clay, plankton and other microorganisms. Size, shape and refractive index of the particles affect the transparency or opacity of water. It is therefore not possible to correlate turbidity and quantity by weight (mg per liter). Two methods of measurement commonly used are 1.) nephelometry—the backscattering principle which measures light reflected by the turbidity in the sample; and 2.) light absorption—the light absorbed by the suspended particles during its passage through a fixed column of water.

Color content of water will interfere with turbidity measurements and the measurement should be calibrated against standards prepared from the water, including the coloring materials.

Temperature. This parameter is reported in °F or °C. Only slow changes in water temperature occur seasonally unless thermal pollution is experienced. Because of the effect of temperature on the solubility of oxygen and other chemical content of the water, temperature is an important parameter.

Specific Ion Activity Measurements

A variety of electrode systems can determine the activity of specific ions, and of interest in water monitoring are those listed in Table 5.1e, which measure the activity rather than the concentration of an ion. In pure solutions and in the absence of interfering ions, the activity reading can be converted to concentration by introducing the ion's activity coefficient into the Nernst equation.

As a specific ion electrode measures the activity of that ion—the rate and extent of participation of that ion in a chemical reaction—it is not a measure of concentration. In dilute solutions, such as may well be found in rivers and streams, activity may approach concentration and, with extensive

Table 5.1d
WATER SOLUBILITY OF OXYGEN AS A FUNCTION OF CHLORIDE ION CONCENTRATION AND OF TEMPERATURE

Temperature °C	*Chloride Ion Concentration in Water (mg/liter) Dissolved Oxygen Concentration in Water (mg/liter)*					*Difference per 100 mg/liter Chloride ion*
	0	*5,000*	*10,000*	*15,000*	*20,000*	
0	14.6	13.8	13.0	12.1	11.3	0.017
1	14.2	13.4	12.6	11.8	11.0	0.016
2	13.8	13.1	12.3	11.5	10.8	0.015
3	13.5	12.7	12.0	11.2	10.5	0.015
4	13.1	12.4	11.7	11.0	10.3	0.014
5	12.8	12.1	11.4	10.7	10.0	0.014
6	12.5	11.8	11.1	10.5	9.8	0.014
7	12.2	11.5	10.9	10.2	9.6	0.013
8	11.9	11.2	10.6	10.0	9.4	0.013
9	11.6	11.0	10.4	9.8	9.2	0.012
10	11.3	10.7	10.1	9.6	9.0	0.012
11	11.1	10.5	9.9	9.4	8.8	0.011
12	10.8	10.3	9.7	9.2	8.6	0.011
13	10.6	10.1	9.5	9.0	8.5	0.011
14	10.4	9.9	9.3	8.8	8.3	0.010
15	10.2	9.7	9.1	8.6	8.1	0.010
16	10.0	9.5	9.0	8.5	8.0	0.010
17	9.7	9.3	8.8	8.3	7.8	0.010
18	9.5	9.1	8.6	8.2	7.7	0.009
19	9.4	8.9	8.5	8.0	7.6	0.009
20	9.2	8.7	8.3	7.9	7.4	0.009
21	9.0	8.6	8.1	7.7	7.3	0.009
22	8.8	8.4	8.0	7.6	7.1	0.008
23	8.7	8.3	7.9	7.4	7.0	0.008
24	8.5	8.1	7.7	7.3	6.9	0.008
25	8.4	8.0	7.6	7.2	6.7	0.008
26	8.2	7.8	7.4	7.0	6.6	0.008
27	8.1	7.7	7.3	6.9	6.5	0.008
28	7.9	7.5	7.1	6.8	6.4	0.008
29	7.8	7.4	7.0	6.6	6.3	0.008
30	7.6	7.3	6.9	6.5	6.1	0.008
31	7.5					
32	7.4					
33	7.3					
34	7.2					
35	7.1					

laboratory investigation, it may well be possible to measure the concentration of the system in mg per liter.

A specific ion activity electrode, like a pH electrode, develops an electrical potential in response to ion activity, with a logarithmic relationship between activity and potential, i.e.,

$$E = Eq + 2.3\frac{RT}{nF}\log A$$

where E = total potential of the system,

Eq = part of total potential due to reference electrode and internal solution,

$2.3\frac{RT}{nF}$ = Nernst factor (54.16/n mv @ 25°C) with R and F being constants, n = the charge of the ion and T = the temperature in °K(°C + 273)

A = activity of the ion being measured

When sensing a cation, potential becomes more positive with increasing activity; when sensing an anion, potential becomes more negative with increasing activity.

A tenfold change in activity of a monovalent ion at 25°C produces an electrode potential change of 59.2 mv, whereas the response change of a divalent ion is 29.6 mv.

Physical Variables

To gain a more complete understanding of the water quality of a large body of water, variables such as solar radiation intensity (sunlight), rainfall, wind speed, wind direction, air temperature and river flow or stage (level), or both, are frequently either automatically monitored or manually logged.

Water Quality Sensing

The simplest albeit not necessarily the most practical method of continuously sensing important variables like dissolved oxygen, pH, conductivity and water temperature is by submersible sensor assemblies (Figure 5.1f). The assembly must be securely anchored in the stream, a measure that frequently makes it difficult to service. In addition, it is subject to damage by floating debris and vandalism. For monitoring water quality parameters in flumes, tanks and other protected areas as well as from boats the submersible assembly may be ideal (Figure 5.1g).

Frequently, however, it will be advantageous to install a permanent monitoring system for ease of maintenance. A submersible pump can be installed—suitably anchored and surrounded with a debris screen—at a site where a representative sample of water can be obtained. Water

Table 5.1e

CHOICES OF SPECIFIC ION MEASUREMENT

Ion	Electrode pH Range	Interfering Ions: Anions													Interfering Ions: Cations											Electrode Type
		ClO_4^{1-}	I^{1-}	ClO_3^{1-}	Br^{1-}	S^{2-}	CN^{1-}	S_2O_3	NH_3	OH^{1-}	Cl^{1-}	NO_3^{1-}	HPO_4^{2-}	HCO_3^{1-}	Cu^{2+}	Mg^{2+}	Ba^{2+}	H^{1+}	Fe^{2+}	Ni^{2+}	Zn^{2+}	Na^{1+}	K^{1+}	Ca^{2+}	Si^{2+}	
Chloride	0–14		√		√	√	√	√		√																Solid State
Sulfide	0–14		none													none										Solid State
Nitrate	2–12	√	√	√	√																					Liquid Membrane
Fluoride	1–8.5		√		√					√	√	√	√	√												Solid State
Divalent cation	7–8															sensitive to most divalent actions										Liquid Membrane
Calcium	5.5–11.0															√	√									Liquid Membrane
Copper Cu^{2+}	2.5															√	√	√	√	√	√	√	√	√	√	Liquid Membrane
Lead	4–7														√					√	√	√		√		Liquid Membrane
Cyanide	0–14		√			√																				Solid State
Hardness	5.5–11														√		√		√	√	√	√				Liquid Membrane

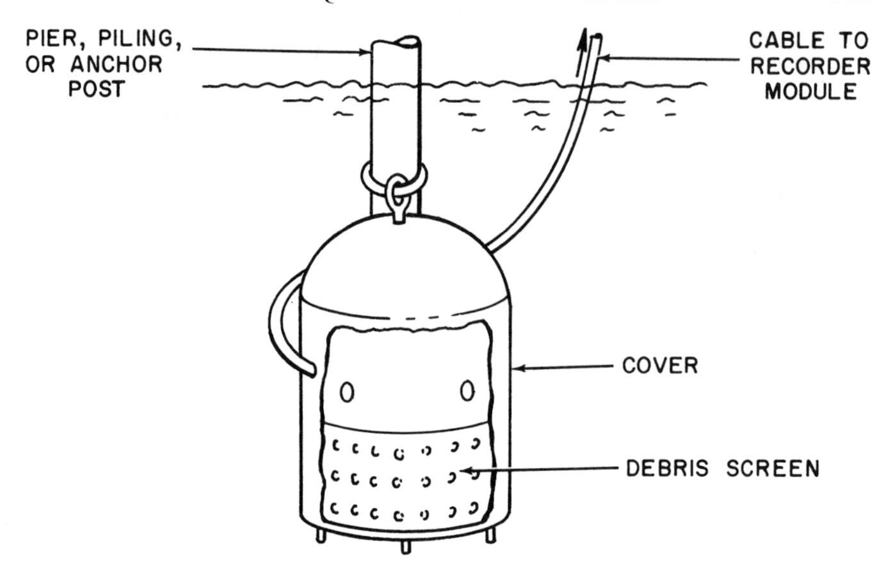

Fig. 5.1f Multiparameter submersible sensor housing

is then pumped continuously to the sampling tank of the water quality monitor (Figure 5.1h). Pump selection is critical, especially when dissolved oxygen is one of the measurements. Peristaltic pumps have the least effect on dissolved oxygen in the sample. According to Henry's law the amount of dissolved oxygen in water at saturation varies linearly with pressure (Table 5.1i).

Unfortunately, there is no assurance that the oxygen level will decrease if suction pressure drops, as indicated by the table. Also, any turbulence through the pump may affect the dissolved oxygen reading especially when the water supply is supersaturated. Sample tank construction affords parallel flow to each sensor sample chamber. Upward flow of water reduces accumulation of sediment as does a common sink and drain (Figure 5.1j).

Sensors are easily removed for cleaning and calibration. Signal conditioners are commonly housed in proximity to the sample tank for easy connection of sensor cables to signal conditioner input terminals. The individual signal conditioners are amplifiers and servomechanisms that accept the varied inputs of the sensors and furnish output signals of the same full range value, such as 10 to 50 or 4 to 20 ma DC and are properly scaled for analog or digital recording equipment.

Recording and Telemetry

The recorders and telemetry transmission and receiving equipment are standard industrial devices (see Chapters V and VI in Volume II).

Data should be in a form which will be most usable at a later date.

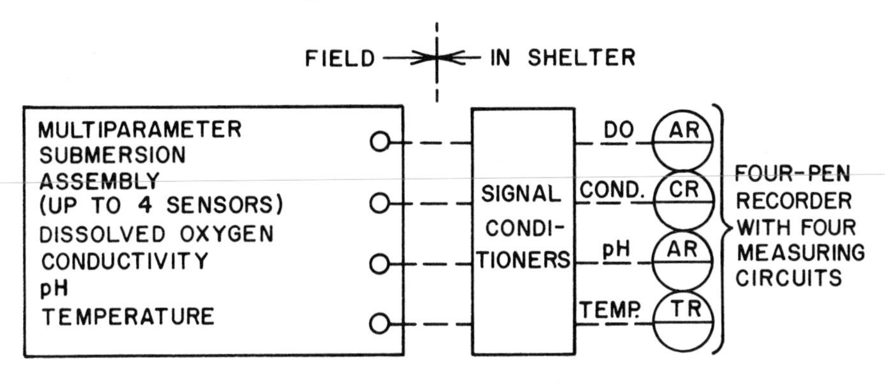

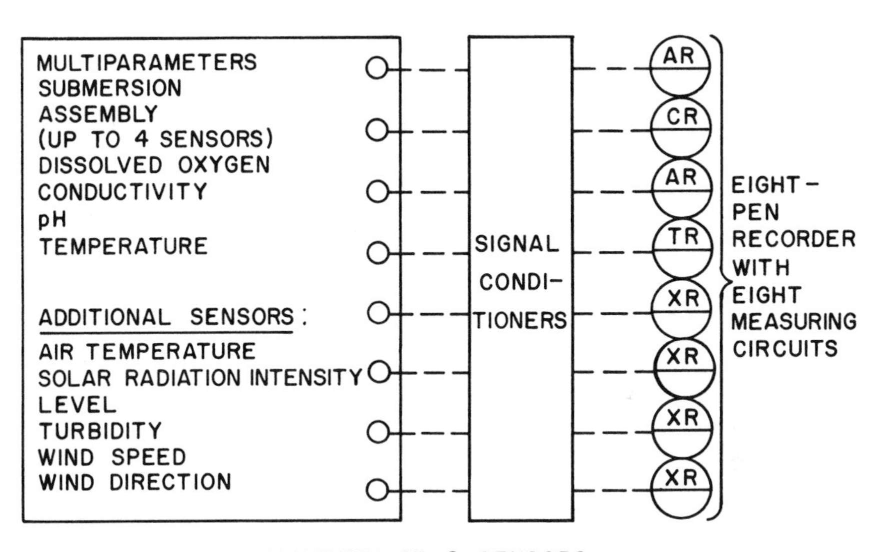

Fig. 5.1g Typical water quality monitoring system for 4 or 8 analog records of water and related parameters

Consideration should thus be given to 1.) determining the most understandable form of data presentation; and 2.) the way in which the data will be used or processed.

The analog or graphic chart record is the most direct reading and certainly the most usable record for observing trends. When all parameters are recorded concurrently on a single chart, it is possible to observe changes of some of the variables with respect to others with time. A record like this is particularly useful when examining long-term trends, short-term excursions of a single parameter or the interrelation of parameters.

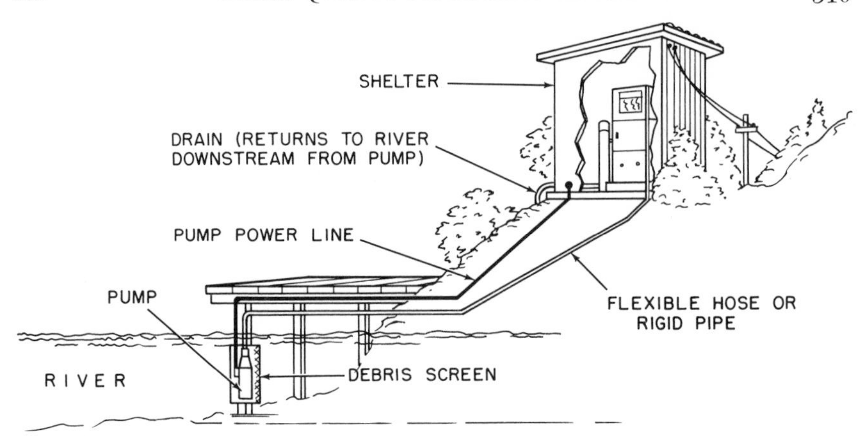

Fig. 5.1h Typical permanent water quality monitoring installation

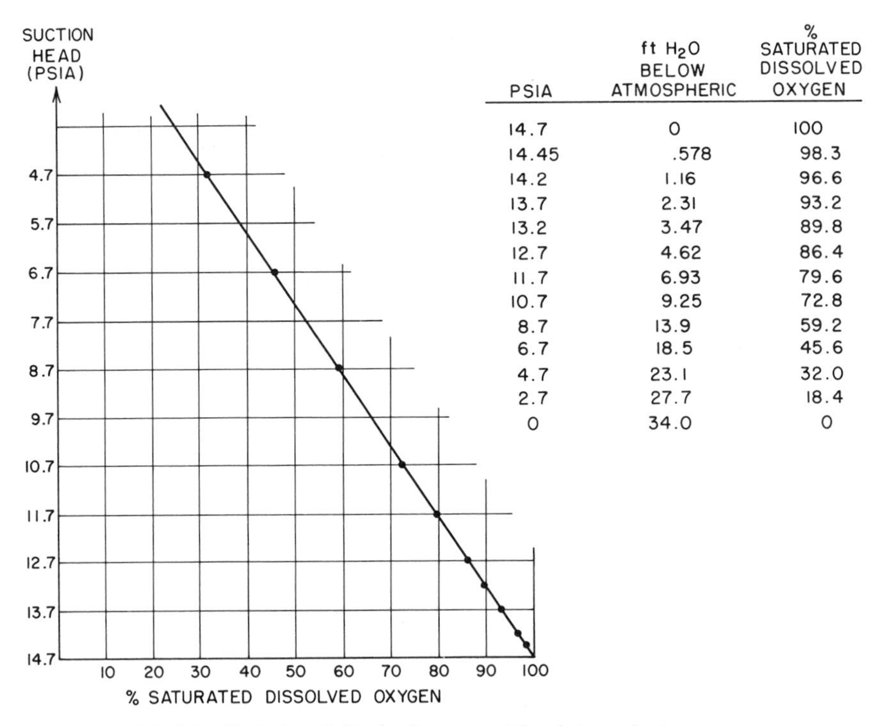

PSIA	ft H_2O BELOW ATMOSPHERIC	% SATURATED DISSOLVED OXYGEN
14.7	0	100
14.45	.578	98.3
14.2	1.16	96.6
13.7	2.31	93.2
13.2	3.47	89.8
12.7	4.62	86.4
11.7	6.93	79.6
10.7	9.25	72.8
8.7	13.9	59.2
6.7	18.5	45.6
4.7	23.1	32.0
2.7	27.7	18.4
0	34.0	0

Table 5.1i Variation of dissolved oxygen with subatmospheric pressures

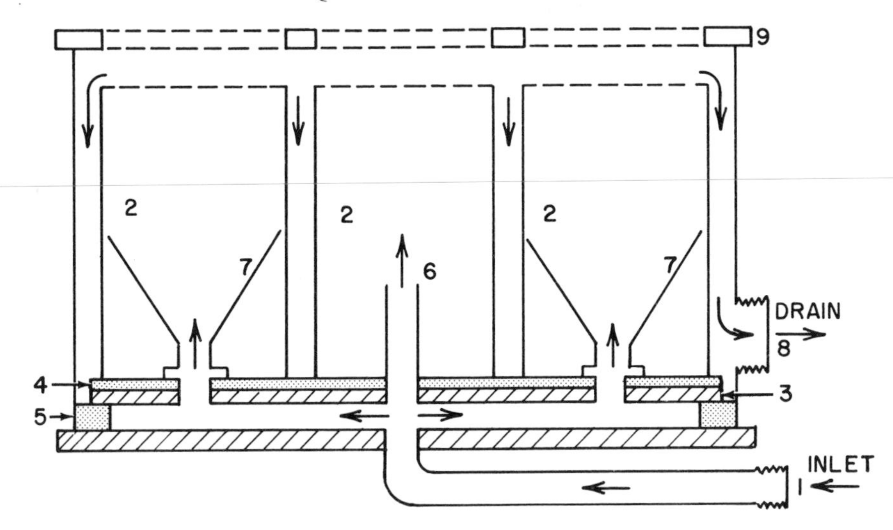

Fig. 5.1j Typical sampling tank. A continuous sample of water enters the tank through inlet (1) and flows into sampling chambers (2) through ports in a distribution plate (3) and rubber gasket plate (4). Sealing ring (5) prevents flow around the perimeter of the distribution plate. Water enters the center sampling chamber through a pipe (6) to provide jet action across the face of the dissolved oxygen sensor. In all other chambers baffle cones (7) keep heavy dirt particles in suspension. Water overflows around outside of each chamber to a large drain (8). Cover (9) provides openings and support for sensors.

When information is required at fixed intervals rather than continuously, it is convenient to log the data with a typewriter. The time and parameter identification can be logged in addition to parameter value so that the information is recorded on a real-time basis and easily identified.

If large amounts of data are to be collected and subsequently processed by a computer, it is desirable to obtain a tape record of the values of the parameters on a programmed time basis. When stored on punched paper tape or magnetic tape, it is subsequently possible to feed the data into a computer for the necessary computational work. Information on tape is convenient for long-term storage and relatively quick retrieval.

Of equal importance to the selection of the data format (type of record) is the location of the recording equipment. The most obvious, and usually the least expensive, location is near the point of measurement. However, it is often more desirable to place the recording equipment at a location farther from the point of measurement and closer to the people who must use the data, in which event it is necessary to telemeter the data from the point of measurement to the point of recording. The main advantages of telemetering are that 1.) information is immediately and continuously available at the point of use; and 2.) equipment malfunctions at the sensing site are quickly detectable and can be responded to "on demand."

A variety of telemetering methods systems exist, and the final choice depends on the accuracy, speed and economic requirements of a particular installation.

Frequency telemetering is reasonably priced and provides satisfactory accuracy and reliability. When accuracy or speed greater than ordinary is desired, digital transmission is used. The parameter values are converted to digital form at the point of measurement and then transmitted to the receiving site and recorded. (For more on water pollution control systems see Section 10.14 in Volume II.)

5.2 SMOKE AND AIR QUALITY MONITORS

Detector Types:

(a) Photometric smoke sensors,
(b) Stack sampling systems for particulates,
(c) Tape samplers for particulates,
(d) Conductometric SO_x sensors,
(e) Colorimetric SO_x sensors,
(f) Coulometric SO_x sensors,
(g) Photometric SO_x sensors,
(h) Saltzman NO_x sensors,
(i) Photometric NO_x sensors,
(j) Hydrocarbon detection by flame ionization,
(k) Infrared hydrocarbon sensors,
(l) Infrared CO sensors,
(m) Total oxidant detectors.

Note: The summary below utilizes the letters (a) through (m) to refer to the listed detector types.

Cost:

About $1,000 (a), $1,500 to $2,500 (b,c,j), $2,000 to $5,000 (d,e,h), $3,000 to $7,000 (f,k,l), about $5,000 (m), $7,000 to $12,000 (g,i).
These price ranges are for order of magnitude estimates of complete systems.

Partial List of On-Line Sensor Suppliers:

Automated Environmental Systems, Inc. (c,e,l), Bacharach Instrument Co. (a), Bailey Meter Co. (a), Barton Co. (f), Beckman Instrument Co. (a,d,f,j,k,l), Bendix Corp. (c,e,j,l,m), Dalmo Victor, A Textron Div. (k), Davis Instruments, Inc. (d,j), DuPont Co. (g,i), Environeers, Inc. (b), Ikor, Inc. (c), Instrument Development Co.

(d), Leeds & Northrup Co. (a,l), Leigh Systems, Inc. (c); Mine Safety Appliance Co. (a,j,k,l), Monitor Laboratories, Inc. (h,m), Peerless Corp. (i), Phillips Electronics Co. (f), Research Appliance Co. (b,c,h), Technicon Instruments Corp. (e,h), Union Carbide Corp. (j), Western Precipitation Co. (b).

Three broad instrumental methods of analyzing the quality of air and determining the concentrations of pollutants include 1.) smoke or opacity monitoring, 2.) source testing and 3.) air quality monitoring. A brief account of the instrumentation systems to measure automobile exhaust emissions is also given.

Smoke Monitoring Methods

Source testing and precise inventory of emissions are necessary in designing pollution control equipment, but once a system is in operation, it is necessary to have means of monitoring the emissions in order to check the operation of the equipment. One method widely used by regulatory agencies is to monitor the smoke from the stack, or more precisely, observe the opacity of the plume. Smoke monitoring methods like this one include 1.) appearance methods, 2.) manual methods and 3.) automatic opacity measurement methods.

Appearance

This method utilizes the well-known Ringlemann smoke charts published by the U.S. Bureau of Mines consisting of four numbered charts. The cross-hatched areas in the charts vary from 20% for No. 1 to 80% for No. 4. There are six Ringlemann numbers, the additions being all white and all black. The observer compares the smoke plume with the Ringlemann charts and assigns a smoke number to the plume based on its resemblance to a particular panel.

The procedure is time-consuming and applicable to various shades of smoke. Also, there is no direct correlation between the opacity of the plume and the quantity of pollutants released to the atmosphere. This method, however, can detect changes in the appearance of the stack and thus determine the trend of the pollution control effort.

Manual

Refinement of the appearance method has led to the Smoke Inspection Guide developed by the U.S. Public Health Service. In this method, the observer compares the plume with film strips of varying shades of grey. An additional improvement is the optical method[1] of sighting the smoke and comparing it with reference film which appears in the same field of view. Methods like these greatly increase reproducibility, even in unskilled hands.

Automatic Opacity

A more sophisticated development is the photometric comparison of the stack effluent with calibrated, carefully controlled black and white smokes. A simpler version, using a photocell in the plant stack, is shown in Figure 5.2a. Various modifications of these methods have been reported.[2,3,4] One source of error in all opacity measurements is the steam plume caused by the condensation of moisture from the exhaust. Condensation of moisture per se is not prohibited by regulations except when visibility is critical or its effect severe. One method of correcting for the steam plume in opacity measurements is to measure the opacity of the stack sample above its dew point[5].

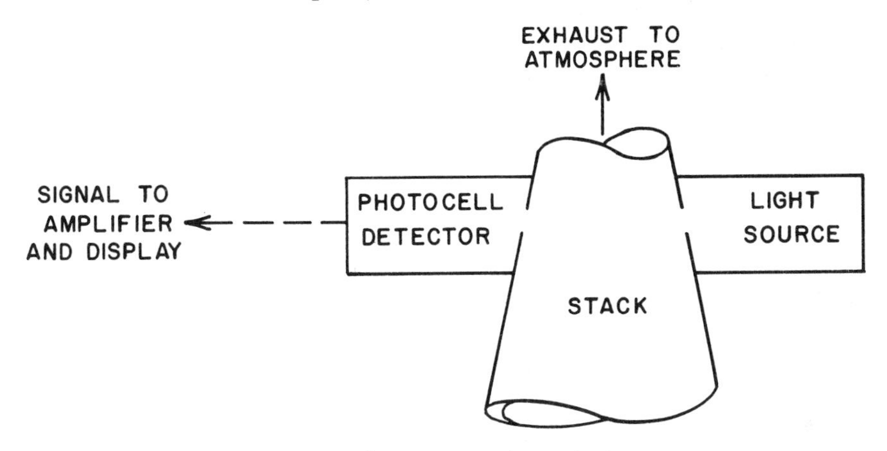

Fig. 5.2a Photometric smoke monitoring

The smoke monitoring methods just described fail to provide quantitative correlation of particulate pollutants; for pollutants other than particulate matter, the methods are of no value at all.

Source Testing Methods

Source testing is a direct method of compiling an emission inventory, and consists of withdrawing a small sample from the stack exhaust and analyzing it for gaseous and particulate pollutants. For proper source testing, 1.) sample should be large enough to be detectable by the analytical method but not large enough to affect collector efficiency; and 2.) sampling time should be short enough to include all significant peaks but long enough to collect a sample of pollutant to indicate minimum level of interest. Very short sampling time will produce insufficient pollutant to be detectable.

Comprehensive reviews[6,7] of an air sampling schematic (Figure 5.2b) define the method in detail. Salient aspects of sampling procedures include:

Location of sampling point. Ports should be located in regions of stable flow—ideally, 10 diameters downstream and 5 diameters upstream of any bends or obstructions. Traverse positions for various round and rectangular ducts are given in the literature.[6]

Gas properties. Provided the stack temperature is below 212°F, conventional wet and dry bulb determinations are adequate. At higher temperatures the gas samples must be cooled to determine dew point.

Gas velocity. This is usually determined by a Pitot tube. Since the standard Pitot tube may become plugged in the duct, S-type Pitot tubes are employed in source testing. For measuring velocities below 10 ft per second Pitot tubes are unsatisfactory and displacement devices are required.

Particulate sampling. Particulate samples should be drawn from the stack under isokinetic conditions, i.e., the average velocity of the gas stream as it enters the sample port should be equal to the average velocity of the gas in the main stream at the sample point. Deviations from the isokinetic rate cause errors in the determinations of dust loading.

Gas flow. Many conventional rate and quantity meters can determine the gas flow.

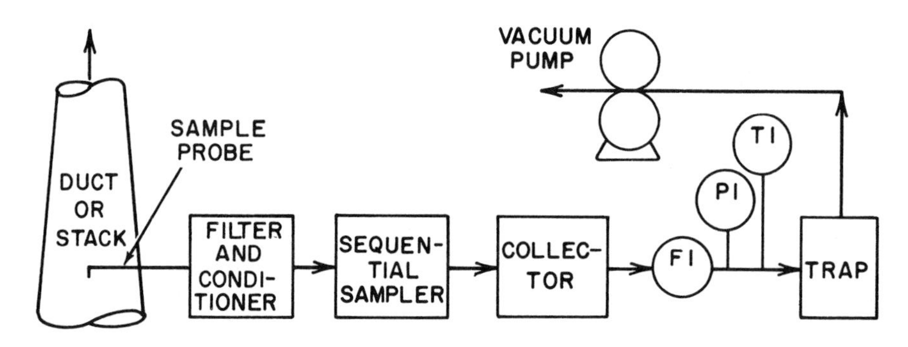

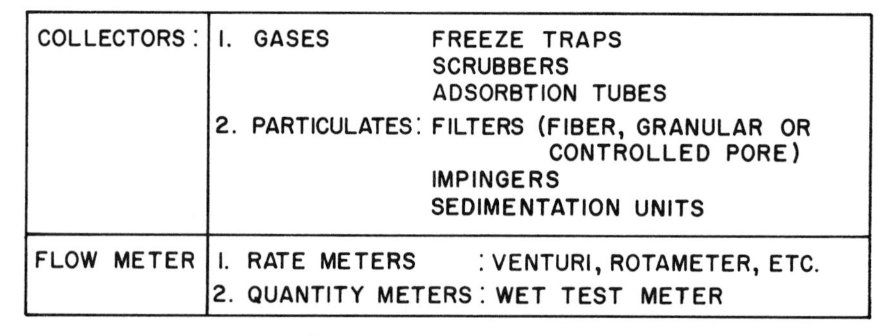

COLLECTORS:	1. GASES	FREEZE TRAPS SCRUBBERS ADSORBTION TUBES
	2. PARTICULATES:	FILTERS (FIBER, GRANULAR OR CONTROLLED PORE) IMPINGERS SEDIMENTATION UNITS
FLOW METER	1. RATE METERS	: VENTURI, ROTAMETER, ETC.
	2. QUANTITY METERS	: WET TEST METER

Fig. 5.2b Stack sampling system components

Sampling for gases. It is only necessary to withdraw a representative sample regardless of flow rate. Various methods of analyzing the sample for gaseous components are reported in the literature.[8,9,10]

Air Quality Monitoring

Particulates

Methods of monitoring particulate matter include

Sedimentation. This is the simplest and the cheapest method, wherein a small bucket containing 1 to 2 liters of water is left in the open for 30 days, at the end of which time the water is analyzed for soluble and insoluble material and the results are reported as tons per square mile per 30 days. This method exposes primary pollution sources and measures long-term changes but cannot be used for continuous monitoring or short-term evaluations.

Filtration. These are among the more widely used techniques. Fiber filters, i.e., paper, and granular filters, i.e., fritted glass, are the most common in source testing and monitoring. Controlled pore filters are used when measurement of particle sizes is desired. Determination of particulates is by weighing, chemical analysis or, in controlled pore filters, sizing methods.

In the high volume sampler, a filtration instrument that monitors radioactive particles, a vacuum pump pulls a large volume of air through a filter of glass fibers. Deposited particulates are analyzed by nuclear radiation detectors.

The filtration device commonly used for continuous analysis is the tape sampler, in which air is drawn through a tape of filter fiber held in a sampling nozzle, and a vacuum pump provides the pressure drop. Particulate matter is deposited on the tape and affects its light transmission characteristics. At preset intervals the tape is advanced to a clean area. Most tape samplers have an automatic and programmable tape advance timer. The tape is analyzed by a photometer for light transmission, the result being reported as coefficient of haze (COH)

$$\text{COH}/1{,}000 \text{ linear ft} = \frac{(\text{OD})\text{A}}{\text{QT}} \times 10^5 \qquad 5.2(1)$$

where optical density (OD) $= \log_{10} \dfrac{I_{100}}{I_m}$

I_{100} = light transmission across clear clean tape (usually set at 100)

I_m = light transmission across measured tape

A = area of sample spot in ft^2

Q = flow rate of air sample in ft^3 per second

T = time of sampling in minutes.

Impingement. When a flowing gas is forced to change direction, particles become deposited, a characteristic utilized in wet and dry monitoring devices. The cascade impactor[11] is in this category and monitors aerosols.

Electrostatic methods. For a brief account on electrostatic precipitation, see Section 5.5 in this volume. The same principle can be employed to monitor particulates.

Centrifugation. Centrifugal methods monitor particles larger than 5μ.

Oxides of Sulfur

The methods of analyzing and monitoring sulfur oxides as pollutants are summarized in Table 5.2c, sulfur dioxide being the predominant pollutant.

Table 5.2c
METHODS OF MEASUREMENT OF SULFUR OXIDES

Method	*Description of Reaction and Analysis*	*Sensitivity (ppm)*
Conductometric	Sulfur dioxide is absorbed in water to form sulfuric acid which is analyzed by change in conductivity; one modification is to titrate sulfuric acid with an alkali reagent.	0.005
Electrolytic	Sulfur dioxide is absorbed in bromine solution and the bromine is titrated by ORP.	0.1
Iodine-thiosulfate	Iodine oxidizes sulfur dioxide to sulfuric acid; remaining iodine is determined by thiosulfate.	0.4
Lead Peroxide	Sulfur dioxide is absorbed by lead dioxide candles to form lead sulfate; lead is determined gravimetrically.	0.02
Colorimetric	Sulfur dioxide reacts with sodium tetrachloromercurate; colorimetric analysis of compound results in sulfur dioxide determination.	0.005
Photometric	Absorption at specific wavelengths by sulfur dioxide is basis for direct measurements.	~10

COLORIMETRIC ANALYZER

The air sample (Figure 5.2d) is wet scrubbed in a rotary vane scrubber by an absorbing agent solution (sodium tetrachloromercurate). The solution is then mixed with formaldehyde-paranosaniline solution and passes through a time delay coil for color development. The solution is photometrically compared with the unreacted paranosaniline solution. The system employs cadmium sulfide photocells, which are usually provided with null point checks. (For a general review of colorimeters, see Section 8.4 in Volume I.)

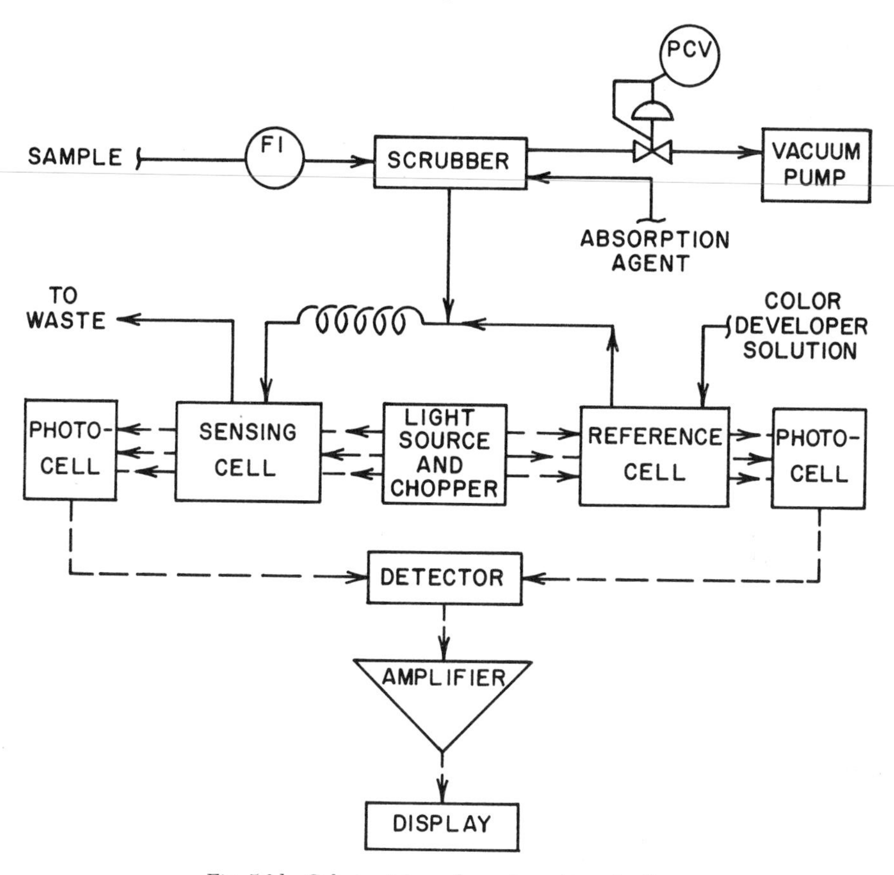

Fig. 5.2d Colorimetric analyzer for oxides of sulfur

ELECTROCONDUCTIVE ANALYZER

Sulfur oxides in the air sample are absorbed by hydrogen peroxide solution in a countercurrent scrubber. Conductivity of the solution is measured to obtain the concentration of sulfuric acid from which sulfur oxides can be determined.

COULOMETRIC ANALYZER

The air sample is drawn through a detector cell containing a buffered solution of potassium iodide (Figure 5.2e). Iodine is generated at the anode, which oxidizes the sulfur dioxide to sulfuric acid, and the unreacted iodine is reduced to iodide at the cathode. Thus, due to the presence of sulfur dioxide, the cathode current is less than the anode current. The difference, measured by the reference electrode, is proportional to the concentration of sulfur dioxide.

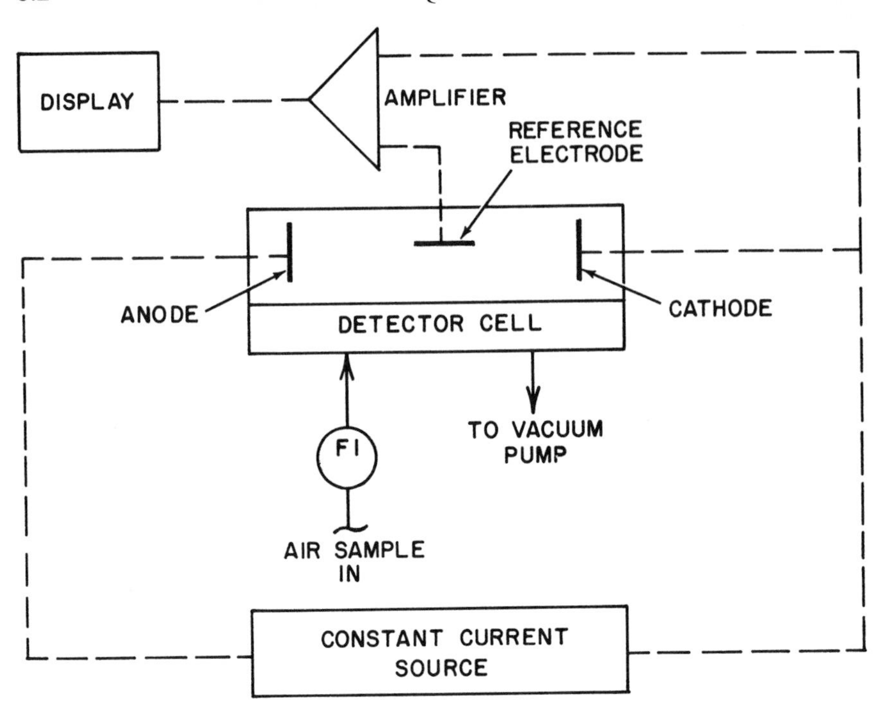

Fig. 5.2e Coulometric analyzer for oxides of sulfur

PHOTOMETRIC ANALYZER

Recently, direct photometric methods have become available for monitoring oxides of sulfur and nitrogen (Figure 5.2f). The pollutant in the sample absorbs certain wavelengths in a beam of light. Analysis of the beam at two particular wavelengths by reference and measurement photocells determines the concentration of sulfur dioxide. The wavelengths are chosen so that the reference wavelength is not absorbed by the pollutant, whereas the measurement wavelength is strongly absorbed. By proper choice of wavelengths, this technique can be extended to a variety of compounds. A variation of this method involves comparison of absorption by the sample with absorption by pure air or nitrogen.

Oxides of Nitrogen

The Saltzman technique is the most common method of NO_x analysis. In this procedure the oxides of nitrogen in the sample are completely oxidized (to nitrogen dioxide) by scrubbing with potassium permanganate solution. A dye complex is formed between the nitrite ion, sulfanilic acid and α-naphthyl amine in an acid medium. Colorimetric determination of the dye

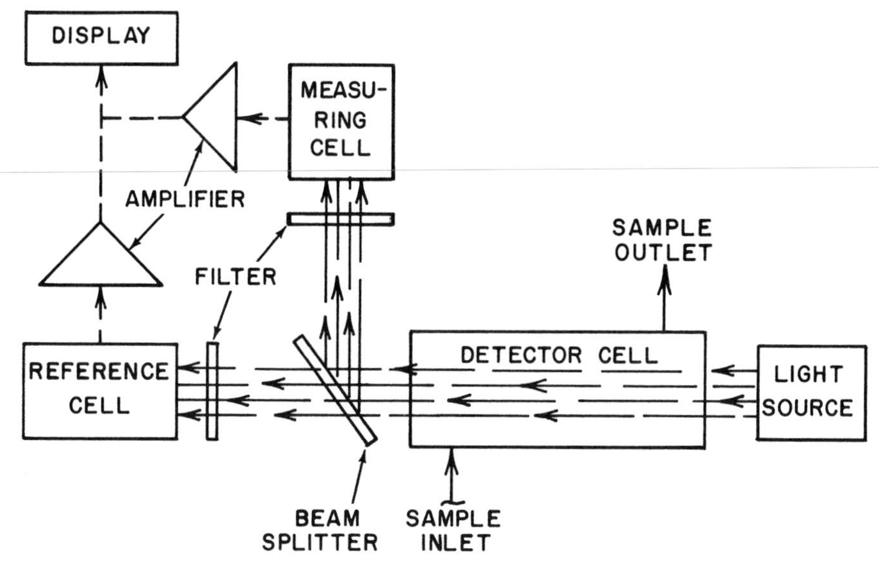

Fig. 5.2f Photometric analysis for oxides of sulfur

complex measures the NO_x content in the sample, usually expressed as parts of nitrogen dioxide, even though other oxides may also be present.

Another method to detect oxides of nitrogen is ultraviolet spectroscopy, in which the sample is first oxidized to convert all NO_x to nitrogen dioxide and then analyzed. (For more on ultraviolet analyzers, see Section 8.2 in Volume I.) This method for NO_x is reported to be sensitive to the ppm range. Recently, direct photometric methods have also become available to monitor oxides of nitrogen. These are similar in principle to those used to monitor sulfur oxides.

Hydrocarbons

The instrumental methods of measuring hydrocarbons in air are the flame ionization and infrared techniques.

FLAME IONIZATION METHODS

The basis of these methods involves measuring the ionization of carbon atoms in a hydrogen flame (see P. 744 in Volume I). A pure hydrogen flame contains negligibly few ions and the presence of even trace amounts of organic compounds will produce substantial numbers of ions.

The sample is mixed with hydrogen and passed to the flame ionization detector (Figure 5.2g). Air is supplied to maintain the flame. The electric potential across the flame jet to an ion collector placed above the flame is measured by an electrometer. The potential is proportional to the concentration of carbon ions in the sample.

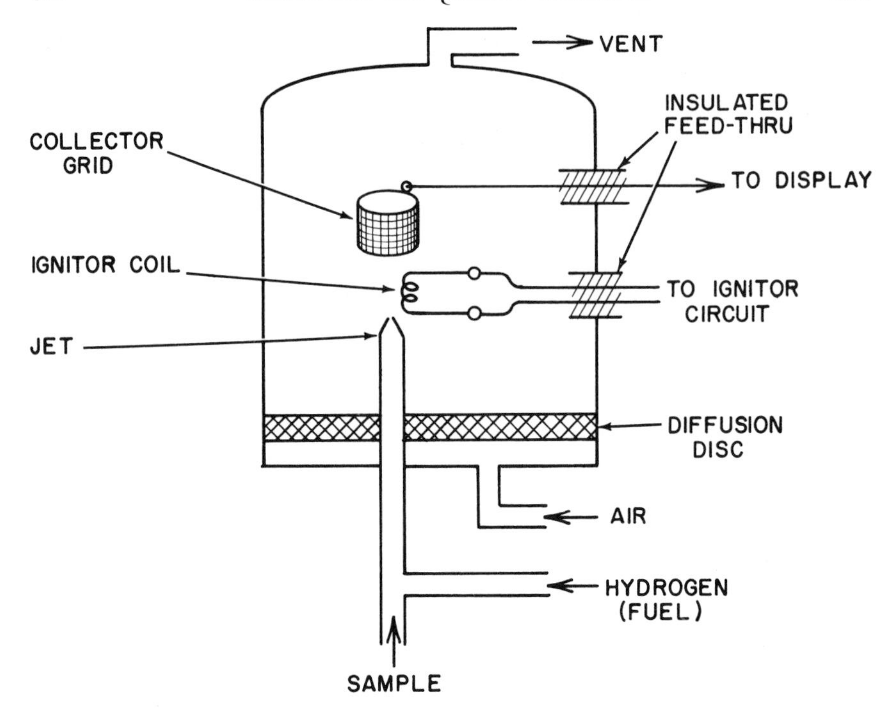

Fig. 5.2g Flame ionization detector for hydrocarbon measurement

INFRARED METHODS

These methods have been used for total hydrocarbon monitoring, particularly from automobile exhausts. The non-dispersive infrared (NDIR) analyzer with hexane-sensitized detectors has been particularly successful (see P. 773 in Volume I.). In this technique, the energy in the IR absorption bands in the air sample is compared to those in normal hexane. The technique requires corrections for carbon dioxide, water vapor and lower hydrocarbons. (For more on IR analyzers, see Section 8.3 in Volume I.)

Total Oxidants

Total oxidants are an approximate index of eye-irritating and plant-damaging pollution caused by an atmosphere containing photochemical smog, and are usually expressed as ppm of ozone but include chlorine, peroxides and other oxidants as well.

In the procedure commonly used to measure total oxidants, the air sample is introduced into a contactor where the oxidants are absorbed by buffered-neutral potassium iodide solution (pH 7.2). The iodide is oxidized to iodine, which can be measured by a dual beam colorimeter or by other spectrophotometric methods. Sampling apparatus is usually a midget impinger. The method is sensitive to a few parts per hundred million (pphm).

In a modification of the procedure, alkaline potassium iodide and phosphoric-sulfamic acids are used. The air sample is absorbed in alkaline potassium iodide and a stable product is formed that can be stored for days. The analysis is completed by the addition of phosphoric-sulfamic acid reagent, liberating the iodine. Thus, a time delay is permissible between sampling and analysis, which is not the case with neutral potassium iodide.

Carbon Monoxide

Of the various methods to monitor carbon monoxide (Table 5.2h), IR analysis is adaptable to continuous monitoring.

In a non-dispersive infrared (NDIR) analyzer for carbon monoxide measurement (Figure 5.2i), two IR sources are used, one for reference and the other for sample. The beams are blocked several times per second by a chopper. In the unblocked condition, the beams pass through the reference and sample cells respectively into the detector. The sample cell receives a continuous stream of sample, whereas the reference cell is a sealed tube containing a reference gas chosen for minimal IR absorption in those wavelengths that are strongly absorbed by carbon monoxide.

The detector assembly consists of two sealed compartments separated by a diaphragm. Both compartments have IR transmitting windows and are filled with pure carbon monoxide. The detector responds to that portion of energy in selected IR bands that are strongly absorbed by carbon monoxide. The differential between the two compartments can be measured by mechanical or electrical means (see P. 773 in Volume I).

Detection from a Distance

Considerable effort has been directed to developing remote sensing instruments that can measure emissions from a distance. Inspection agencies can monitor plant emissions with these devices. Development has also been aided by space research and techniques for detecting and analyzing chemicals.

Remote sensing instruments employ spectrum analysis to detect and measure concentration of pollutants, and correlation spectrometers for remote sensing are available[12, 13].

The principle of remote sensing is illustrated in Figure 5.2j. Scattered and reflected radiation from a distant source is collected in a telescope and dispersed through a grating or prism spectrometer. The spectrum of the radiation can be compared photometrically with a photographic replica of the spectrum of the pollutant.

Problems associated with these instruments include 1.) sensitivity—it changes according to light conditions and to scattering in the lightpath from the source to the instrument; 2.) temperature of the gas at the source—it broadens the absorption spectrum and affects sensitivity; and 3.) wind conditions and particulate turbidity—they cause errors.

Table 5.2h
METHODS FOR DETERMINATION OF CARBON MONOXIDE

Method	*Sampling*	*Sensitivity (ppm)*	*Reaction*	*Analytical Method*
Iodine Pentoxide	Heated I_2O_5 tube after purification train	10 or better	$5\ CO + I_2O_5 \longrightarrow 5\ CO_2 + I_2$	Photometric or titrimetric determination of I_2
Gas Chromatography	Special columns in series	2	CO is separated in first column and catalytically reduced by H_2 to CH_4 in second column	CH_4 determined by flame ionization detector
Infrared Analysis	Pressurized. Nondispersive infrared spectrophotometer	100 or better	—	CO determined by IR absorption

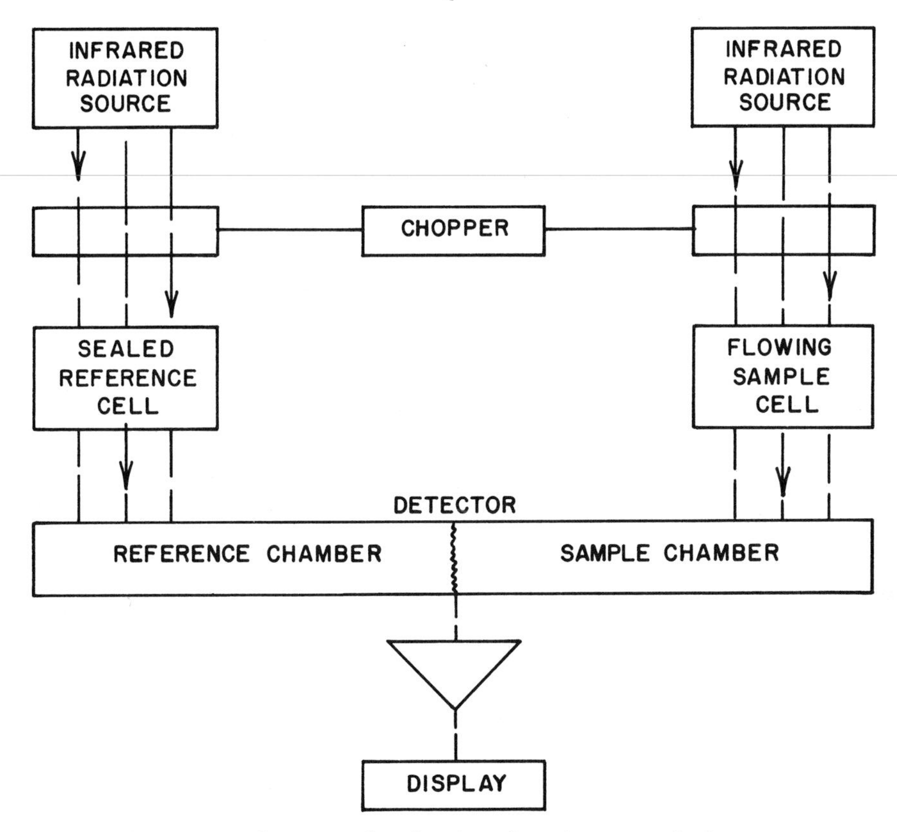

Fig. 5.2i Nondispersive infrared analyzer for carbon monoxide detection

Automobile Exhaust Emission Measurement

By 1975, all new vehicles will have to meet Federal emission standards. These standards require a 90% reduction in emissions for 1976 automobiles compared to those of a few years ago; some states have already passed stringent automobile emission standards—hence, the current interest in instrumentation systems to measure pollutants in automobile exhausts.

Automobile emission standards focus on four types of pollutants including 1.) particulate matter, 2.) hydrocarbons, 3.) carbon monoxide and 4.) nitrogen oxides.

One important aspect of automobile emission instrumentation is the method of sampling. Frequently, a variable dilution sampling system is used to minimize condensation and other problems. A typical instrumentation scheme for automobile exhaust emission measurement is shown in Figure 5.2k.

Smoke is usually measured by a light transmission type photometer, and total hydrocarbons are generally measured by flame ionization methods.

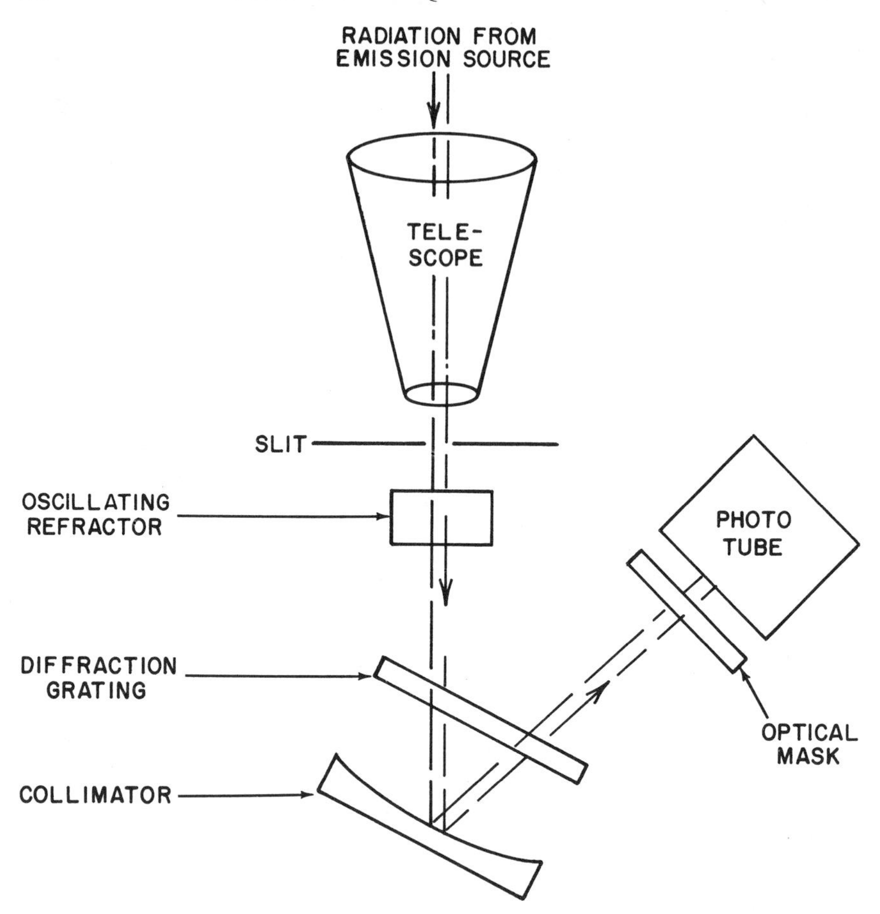

Fig. 5.2j Remote sensing spectrometer

Non-dispersive infrared methods are the most widely used for carbon monoxide and carbon dioxide. Measurement of NO_x in exhausts is relatively new. Presumably, the Saltzman technique and other procedures can be employed.

Future Trends

Rapid changes in laws and public attitudes concerning air pollution have provided a major impetus for the development of instrumentation to monitor air pollutants. Major improvements are likely to be made in sensitivity, reliability and ease of use of monitoring devices. The more sophisticated instruments are still in the early stages of development, and problems do arise in field use.

Major metropolitan areas are likely to install extensive telemetering systems for pollutant concentration surveys. These, together with extensive

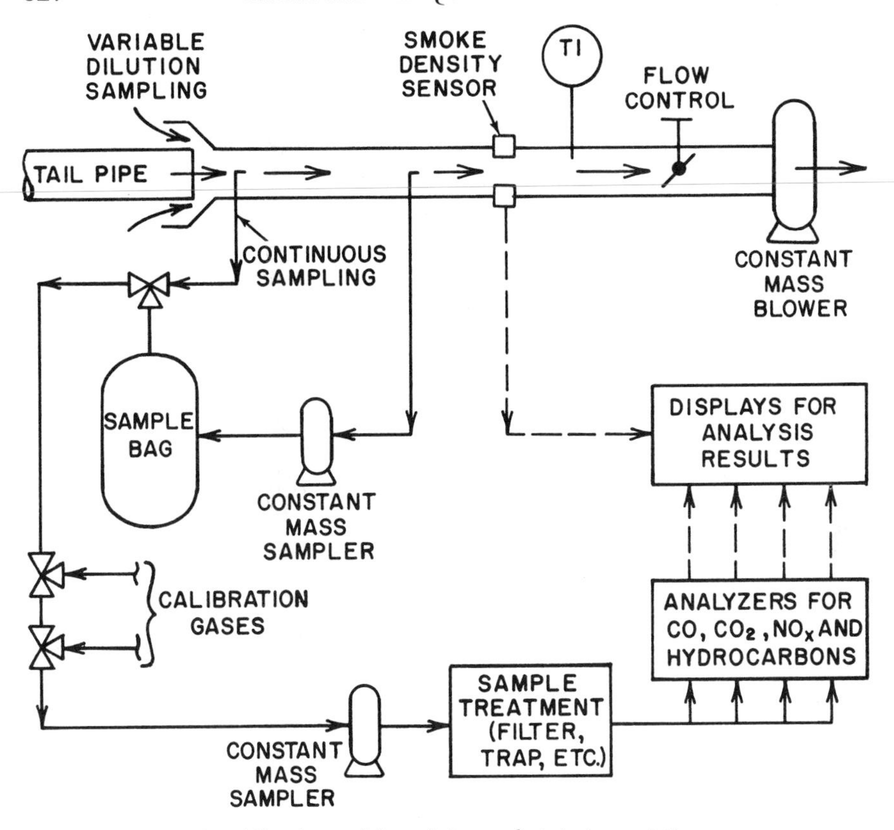

Fig. 5.2k Automobile emission analysis instrumentation

meteorological data, will permit continuous and comprehensive monitoring of air quality and rapid anticipation of potentially unsafe pollution levels.

REFERENCES

1. "Air Pollution Handbook," New York: McGraw-Hill, 1956.
2. Conner, W. D. and Hodkinson, J. R., "Optical Properties and Visual Effects of Smoke Stack Plumes," U.S. Department of Health, Education and Welfare, Bulletin 999-AP-30.
3. Coons, J. P., et al., "Development, Calibration and Use of a Plume Evaluation Training Unit," *J. APCA*, *15*: 199, May, 1965.
4. "Smoke Density Indicators and Recorders," British Standard 2811, British Standards Institute, London, 1957.
5. Jacom, J. E., "Problems in Judging Plume Opacity," *J. APCA*, *13*: 36, Jan., 1963.
6. "Methods for the Determination of Velocity, Volume, Dust and Mist Content of Gases," Bulletin WP-50, Western Precipitation Division, Joy Manufacturing Co., Los Angeles, Calif., 1968.
7. Stern, A. C. (ed.), "Air Pollution," Vol. II, New York: Academic, 1968.
8. "Selected Methods for the Measurement of Air Pollutants," Public Health Service Bulletin 999-AP-11, APCO Technical Information Center, Research Triangle Park, N.C.

9. "Methods of Measuring Air Pollution," Organization for Economic Cooperation and Development, Paris, 1964.
10. Blosser, R. O. and Cooper, H. B. H., "Analytical Equipment and Monitoring Devices for Gases and Pollutants," National Council for Air and Stream Improvement, Technical Bulletin No. 35, 1968.
11. Brink, J., "Cascade Impactor for Adiabatic Measurements," *I&EC,* 50: No. 4, 645, April, 1958.
12. Barringer, A. R. and Schock, J. P., "Progress in Remote Sensing of Vapors for Air Pollution, Geologic and Oceanographic Applications," Proceedings of the IVth Symposium on Remote Sensing of the Environment, University of Michigan, Ann Arbor, Mich., 1966.
13. Williams, D. T. and Roltig, B. L., "Molecular Correlation Spectroscopy," *Applied Optics,* 7: No. 4, 607, 1968.

5.3 AIR POLLUTION CONTROLS

Many of the control systems in air pollution abatement are ill suited to feedback control. Often, they are also unnecessary, due partly to the nature of the system itself and partly to the unavailability of reliable continuous analyzers.

Legal Aspects

The design of all air pollution control systems has to recognize local, state and federal standards on permissible emission of pollutants. The various legal standards fall into two broad categories including ambient standards and emission standards.

Concerning legal standards on air quality, the U.S. Government through the Environmental Protection Agency (EPA) establishes the ambient air quality standards. These limits define the permissible level of pollutants in ambient air. Taking this basic standard, all 50 states are to develop emission standards and implementation plans with which process industries will be increasingly concerned. In addition, highly populated areas have special local standards.

Federal Ambient Air Standards

The National Primary Ambient Air Quality Standards establish minimum standards in the United States, which in the judgment of the EPA are a requisite to protect public health. The more stringent National Secondary Air Quality Standards were established to protect the public from any known or anticipated ill effects from pollutants in ambient air.

In April 1971, primary and secondary ambient air quality standards[1] were published for six air pollutants including oxides of sulfur, particulate matter, carbon monoxide, photochemical oxidants, hydrocarbons and oxides of nitrogen (Table 5.3a). Ambient standards are not specified for many chemicals including those that are toxic. These are still to be established, based on the philosophy that the permissible levels shall be set to protect public welfare from known or anticipated adverse effects.

Table 5.3a
NATIONAL AMBIENT AIR STANDARDS*

Pollutant	Primary Standard	Secondary Standard
Sulfur Dioxide (in micrograms/m^3)		
annual arithmetic mean	80	60
maximum 24-hour concentration	365	260
maximum 3-hour concentration	—	1300
Particulates (in micrograms/m^3)		
annual geometric mean**	75	60
maximum 24-hour concentration	260	150
Carbon Monoxide (in milligrams/m^3)		
maximum 8-hour concentration	10	10
maximum 1-hour concentration	40	40
Photochemical Oxidants (in micrograms/m^3)		
maximum 1-hour concentration	160	160
Hydrocarbons (in micrograms/m^3)		
maximum 3-hour concentration	160	160
Nitrogen Dioxide (in micrograms/m^3)		
annual arithmetic mean	100	100

* All values are so specified that the maximum concentration cannot be exceeded more than once per year.
** Geometric mean is the nth root of the product of n numbers.

Emission Standards

Emission standards govern the permissible levels of pollutants from all stacks and are set by the states. The theory underlying emission standards is that by specifying exhaust standards, the general level of pollutants can be controlled to meet ambient air standards. In order to make the emission standards equitable to large and small units, a graduated emission standard is coming into general use. On this basis, the permissible concentration of pollutants decreases with increasing quantities of exhaust from a stack, i.e., the larger the stack, the more stringent the requirements.

At present, emission standards and procedures vary widely. New Jersey, for example, defines permissible emission standards[2] on the basis of 1.) effect factors for coarse and fine particulates, 2.) stack height and, 3.) distance of stack to plant property line.

Other states, like Pennsylvania[3], define allowable emission of particulates as a fraction of potential emission (without air pollution abatement equipment), taking into account the location.

New emission standards are likely to be established in most states. The reader should refer to the latest regulations from the particular state and locality.

Special Local Standards

Besides ambient air and emission standards set at federal and state levels, there are pollution programs initiated by local air pollution control districts, as in highly urbanized areas like Los Angeles and New York, with standards that must be met by a manufacturing facility if it wants to operate within these areas. The standards are often more stringent than the state or federal requirements. In Los Angeles, for example, organic solvent emissions are severely restricted.

Pollution Control Systems

To design a system for pollution abatement, it is necessary to know the type and quantity of pollutants in the exhaust. Three methods that supply this emission inventory include 1.) a comprehensive and accurate material balance over the whole process plant. The computed losses can be attributed to emissions; 2.) use of emission factors, compilations of which can be found in reference 6. These factors often permit the approximate estimating of pollutant emissions; and 3.) source testing at the stack. This is the most accurate method.

Gravity Chambers and Cyclones

In gravity chambers, the dust-laden gas enters a large chamber where the gas velocity drops sharply and the dust or droplets settle out by gravity. The chamber is usually a horizontal rectangular chamber with an inlet at one end and an outlet at the top or at the other end. Their inherent simplicity makes the chambers applicable to many types of construction, but they are rarely useful for removing particles less than 50μ in diameter. The potentials for process control in a gravity chamber are limited. Although the pressure drop can be measured across the unit, the information is not used for control.

Cyclones are settling chambers in which centrifugal acceleration replaces gravity. The dust-laden gases tangentially enter a cylindrical control chamber and leave through a central opening on the top (Figure 5.3b). By virtue of inertia, the dust particles move to the outside wall, from which they fall to a receiver. Cyclones operate at equivalent force levels of as low as 1 G to as much as 2000 G's.

Cyclones can effectively separate particles which are more than 5μ in diameter. Where dust agglomeration is substantial, cyclones can be efficient at the 2μ particle size level. Cyclones can be classified as conventional, high efficiency and multiple-tube.

High efficiency cyclones have smaller inlets and produce larger separating forces than the other two types, and their body diameters are usually less than one foot. When these small diameter units are employed, many of them are run in parallel to increase capacity. A common inlet plenum chamber,

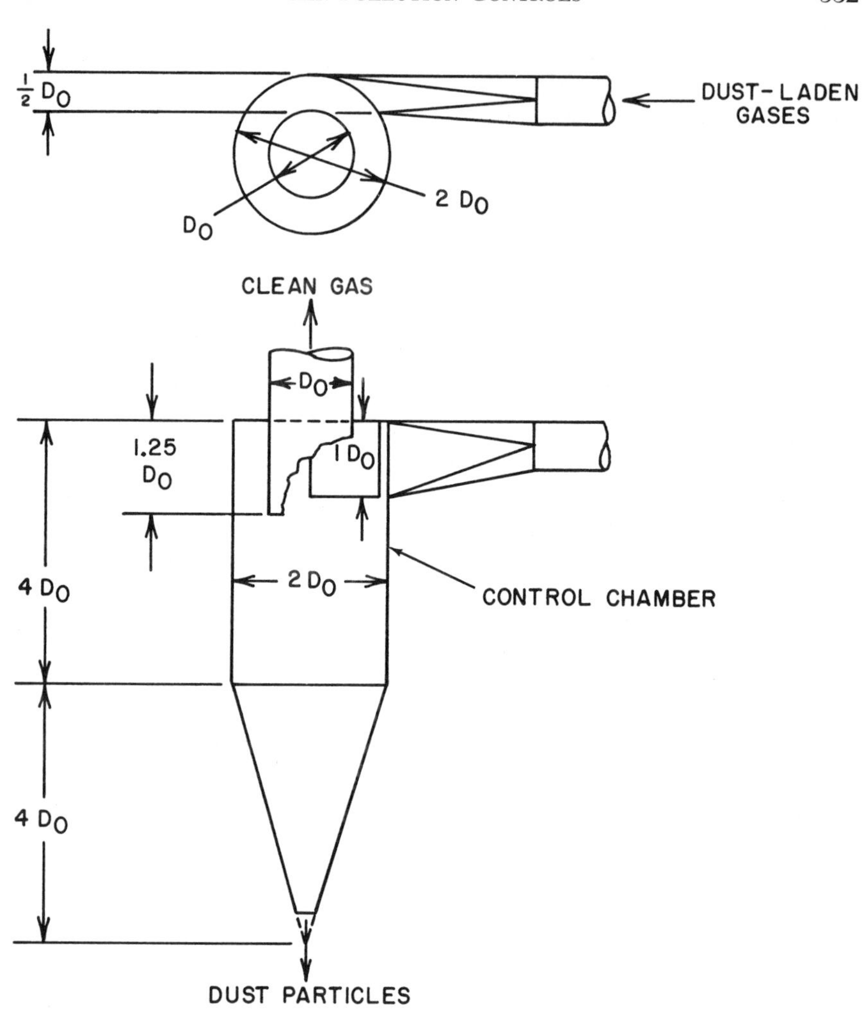

Fig. 5.3b Typical dimensions of a cyclone

dust receiver and outlet plenum chamber usually serve a number of high efficiency cyclones in parallel. In such systems, it is necessary to equalize the vapor flow distribution to the various cyclones so as to prevent backflow, re-entrainment and plugging.

Characteristics of the various cyclone designs are listed in Table 5.3c. The following operating variables are pertinent to cyclone operation and control:

Flow rate. As inlet velocity increases, there is a rapid rise in dust removal efficiency followed by a flat level-off region.

Table 5.3c
CYCLONE CHARACTERISTICS

Type	*Space Required*	*Efficiency by Weight*	*Pressure Differential ("H_2O)*	*Power HP/1000 ft^3 of Gas*
Conventional	large	50% on 20μ	1 to 3	0.2 to 0.7
High Efficiency	medium	85% on 10μ	3 to 6	0.6 to 1.5
Multiple-tube	small	90% on 5 to 7μ	4 to 6	1 to 1.5

Gas properties. Gas density and viscosity are both important. Gas viscosity increases with temperature and hence, collection efficiency falls.

Dust properties. Size and density of dust are the principal factors.

Cyclone efficiency. This variable is increased by letting as much as 10% of gas escape through the apex at the bottom.

Cyclones are frequently operated without automatic controls, and pressure drop across the unit is measured by manometers and periodic checks on dust loading are made. Because the vapor flow rate is a function of plant operation, cyclones are usually sized for maximum capacity and then allowed to operate at a lower load and efficiency.

Bag Filters

Bag filters or "baghouses" offer high separation efficiency at small particle sizes. They produce dry dust which is easy to handle and which can frequently be reused. The exhaust gases are not saturated with moisture (as with scrubbers) and will not cause plume formation. In Figure 5.3d dust-laden gas flows through a porous medium and deposits dust in the voids. As the voids fill, pressure drop increases, and eventually a point is reached when the fabric must be cleaned.

The factors in designing a baghouse include cloth form, cloth type, cleaning method and housing.

CLOTH FORM

The two basic forms are the 1.) envelope type, i.e., flat bags. These are basically sheets held in place by a frame; and 2.) tubular type, i.e., round bags. Multi-bags are tubular bags sewn together in groups. Direction of gas flow through the tubular bags can be inside-out or outside-in. In the latter case, a frame is inserted inside the bag to prevent collapse.

CLOTH TYPE

Filter cloth may be woven or felt fabric. When woven cloth is used, a residual load of dust is usually left on the cloth after cleaning and serves as a grid for the next operating cycle. Air-to-cloth ratios in woven cloth are

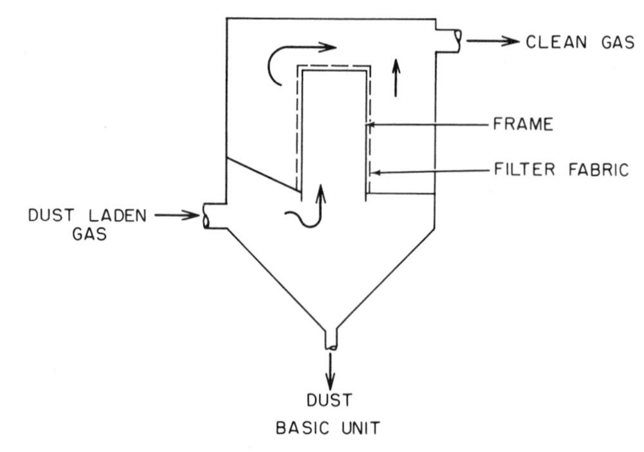

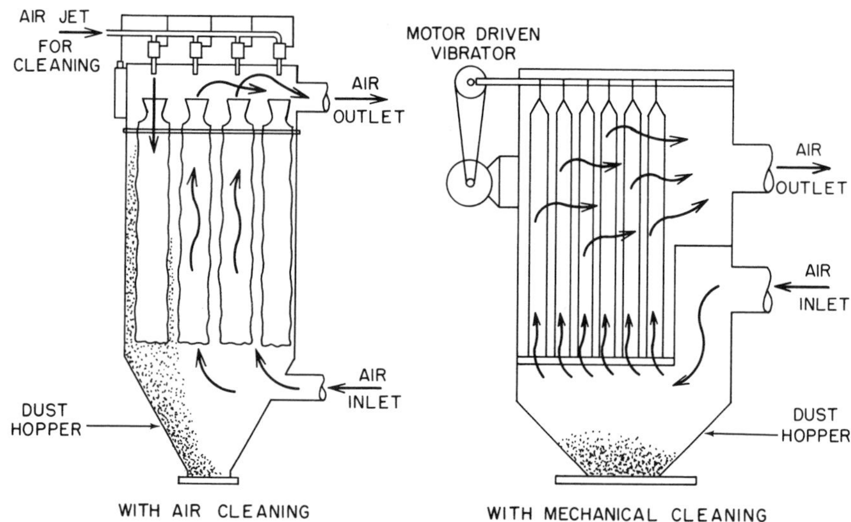

Fig. 5.3d Bag filter designs

usually low and range from 1 to 6 ACFM per sq. ft. Woven fabrics are limited to a maximum of 8 ACFM per sq. ft of cloth surface by the compacting action of higher velocities. These are not efficient for small size dust, i.e., $<5\mu$.

Felt fabrics are more expensive, but permit higher velocities. Air-to-cloth ratios for felt fabrics range from 4 to 10. In air pollution control duty, some regulatory agencies have suggested air-to-cloth ratios of 5 to 7. The properties of some filter cloth materials are summarized in Table 5.3e. The major drawback of all bag filters is the temperature limitation on fabrics. It is often desirable to cool a hot gas to permit the use of bag collectors; otherwise they are very efficient.

Table 5.3e
FILTER CLOTH MATERIAL PROPERTIES

Fabric	Max. Temperature, (°F) Long Exposure	Abrasion Resistance	Tensile Strength	Resistance to		Cost
				Acids	Alkalies	
Acrylics	250	G*	S	G	S	Moderate
Cotton	160	G	S	P	S	Low
Nylon	220	E	E	S	G	Moderate
Nomex	400	G	E	G	G	High
Teflon	450	P	G	E	E	High
Polypropylene	150	G	E	G	S	Low
Polyesters	275	G	E	G	S	Moderate

*E = excellent G = good S = satisfactory P = poor

CLOTH CLEANING

The cleaning methods apply shaking, reverse flow or a combination of the two. Shaking techniques can be mechanical, air shaking, bubble cleaning, jet pulse, reverse air flex and sonic.

Reverse flow techniques include repressurizing, atmospheric and reverse jet methods.

For woven fabrics, shaking methods are the most common. In addition to mechanical shakers, bags can be whipped by air jets between rows of bags. In the bubble method, air bubbles impart a wavy motion, thus shaking the dust from the fabric. The jet action of compressed air through a Venturi section at the top of bags is also utilized. Sonic methods are not fully proven and are commonly employed with the bags collapsed.

Reverse flow cleaning by repressurizing is very successful if bags are supported by a frame to prevent collapse. Repressurizing is by a separate backwash fan. Similarly, a suction fan can be used if the filter operates at atmospheric pressure. In the reverse jet method, a moving ring blows compressed air to remove the dust.

CONTROL OF BAGHOUSES

Bag filter control is concerned with cleaning cycle rather than with normal operation. Cleaning cycles can be controlled in a manual, semiautomatic, automatic or continuous manner.

In small bag filters, or if cleaning once a shift is adequate, the electrical or pneumatic shakers are *manually* activated. Interlocks ensure that the fan is shut off before the shaker can be activated. In intermittent processes in which the *semiautomatic* control frequently employed as the blower is turned off, a timer is activated. The shaking cycle is automatically started after a delay to permit the blower to come to rest. An interlock prevents the fan from being started until the cleaning cycle is over.

In *automatic* control, compartmented baghouses are automated on a programmed cycle. The operation is shown in Figure 5.3f for a two-compartment unit. At a predetermined pressure drop across the units detected by dPIS-1 or dPIS-2, the cleaning cycle is initiated. The section with the high pressure drop is isolated from the process and the shaker drive is started. Cleaning action may be assisted by reverse air flow through UV-5 or UV-6.

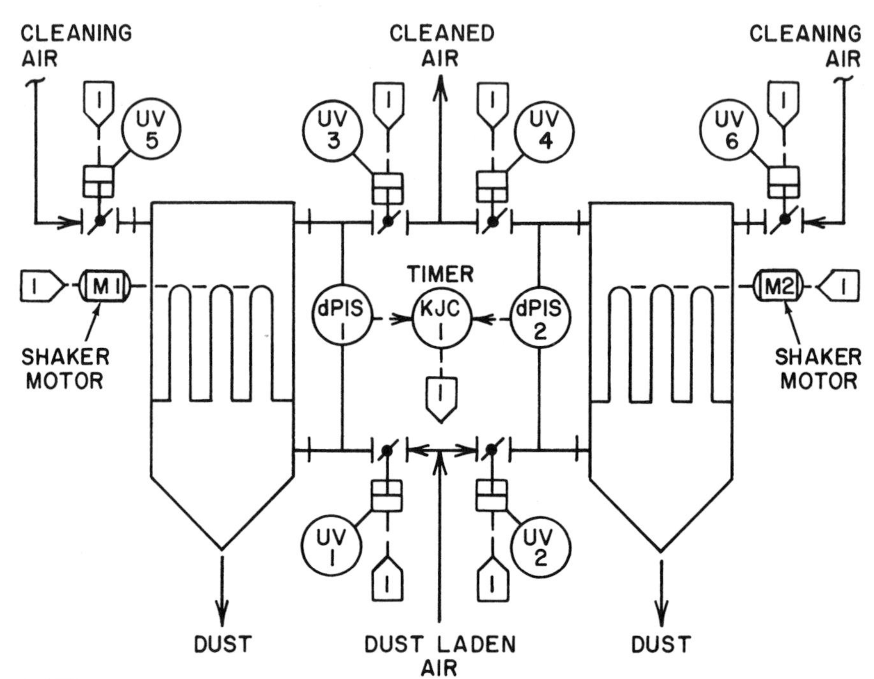

NOTE:
INTERLOCK FLAG [I] REFERS TO THE FOLLOWING OPERATING LOGIC:

OPERATING STATES \ CONTROLLED DEVICES	dPIS-1	dPIS-2	KJC-1	M-1	M-2	UV-1	UV-2	UV-3	UV-4	UV-5	UV-6
dPIS-1 AND-2 ARE NOT HIGH, KJC-1 IS OFF	LOW	LOW	OFF	OFF	OFF	O	O	O	O	C	C
dPIS-1 STARTED KJC-1 CYCLE BECAUSE HIGH DIFFERENTIAL	HIGH	LOW	ON	ON	OFF	C	O	C	O	O	C
dPIS-2 STARTED KJC-1 CYCLE BECAUSE HIGH DIFFERENTIAL	LOW	HIGH	ON	OFF	ON	O	C	O	C	C	O

O – OPEN, C – CLOSED

Fig. 5.3f Automatic baghouse cleaning cycle

At the end of the cycle, the section is automatically returned to the operating cycle. The time required for cleaning one section is from 3 to 6 minutes, whereas the operating time may be 15 minutes to several hours.

Compartmented baghouses should be designed so as to furnish adequate filtering area during all phases of operation. For efficient operation without excessive over-design six or more compartment sections are frequently needed.

Felt fabric filters of the Hershey-type using reverse jet cleaning can be fully *continuous*. Compressed air is blown into the fabric by a moving ring (Figure 5.3g). Indenting the felt, the blow ring is sealed so that the jet passes directly into the felt, dislodging the dust. These filters are thorough in cleaning and require felt bags.

WET SCRUBBERS

Wet scrubbers can remove gases, mists or particulate matter by gas absorption or impingement. A third collection method, nucleation, is feasible for fume or dust of submicron size.

The major types of wet scrubbers include cross flow, counter-current, wet cyclone, venturi and vertical air washers.

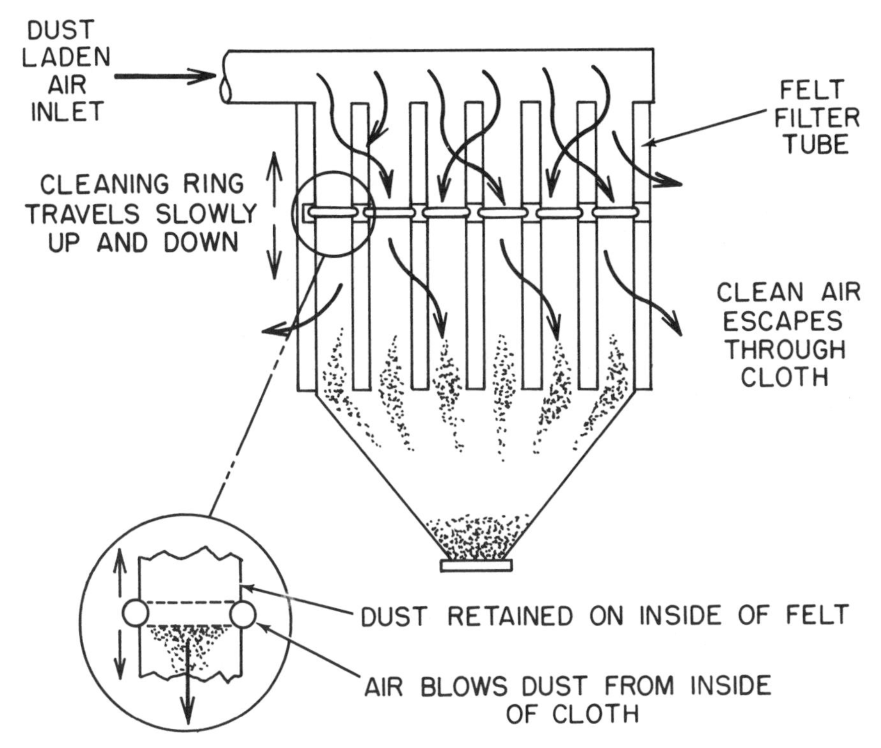

Fig. 5.3g Reverse jet cleaning of bag filters

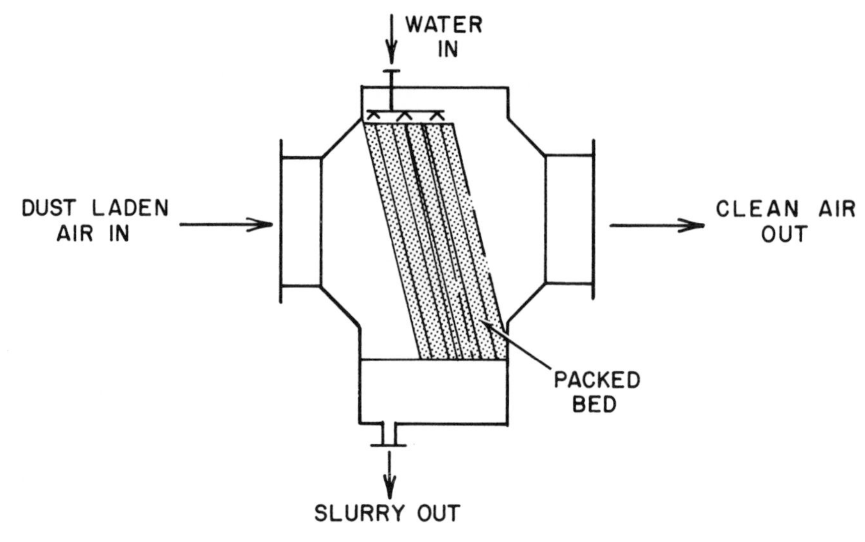

Fig. 5.3h Crossflow scrubber

In the *cross-flow* type, an air stream moves horizontally through a packed bed while the liquid flows downward (Figure 5.3h). This design will effectively remove particles above 3μ in size. The water consumption of these units is low and they can handle high air capacities at low pressure drops.

A modification of the cross-flow design is the parallel-flow type in which both gas and liquid flow through the bed horizontally. Liquid flow prevents blinding of the bed. This design is applicable to gas streams with high dust loadings.

In the *counter-flow* scrubber the gas stream moves upward to meet a descending liquid stream (Figure 5.3i). The methods to increase contact between liquid and gas include plate-type contacts including valve trays, sieve trays and impingement plates; and various types of packings. The design provides a reasonably constant driving force to move the gaseous or solid contaminants into the liquid.

A modification of the counter-current design is the concurrent type, in which both the gas and liquid move down through the bed, a useful maneuver if, owing to space limitations, high gas velocities must be handled.

As shown in Figure 5.3j, in the *wet cyclone* scrubber the gas stream is forced into a spiral pattern, generating high accelerational forces. High pressure spray nozzles provide the liquid spray mist which intercepts the solids to increase collection efficiency. Wet cyclones are very efficient in removing fine dust.

Venturi scrubbers are high performance devices for the removal of particulates in low concentrations, where high pressure drops (10 to

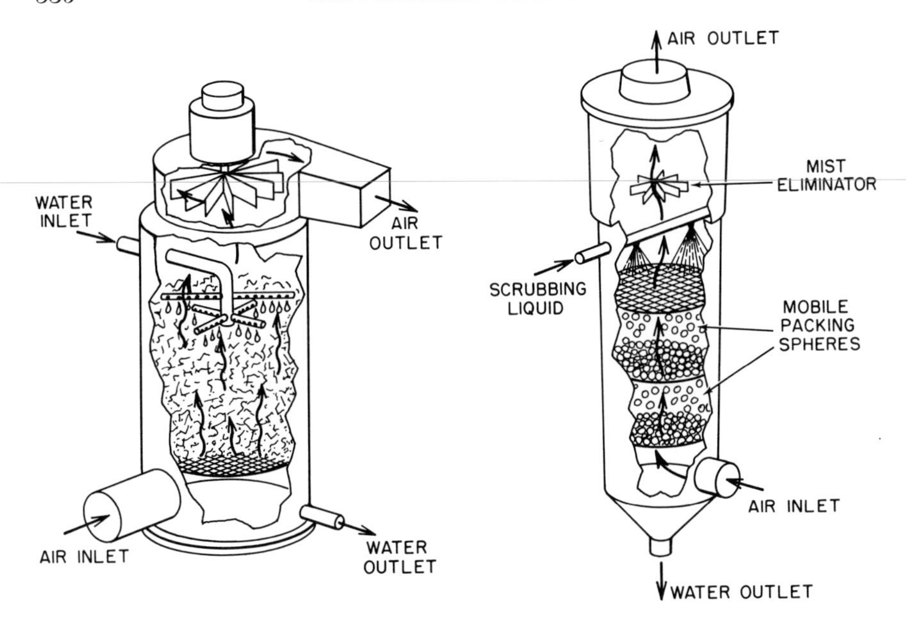

Fig. 5.3i Counterflow packed tower designs

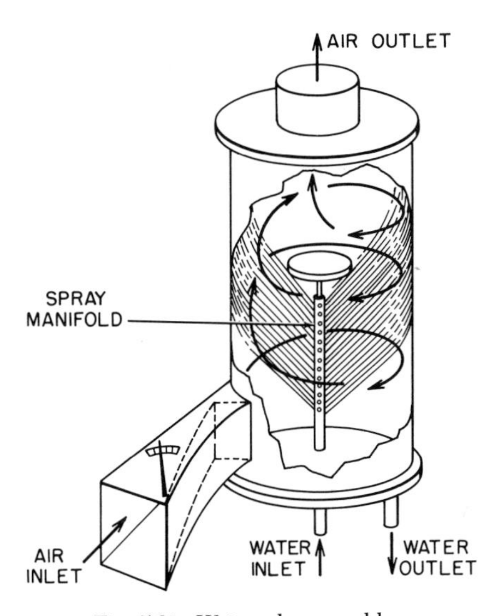

Fig. 5.3j Wet cyclone scrubber

80 inches water) are available. Gas and liquid pass through the small throat section of a wet venturi (Figure 5.3k), and the velocity breaks the liquid into small particles, which in turn collect the dust.

VERTICAL AIR WASHER

This is an inexpensive scrubber designed for removing liquid particles from exhaust gases (Figure 5.3 l).

CONTROL OF WET SCRUBBERS

The gas being scrubbed ordinarily contains solid or liquid particulate matter, and the scrubbing liquid is usually water. Because the outlet gas is cooled and saturated with moisture, some of the water vaporizes and leaves with the clean gas, thereby increasing the water consumption. The amount of liquid required to remove the solid particulates is determined by the concentration of slurry that can be handled—between 2 and 10% solids in the slurry. Pressure drop on the gas side varies from 3 to 15 inches or more of water, whereas liquid is often sprayed and therefore requires a 20 PSIG supply.

The gas stream is usually uncontrolled, its flow rate being a function of the process, whereas the liquid side is automatically controlled.

In the control system shown in Figure 5.3m, the make-up water is on flow control whereas the removed slurry is on level control. The blower is on the exhaust side so that it can handle a cleaner gas, and the pressure

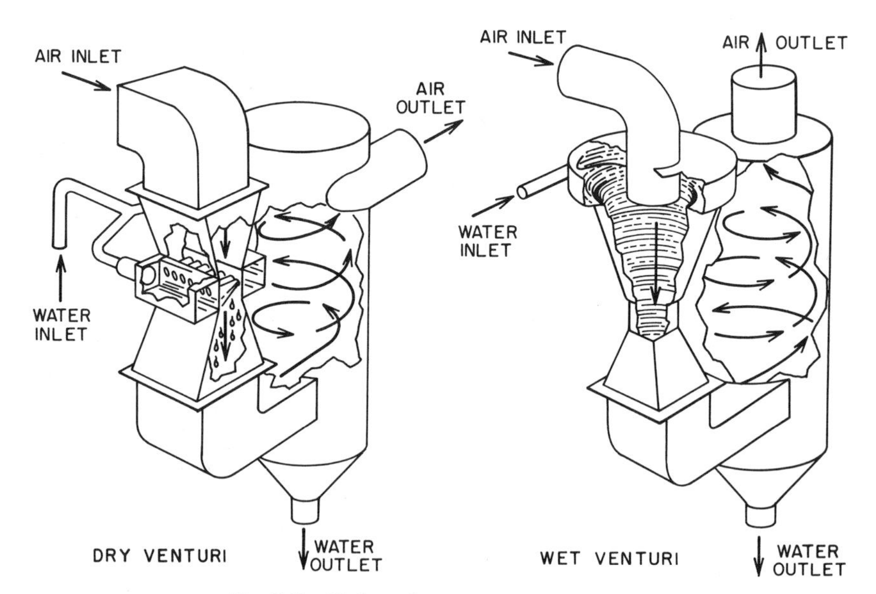

Fig. 5.3k High performance venturi scrubbers

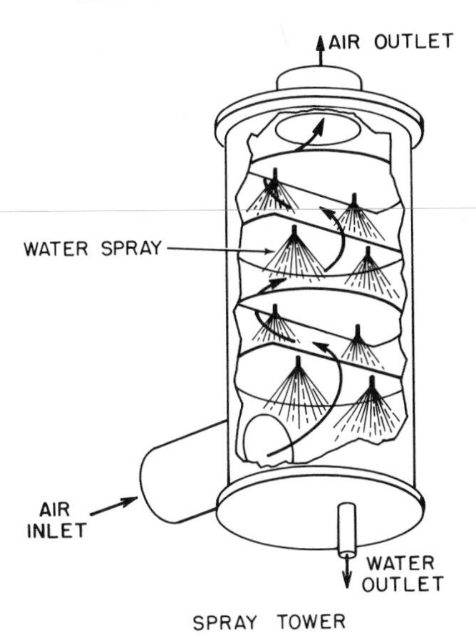

Fig. 5.3l Vertical air washer

indicator and butterfly valve permit manual adjustment of operating pressure.

In another variation the make-up water may be on level control with the outflow throttled so as to maintain constant slurry concentration. In this system a density detector is installed to measure % solids.

PLUMES

The exhaust gases leaving the scrubber are saturated with water. Particularly if they are hot, this causes severe opacity owing to formation of a plume. Ordinarily, the white plume is only water vapor but it can still cause public concern. Therefore, devices have been installed eliminating the plume by 1.) diluting it with atmospheric air, 2.) cooling the exhaust before it leaves the stack and 3.) superheating the exhaust gas by duct burners.

Electrostatic Precipitators

When suspended particulate matter in a gas is exposed to gas ions in an electrostatic field, it becomes charged and migrates to the opposite electrode. Electrostatic precipitators employ this method to separate solid particulate matter or aerosols from gas streams. Precipitators (Figure 5.3n) are of the plate type—mainly used to remove solids from gas streams, and of the tube type—utilized to remove aerosols (solid or liquid particles in a colloidal system with gas) and fumes from gas streams.

The collecting surface in a precipitator also acts as the storage area for the collected dust. Discharge electrodes are usually barbed or twisted steel

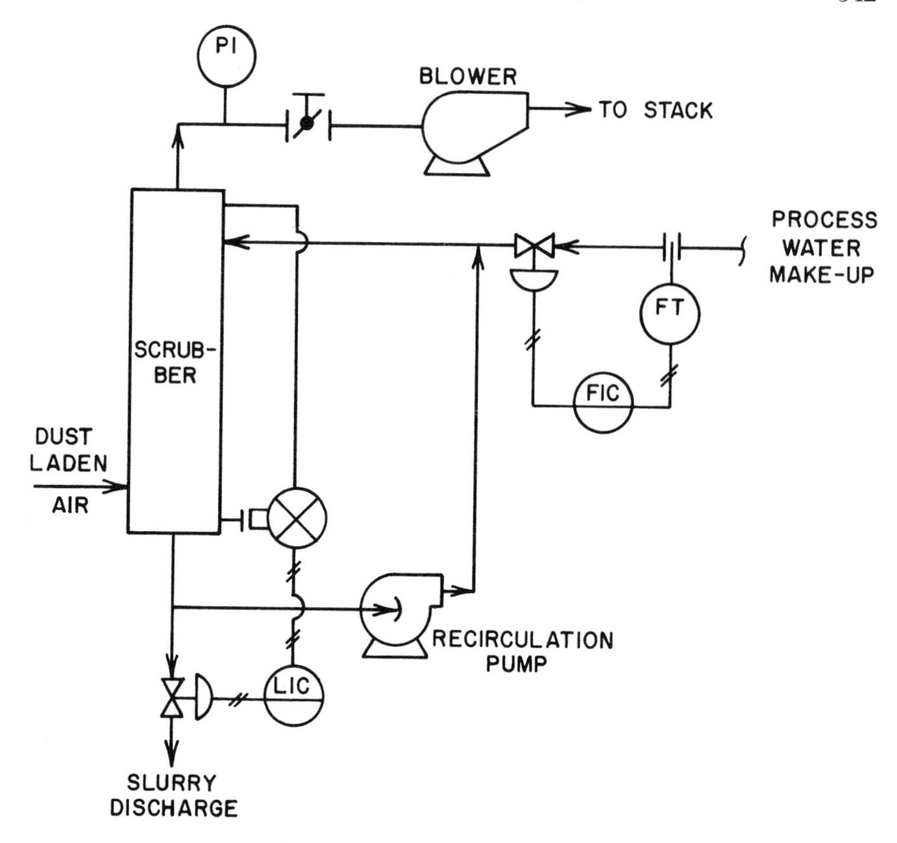

Fig. 5.3m Control of a wet scrubber

wire, maintained under tension because precise alignment of the wire and the collecting surface is necessary for efficient operation. Dust must be frequently removed from the collecting surface by water washing or mechanical rapping. The latter is more common and is applied by a number of collecting surfaces being connected to a rapping bar. A hammer strikes the bar at a frequency of 60 or more times per minute to dislodge the collected dust. Hoppers located beneath the collecting surfaces accumulate the dust until it is disposed of.

AC power supply to electrostatic precipitators is rectified and transformed to a suitable DC level. Voltage between electrodes is limited by their spacing because sparkover is likely to occur at 10 to 15,000 volts per inch. Although either positive or negative potential can be applied to the discharge electrode, the latter is more common.

Electrostatic precipitators have very high efficiency, in the range of 95 to 99.9%. Pressure drop across the unit is low, usually ranging from $\frac{1}{4}$ to 1 inch of water. Electric power consumption ranges from 0.2 to 0.5 kW per

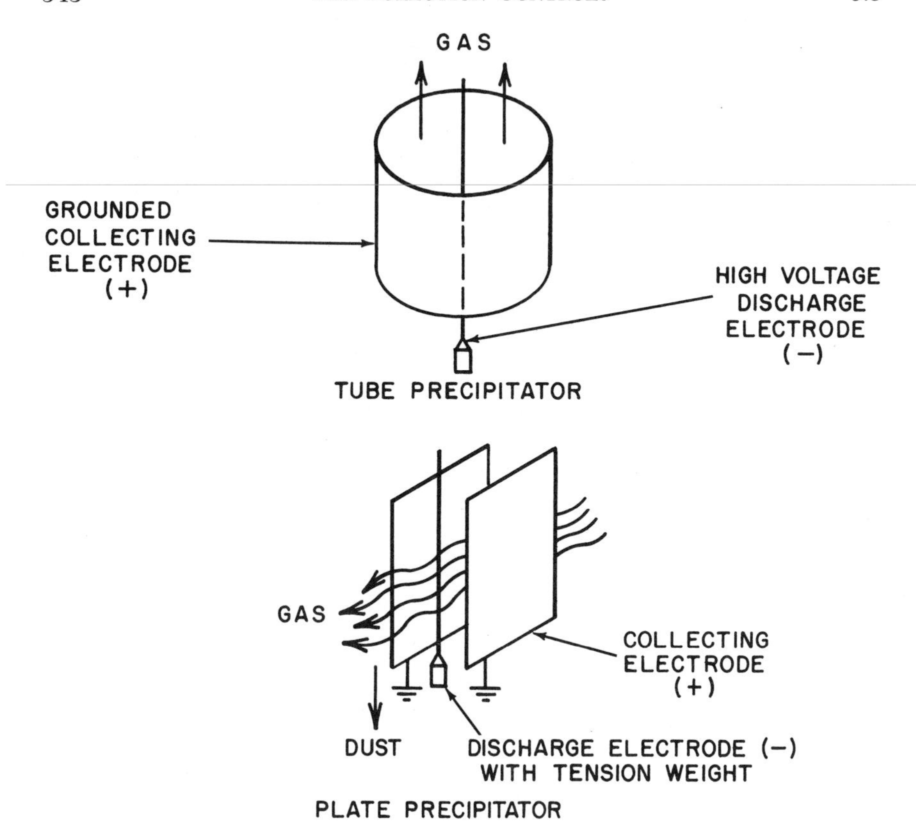

Fig. 5.3n Electrostatic precipitators

1,000 CFM gas. Disadvantages include the high initial cost and the careful maintenance that is required. A special type of precipitator in air conditioning is the two-stage unit in which corona discharge occurs in the first stage and collection is effected in the second under a steady field. Such units have also been used for fly ash collection.

CONTROL OF PRECIPITATORS

There is little feedback control on the gas side. Gas distribution at the inlet is important, but is usually achieved by the entrance configuration. The power supply consists of a transformer to step up the commercial line voltage and a rectifier to convert it to DC current. Maximum output voltages approach the 40 to 65,000 volt range. Taps on the transformer primary permit adjustments in step-up ratio. Ballast resistance in the primary transformer suppresses sparking; the rapping mechanism is usually controlled by a timer with the frequency and force of rapping adjusted so that it dislodges the dust without leaving the electrode naked.

Adsorption Systems

Any gas, vapor or liquid will partially adhere to a surface due to the binding effect of "residual" forces on surfaces. This phenomenon, called adsorption, is useful in air pollution control in concentrating gaseous pollutants. The quantity of material that can be adsorbed on a given weight of adsorbent depends largely on the concentration of the material in the space around the adsorbent, on the total surface area of the adsorbent and on temperature.

ADSORBENTS

Common materials used as adsorbents include activated carbon, silica and some metallic oxides. *Activated carbon* in the moderately to very pure grade is used in the 4 to 20 mesh size range. *Silica* as silica gel, diatomaceous earths and synthetic zeolite sands exhibits far greater specificity than carbon for polar molecules. Molecular sieve dryers are an example of silica-based systems. *Metallic oxides* such as activated alumina are widely used as desiccants and as catalyst carriers.

ADSORPTION EQUIPMENT

Among the considerations in designing adsorption equipment are 1.) contact time required between airstream and bed, 2.) adsorption capacity for the desired service life, 3.) resistance to air flow and 4.) uniform distribution of air flow. Adsorption equipment may be carrier-based, fixed-bed or fluid-bed.

Carrier-based adsorption equipment is uncommon in air pollution control. In this system the adsorbent is deposited on an inert carrier such as paper or plastic pellets. *Fixed-bed* adsorbers are the most widely used. The two types are the thin-bed adsorbers, with beds up to a few inches in depth for light duty work, and the thick-bed adsorbers, with beds 1 to 3 feet deep, for larger capacity. *Fluid-bed* adsorbers offer the advantages of uniform contact through fluidization, but are not used in pollution control.

ADSORPTION CONTROL

Adsorption systems can be regenerative or non-regenerative. In *non-regenerative* systems the adsorbent is discarded after use. Frequently the service life of the adsorbent is determined by experience; consequently, no instrumentation monitors the degree of saturation or the efficiency of performance. Alarm detectors can be installed on the downstream side to warn the operator of a premature breakthrough. These systems are particularly economical for odor control, because odorous vapors in low concentrations require only small units and still have substantial service life.

Regenerative systems can automatically regenerate the adsorbent after it is saturated. Such regeneration can be achieved together with destruction or recovery of the adsorbed substance.

Figure 5.3o shows a control system which incorporates the recovery of the adsorbate. The polluted air is fed (usually downward) through adsorber 1 or 2 with the pollutant being adsorbed while clean air is exhausted. While adsorber 1 is in operation, adsorber 2 can be regenerated by steam purging

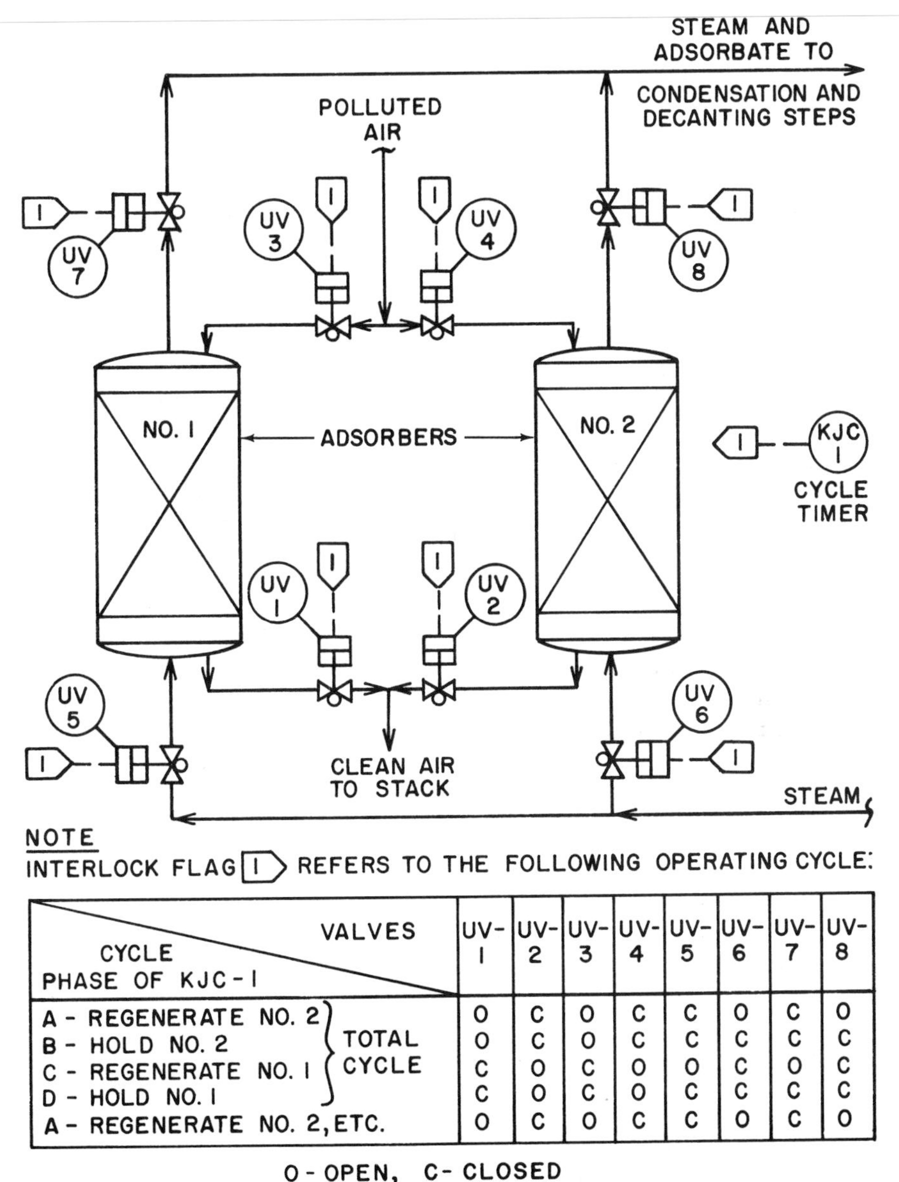

CYCLE PHASE OF KJC-1 \ VALVES	UV-1	UV-2	UV-3	UV-4	UV-5	UV-6	UV-7	UV-8
A - REGENERATE NO. 2 (TOTAL CYCLE)	O	C	O	C	C	O	C	O
B - HOLD NO. 2 (TOTAL CYCLE)	O	C	O	C	C	C	C	C
C - REGENERATE NO. 1 (TOTAL CYCLE)	C	O	C	O	O	C	O	C
D - HOLD NO. 1 (TOTAL CYCLE)	C	O	C	O	C	C	C	C
A - REGENERATE NO. 2, ETC.	O	C	O	C	C	O	C	O

Fig. 5.3o Automatic regeneration of adsorption systems

of the bed in the upward direction. The adsorbate is carried away by the steam and condensed. If the adsorbate is immiscible with water, decantation can produce crude adsorbate.

The control valves around the two beds are interlocked to permit sequential operation (Figure 5.3o). A variation of the method described is used in molecular sieve dryers. Regeneration of the bed is accomplished by heated air, and no attempt is made to recover the adsorbate. Following regeneration, the bed is cooled by dry nitrogen or other inert gas. Regeneration of the bed can also be done by oxidation of the adsorbate, resulting in partial oxidation of the carbon (if that is the adsorbent medium). In small sizes, this may be more convenient and less expensive.

Incinerators

Incineration is the thermal oxidation of organic compounds by exposure to high temperature in the presence of air. Because the oxidation is carried to the ultimate products (CO_2, H_2O, and N_2), toxicity, if due to organic material, is neutralized. Reference 7 gives a review on the design and application of incinerators for gases and vapors. Incinerators are either direct flame or catalytic.

In the *direct flame* incinerator the organic material is oxidized at 1,100°F to 1,800°F in the presence of a direct flame. The flame increases the efficiency of the oxidation process. Due to the high temperature requirement, and the consequent fuel cost, heat recovery is usually justifiable in the form of either preheating the exhaust (Figure 5.3p) or using a waste heat boiler. Direct flame incinerators usually operate at high efficiencies, and 90 to 99% destruction of organic matter is achieved.

Direct flame (direct fired) incinerators are designed to prevent accumulation of combustible vapors at concentrations within explosive limits. Most direct fired incinerators act on fumes at or below 30% of their lower explosive

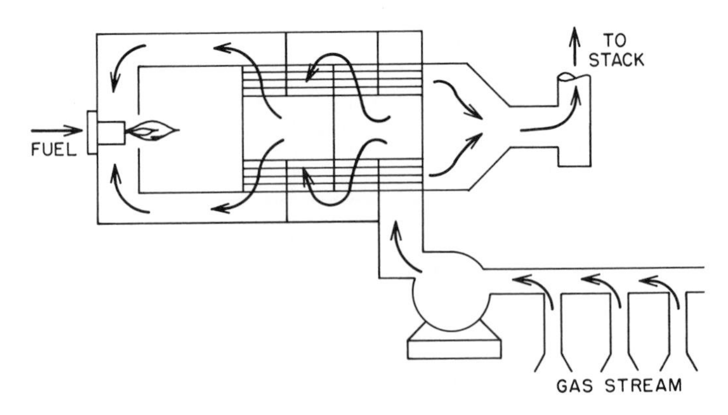

Fig. 5.3p Direct flame incineration

limit (LEL) and are minimally hazardous. Fumes at levels of 50 to 100% of LEL, however, have the inherent danger of flashback. Should flashback occur, the fuel gas supply is shut off after initial heat-up and combustion becomes self-sustaining. This operation requires a minimum amount of fuel but care is needed against explosion hazards. (For more on both continuous and safety controls for the units described, see Section 10.7 in Volume II.)

In *catalytic* incinerators, the oxidation process takes place at 600°F to 1,100°F in the presence of a platinum catalyst. Thus, catalytic systems greatly reduce the heat and fuel requirements in oxidizing fumes (Figure 5.3q). Although efficiencies of 85% or more are reported, the systems are very sensitive to catalyst poisons, suppressants and foulants. Some air pollution control districts require incineration efficiency of no less than 90%, which is difficult to achieve with present catalytic systems. The inlet temperature to the bed is usually controlled, and the catalyst bed is generally changed at intervals determined by experience.

Stacks

After pollution abatement by any of the systems discussed, the exhaust is dispersed by tall stacks to ensure that only negligible concentrations of residual pollutants reach the ground. Factors influencing dispersion from a stack include wind velocity, effective stack height, quantity of pollutants and atmospheric conditions.

To estimate ground level pollutant concentration, several correlations[5] have been developed by Briggs[8], Bosanquet, Pearson and Sutton. In stack design, stack height should be greater than any surrounding building or obstacle; stacks should be located so that the gases escape the wake of nearby structures and of the stack; and exhaust velocity of 100 ft per second is desirable to disperse the material and minimize plume.

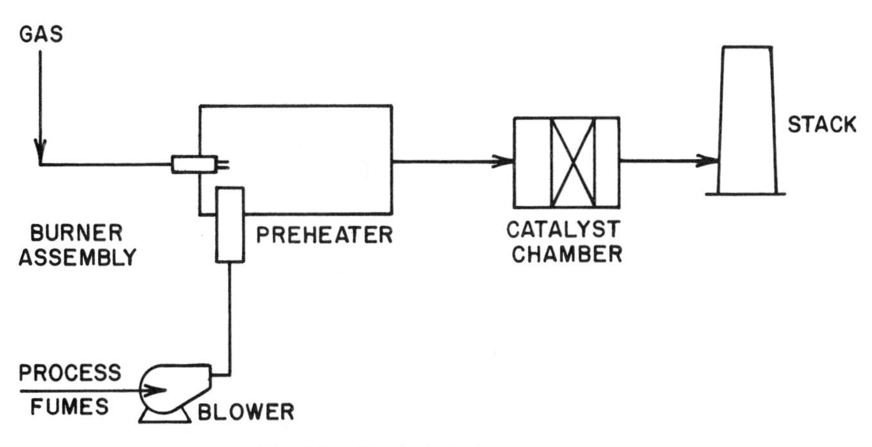

Fig. 5.3q Catalytic incineration system

Maximum ground level concentration of pollutants is at 5 to 10 stack heights from the stack. Stack design involves a value judgment on the extent of dispersion desired, but no stack can ensure dispersion under all atmospheric conditions.

Removal of Oxides of Sulfur

By 1975, permissible concentrations of sulfur oxides in ambient air will be limited to a maximum of 0.14 ppm and to an annual average of 0.03 ppm. In order to achieve this goal, larger sources will have to limit emission concentrations from their stacks to 200 to 400 ppm. Processes of sulfur removal presently available or under investigation include[9] dry absorption, wet absorption, adsorption and catalytic oxidation.

Current SO_x removal processes are given in Table 5.3r. The available processes are in various stages of development or in the very early stages of commercialization, but a completely reliable processing method that meets 1975 requirements is not yet available.

Removal of Oxides of Nitrogen

Nitric oxide (NO) formed in all stationary and mobile combustion sources is converted to nitrogen dioxide (NO_2) in the atmosphere. These two compounds are important in causing photochemical smog and other deleterious effects. Federal standards by 1975 will limit the concentration of nitrogen oxides to 0.05 ppm. This refers to all oxides of nitrogen but is expressed in terms of an equivalent quantity of NO_2. To accomplish this requires emission control at the source in the range of 100 to 300 ppm.

Methods under study to reduce emissions of nitrogen oxides and remove them prior to exhausting to the atmosphere include

Controlling NO_x formation. Combustion techniques can be modified to produce less nitrogen oxides.[10] The important parameters available are low excess-air firing, flue gas recirculation, water injection and two-stage combustion.

Reduction of already formed NO_x. If the concentration of oxides of nitrogen is relatively high, it can easily be reduced by using this polluted air to burn a hydrocarbon. If the available oxygen in the air is insufficient to burn the hydrocarbon, the nitrogen oxides will be reduced to nitrogen as the oxygen is used to oxidize the hydrocarbon. This technique can be adapted to conventional burners in fume incinerators.

Recovery of already formed NO_x. Adsorption methods are under development to remove oxides of nitrogen selectively from exhaust air. Methods to remove nitrogen oxides from emissions are still in preliminary stages, and much development has to be done before techniques are available to meet 1975 air quality criteria.

Table 5.3r
SULFUR OXIDE REMOVAL PROCESSES

Process	Basic Scheme
Dry Absorption Methods	
Alkalized Alumina Process (U.S. Bureau of Mines)	Sulfur dioxide absorbed by alkalized alumina; spent absorbent regenerated by a reducing gas that converts sulfur dioxide to hydrogen sulfide, which is sent to Claus Unit.
DAP-Mn Process (Mitsubishi Heavy Industries)	Sulfur dioxide absorbed by manganese dioxide; regeneration of spent absorbent by ammonia and air producing ammonium sulfate.
Wet Absorption Methods	
Combustion Engineering Process	Furnace injection of an alkaline earth additive followed by wet scrubbing.
Wellman-Lord Process	Sulfur dioxide is absorbed in potassium sulfite solution producing potassium bisulfite; spent liquor steam stripped to give sulfur dioxide and absorbent.
New Lime Process (Mitsubishi Heavy Industries)	After cooling, sulfur dioxide is absorbed by lime and calcium sulfite slurry; spent liquor oxidized by air to form gypsum.
Adsorption Methods	
Sulfacid Process (Lurgi)	Wet carbon adsorption in fixed bed adsorber to produce dilute sulfuric acid.
Hitachi Process	Wet carbon adsorption in a series of towers; water wash produces sulfuric acid.
Reinluft Process	Coke adsorbs sulfur dioxide and sulfur trioxide mist; regeneration by reducing gas.
Oxidation Methods	
Cat-Ox Process (Monsanto)	After complete dust removal, sulfur dioxide in flue gas is oxidized catalytically to sulfur trioxide and absorbed in a packed tower to form sulfuric acid.

Conclusion

Preserving the quality of the environment has become a matter of vital concern to everyone. Comprehensive laws are both on the books and under consideration, and elaborate enforcement methods are being devised. There is little doubt that environmental control has become a major aspect of process plant design.

Optimization and control of a process plant is no longer a function of product economics alone. The objective of leaving minimum impact on the environment has become mandatory.

REFERENCES

1. "National Primary and Secondary Air Quality Standards," Federal Register, April 30, 1971.
2. "Control and Prohibition of Air Pollution from Solid Particles," N.J. Air Pollution Code, Chapter VII.
3. "To Control Area (Air Basin) Air Pollution," Commonwealth of Pennsylvania, Regulation V, January 28, 1969.
4. Perry's Chemical Engineers' Handbook (4th ed.), New York: McGraw-Hill, 1963.
5. Stern, A. C. (ed.), Air Pollution (Vol. I), New York: Academic, 1968.
6. "Compilation of Air Pollution Emission Factors," Public Health Service Publication 999-AP-42. APCO Technical Information Center, Research Triangle Park, N.C.
7. Danielson, J. A. (ed.), Air Pollution Engineering Manual. Publication PB 190 243, Public Health Service, Cincinnati, 1967.
8. Briggs, G. A., "Plume Rise," U.S. Atomic Energy Commission Division of Technical Information, 1969.
9. "Control Techniques for Sulfur Oxide Pollution," Public Health Service Publication PHS-AP-52, APCO Technical Information Center, Research Triangle Park, N.C.
10. Bartok, W., et al., "Systems Study of Nitrogen Oxide Control Methods for Stationary Sources." Final Report GR-2-NOS-69 under NAPCA Contract PH 22-68-55 to Esso Research and Engineering.

5.4 NOISE MEASUREMENT

Available Instruments:	Sound level meters for measurement of RMS pressures; octave-band and ⅓ octave-band filters for frequency analysis; narrow-band analyzers; graphic level recorders; impact noise level meters; loudness-evaluating instruments; and devices for monitoring sound and noise levels.
Ranges:	From about 20 to 20,000 Hz; usually from 40 to 140 db, higher for special purposes.
Cost:	From $200 up.
Partial List of Suppliers:	B & K Instruments, Inc., Conwed Corp., General Radio Corp., Hewlett-Packard Co., H. H. Scott Co., Inc., Mine Safety Appliance Co.

Sound—or noise—as a pollutant has both physiological and psychological effects. Its units and definitions subsume physical, physiological and psychological factors. The basic units are the objective or impersonal units of intensity and quantity, and the subjective or personal units of loudness or annoyance. (See also Appendix 3 in Volume II.)

Objective or Physical Magnitudes of Noise

Audible sound is a pressure vibration in the atmosphere expressed in units of pressure, usually dynes per square centimeter, or microbars. Sound pressure vibrations alternate above and below the steady state atmospheric pressure, and therefore are given as RMS (square root of mean square) values. (Sound pressure is analogous to voltage in the electrical system.)

The threshold of audibility is taken as 2×10^{-4} microbar at 1,000 Hz, or about 2×10^{-10} atmosphere. The threshold of discomfort is about 200 microbars, and at 2,000 microbars the threshold of pain has been passed. It is convenient to establish levels within this very wide range by a logarithmic scale. The unit of sound pressure level (SPL) is the decibel (db), and

$$SPL = 20 \log_{10} (P/P_o) \text{ db} \qquad 5.4(1)$$

where P is "observed" sound pressure and P_o is the reference pressure, at the threshold of hearing, 2×10^{-4} microbars. (This reference pressure, in English units, is about 0.3×10^{-8} PSI. In acoustics, however, metric units are universally used.)

Sound intensity (analogous to electric power) is the average rate of propagation of sound energy through unit area normal to the sound path. Sound intensity is defined mathematically as the product of sound pressure and particle velocity, and therefore sound intensity, (I)

$$I = P^2/(\rho C) \text{ watts/meter}^2 \qquad 5.4(2)$$

where P is the sound pressure, ρ is the density of the medium and C is the velocity of sound in the medium. (The term ρC is the acoustic impedance of the medium.) Sound intensity level (SIL) is also expressed in db

$$SIL = 10 \log_{10} I/I_o \text{ db} \qquad 5.4(3)$$

where I is "observed" intensity, and I_o is the reference intensity (at the threshold of audibility) of 10^{-16} watts per cm^2.

Another unit, the acoustic power of a noise source, is useful in machine design where the total amount of noise produced is of interest. The unit of acoustic power is the watt, and sound power levels (PWL) are expressed as

$$PWL = 10 \log_{10} W/W_o \text{ db} \qquad 5.4(4)$$

where W is the sound-power output of the machine, and W_o is a reference level, 10^{-12} watt. (An earlier reference level was 10^{-13} watt.)

All of the physical magnitudes listed can be objectively measured with electronic instruments. The effect of noise on human beings, however, involves physiological and psychological reactions as well as the purely physical; psychological reactions can produce very real physiological damage. No purely instrumental method of measuring the latter has yet been devised, but effective research has been done through comparison tests using human subjects as listeners.

Subjective Noise Magnitudes

Roughly speaking, the sensation of loudness is represented by sound pressure levels, with each 10 db increase in level approximately doubling the loudness. However, the ear is not uniformly sensitive at all frequencies, and its frequency discrimination changes as the sound level changes. This response of the ear is illustrated in Figure 5.4a; shown are contours of equal loudness

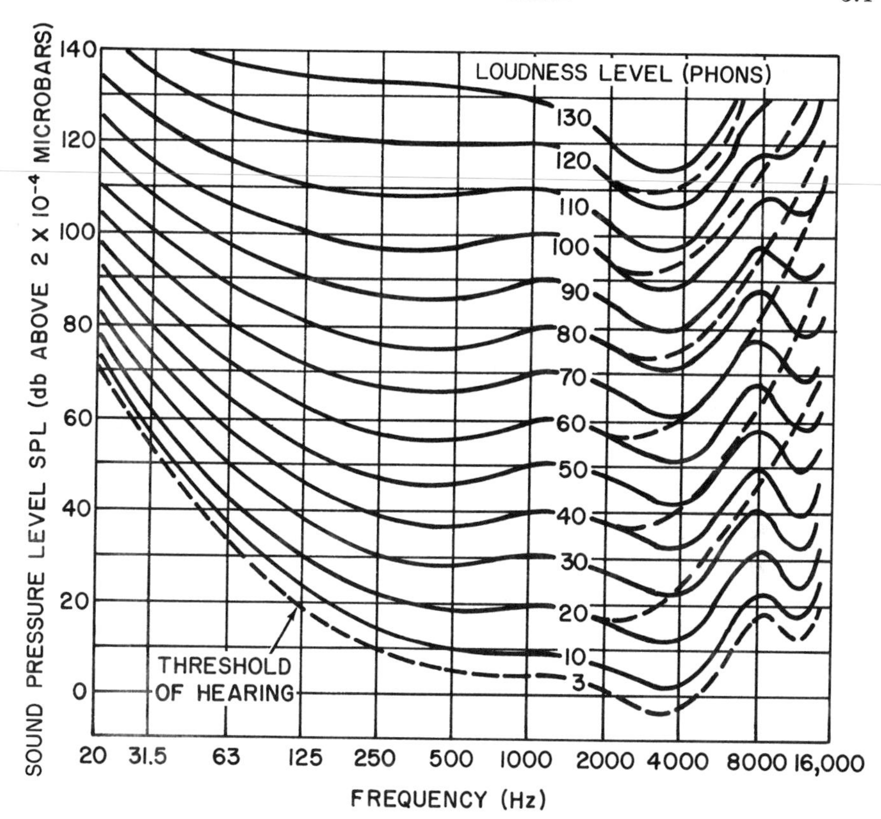

Fig. 5.4a Curves showing equal loudness for pure tones in frontal field. (After ISO recommendation.) The interrupted lines at the high frequencies indicate typical presbycusis for ages between 55 and 60 years.

for the average normal young ear (solid lines) and for ears suffering from progressive perceptive hearing loss occurring with age—presbycusis (dotted lines)—typical of an age of 55 to 60 years. The normal ear is most sensitive at frequencies of about 4,000 Hz, and the differences in sensitivity decrease as the sounds become louder.

The unit of loudness level is the *phon*, related to the sound pressure level through listening tests; a 40-phon sound is one judged by a group of listeners to be as loud as a 40-db sound at 1,000 Hz; 60 phons sound as loud as a 60-db 1,000 Hz sound, and so on. The phon, like the db of sound pressure level, is a logarithmic unit; it is not subject to easy addition and it does not seem to represent subjective sensations.

The *sone*, the unit of loudness, avoids these inconveniences. A 40-phon loudness level equals one sone; each 10-phon increase doubles the value in sones. Some representative values include:

Loudness Level (phons)		*Loudness (sones)*
140	Threshold of pain	1,024
120	Jet plane, landing	256
100	Dump truck	64
80	Speech (oration)	16
60	Conversational speech	4
40	Quiet living-room	1

Complaints about noise are based on its quality of annoyance as often as they are on loudness alone. At moderate levels sone values usually describe noise adequately, but at loudness levels above 90 phons—particularly for high frequency noise—the concept of "perceived noise level" expressed in decibels (PNdb) is becoming popular. The "perceived noise" value recognizes the annoyance factor as well as the loudness factor alone; it is therefore used to describe noise with considerable high-frequency components—jet aircraft, high-speed pumps, gas turbines or gas-turbulence noise and computer printers.

The frequency of uniformly repetitive sounds is defined in cycles per second (Hz); the normal hearing range is from about 20 Hz to 20,000 Hz.

Intermittent or interrupted sounds produce physiological effects somewhat different from uniform sounds, because the ear requires a few milliseconds to adjust to change in noise level, and because of the "startle effect." Criteria suitable for judging continuous sounds are not equally good for impulsive or impact sounds, since the protective reflex of the ear requires variable amounts of time—longer time if the ear is fatigued.

Standard definitions, standard calibration procedures and standards for measuring instruments have been established by national and international agreement. For field measurements, official standards have not been set up but recommendations are available in the literature, and some technical organizations have agreed on procedures within their own specific areas. Accurate work requires experience and judgment.

Sound Level Meters

Sound level is the parameter most often measured in noise surveys, and ordinarily the measurements are made on continuous, more-or-less uniform sounds. The measurements do not take detailed account of frequency characteristics, but standard sound level meters do have weighting devices to modify the frequency response of the instrument so that its sensitivity approximates that of the ear. (True sound pressure level is for flat or uniform frequency response; "sound level" or "noise level" are terms often used to indicate a non-uniform or weighted response; the weighting network used should always be specified, as "db[A]" or "db[B]".)

The frequency response supplied by the "A" weighting network approximates the ear's response at a 40-db average level; the "B" weighting is for about 70-db levels, and the "C" weighting matches the ear at about 100 db levels. The curves describing these responses are shown in Figure 5.4b and roughly correspond, to the 40-db, 70-db and 100-db equal loudness contours shown in Figure 5.4a. These three weighting networks appear in all standard sound level meters. A fourth, the "D" weighting, has been proposed but not yet adopted; it is seen occasionally and is also shown in Figure 5.4b. Its frequency response is designed to reflect readings approximating perceived noise level values in PNdb. Perceived noise levels can be approximated for jet-aircraft noise above 90 dbA by taking a sound-level meter reading on the "A" scale and adding 12 db.

All standard sound level meters employ a high quality, wide-range microphone, electronic amplifier, weighting networks, attenuator in 10-db steps and an indicating meter graduated in decibels. Frequently, various types of microphones are available, and occasionally provisions are made for accessories such as pre-amplifiers, octave-band filter sets or graphic level recorders.

For industrial noise the energy is distributed uniformly over a wide frequency range. If this is not the case, octave-band analysis is desirable. This

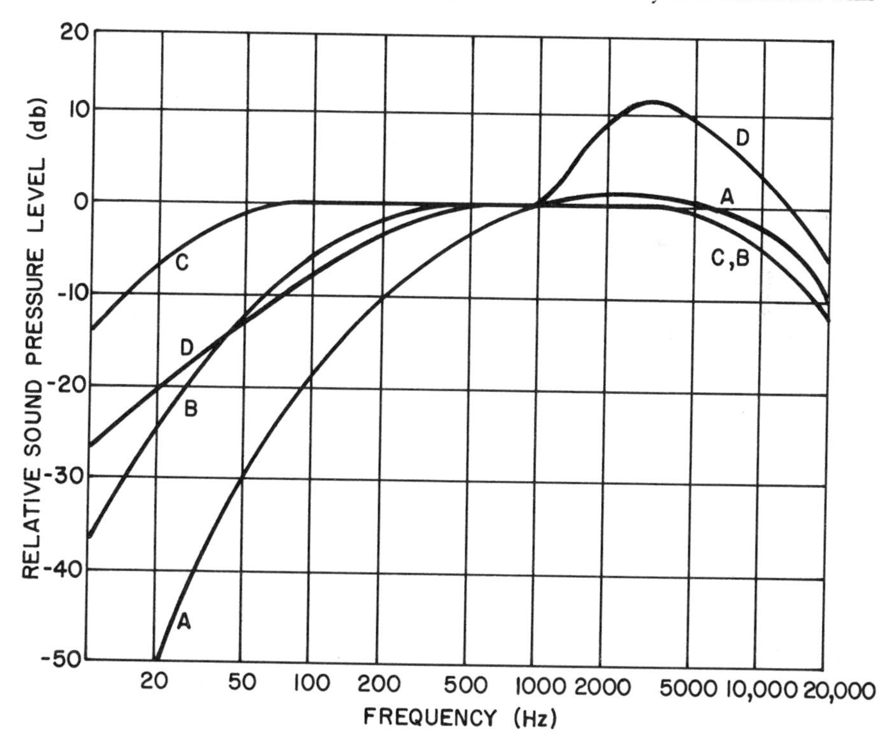

Fig. 5.4b Weighting curves for sound level meters. (After ISO recommendation.) The A, B and C weightings are standards; the D curve has been proposed for monitoring jet aircraft noise.

is done with a sound-level meter fitted with filters, each of which has a pass-band of one octave (or ⅓ octave). Such filters are provided in sufficient numbers to cover the entire range. The sound level for each octave is read separately, and the strongest frequency components are immediately apparent. For machine design, even more detailed frequency analyses are used.

Since hearing hazard is greater for narrow-band than for wide-band noise, it is valid practice to make octave-band analyses of noise and convert them to "equivalent A-weighted noise" levels in applying hearing-hazard criteria. This is often done by plotting the octave-band data on top of a set of curves like those in Figure 5.4c and reading the highest numbered contour penetrated by the octave-band data, i.e., the "equivalent A-weighted level."

One-third-octave band data are also used for the two standardized methods of calculating loudness. Stephens' method and Zwicker's method. The details differ, but both methods use narrow-band measurements (which can be ⅓ octave) to obtain information which is then used in an essentially nomographic procedure to give loudness figures in phons.

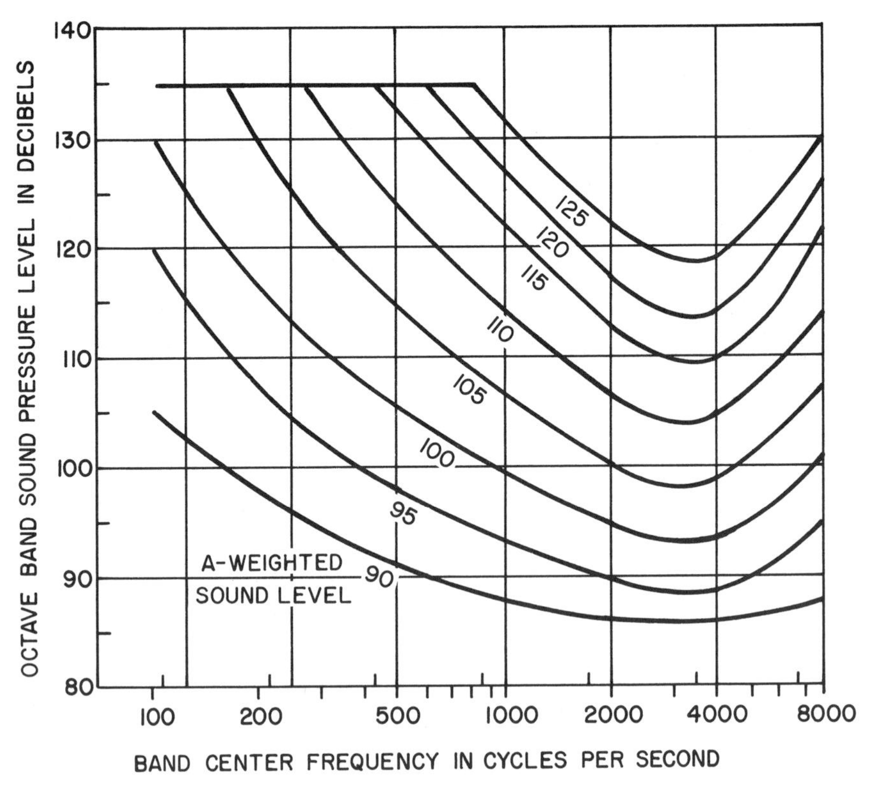

Fig. 5.4c "A-weighted equivalent sound pressures" for hearing-hazard evaluation can be reached by plotting octave-band data on these curves.

Zwicker's method is commonly used in electronic equipment to make loudness computations instantly available. Higher accuracy can be expected—at higher cost—than by sound-level meter procedures.

Impact, impulsive or explosive noises must be measured with special equipment, the duration of the sound being so brief. These instruments show the peak pressure, an RMS value derived by an arbitrary procedure or even by frequency analyses.

Special-purpose instruments are available for continual monitoring of noise levels. Called noise monitors, noise exposure meters, noise limit indicators, sound hazard indicator or noise hazard integrators, they continually observe the noise level and signal or record when noise levels exceed prescribed values. One form intended especially for hearing hazard monitoring gives as its output an "integrated noise exposure value" according to the Walsh-Healey criteria. (For more on noise pollution control techniques, see Section 5.5 in this volume.)

REFERENCES

1. Acoustics Handbook: Application Note 100. Hewlett-Packard Company, Palo Alto, Calif.
2. B & K Technical Review. Bruel and Kjaer Instruments Company, Denmark.
3. Beranek, L. L., "Acoustic Measurements," New York: Wiley, 1949.
4. Kryter, K., "The Effects of Noise on Man," New York: Academic, 1970.
5. Peterson, A. P. G. and Gross, E. E., Jr., "Handbook of Noise Measurements," General Radio Company, Concord, Mass.
6. Stephens, R. W. B. and Bate, A. E., "Acoustics and Vibrational Physics," London: Arnold, 1966.

International Standards are available from International Organization for Standardization, 1 Rue de Narembé, Geneva, Switzerland.

U.S. Standards are available from American National Standards Institute, 1430 Broadway, New York, N.Y. 10013.

5.5 NOISE POLLUTION CONTROLS

Heightened noise levels in industry have caused serious problems both to the efficiency and health of workers, and have affected the environment within and surrounding industrial plants. Some legal restrictions are currently being imposed by federal and local governing bodies to correct the excesses.

To assist in reducing and controlling noise and in affording better working and living conditions, some causes and effects of noise are discussed, together with commonly used criteria for legislative control. The general principles of physical control of noise are listed, although usually each problem must be individually treated. (For definitions, terminology and methods of measurement, see Section 5.4 in this volume.)

Sources and Levels of Noise

Almost every moving object produces noise: machines, flowing gas or liquid or turbulent flames—anything for that matter which can produce a pressure vibration in the air. In a metal-working plant the primary source of noise may be mechanical vibration; in a chemical plant it may be turbulent flow of gas or liquid in ducts, pipes or valves. Energy may be transmitted through mechanical vibration, finally becoming audible at a point some distance from the source.

Both the intensity and the character of noise vary widely. Some typical noise levels are summarized in Table 5.5a. Noise levels at individual workers' locations vary because of local conditions and because more than one source of noise is ordinarily present.

Harmful Effects of Noise

In this context, noise is simply sound that is harmful or objectionable because of its level or intensity, its character or perhaps because it occurs at the wrong time or place. Obviously, noise produced in an industrial plant is objectionable on the plant premises, but it may also pass to neighboring residential areas; there the level may be lower but the annoyance factor is far greater, especially if it occurs during sleeping hours. Aircraft and other traffic noises are often particularly troublesome in this respect.

The adverse effects of noise on people may be grouped under three

Table 5.5a
TYPICAL NOISE LEVELS

Noise Source	*Level in dbA*
Noise in high-pressure coolant pipes	as high as 200
Saturn rocket motor, close by	195
Noise in propeller tunnel of VTOL aircraft	165–170
Jet plane take-off; artillery fire	140
Thunder	as high as 135
Orchestra, 100 piece, fortissimo	130
"Discotheque"	as high as 135
Chain saw at operator's ear	120
Textile mill	110
Farm tractor at driver's seat	110
Saturn rocket at 10 miles	110
Dump truck at driver's seat	110
Automobile horn at 10 feet	105
Pneumatic jackhammer at 25 feet	95–100
Diesel truck at 50 feet	88
Oration, front row	80
Cooling tower at 50 feet	80–95
Passenger car, 60 MPH, smooth highway, inside	78–80
Office with typewriters	80
Face-to-face conversation at 3 feet	65

headings: impairment of hearing, general damage to health and indirect effects such as decrease in efficiency and increase in accidents.

It has long been known that workers in noisy industries suffer hearing loss; weaver's deafness and blacksmith's deafness have long been recognized as occupational diseases, as have the ruptured eardrum associated with cannoneer's and miner's deafness. The exact noise levels at which diverse injuries occur cannot be explicated, owing to the differences among individuals. In general, lung damage may be expected above 175 db, eardrum rupture above 150 db and headaches and dizziness connected with permanent hearing impairment possible, even for brief exposures above 140 db. Noise levels above 150 db are usually produced only by explosions, rocket motors and the like.

At lower levels the damage is less drastic but no less real. For example, 10 minutes' exposure at 120 db will usually cause ringing in the ears, headaches, dizziness and a temporary loss in hearing sensitivity (temporary threshold shift, or TTS) of 15 to 25 db; the duration of the loss is roughly proportional to the severity of exposure. Some permanent loss is also probable, the amount of which is also proportional to exposure.

It is sometimes thought that workers in noisy locations become "acclimated" to the high level of noise and no longer notice it. Careful examination seems to indicate that the acclimatization is no more than a temporary loss in hearing acuity, which if continued becomes permanent.

At still lower levels, down to perhaps 85 db, longer exposure will also produce temporary—and possibly permanent—hearing impairment. Other common physiological effects include, for example, a 25% loss in visual acuity accompanying noise exposure at 90 db.

At levels below 85 db, permanent hearing loss is not probable for normal periods of exposure, but physiological and psychological effects still exist. Noise at 75 db can cause vasoconstriction, nervous tension and excessive fatigue.

Adverse effects occur at levels below 75 db, although in the modern world it is only in residential areas and some offices that such levels can be expected. Sudden changes in the level often cause strain and annoyance, and impair general health. Many medical authorities feel that restful sleep requires a sound level no higher than 35 db; a passing vehicle producing 50 db will waken many people, and 70 db will waken most. Even though the individual may not fully awaken, the restful "deep sleep" stage can easily be disturbed by relatively low level intermittent noises.

Noise has an adverse effect on general health and welfare, since it produces fatigue and strain; the degree varies with the individual. Social disadvantages arise from noise, too. Education is hampered by noisy homes and schools.

At even moderately high noise levels nervous strain and fatigue are heightened, and there is usually a reduction in work efficiency. Where there is pressure to keep the work rate constant, breakage and errors usually increase; when the worker can control this rate, breakage and errors decrease. Absenteeism and employee turnover increase rapidly when noise levels are high.

Accident rates often rise because of the fatigue effect of "permissible" noise levels. Judgment can be impaired, and as already noted there can even be impairment of vision in high noise fields. Hazards are greatly increased by the masking of needed signals by noise.

Interference with communication is an ever-present disadvantage, causing errors, loss and accidents in plant or office. Speech which is easy at 65 to 70 db grows increasingly difficult until at a little above 100 db one depends more on lip-reading than on hearing. Speech interference is common in factories but also occurs in offices—and even in courtrooms.

Criteria of speech interference are concerned more with operating requirements for efficiency than with legalistic ones for safety and health; safety is obviously involved, however since impaired communication impairs safety. A set of noise criteria has been developed, seen most graphically in

Figure 5.5b. Each of these "NC-curves" represents a noise environment in which speech has to be at a certain level in order to be heard; for each purpose, one of these environments may be considered satisfactory. Background levels lower than NC-30 are considered to be "quiet" by most people. NC-20 is often specified for theaters, recording studios and music halls. Background

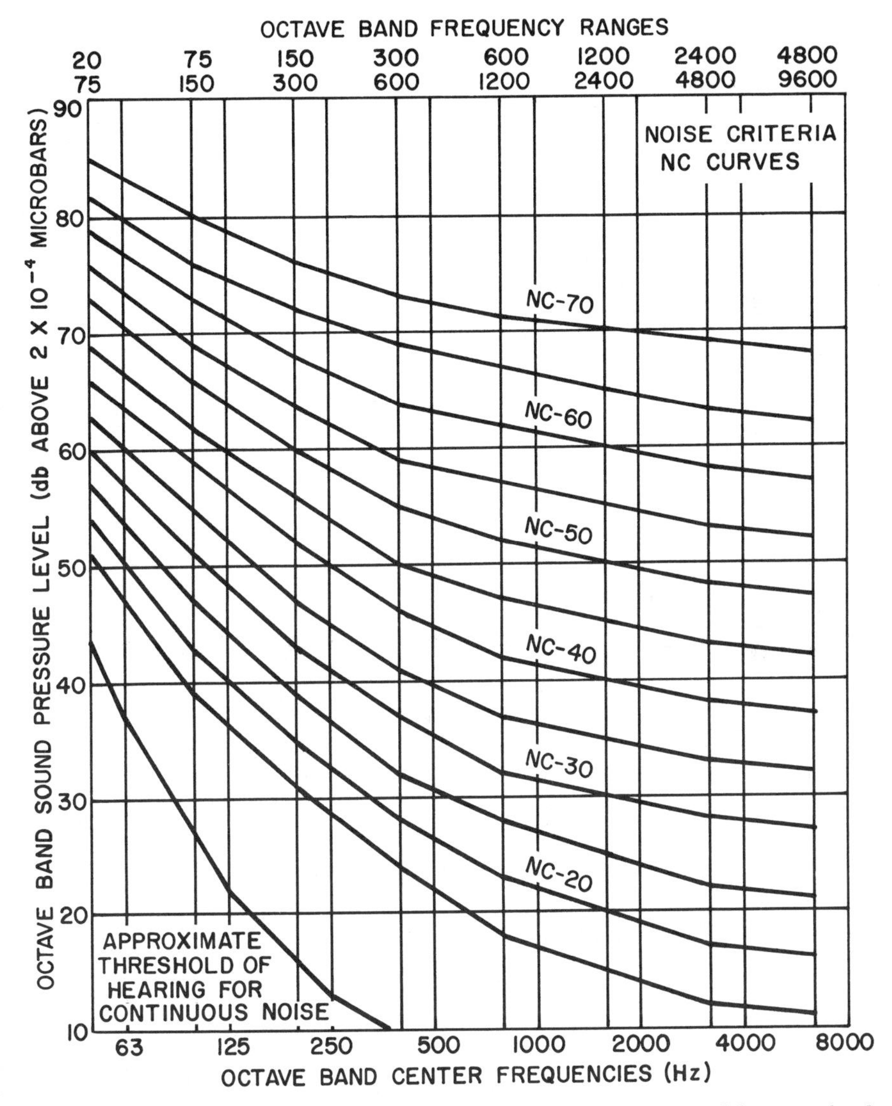

Fig. 5.5b NC-curves. An environment conforms to a specified noise criterion if the octave-band data, when plotted on this chart, fall entirely below the NC-curve.

noise above NC-50 are regarded as noisy by most people, although some large offices reflect higher levels.

It is to be expected that these NC-curves would resemble the equal-loudness curves, since the same physiological reactions are involved. For rough checks an NC-criterion can be approximated by taking a dbA reading and subtracting 7 to 10 from the observed value. The NC-curves are used only when the intensity of noise is relatively constant, without sudden pulses; when these are present impulsive or impact criteria should be used.

Vibration, which causes and is caused by noise, may disturb the operation of some equipment (as well as people); if persistent it may cause fatigue breaks. Vibrations caused by turbulent flow around aircraft or in pipes are often difficult to find and correct.

Hearing Hazard Criteria

Criteria recommended for relative safety from permanent hearing impairment have necessarily been established largely by statistical studies, since test animals cannot be used to predict effects on human subjects; over a period of years, however, the former have adequately predicted such losses.

For noise that is relatively constant in intensity and uniform in character, the maximum period of exposure permissible for each of a range of levels can be stated. This body of data is given in Table 5.5c. If the energy is distributed randomly over the entire frequency spectrum, the first column ("white noise") of the table applies. If the frequency range is restricted, the energy reaching individual portions of the hearing mechanism is increased;

Table 5.5c
DAMAGE RISK CRITERIA

Duration (daily exposure in hours)*	*White Noise (dbA)*	*One-octave Bandwidth (dbA)*	*Pure Tones (dbA)*
8	90	85	80
4	90	85	80
2	92	87	82
1	95	90	85
½	98	93	88
15 minutes	102	97	92
7 minutes	108	103	98
3 minutes	115	110	105
1½ minutes	125	120	115

* No exposure is to be permitted when ear protectors are being worn to levels above 150 dbA. When ear protectors are not worn, no exposure is to be permitted above 135 dbA. These criteria assume one period of exposure per day, and that a noise environment is to be considered acceptable if after 10 years the hearing loss is no greater than 10 db up to 1,000 Hz or 15 db up to 2,000 Hz or 20 db up to 3,000 Hz.

if the noise falls within about a 1-octave range, the second column should be used; and if the frequency range is even narrower, the third column should be used. Each column is uniformly 5 db lower than the preceding one; permissible exposure times for narrow-band noise are shorter than for wide-band noise.

These criteria assume that a noise environment may be considered acceptable for the average person if after 10 years of everyday exposure the hearing loss is no greater than 10 db up to 1,000 Hz, 15 db up to 2,000 Hz or 20 db up to 3,000 Hz; note that there is no assurance that hearing loss will be zero.

It is also assumed that no exposure is to be permitted, even if hearing protectors are worn, to noise levels above 150 db (since if the protectors were inadvertently removed, irreparable damage might occur before they could be replaced) and that without hearing protectors no exposure whatever is to be permitted to levels above 135 db.

These criteria assume constant noise levels, which often is not the case; an alternative means is available where noise levels are not constant. This requires that the noise level and time of exposure for each level be noted, with the partial exposures added in this manner: the actual exposure time divided by the permissible time is taken for each level, and the resulting fractions are added. If the sum of the fractions is greater than unity, the total permissible exposure has been exceeded.

Current and Probable Noise Regulations

Although for some years the Federal Aviation Administration (FAA) has compiled rules to regulate the noise emitted by aircraft in and near airports, the first federal regulations applying directly to a large segment of the public was the noise amendment to the Walsh-Healey Public Contracts Act, which became effective May 20, 1969. More recently, the federal Occupational Safety and Health Act was enacted, containing noise restrictions similar to those of the Walsh-Healey Act. Its enforcement began August 15, 1971. Since the provisions of these two acts are similar, they will be described together.

The Walsh-Healey Act applies to industries or commercial organizations that have contracts with any government agency for $10,000 or more. The Occupational Safety and Health Act is intended to apply to most privately owned industrial or commercial organizations; exempted from its provisions are governmental establishments, railroads, mines and the Atomic Energy Commission.

The noise exposures permitted by both these Acts are listed in Table 5.5d. The values are similar to those in column 1 of Table 5.5c but somewhat less restrictive at the higher levels. Impact noise exposure is permitted up to 140 db peak sound pressure level, instead of the 135 db limit shown in Table 5.5c. Accurate measurement of these peak levels requires proper instru-

Table 5.5d
PERMISSIBLE NOISE EXPOSURES
(Walsh-Healey Act)

Duration per Day (hours)	*Sound Level (dbA) Slow Response°*
8	90
6	92
4	95
3	97
2	100
1½	102
1	105
½	110
¼ or less	115

° Exposure to impulsive or impact noise shall not exceed 140 dbA peak sound pressure level. (Daily noise exposure may also be computed by octave-band data or effective partial daily exposures, or both.)

mentation, usually of the sample-and-hold type. If this apparatus is not available, an approximation can be used if the duration of the impulsive noise is longer than 5 but shorter than 50 milliseconds.

A standard sound-level meter is set on the "C" weighting (see Section 5.4 of this volume) and fast response of the meter; the peak deflection is noted and if it is less than 125 dbC, the peak level is presumed to be below 140 dbA. The procedure is at best not highly accurate, and tends to give low readings for noise pulses shorter than 5 ms and high readings for pulses longer than 50 ms.

Similarly, if it is not known that noise is wide-band—essentially white noise—an octave-band analysis should be made; the procedure described in Section 5.4 and Figure 5.4c in this volume should be used. Provision for use of this method is a part of both the Walsh-Healey and Occupational Safety and Health Acts. An additional provision is that when noise is intermittent, noise pulses shorter than 1 second shall be counted as 1 second, whereas interruptions in continuous noise of shorter duration than 1 second shall not be counted as interruptions.

The regulations provide that when sound levels exceed these values, an effective, continuing hearing conservation program shall be administered. Penalties are established for continued non-compliance. (As of this writing little information is available as to how enforcement is to be implemented.) A minimum hearing-conservation program would include audiometer tests on all employees at the beginning and the termination of employment, with similar tests at 6-month (preferably 3-month) intervals during employment; personal hearing protectors (ear-plugs or ear-muffs) should be available when needed, and continuing records should be kept.

The FAA regulations regarding aircraft set the permissible levels in PNdb which may occur at specified distances around established airports, but have little applicability outside that specialized field. Federal regulations on construction noise generally follow the same criteria as the Walsh-Healey Act.

In the past, various state and local regulations have been set up to restrict noise; usually they have been directed at traffic or construction noise, or both. Most of the provisions have either been ignored or repealed, but new ones are being prepared at the federal level. It is to be expected that the regulations will apply to all transportation noises except those from aircraft, and that additional regulations will be issued to restrict noise levels in public buildings (such as schools and hospitals) and the noise transmission characteristics of building components; perhaps there will be others as well.

Although there is good reason to anticipate more federal, state and local noise regulations in the future, the specifics of such regulations cannot even be surmised at this time. It seems likely that the current criteria in the Walsh-Healey Act and the Occupational Safety and Health Act will be made stricter, since comparison shows that both acts now permit higher noise levels than those listed in Table 5.5c; changes in workmen's compensation standards may make such changes necessary. It also appears likely that specific noise-emission maxima will be applied to many machines and appliances not presently regulated in any way.

One area not yet protected by federal legislation is the contamination of adjacent residential areas by industrial noise. The trend appears to be toward applying regulation to contaminants having effects at remote distances as well as to areas immediately adjacent. It is possible that there will be restrictions limiting the noise contribution of an industrial installation to residential-area noise—possibly a prohibition of the raising of any residential noise level by more than say 5 db over its 1965 value. Since each source of noise makes its own contribution—large or small—to the whole, these restrictions might even limit the size of an installation and compel dispersion of industry outside heavily populated areas.

Sources of Noise

Audible noise is a pressure vibration in the air produced by anything that can set air into vibration at frequencies within the range of hearing of the human ear. (Frequencies above and below this range are produced [ultrasounds and infrasounds] but their effects will not be discussed here.) A large area in vibration, a wall or a vibrating pipe perhaps, might constitute an extended source. High-amplitude noise can set elastic members into vibration, producing noise at a remote location, and perhaps (depending on the resonance characteristics of the elastic member) at a frequency other than the one exciting the vibration.

In the mechanical industries, rotating and reciprocating machines are doubtless the primary sources of noise. Power plants and generating stations also contain rotating machines, as do shipboard engine-rooms, trucks and locomotives and the mechanical-equipment rooms of buildings. Electrical distributing substations produce noise at 60 and 120 Hz (with some higher harmonics) largely as a result of magnetic forces causing vibration of transformer cores.

The noise from rotating machines may be caused by vibration from unbalance of rotating parts; it may be tooth noise from gears or noise from belts, chain drives or flexible couplings; electric motors commonly produce vibration and noise at the power-line frequency because of magnetic forces. There is wind and bearing noise from any rotating machine, and cooling fans and cooling air flow contribute to the total.

The character of noise from rotating machines varies but in general is relatively broad-band; some types of machines will have peaks at certain frequencies characterizing these types. Figure 5.5e shows the noise spectra for a rotating machine—a gear-driven high-speed pump. Two sets of data are

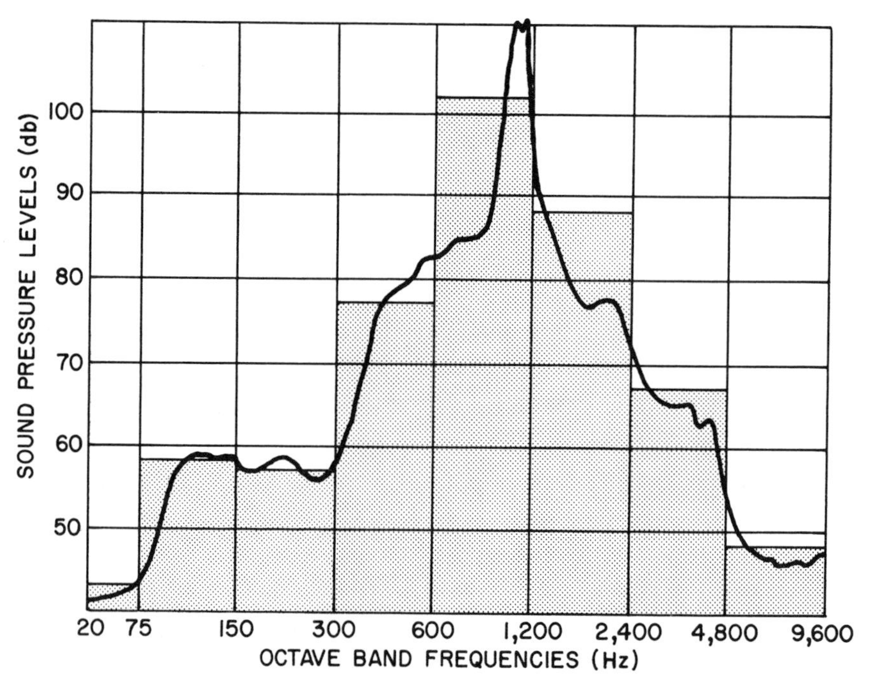

Fig. 5.5e Noise spectra for pump. The cross-hatched area represents octave-band data; the solid line, narrow-band (1% wide) data.

superimposed in this figure; one is the octave-band spectrum, useful in assessing the hearing damage hazard; the other is a continuous spectrum which shows peaks. One of these peaks is caused by a malfunction—a chattering valve. Continuous spectrum analysis is effective in diagnosing defects or malfunctioning, or in reducing noise during design. In this example, the valve noise is a narrow-band vibration at the natural frequency of the metal valve, modified by the acoustic resonance of a column of fluid in a pipe, and the noise radiated into the air is a combination of these. In comparison, a noisy gear-tooth would show as a noise-pulse appearing each time the defective tooth made contact with a mismating surface; its pulse frequency could identify the gear by its speed.

Plant noises that are sometimes overlooked in surveys are caused by apparatus operated intermittently—machines like conveyors and lift trucks (both of which may produce different noise when empty or loaded, and may not be operated continually), ventilators or unit-heating devices, and infrequently used emergency or repair equipment.

In metal-fabricating and assembly plants, a very serious source of noise is the clanging and clattering of metal parts striking one another. Unlike the noises of rotating machines, conveyors, welders, riveters or other fastening machines, this noise is largely unpredictable in plant design, since it depends in part on methods of operation and employee care, factors that change from time to time.

The flow of fluids in pipes or ducts, especially turbulent high-pressure flow such as in pressure reducers, can be very noisy. A considerable amount of energy is being exchanged, and if only a minute fraction is converted into acoustic energy the sound level will be high. In large ducts (such as heating and ventilating ducts) noise can be troublesome even when the pressure is low; noise is produced as a result of turbulence (as at a change in cross-section or direction, or at a damper) and is transmitted along the duct. Thin-walled ducts may vibrate in response to machine vibration or to air flow, producing noise from a new source. The noise produced by fluid streams and control valves are commonly of rather high frequency: gas-turbulence noise may range from 1,000 to 10,000 Hz, depending on conditions; liquid turbulence is frequently due to cavitation and is likely to produce broad-band noise above 1,000 Hz; many of the mechanical noises produced in pipes and valves are also above 1,000 Hz and are particularly annoying and potentially harmful because of the higher sensitivity of the ear at these frequencies. A sample control-valve-noise spectrum is given in Figure 5.5f.

The noise in ventilating ducts is very different; fan rumble and line-frequency noise can be an appreciable fraction of the total, and the total noise pattern may resemble the example shown in Figure 5.5g, in which the noise-spectrum can also be used to establish probable source of noise, or to diagnosis malfunctions. The causes and the treatment of valve and flow-stream noise

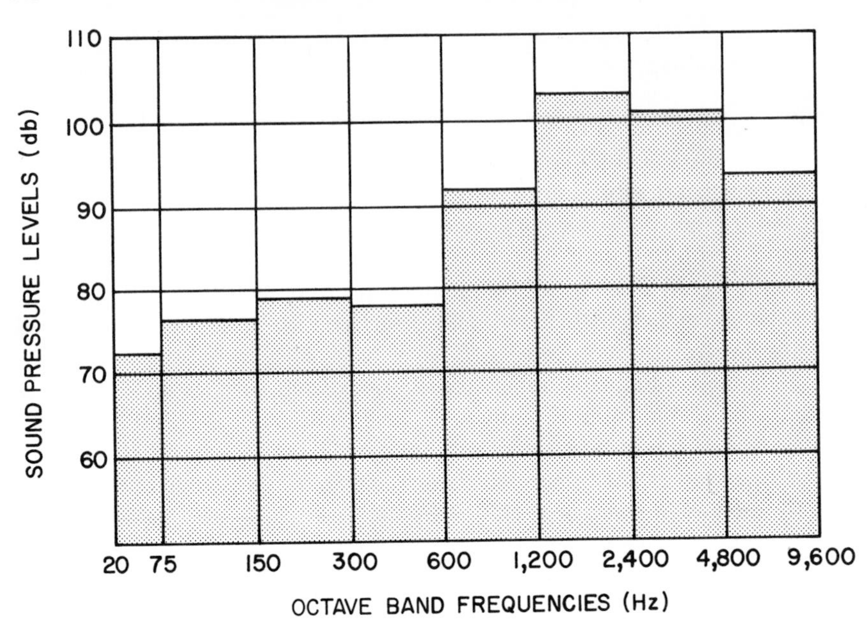

Fig. 5.5f Typical octave-band spectrum of control-valve noise

are discussed in greater detail in Section 1.21 in Volume II and Appendix 3.

Turbulence in flames is often a powerful noise source; it is broad-band, its spectral characteristics modified by the local conditions—resonance characteristics of pipes, for example. Its elimination is difficult, but control of immediate surroundings offers a useful method. This noise is produced primarily by two phenomena: the turbulent mixing of gases during combustion, and the turbulent mixing of gases during emission into the atmosphere. Unsteady flow induces turbulence, and combustion produces unsteady flow; thus some noise is inevitable. The most detailed theoretical work in this area has been with regard to jet turbulence noise in the open (not within pipes); recent work, however, has dealt with restricted streams.

Machines for office use rather than in the plant also make noise. In recent years their numbers—and noise output—have increased strikingly.

Ordinary office machines like typewriters, telephones and adding machines create noise simply because of their numbers; a large office may have scores or even hundreds of them. Even though each machine is a relatively small source of noise the sounds are additive and may seem unbearable. It is not uncommon for the noise level in large offices to approach 80 dbA. Some duplicating machines are very noisy, and for convenience are installed close to desk workers. In newspaper offices, teletypewriters and wirephoto machines are ordinarily located in the newsroom. Both there and in engineer-

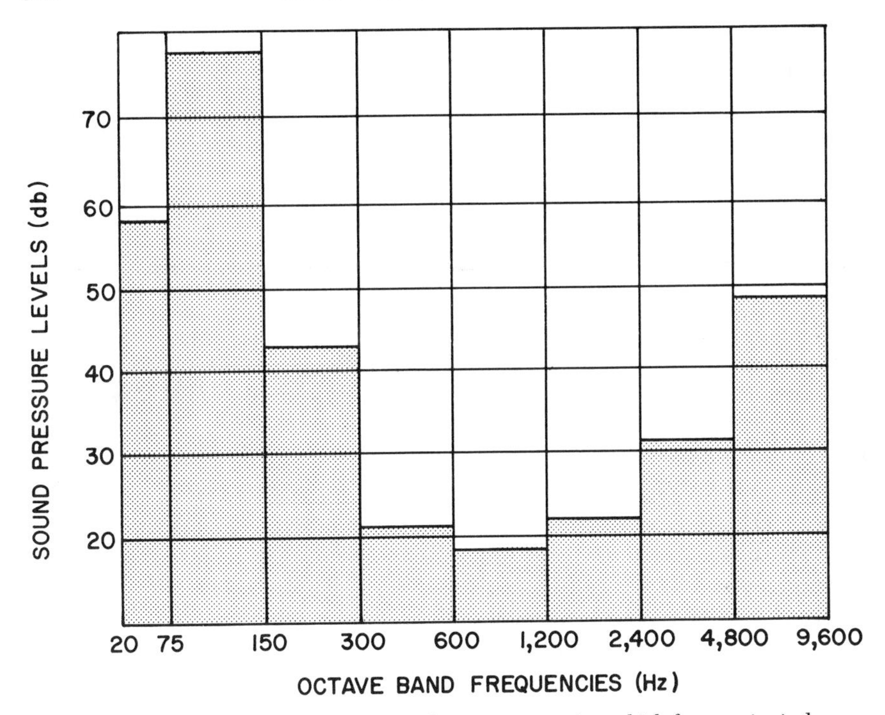

Fig. 5.5g Noise at ventilating duct outlet. The increasing noise at high frequencies is damper noise; that between 75 and 300 Hz. is probably fan rumble.

ing offices, tape-punching typewriters and reperforators may be used, all of which are noisy.

A particular source of noise is the high-speed computer—especially its interface equipment. Keypunch stations produce noise (although less than in the past). Collators and card-verifiers are often very noisy, as are devices for rapid reading-in of decks of cards. Even worse are high-speed output printers. Magnetic-tape-reader clutches, because of their rapid action, are often extremely noisy, and are ordinarily enclosed in cabinets for protection as well as to reduce noise, but most of the other items mentioned are not enclosed. In contrast, the new electronic desk-top computers make less noise than their mechanical predecessors.

Many pieces of ordinary equipment also add to the total noise, and where noise levels should be kept low (as in offices, laboratories, first-aid or medical stations, audiometric rooms, instrument calibration and repair shops and so forth) the smallest sources may be too much. Thus, noise from ventilating ducts or diffusers, door-openers, unit-heaters or even plumbing outflow pipes must be monitored.

Control of Noise

Basically, noise may be controlled by 1.) reduction at the source, 2.) isolation of the source from other areas or 3.) isolation of sensitive areas and protection of workers—or by all three methods simultaneously. Of these, reduction at the source is best.

The greatest benefit in reducing noise comes from selecting equipment which makes the least noise in its operation. It may be possible to reduce the noise produced through adjustment of the machine or other device. (One necessarily depends on the manufacturer of the machine for both these procedures.) In the last few years manufacturers have begun to design equipment that makes less noise, and to supply data describing the amount of noise produced in normal operation. As yet, few manufacturers supply much information on techniques for reducing noise in equipment already installed, except through their own installation engineers, but various devices are available for reducing noise at or near its source. The most useful of these are flow silencers, or inline silencers, which reduce the noise transmitted from an area of turbulent flow along the fluid stream, thus preventing a long length of pipe from becoming a source of airborne sound. Outlet silencers or mufflers are designed to prevent a pulsating or turbulent stream from making too much noise as it enters the atmosphere, and various elastic and viscous devices are designed to absorb mechanical vibration and to prevent it from setting the surrounding air or solid objects in vibratory motion.

Isolation of a machine, or even a group of machines, is an effective procedure since it prevents noise from being freely radiated even though the machine itself may still vibrate. A machine alone can be placed on elastic mounts, a heavy base ("seismic mass") carrying the machine may be suspended or an entire section of floor ("floating floor") may be mounted on soft material which combines elasticity and damping. Large and heavy machines are frequently placed on foundations separated from the building and its foundation, so that the earth acts as a vibration isolator.

The physical principle is the same for all of these, i.e., if the natural frequency of the mass on its elastic suspension is less than ¼ of the vibration frequency, isolation is efficient. It is convenient to describe the resonant frequency of the system in terms of its static deflection under load. In this equation

$$F_n = 3.13\,(1/d_{st})^{1/2} \qquad 5.5(1)$$

in which the natural frequency, F_n, is in cycles per second, the d_{st} signifies the static deflection (loaded) in inches. Figure 5.5h is a diagram of a floating floor carrying two machines, one of which is itself on shock-isolating mounts. This procedure can combine good results with moderate cost, particularly for low frequencies, which are much more difficult to isolate.

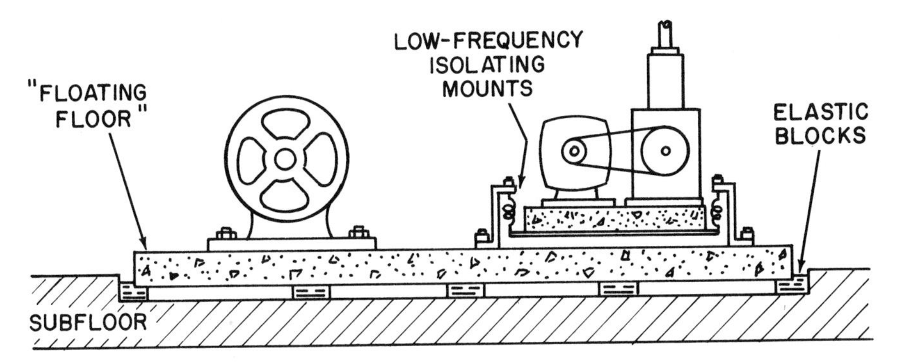

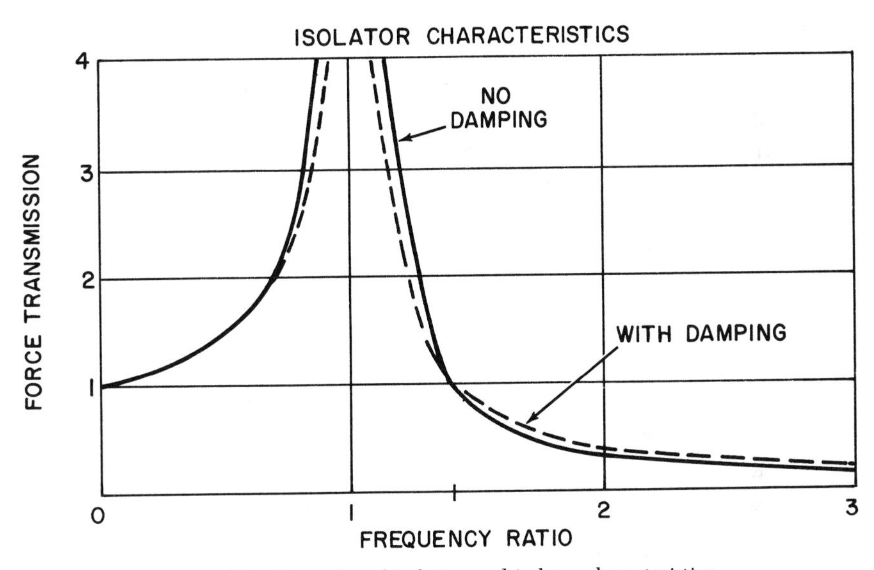

Fig. 5.5h Examples of isolation and isolator characteristics

The lower part of Figure 5.5h shows the curve describing isolation characteristics. The amplitude of the driving force is plotted against the ratio of the two frequencies: that of the resonant suspension and that of the driving force. It is important to note that if the frequency of the machine vibration is *less than 1.4 times* that of the resonant suspension, there will be an increase in transmission rather than a decrease. If there were no damping (frequency ratio of 1), the amplitude of the driving force would tend to become infinite. Since the system must pass through this critical frequency to reach its operating speed, partial damping is necessary. It is seldom a problem, however, being easily furnished; in fact, the elastic blocks that suspend floating floors usually have adequate damping capacity because of their composition. Typical

Table 5.5i
VIBRATION ISOLATOR
CHARACTERISTICS

Revolutions per Minute	*Static Deflection (inches)*°	
	5 percent "good"	*20 percent "fair"*
3,600	0.06	0.02
2,400	0.14	0.04
1,800	0.26	0.07
1,200	0.6	0.15
600	2.2	0.6
300	7.0	2.5

°Typical values of static deflections to give "good" isolation at various speeds (5 percent force transmission), and "fair" isolation (20 percent force transmission).

values of static deflections suitable for machines running at various rotational speeds are listed in Table 5.5i.

Vibration-isolating mountings for machines reduce the vibrational energy passing to the floor and building structure, but similar precautions must be taken with pipes, ducts and electrical conduits connected to the machine; otherwise, these will both radiate noise directly into the air and also conduct vibratory energy elsewhere. This energy can be conducted through a liquid within the pipe as well as along the metal pipe itself. Vibration can be damped in the liquid by inline silencers, by surge chambers or to some extent by flexible joints. Vibrational energy traveling along the pipe itself can be isolated by flexible joints, flexible hangers (which also isolate the pipe from building structure), U-bends and 90° turns in the pipe (to afford flexibility from longitudinal force) and elastomeric bushings where pipes pass through hangers or walls. Small-capacity damping devices protect gauges or other delicate devices from fluid shock. Rapidly closing valves, or oscillating control valves, can induce severe hydraulic hammer and cause damage.

Many noise pollution problems in industrial plants are the result of vibration transmitted through building structures by overlooked paths. Often-ignored pathways are through mounting bolts or braces to building structure; along fluid paths within pipes or ducts; and even through resonant acoustic couplings between thin walls, partitions or doors. Although these paths may exist anywhere, they are most troublesome very close to the noise source, and thus should be considered as subject to source treatment.

Isolation of Noise Sources

Noise reduction, when it can be done at the source, is the most efficient procedure, but it is not always sufficient. Some operations cannot be silenced, and in others economic considerations suggest other means.

Operations that are extremely noisy must be located in isolated areas. Occasionally they can be sealed within massive enclosures that do not require the presence of personnel; robot material-handlers are being used in such instances.

More commonly, however, the noisier operations are housed together in separate buildings or wings, reducing the personnel needed there and providing protection for them. In these circumstances, it is practical to afford as much noise isolation as possible by locating corridors and storage areas where they can serve as buffers between noisy areas and those in which quiet is requisite.

Walls, partitions and doors may possess good noise-isolation properties if they are properly designed and installed. Although this is usually an architectural and construction problem rather than one of operation or maintenance, the same principles can be applied to design noise-isolating enclosures around especially noisy machines, or around specific areas (as the audiometric test room). The sound-attenuating property of a partition or wall depends on its mass, on its elastic properties and also on its lack of porosity or openings.

The sound attenuation of a wall or door is described in terms of transmission loss (TL)—the ratio of the sound energy incident on one side of the wall to the sound energy radiated from the other side expressed in decibels. Transmission losses vary with frequency; if the partition is set into resonant vibration its transmission loss will be quite low, i.e., its ability to impede noise is poor.

It is customary, for the sake of simplicity, to describe the characteristics of walls, windows and doors by their sound transmission class (STC). This is a single-number value which describes the integrated TL values for a typical range of frequencies; some typical STC ratings are given in Table 5.5j. As

Table 5.5j
SOUND TRANSMISSION RATINGS OF WALLS AND DOORS

Material or Component	*Sound Transmission Class (STC) Rating in dbs*
½-inch thick steel plate	35
¾-inch plywood	28
4-inch brick wall	41
4-inch lightweight block wall	36
8-inch dense block wall	48
8-inch dense block wall, plastered both sides	51
12-inch poured concrete wall	56
¼-inch plate glass	26
1¾-inch hollow metal door	30
double-strength window glass	21

a rule of thumb, the transmission loss of a wall will be increased 5 to 6 db if the mass per unit area is doubled. Two walls or doors separated by a space are usually much better than simply making a wall thicker. Acoustic absorbing material is of little value in reducing transmission of noise through walls or ceilings, although it can reduce the noise level within a room to a limited extent.

Partial-height partitions permit sound to travel easily over their tops and can be moderately effective for small areas if heavily lined with high-absorption material. Thus, a three-sided telephone booth becomes possible, and similarly, a noisy small machine (as a teleprinter in an office) can be partially isolated.

Noise "leaks" are common through suspended ceilings; the noise travels up on one side of a wall, crosses the open space and descends the other side into adjacent rooms. Ventilating ducts provide similar weak points ("flanking") for passage of noise from room to room.

Hearing Conservation Programs: Personal Hearing Protectors

A hearing conservation program is an essential part of any adequate medical and safety program, and should include monitoring of noise levels, inspection of employees, scheduling of work periods and provision for personal hearing protectors. In the light of recent legislation such a program is probably best accomplished as a completely inplant program.*

A limited hearing conservation program can be conducted through a company personnel office or first-aid department; in a small plant this may be economically necessary. Audiometric tests may be performed by an outside medical agency on a definite schedule, detailed records being kept in the plant premises. Noise levels must be measured at all work locations; again, outside help may be necessary to insure that testing personnel are competent. With the data obtained, work schedules may be adjusted and personal hearing protection supplied as necessary.

A minimal audiometric test schedule should include tests at the beginning of employment for all personnel, tests at 3-month or 6-month intervals and a final test when employment is terminated. Detailed records should be kept and made available to the employee. Both office and plant personnel should be tested.

Continuous monitoring equipment may occasionally be needed, but in many sites spot checks with a sound level meter will suffice. If equipment

* "When employees are subjected to sound exceeding those listed in Table 1 of this section (Table 5.5d), feasible administrative or engineering controls shall be utilized. If such controls fail to reduce sound levels within the levels of the table, personal protective equipment shall be provided and used to reduce sound levels within the levels of the table." From Sec. 50-204, Safety and Health Standards, Walsh-Healey Public Contracts Act.

Table 5.5k
CHARACTERISTICS OF EAR PROTECTORS

Frequency Range (Hz)	*Nominal Sound Attenuation*		
	Earplugs (db)	*Earmuffs (db)*	*Combination of Both (db)*
50–200	24	20	35
200–800	25	32	36
800–2,000	27	36	40
2,000–5,000	36	41	48
5,000–10,000	40	38	45

or operation changes are made, noise levels must be rechecked closely. Standard calibrated instruments must be used—by operators who are competent.

Personal hearing protectors (ear-plugs and ear-muffs) should be properly selected and carefully fitted. Ear-plugs are less popular than ear-muffs: they are less comfortable and require periodic refitting; ear-muffs are more easily donned, with less danger of their failing to give protection if not carefully applied.

Ear-plugs and ear-muffs are available from several suppliers, and Table 5.5k lists some characteristics for both types of protectors. The attenuation values are for essentially ideal conditions; in practice, because of imperfect fitting, carelessness and haste, the attenuation values may be as much as 10 to 20 db lower than those shown.

REFERENCES

1. Crede, C., "Vibration and Shock Isolation," New York: Wiley, 1951.
2. Harris, C., "Handbook of Noise Control," New York: McGraw-Hill, 1957.
3. Lightbill, M. J., "On Sound Generated Aerodynamically. I. General Theory," Proc. Roy. Soc., Ser. A: Vol. 211, No. 1107, Mar. 20, 1952, pp. 654–687.
4. Lightbill, M. J., "On Sound Generated Aerodynamically. II. Turbulence as a Source of Sound," Proc. Roy. Soc. Ser. A: Vol. 222, 1954. Pp. 1–32.
5. Roberts, H. C., "Quiet Please! What M/E Engineers Should Know About Noise Control. I, II, and III," Actual Specifying Engineer, May, July, and September, 1970.
6. Yerges, L., "Sound, Noise, and Vibration Control," New York: Van Nostrand Reinhold, 1969.

Chapter VI

INSTRUMENT INSTALLATION MATERIALS

L. V. Corsetti
P. M. Glattstein

CONTENTS OF CHAPTER VI

6.1 TUBES AND TUBE BUNDLES

Tube Types:	(a) SINGLE TUBE (a1) Metallic (a2) Plastic (a3) Coated metallic (b) MULTI-TUBE BUNDLE (b1) Metal (b2) Plastic. *Note:* In the feature summary below, the letters (a1) to (b2) refer to the listed tube constructions.
Individual Tube Sizes (OD):	$\frac{1}{8}''$, $\frac{3}{16}''$, $\frac{1}{4}''$, $\frac{3}{8}''$, $\frac{1}{2}''$, $\frac{3}{4}''$ and 1″ ($\frac{1}{4}''$ is the most frequently used size).
Individual Tube Wall Thicknesses:	0.030″ (a1,a3), 0.032″ (a1,a3), 0.035″ (a1,a2,a3), 0.040″ (a2), 0.049″ (a1,a3), 0.050″ (a2), 0.062″ (a2), 0.065″ (a1,a3), with 0.030″ (a1,a3) and 0.040″ (a2) being most popular on instrument air service.
Materials:	*Single Tubes:* aluminum (a1,a3), carbon steel (a1,a3), copper (a1,a3), nylon (a2), polyethylene (a2), polyvinyl chloride (a2), stainless steel (a1,a3) and teflon (a2). *Coatings for Single Aluminum or Copper Tubes:* polyethylene (a3), PVC (a3). *Multi-tube Bundle Jacket:* polyethylene (b1,b2), PVC (b1,b2) and steel outer armor (b1,b2).
Number of Tubes per Multi-tube Bundle:	2 to 5, 7, 8, 10, 12, 14, 19 and 37.
Cost of $\frac{1}{4}''$ OD Tubing in 100-ft Length:	\$6 to \$8 (a1, aluminum), \$13 to \$15 (a1, copper), \$60 to \$80 (a1, steel), \$120 to \$150 (a1, stainless steel), \$500

(a1, monel), \$3 to \$4 (a2, polyethylene), \$20 to \$30 additional (a3), \$60 to \$70 (b2, 10 plastic tubes in plastic jacket), \$100 to \$110 (b2, 10 plastic tubes in galvanized steel armor), \$500 (b1, 10 copper tubes in fireproof double plastic jacket).

Partial List of Suppliers: Astubeco Inc. (a1,a3), Crescent Insulated Wire and Cable Co. (a2,b1,b2), Dekoron Div., Samuel Moore and Co. (a2,a3,b1,b2), Gulf Supply Co., Inc. (a1,a3), Charles F. Guyon, Inc. (a1,a3), Okonite Co., Wire and Cable Div. (b1,b2), Parker Hannifin (a2), Pyramid Plastics, Inc. (b2), Ryerson, Inc. (a1,a3), Whitehead Metals, Inc. (a1,a3).

Single Tubing

Although the variety of tube sizes, materials and constructions commercially available is wide, the tubing used in a process plant should be standardized in order to reduce inventory and simplify maintenance.

Criteria in selecting single tubing include the service (process fluid, pressure and temperature) for which the tubing is intended; the environment (outdoors, corrosive atmosphere, heat or humidity) to which the tubing is exposed; and the distances involved.

Copper Tubing

Soft copper (annealed) seamless tubing, $\frac{1}{4}''$ OD, is the most popular type in instrument air service for both controlled air signals and instrument air supplies. It is moderately resistant to corrosion, is easy to install and maintain and has a mechanical strength superior to that of single plastic tubing. Copper tubing comes in coils of 50 to 1,000 ft and should always be straightened and stretched before installing. Single tubing rather than tube bundles are used if no more than three tubes are laid in the same direction, and the distance is less than 25 ft.

The time constant of 200 ft of $\frac{1}{4}''$ OD tubing into a receiving bellows is about 1.4 seconds, the constant increasing about 2 seconds for each 100 ft to a maximum transmission distance of 1,000 ft. If this lag cannot be tolerated on fast control loops, 1. a larger diameter tube can be installed, 2. the controller can be left to be locally mounted as in a 4-pipe system, 3. pressure booster relays can be installed along the transmission path or 4. electronic instrumentation can be employed. (For more on tubing, see Sections 4.1, 4.3 and 6.1 in Volume II.)

Plastic Tubing

Plastic tubing is subject to mechanical and heat damage and consequently is inferior to metallic tubing. Less expensive than copper tubing, it nevertheless requires more elaborate supports, thus making the total installation as costly as if copper tubing were used. It is not fire resistant and loses strength when exposed to temperatures above 200°F.

Plastic coating is used in highly corrosive atmospheres; the latter cause the outside diameter of the tubing to be increased and therefore wherever a tubing fitting is used, the jacket must be removed, after which it must be resealed either by a protective plastic spray or by tape.

Aluminum Tubing

Aluminum tubing is the cheapest of the metal tubes. Its corrosion resistance in sulfur-containing atmospheres is superior to that of copper, but it is attacked in atmospheres containing hydrochloric acid or alkalies. Anodizing, which produces a tough coating of aluminum oxide on the surface, improves the metal's corrosion resistance. Corrosion from electrolysis (which may be caused by arc-welding in the area) may be minimized by thorough insulation, grounding all metallic conduits and avoiding combinations of dissimilar metals.

Another disadvantage of aluminum tubing is its low vibration resistance.

Multi-tube Bundles

If the tube runs exceed 25 ft, multi-tube bundles are more economical than single tubes.

Bundles come in many different constructions with two or more layers of protective material surrounding the core of tubes (Figure 6.1a). The larger the bundle, the heavier and less flexible it is and the more time consuming

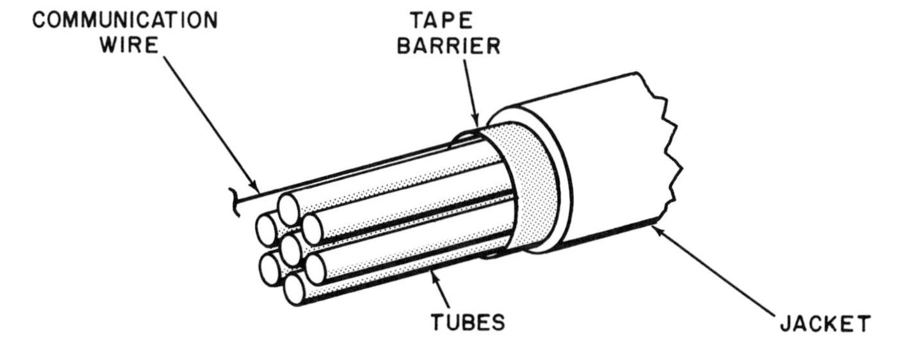

Fig. 6.1a Multitube bundle

its installation. For standardization, one bundle size should be used throughout a given plant and should be between 8 and 14 tubes per bundle for ease of handling.

Within the core the tubes are either in a parallel or in a spiral cabling arrangement. The latter permits greater flexibility and smaller bending radius. Table 6.1b summarizes the minimum bending radius requirements for spiral

Table 6.1b
MINIMUM BENDING RADIUS REQUIRED FOR MULTI-TUBE BUNDLES WITH SPIRAL TUBE ARRANGEMENT

Construction	*Number of 1/4" OD Tubes (in inches)*		
	5	*10*	*19*
Copper with PVC jacket	1.4	2.0	3.5
Polyethylene with PVC jacket	1.4	2.0	3.5
Copper with heavy duty PVC jacket	1.5	2.5	4.0
Polyethylene with heavy duty PVC jacket	3.0	5.0	9.0

bundles with 1/4" OD tubes. The minimum bending radius for parallel tube construction is about four times the values for spiral tube construction. Spiral construction should be used when the tube routing has sharp bends and turns.

There should be 10 to 20% spare tubing per run for future installation requirements and replacement.

Tube Bundle Selection

The most commonly used tube bundle has a copper core with either a polyvinyl chloride or polyethylene jacket and high mechanical strength and flexibility in the smaller sizes, but is more expensive than the plastic core bundles. Metal core tubing needs supports every 8 ft, whereas plastic core tubing needs supports every 4 ft.

The jacketing selected depends on mechanical strength requirements. Designs range from a single layer of plastic to two plastic layers plus an overall armored steel jacket. Within these layers additional moisture protection tapes and asbestos fire protection sheets can be placed (Figure 6.1c). Most bundles are designed either for direct burial or to be set within cement casings. Metal core tubing can withstand the heat of fire longer than plastic tubing can. When

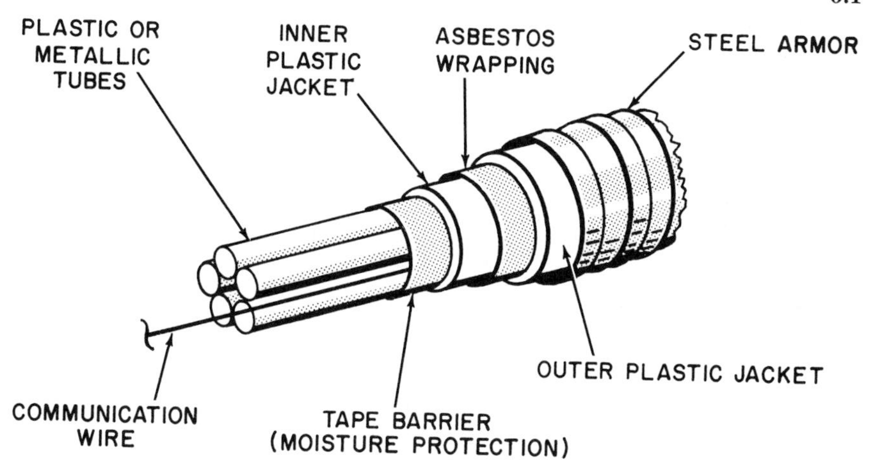

Fig. 6.1c Multitube bundle construction

economy is of primary importance, plastic core bundles are the structures of choice. They can be supported by existing structural members or if long, unsupported spans are being used, support cables are economical (Figure 6.1d).

Thermal expansion and contraction due to changes in ambient temperatures create compressive and tensile forces within the tubing material. Plastics do not resist tension as well as metals, a characteristic which may cause plastic tubes to pull out of their fittings. One ft of slack per 100 ft of tubing should be allowed for the contraction of plastic tubes due to a 100°F temperature drop.

Weatherproofing is required if the bundle is buried in earth, and additional reinforcement against crushing is needed if it is buried under a roadway.

Aluminum core bundle tubing is used in sulfur-containing atmospheres; steel bundle cores are inapplicable because of their inability to bend. Unlike copper or aluminum, steel tubing can be furnished only in straight random lengths.

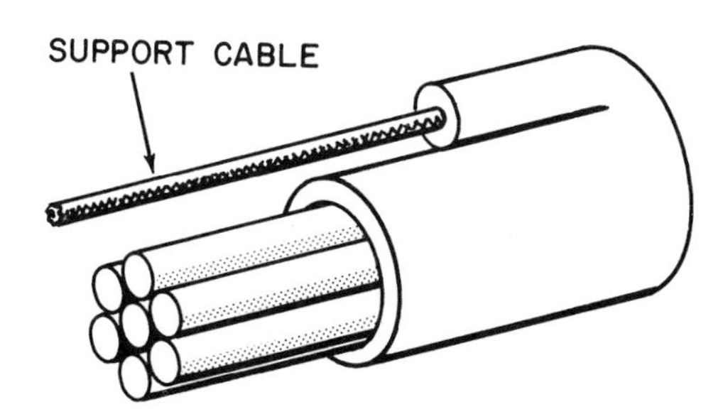

Fig. 6.1d Integral cable support for long unsupported tube bundle runs

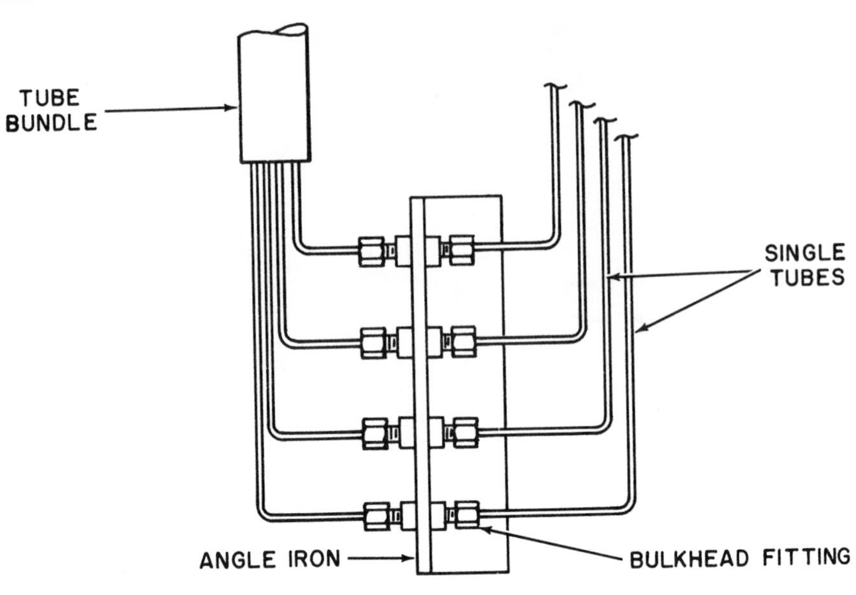

Fig. 6.1e Terminal strip

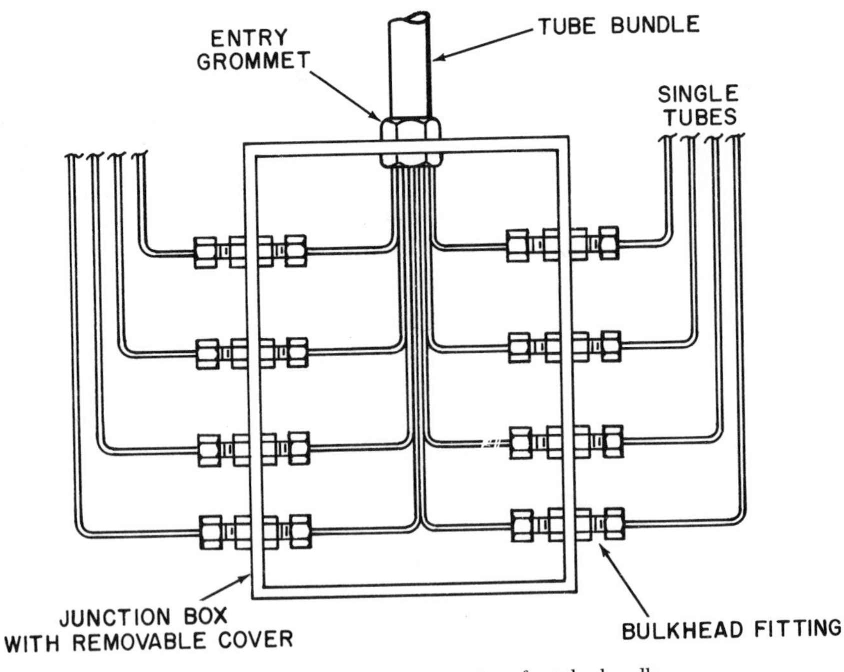

Fig. 6.1f Junction box termination of a tube bundle

Tube Bundle Termination

Bundle tubing ends either at a terminal strip (Figure 6.1e) or in a junction box (Figure 6.1f). The tubes after termination are routed individually to their destinations.

Plastic tubes should always end in a junction box so as to afford them maximum protection. The tube terminations should be identified by a metal or plastic tag, indicating the equipment or instrument they serve. In terminating tubing runs at a panel, it is preferable to have the terminal strips as part of the panel. (For more on recommended control panel tubing practices, see Section 5.1 in Volume II.)

In terminating tubing bundles, the jacketing must be removed. Once the tubes have been shaped to the desired termination point, they can be regrouped and bonded together for the sake of appearance.

Spare tubes should be coiled with the ends taped to prevent contamination.

6.2 TUBE FITTINGS

Fitting Designs:	Compression (flareless) and flared type.
Sizes:	Most commonly used is 1/4" OD × 1/4" NPT. Other sizes are from 1/8" to 1".
Materials:	Brass, aluminum, steel, 316SS, polyvinyl chloride, polyethylene or any machinable metal or plastic on special orders.
Costs:	In brass construction the 1/4" fittings are priced at about $1 each. Stainless items cost three times that amount.
Partial List of Suppliers:	Afco Fitting Co., C. B. Crawford Co., Crawford Fitting Co., Dekoron Product Div., Fast & Tite Products, Flodor Corp., Hoke Valve Co., Parker-Hannifin, Weatherhead Co.

The two basic tube fitting designs are the compression type (flareless) and the flared type. The former develops a pressure seal against the tube wall (Figure 6.2a). The flared design uses the tube end as a seating surface (Figure 6.2b). Compression fittings are used on both metallic and plastic tubing, whereas flare type fittings are limited to metallic tubing.

Metallic, Flareless (Compression) Fittings

The flareless (compression type) tube fitting is the newer of the two basic designs. It improves tube alignment and furnishes effective sealing against leaks.

When joining the tube to the tube fitting, it is important that the tube be evenly cut and the metal burrs removed after cutting. There are no special tools for preparing the tube or the tube fitting, which does not have to be disassembled for installation. The tube is pushed into the fitting opening and

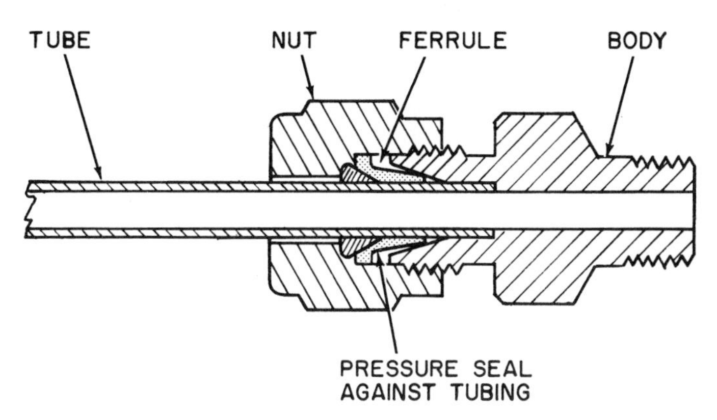

Fig. 6.2a Compression tube fitting (flareless)

the hex nut is tightened to the desired torque, forcing the ferrules to grasp the tubing without damaging or biting the tube walls. The seal developed will withstand pressures in excess of the tube design pressure and will also be reasonably effective on vacuum service.

Variations in flareless fitting designs are the result of modifications in the ferrule. Some manufacturers use two separate ferrules; others use a single ferrule. The main disadvantage of the two-ferrule design is the possibility of installing one of the ferrules backward.

Flared Tube Fittings

The earliest tube fitting design, the flared type, required both skill and time to make a proper installation. The tube must first be evenly cut, and all burrs and metal filings must be removed. Next, the tube end is shaped with a flaring tool. A 37° flare is most commonly used. Too much pressure against the tube when compressing the end into a flared shape will change the wall thickness and distort the ends. The main advantage of the flare fitting is its ability to withstand more vibration than the flareless types.

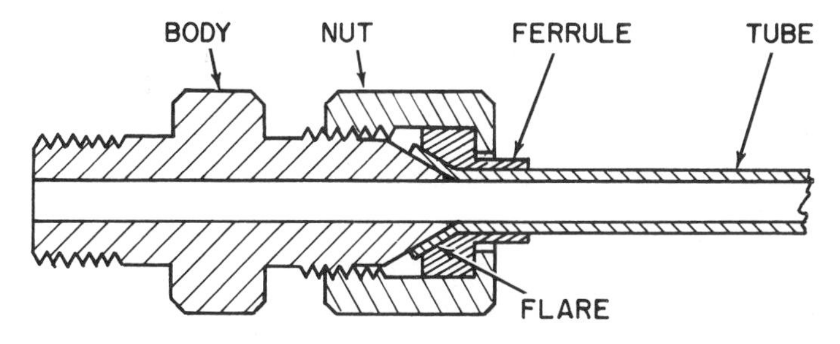

Fig. 6.2b Flared tube fitting

Plastic Flareless Fittings

Plastic fittings are relatively new and chiefly designed for plastic tubing; they should never be installed in metallic tube lines. They are available only in the flareless design and must be protected from mechanical damage, heat and extreme cold. One use for plastic fittings and tubing is on the back of a control panel (Figure 6.2c).

The ferrule of a plastic tube fitting is also plastic, and therefore cannot grip the tube strongly. Metal fittings for plastic tubing give a tighter seal (Figure 6.2d). They can be designed with a metal sleeve that is forced inside the plastic tube, thereby strengthening it (Figure 6.2e). In this case, the metal ferrule forces the plastic tube against the metal sleeve, giving a more secure seal.

Tube Fitting Selection

A fitting is used to connect a pipe header or a shut-off valve to a tube. Most manufacturers use standard designations for the various fittings available. Table 6.2f lists the more common fitting designations which in fact are abbreviations for their descriptive names. The ISA's standard nomenclature is given in Figure 6.2g.

Tube fittings can be obtained with either screwed or welded ends, can

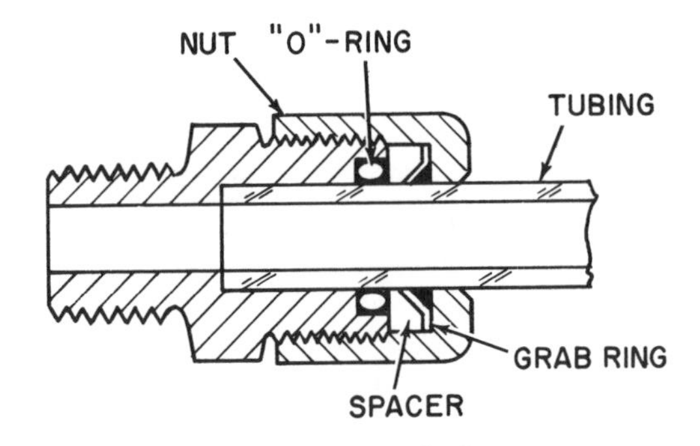

Fig. 6.2c Plastic tube fitting

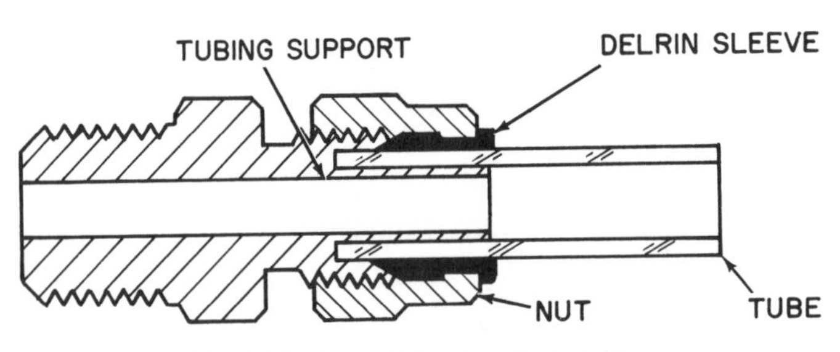

Fig. 6.2d Metal fitting for plastic tubing

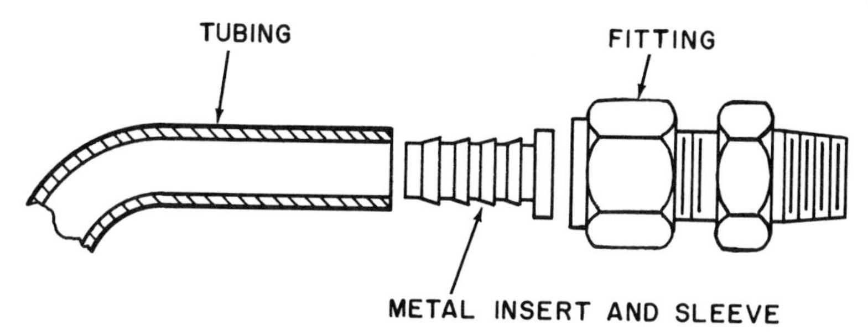

Fig. 6.2e Metal insert for plastic tubing and metallic fittings

Table 6.2f
TUBE FITTING
DESIGNATIONS

Fitting Type	*Designation*
Connector, union, male	CUM
Bulkhead, union	BU
Connector, female	CF
Cross, tube	C
Elbow, male	LM
Tee, all tube	TTT
Tee,* with two tube and one pipe connections	TTP

*Tees are described by first giving the inline connections and then the branch connections.

be used on either light or heavy wall tubing and can withstand pressures in excess of the pressure rating for the tubing being used.

Special Fittings

The quick-connect fitting is ideal when a large number of tubes are frequently connected and disconnected. These fittings have a single or double end shut-off valve, allowing the fitting to be broken with no loss of process material in the line. Because these lines are being transferred from one to another (similar to a telephone switchboard), plastic tubing is generally used, due to its flexibility. The quick-connect fittings cannot be used at high temperatures because of the O-rings and bushings used in their design.

The quick-crimp fitting is constructed of plastic and has to be crimped to fasten it to the tubing, and a special crimping tool joins one tube to another. The fitting does not have screwed ends, but threaded adapters are available to maintain versatility in integrating with other components. This fitting should be used only in protected areas where it is not subject to mechanical damage.

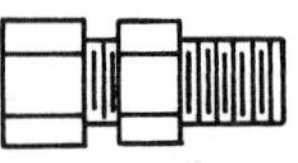

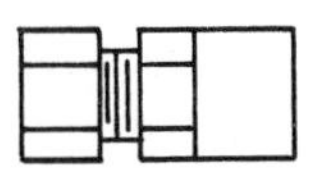

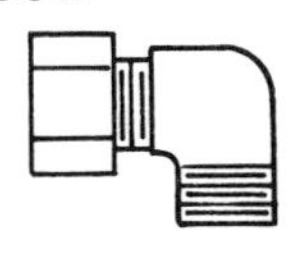

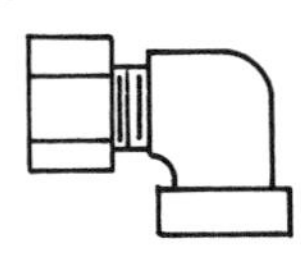

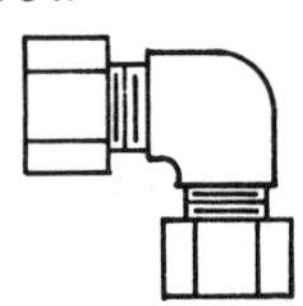

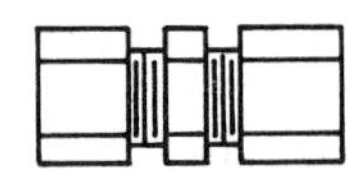

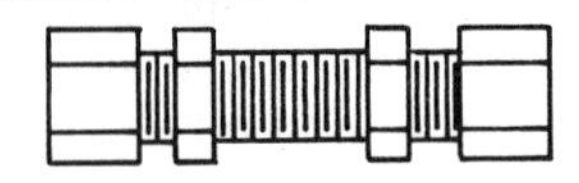

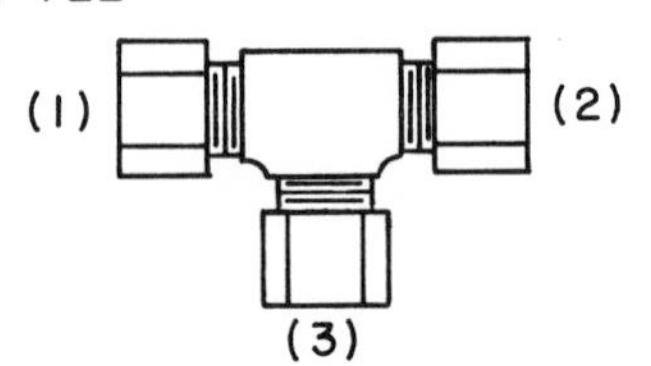

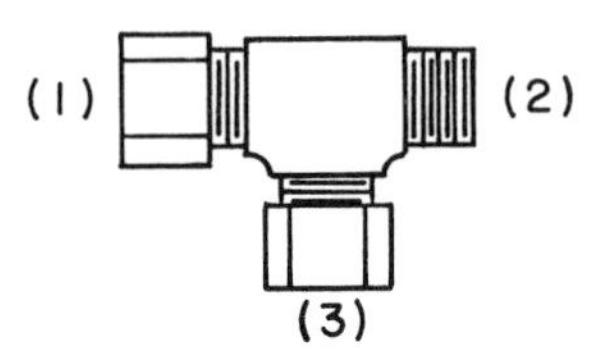

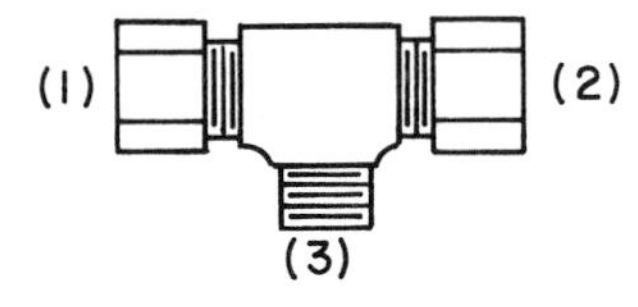

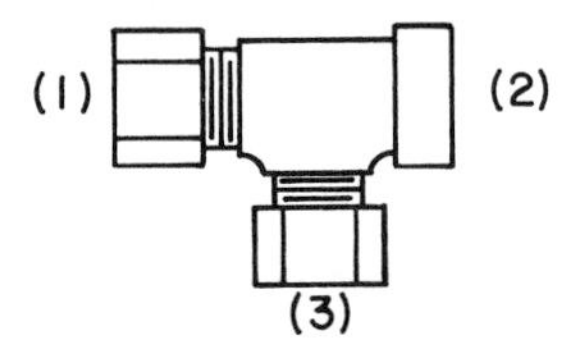

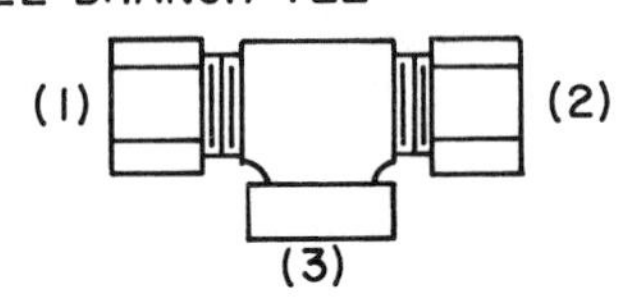

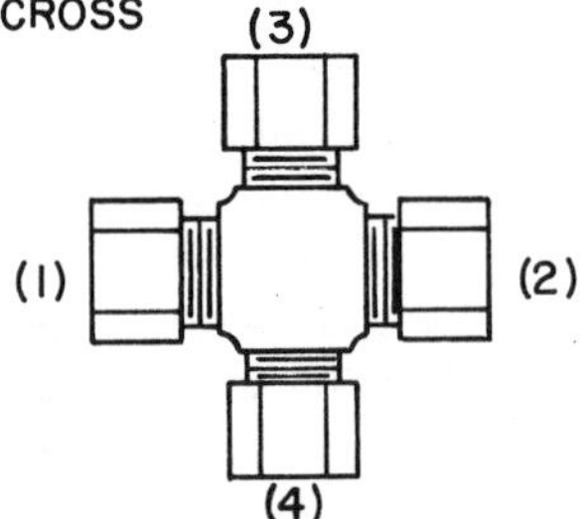

Fig. 6.2g Nomenclature for instrument tube fittings

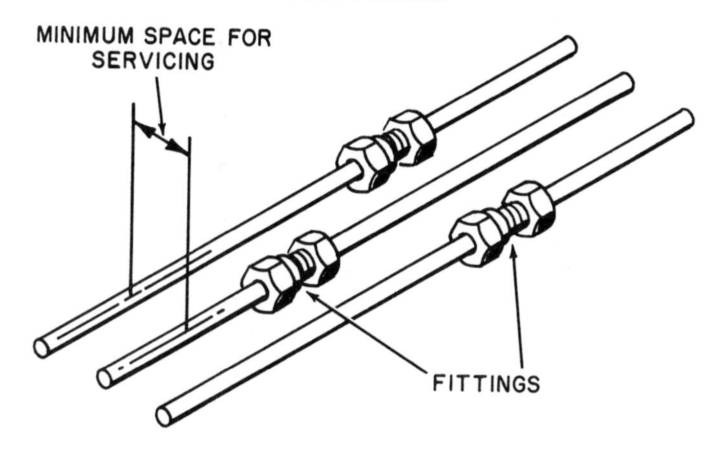

Fig. 6.2h Staggering of fittings in parallel tube runs

Miniature valves for use in tube lines are available with tube fitting ends. They were designed to eliminate the need for extra fittings when installing valves in tube lines. Available in all tube fitting sizes and in a wide variety of materials, they can be obtained from the same manufacturers of the tube fittings.

Tube fittings for high pressures in the 10,000 to 20,000 PSIG range are also available. Equipped with a threaded ferrule which is then screwed to the tube, they can withstand pressures of as much as 60,000 PSIG.

Tube Fitting Installation

Fittings are packaged and shipped assembled and ready to use. They should not be disassembled because contamination with foreign matter can cause leaks.

Fittings are installed by inserting the tubing into the tube fitting, so that the tubing rests firmly on the shoulder of the fitting and the nut is finger tight. With the fitting body held steady with a back-up wrench, the nut is tightened $1\frac{1}{4}$ quarter turns, producing a leaktight joint.

In certain fitting materials such as 316 stainless steel, Inconel, Monel and other special alloys, galling can occur (the nut becomes welded to the body) because of over torque and periodic making and breaking of the fitting. Whenever these special alloys are used, a lubricant should be applied to the threaded portion of the fitting.

In installing fittings in a parallel tubing layout, it is recommended that the fitting be staggered and offset (Figure 6.2h). This allows easy access to the fitting for servicing or leak testing. When bending the tubing, have some straight run beyond the fitting so that the tubing is not offset inside it. There should be expansion loops between fittings to prevent thermal contraction tension stresses from developing.

6.3 JUNCTION BOXES AND INSTRUMENT HOUSINGS

Types:	(a) Junction boxes (b) Instrument housings *Note:* In the feature summary below, the letters (a) and (b) refer to the above listed types.
Standard Dimensions:	Height: 10″ (a), 20″ (a), 24″ (b), 30″ (a,b), 36″ (b) and 40″ (a) Width: 10″ (a), 12″ (a), 16″ (a), 18″ (a), 19″ (b) and 24″ (b) Depth: 4″ (a), 6″ (a), 8″ (a), 16″ (b), 18″ (b) and 20″ (b) Custom built units available in any desired size.
Construction Materials:	Aluminum or steel
Cost Range:	\$30 to \$180 (a), \$70 to \$200 (b)
Partial List of Suppliers:	Columbus Metal Products Co., Inc. (a), Crescent Insulated Wire and Cable Co. (a), Dekoron Div., Samuel Moore and Co. (a), Winston Co. (a,b), John Zink Co. (b)

Junction Boxes

Figure 6.1f shows a typical junction box serving as an enclosure wherein tubing bundles may be separated into individual tubes. The requirements for weather resistance and for mechanical strength are major considerations in selecting junction boxes, and the ruggedness of the box will depend on the thickness of the metal used. Table 6.3a provides guidelines for relating size to desirable wall thickness. Junction boxes have either hinged or removable doors. For greater accessibility, a box with removable door and wing-nut fasteners should be used.

The box can be painted or galvanized (zinc is applied to the metal surfaces) or both. For indoor use, painting is sufficient. If the box is exposed

Table 6.3a
RECOMMENDED JUNCTION BOX WALL THICKNESSES

*Recommended Wall Thickness (Gauge)**	*Largest Wall Surface is under this Value (in²)*	*Largest Dimension is under this Value (in)*
16 (0.062″)	360	24
14 (0.078″)	1,000	40
12 (0.109″)	1,500	60
10 (0.141″)	Over	Over

*This unit is based on the weight (in pounds) per square foot of a steel sheet.

to the outdoors, to water or steam from hose stations or to oil dripping from equipment, it should be galvanized. Mounting supports should be an integral part of the box.

Junction boxes are available with standard knock-outs. (A knockout is an almost completely precut hole which eliminates the need for drilling and allows one to punch out the hole as required for locating terminal fittings.) The pre-determined location of the knock-outs, however, may restrict design flexibility. Fittings on the walls of the box should be located on 2½″ minimum centers. If more than one box has to accommodate the tubes, they can be stacked with conduit(s) as the interconnection between the boxes.

Most standard junction boxes meet the NEMA 12 classification, which requires the enclosure to exclude dust, dirt, oil, lint, fibers, flying's or coolant seepage.

Junction boxes should be located at strategic points so as to serve particular groups of instruments, and the instrument being served should be identified on the corresponding fitting within the box. Junction boxes are ordinarily mounted on steel support columns or walls and their center line should be set at approximately 5′ 2″ from the floor.

Instrument Housings

Instrument enclosures are designed to protect the instrument from inclement weather, atmospheric pollutants, insects, moisture, dust and freezing temperatures. Unlike junction boxes, instrument housings must be designed for a particular type of instrument. Owing to the large number of requirements, housing suppliers find it difficult to standardize on housing sizes. The housings, usually galvanized and painted, should be large enough to afford free space around the instrument for routine maintenance and removal. Properly sized and positioned access doors are necessary. Figure 6.3b illustrates several available design choices.

If the instrument requires visual inspection, the housing should have a shatterproof safety glass window in the front access door.

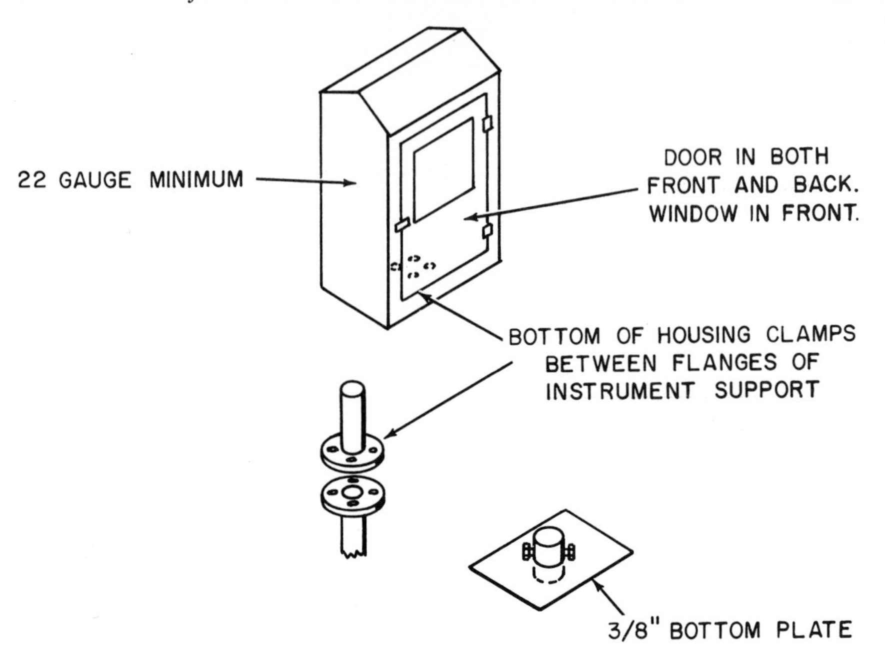

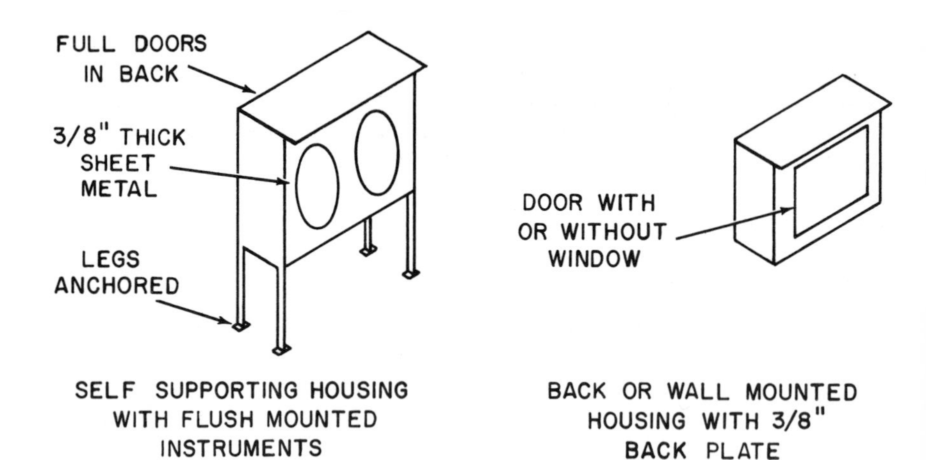

Fig. 6.3b Typical instrument housings

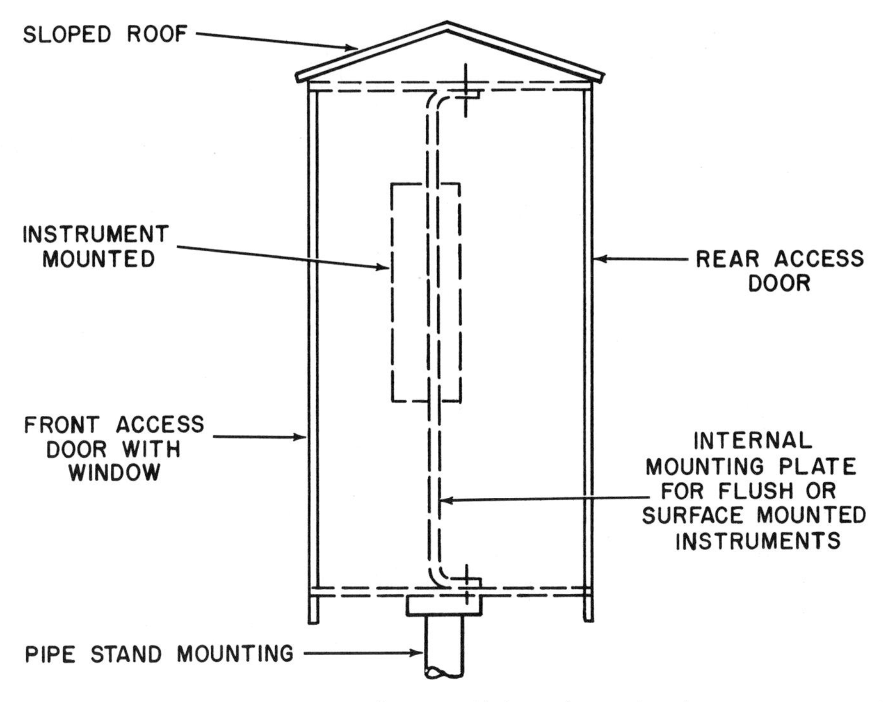

Fig. 6.3c Instrument housing with internal mounting plate

For a flush or surface mounted indicator, an internal mounting plate may be needed (Figure 6.3c), although it is also possible to use the housing walls for supporting the instruments. All tubing, wiring or piping enter through the side or bottom of the box.

For protection against winter weather, the box can be furnished with either a steam coil or an electric heater (Figure 6.3d). Electrical heaters need less maintenance and are easier to control. In some hazardous locations, the instrument housing is purged to pressurize the enclosure, thereby allowing no explosive gases into the box. When purging is required, the housing should also be protected against overpressures.

Housings can be wall or pipe stand mounted (Figure 6.3e). The use of legs with the smaller housings is impractical, and pipe stand mounting is preferable. Houses designed for pipe stand mounting require less floor space and can also be used for wall mounting after plugging the pipe yoke opening. Pipe stands can be supported by inserting the pipe in a cement encasement in the ground. On hard surfaces like concrete or steel platforms, the pipe can be mounted in a flange and bolted to the surface. The centerline of instrument housing windows should be at 5′ 2″ from grade.

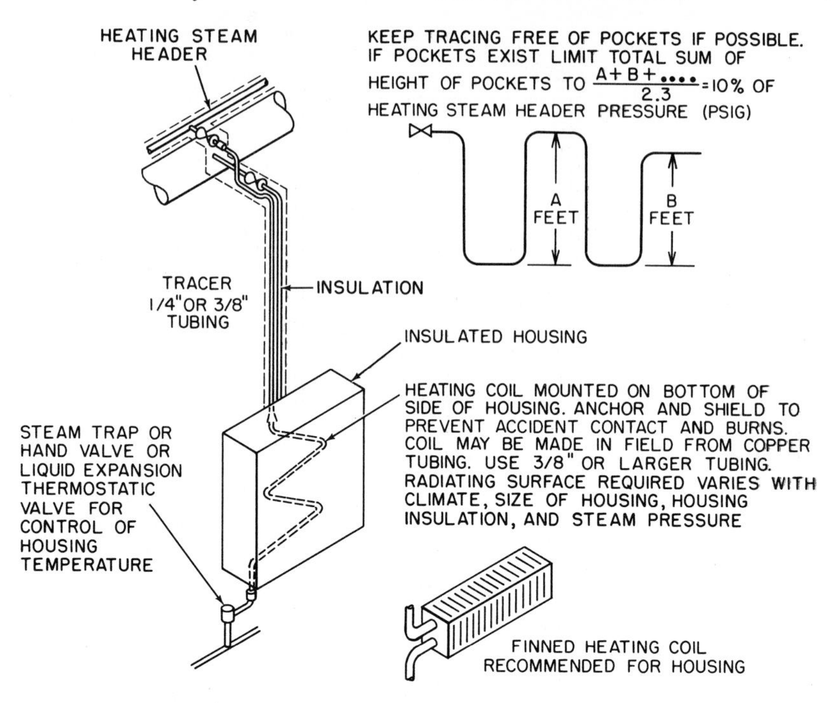

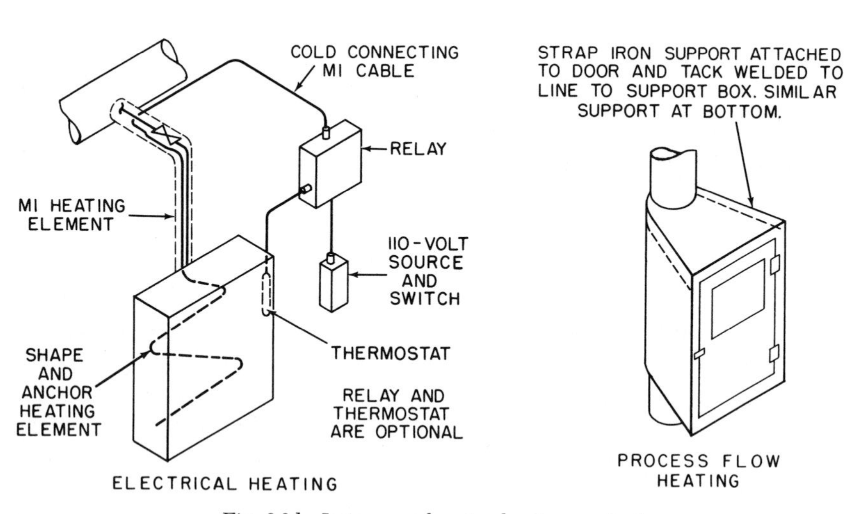

Fig. 6.3d Instrument housing heating methods

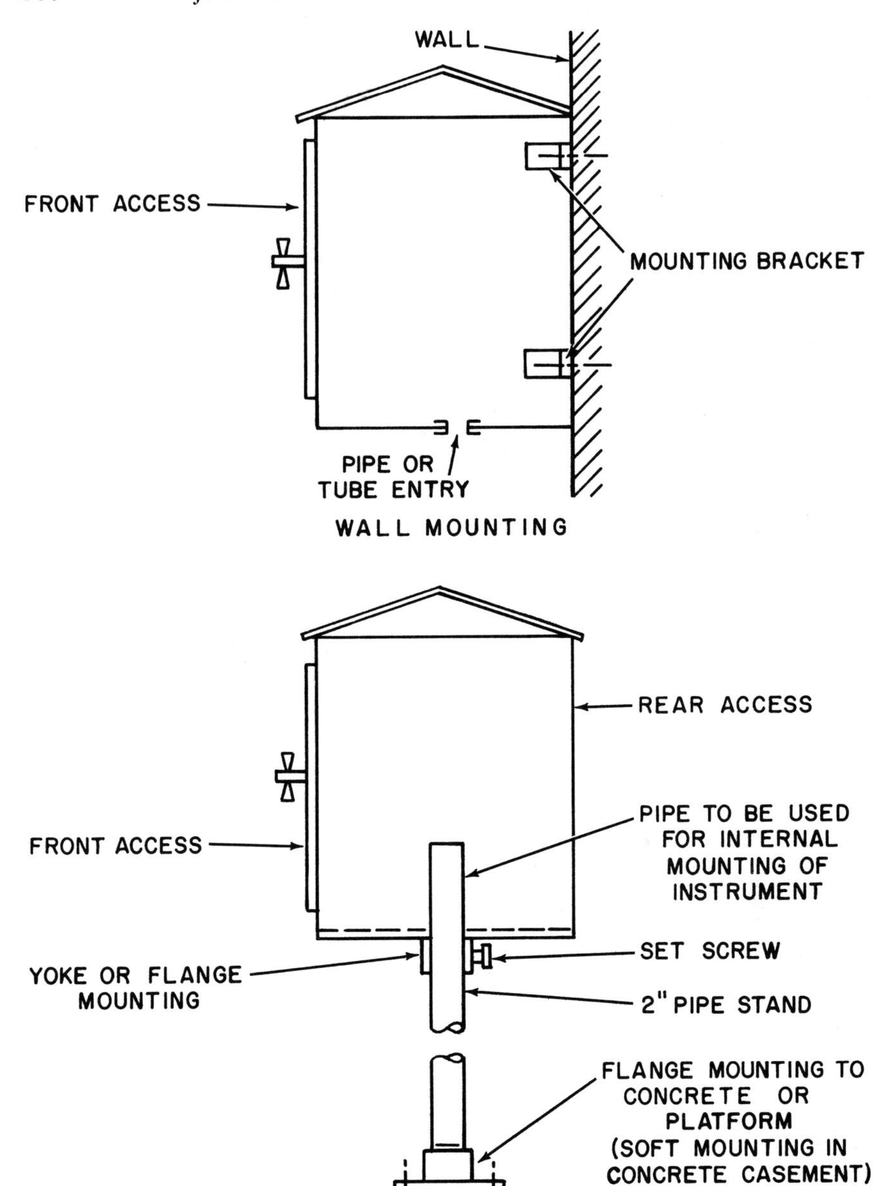

Fig. 6.3e Instrument house mounting methods

Walk-in instrument housings are also available in various sizes, materials and construction, and house complete systems, analyzers and local control panels. They are obtainable as prefabricated structures in various standard sizes with options like type and number of windows, type of doors and door locations, type and number of louvers and peaked or sloped roofs.

6.4 TUBE SUPPORTS

Types and Shapes:	(a) Ladder, trough and channel type (b) angle, solid or prepunched (c) Flat steel, solid or prepunched
Sizes and Dimensions:	(a) Lengths, 12′, 16′ and 24′, Widths, 1″, 1½″, 2½″, 3″, 4″, 6″, 9″, 12″, 18″, 24″ and 30″ Depths, ¾″, 3½″, 4½″ and 6″ (b) 2″ × 1½″ × ⅛″ × 20′ long, or 2″ × 2″ × ⅛″ × 20′ long and (c) 2″ × ⅛″ × 20′ long.
Standard Materials:	Steel, stainless steel, aluminum.
Cost:	2% to 5% of the total instrument cost should be allotted for a well designed support system.
Partial List of Suppliers:	Channel-Way Industries, Inc., Daburn Electronics & Cable Corp., Girard Development Inc., Husky Products, Inc., Imperial-Eastman Corp., Jonathan Mfg. Co., Norrich Plastics Corp., Panduit Corp., Parker Hannifin Corp., Premier Metal Products Corp., P-W Industries, Inc., J. C. White Co., Zero Mfg. Co.

Tubing runs are commonly supported on racks or on expanded metal trays. Both approaches are fairly expensive, contribute to a well organized installation and provide accessibility for maintenance and efficient mechanical protection, because the routing of the trays or racks is selected with these considerations in mind. Existing structural steel along the route can be used to support the trays.

Support on Racks

Rack support systems are costly because they require large amounts of field labor (Figure 6.4a) and must be pre-assembled before erection. Constructed from various steel members (angles, channels and structural tees) racks are all assembled at ground level and then transported to the site of installation. The rack system affords limited mechanical protection for the tubes because they are more exposed. The distance between supports, particularly for plastic tubing, must be kept sufficiently close to prevent the tubing from sagging.

Support on Trays

A tray support system (Figure 6.4b) is made up of prefabricated sections joined together at the installation site and require little pre-erection time. Tray systems are less costly than rack systems and can span a greater unsupported distance. They offer greater mechanical protection because the tubes are not exposed, but on the contrary are confined within the structure of the tray. For added protection, trays have protective covers. Various tray fittings furnish good support and change of direction.

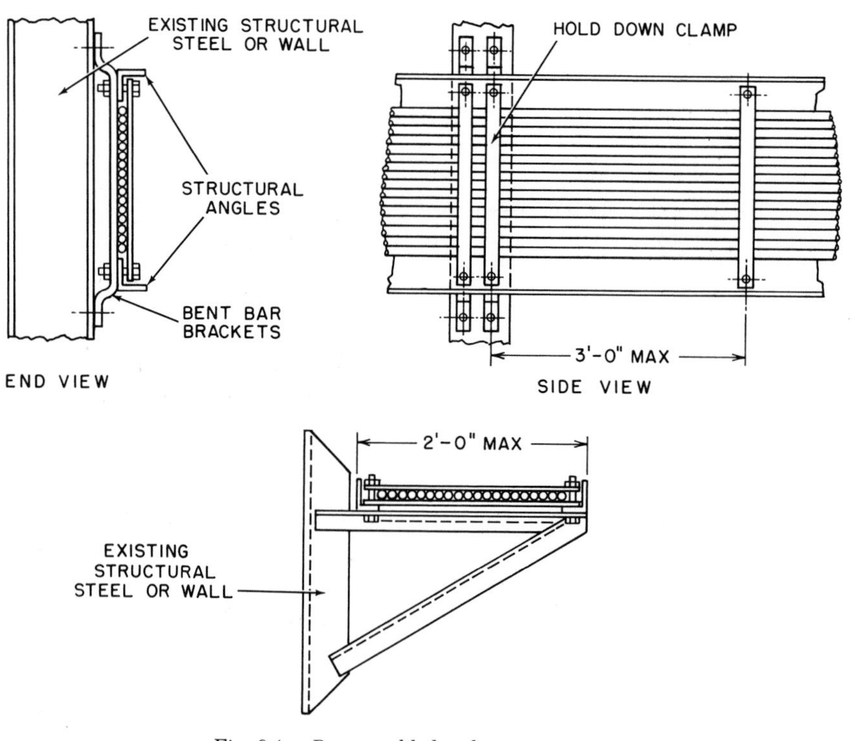

Fig. 6.4a Pre-assembled rack support system

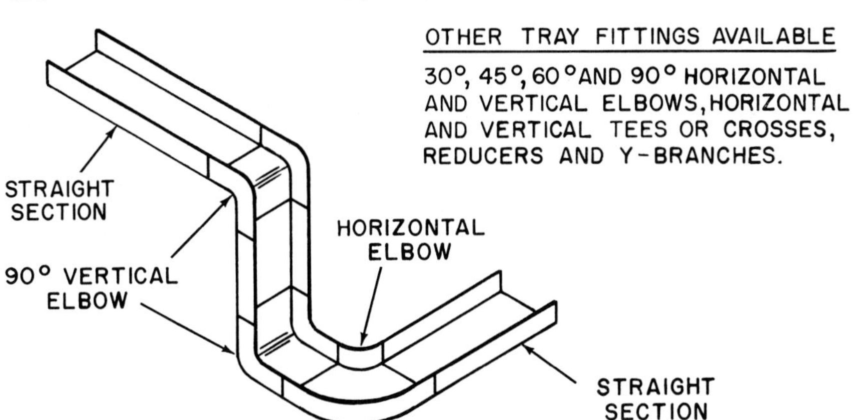

Fig. 6.4b Prefabricated tray support system. Other tray fittings include 30°, 45°, 60° and 90° horizontal and vertical elbows, horizontal and vertical tees or crosses, reducers and Y-branches.

Tray support systems should be sized so that space is available for subsequent addition of either electrical cabling or pneumatic tubing. Standard designations (Figure 6.4c) to identify the components in a tray support system include:

Ladder type trays. A prefabricated metal structure of two longitudinal side rails connected by individual traverse members.

Trough type trays. A prefabricated metal structure more than 4″ wide, with an open bottom with closely spaced or mesh support within the longitudinal side rails.

Channel type trays. A prefabricated metal channel not more than 4″ wide.

Fittings. Sections joined to other tray sections to change the size or direction of the tray system.

Connectors. A fitting that joins together the straight tray sections or fittings.

Figure 6.4d illustrates typical tray support methods including the cantilever bracket, trapeze and individual rod suspension. Accessories include dropouts, covers, conduit adaptors and hold-down devices. Tray system suppliers usually furnish installation instructions on spacing supports, accessories and permissible load capacities. It is not uncommon for maintenance workers to use a tray system with covers as a walkway. Supports should be located so that connectors between horizontal straight sections of tray runs will fall between the support point and the quarter point of the span.

Tray and rack systems are mainly for supporting bundle tubing which is laid in the tray on horizontal runs, with the weight of the bundle itself holding it in place. In vertical runs, the tube bundle must be supported, especially if it is all plastic; plastic tube bundles under tension are likely to

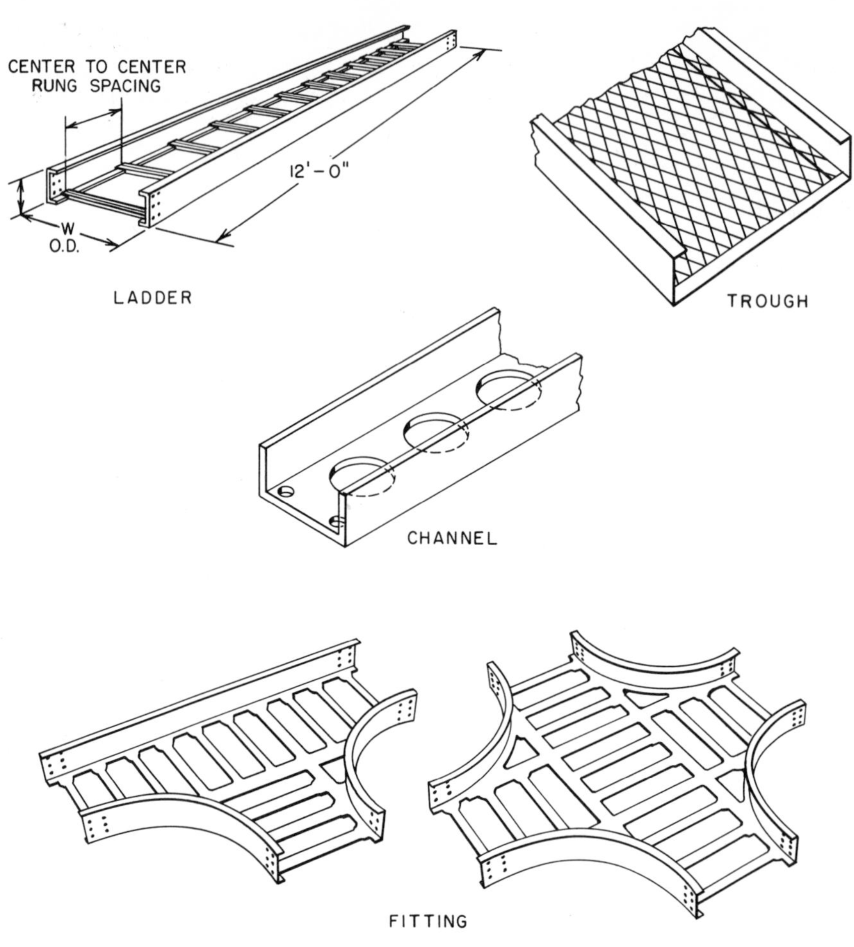

Fig. 6.4c Support system terminology

stretch. To relieve the load on vertical plastic tube bundles, they should be supported one quarter of the way down the vertical run (Figure 6.4e).

For metal core tube bundles, clamping devices suffice for vertical support (Figure 6.4e). If the bundles in the tray are parallel, it is possible to splice single tubes into the bundle by removing the outer protective jacket and teeing into the individual tube (Figure 6.4f). Tube bundles are installed in trays or racks either by lifting them into place or by pulling the bundles along the trays or racks. Pulling devices have rollers and pulleys for negotiating around bends and vertical runs. Prior to installation, the support racks should be inspected to protect the tubing against sharp edges.

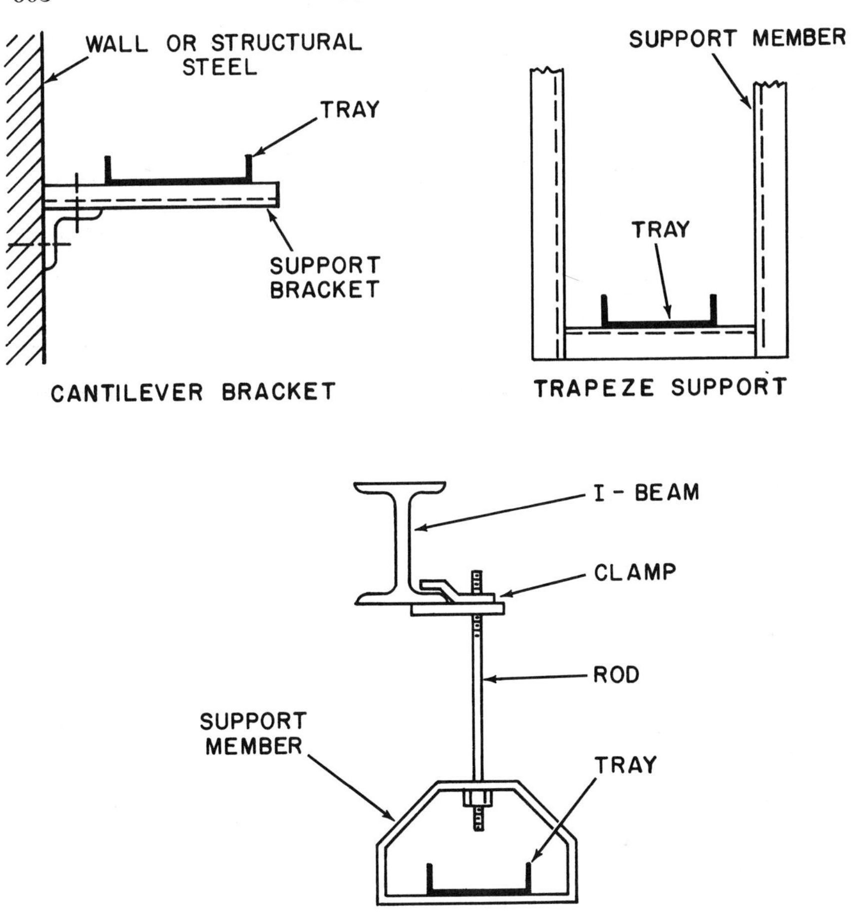

Fig. 6.4d Tray support methods

Single Tube Supports

Protection against mechanical damage is the primary consideration in supporting single tubes. Unlike tube bundles, which are generally run at higher elevations, the single tube is brought to the instrument at a low elevation. Therefore, metal tubes with minimum exposure should be used. Maximum protection for single tubing is the angle support with a protective clip (Figure 6.4g). The angle gives full protection from two directions with the clip and partial protection from two other directions. The angles with prepunched holes can be mounted to a wall or to an existing steel member or can join the angles together.

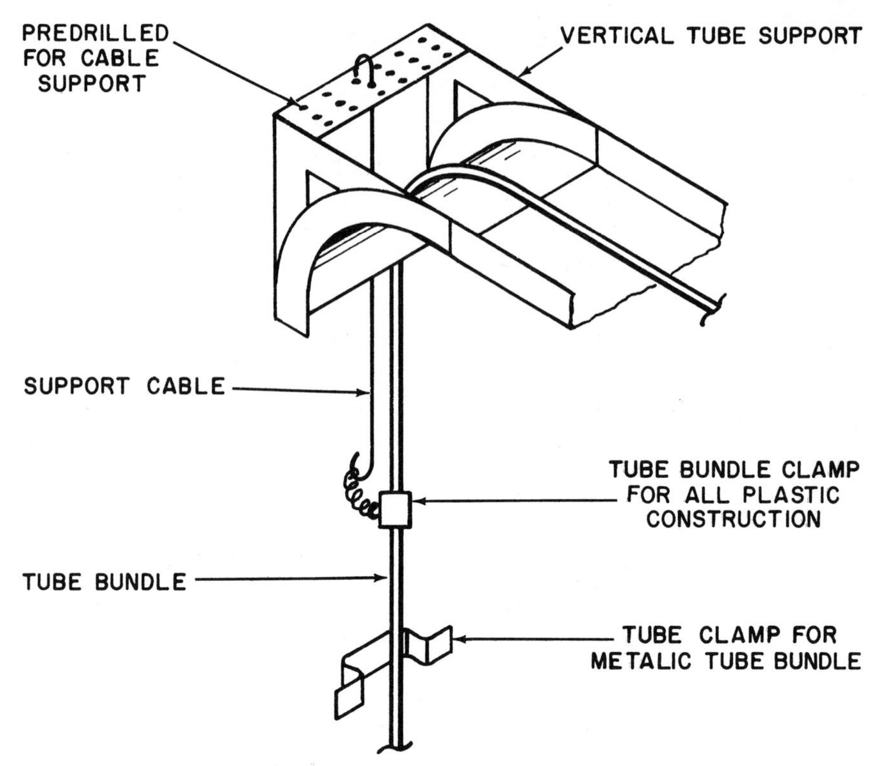

Fig. 6.4e Methods of supporting vertical tube bundle runs

An economical way to support a group of single tubes running in the same direction is by raceways. Raceways are prefabricated from galvanized steel in 16-foot straight channel sections with tube clips affixed to the inside of the channel (Figure 6.4h). Raceways accommodate tube sizes of $\frac{1}{4}''$, $\frac{9}{32}''$, $\frac{3}{8}''$ or $\frac{1}{2}''$ OD. Because the tube clips are an integral part of the raceway and because the tubes are held by the spring tension produced by these tube clips, the raceway can be mounted at an angle. Efficient mechanical protection is obtained by installing the raceway with the channel facing down, which also allows the uncoiled and stretched tube to be clipped easily into place.

Single tubes can also be supported by the existing steel structure, by far the most economical technique because the only support material required are tube or pipe clamps. This approach offers little mechanical protection for the tubing and usually necessitates that the routing be decided in the field because of the limited information on the location of steel members at the time of design preparation.

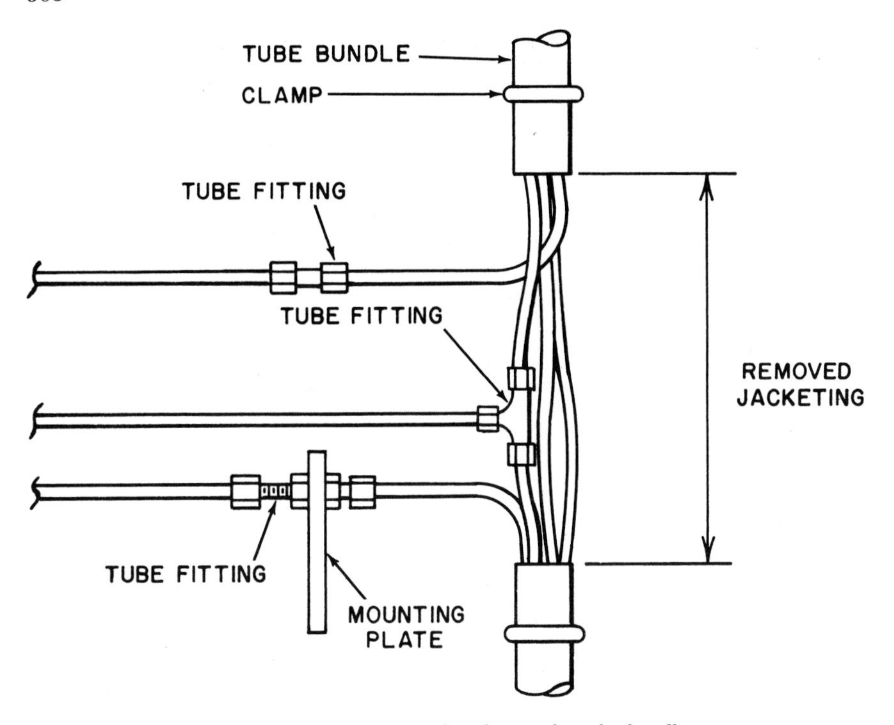

Fig. 6.4f Connecting single tubes to the tube bundle

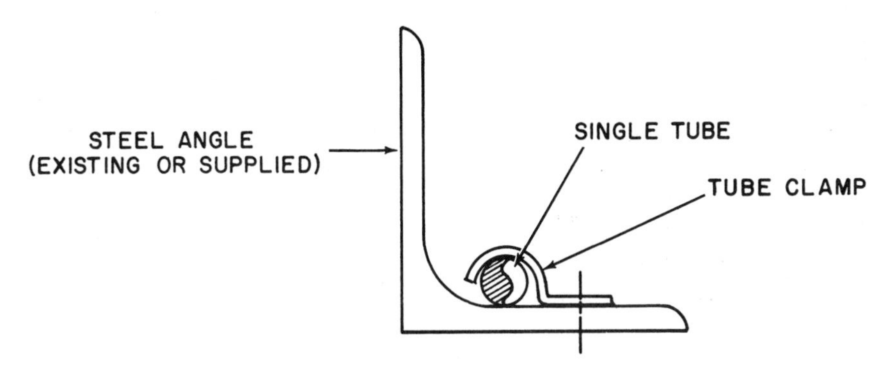

Fig. 6.4g Supporting single tube off steel angle

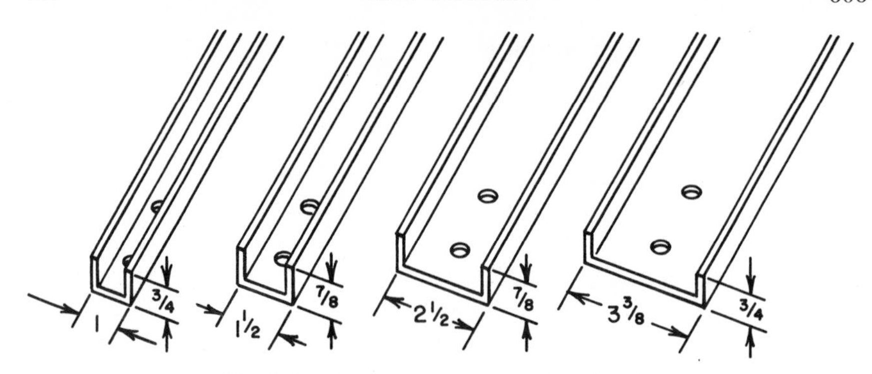

Fig. 6.4h Raceways support several single tubes

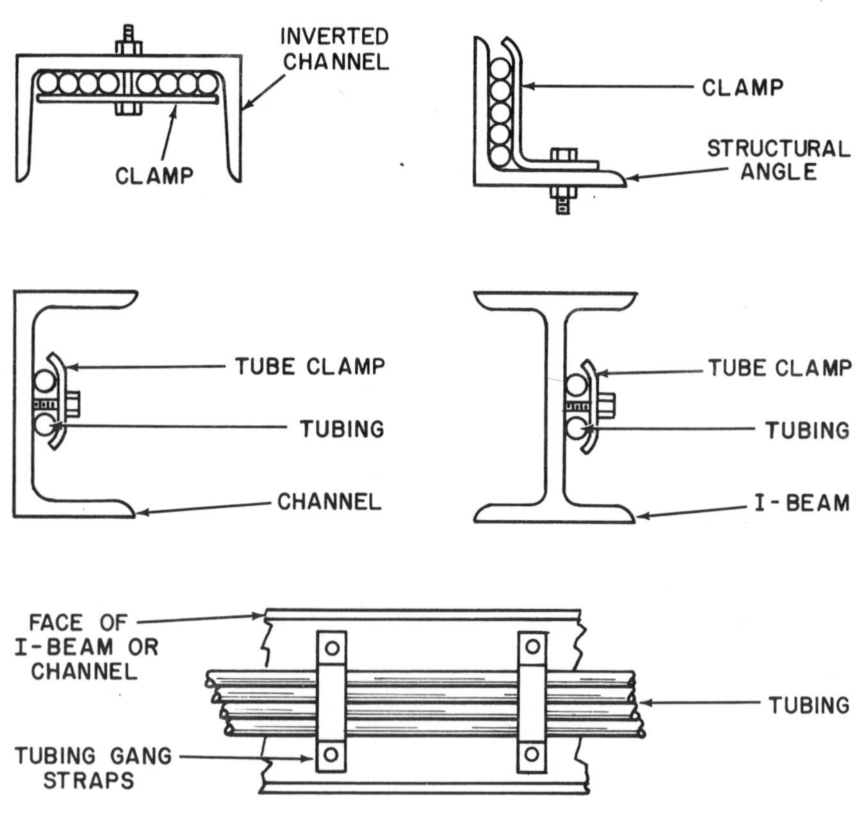

Fig. 6.4i Single tubes supported off existing steel members

The tubes should always be run inside the frame of the support member selected such as the running of the tube in the web of an I-beam. Figure 6.4i shows some typical supports for single tubing, using existing structural steel.

Thin wall conduit also protects and supports all-plastic tube bundles, in which application whenever the conduit run changes direction, a pull box is required to pull the tube bundles through the conduit in straight runs. Maintenance of the tubing or splicing it into the bundles is not practical with this design. The main advantage of the conduit system is the maximum mechanical protection that it offers.

6.5 AIR FILTERS AIR SUPPLIES

Filter Connection Sizes:	1/4″, 3/8″, 1/2″, 3/4″ and 1″ NPT.
Design Pressure:	Plastic bowl, up to 250 PSIG at 150°F. Metal bowls available for higher pressures.
Standard Particle Stoppage Rating:	2, 5, 10 and 40 microns.
Materials:	*Housing*—aluminum, stainless steel, brass and zinc alloy. *Bowl*—plastic or metallic. *Filter element*—stainless steel wire mesh, phenolic impregnated cellulose, woven cloth and paper.
Cost:	$18 for 1/4″ aluminum unit with plastic bowl, $300 for 1″ all-stainless steel filter.
Partial List of Suppliers:	ACS Industries, American Meter Co., Conoflow Corp., Deltech Engineering, Inc., Fisher Controls Co., Mine Safety Appliances, Moore Products Co., Wilkerson Corp., Zeks Industrial, Inc.

Air Supply Systems

A typical air supply system is shown in Figure 6.5a, with all of the main components up to the level of the main instrument air distribution header. Headers and subheaders are run to the sites in the plant where air is required, present and future requirements determining the extent of the air distribution system.

A loop pattern (Figure 6.5b) is a common design, allows the use of small pipe and assures equal distribution throughout the plant. A manifold type (Figure 6.5c) allows the pipelines to go only to those areas needing air. Subheader outlets should always be removed from the top of the header so

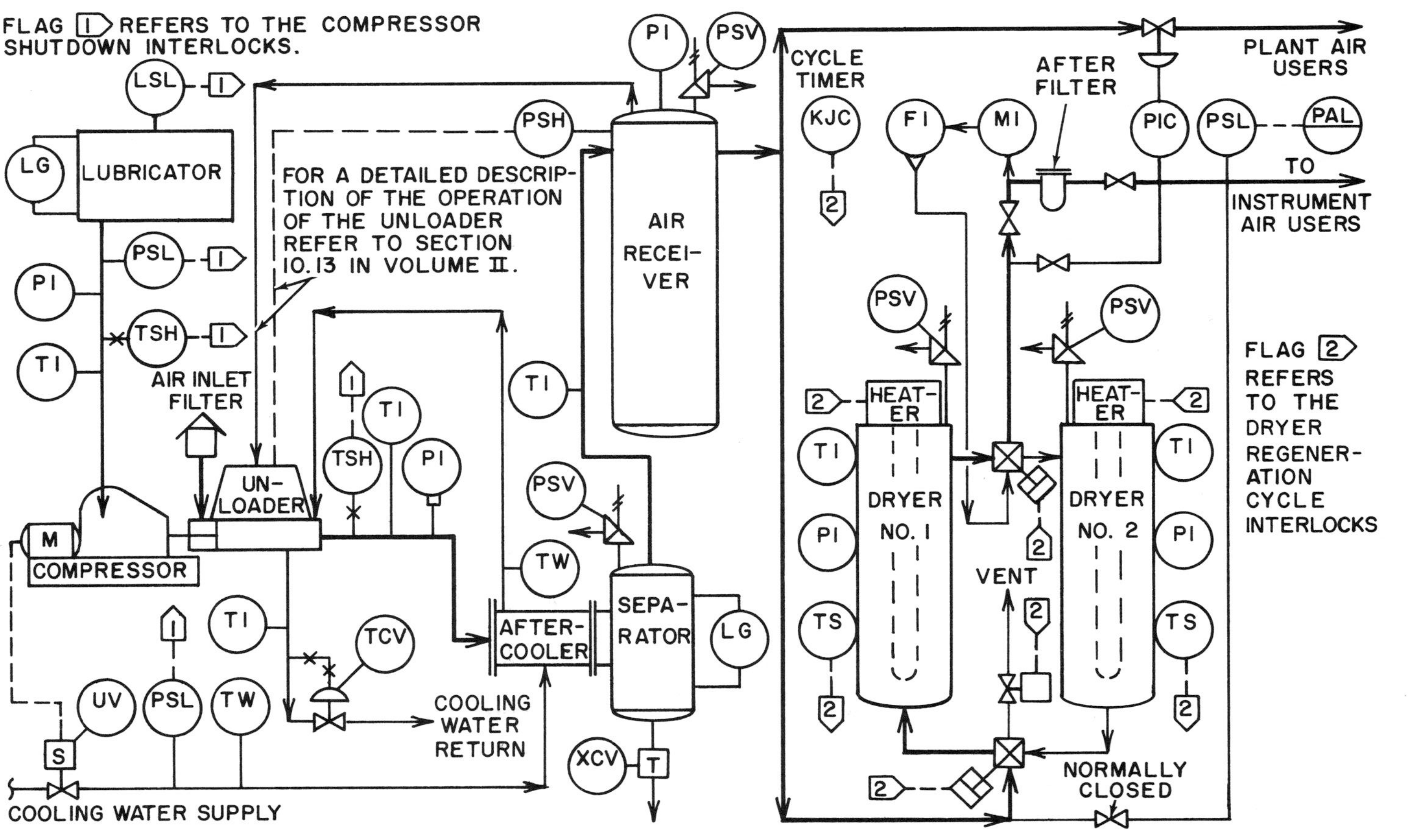

Fig. 6.5a Integrated instrument and plant air supply system utilizing a single stage compressor

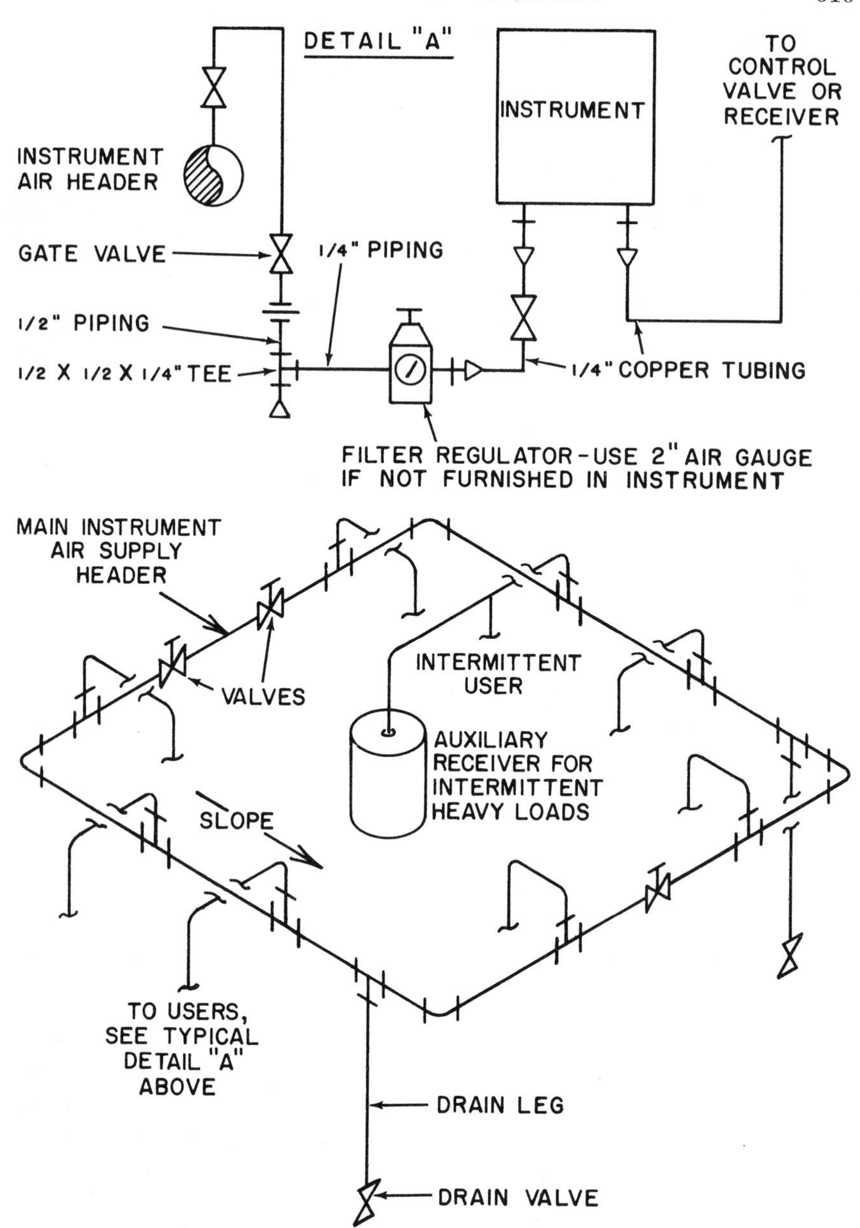

Fig. 6.5b Air supply distribution loop

that accumulated oil or moisture is not sent to the instruments. The lines should slope approximately 1 inch per 10 feet in the direction of the air flow, with drains at the low points.

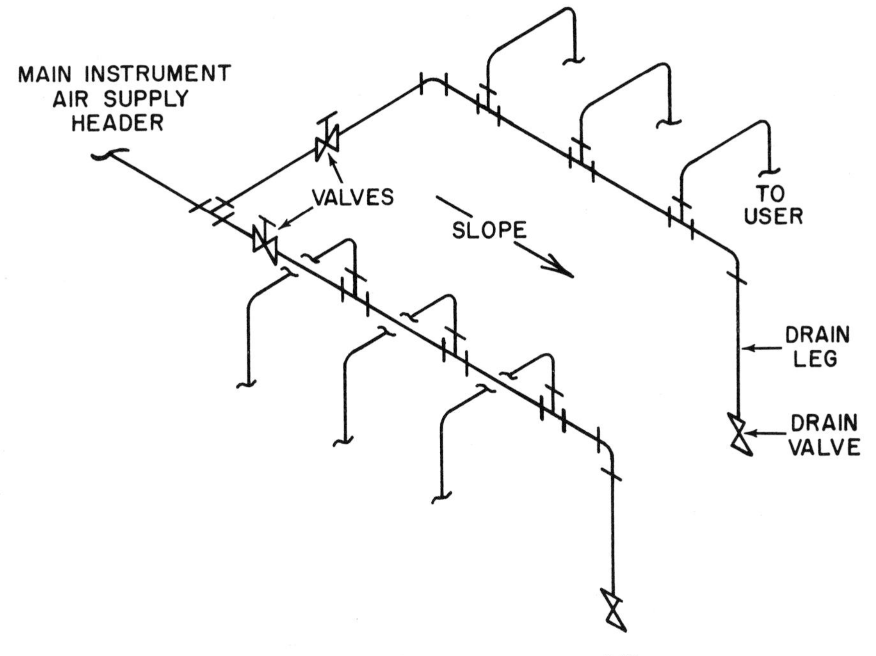

Fig. 6.5c Air supply distribution manifold

Air headers should be made of galvanized pipe, and all foreign matter should be blown from the pipes, tubing and hoses before installation.

Air Filters

An air filter is usually installed before each instrument air user to remove the contaminants which otherwise would reduce the life of instruments (and operators), clog orifices and cause malfunctions. Most inline air filters are of the two-stage, mechanical separation type. The air entering the unit is deflected into a swirling path, and the centrifugal force causes larger solid particles and liquid droplets to collect on the wall of the bowl.

A baffle plate in the lower part of the filter reverses the direction of the air flow, providing a quiet zone for the separated solids and liquid. After this cyclonic separation, the air flows through the filter element which removes the finer particles (Figure 6.5d). Filter bowls are normally made from a transparent polycarbonate plastic, which allows visual inspection of the filter element and of the amount of material collected in the bowl without disassembling the unit. The plastic bowls can have protective guards against mechanical damage, but for full protection against mechanical damage, a metal bowl must be used.

In most designs, the filtering element needs only to be cleaned, not replaced, and therefore units are available with a quick release collar, which

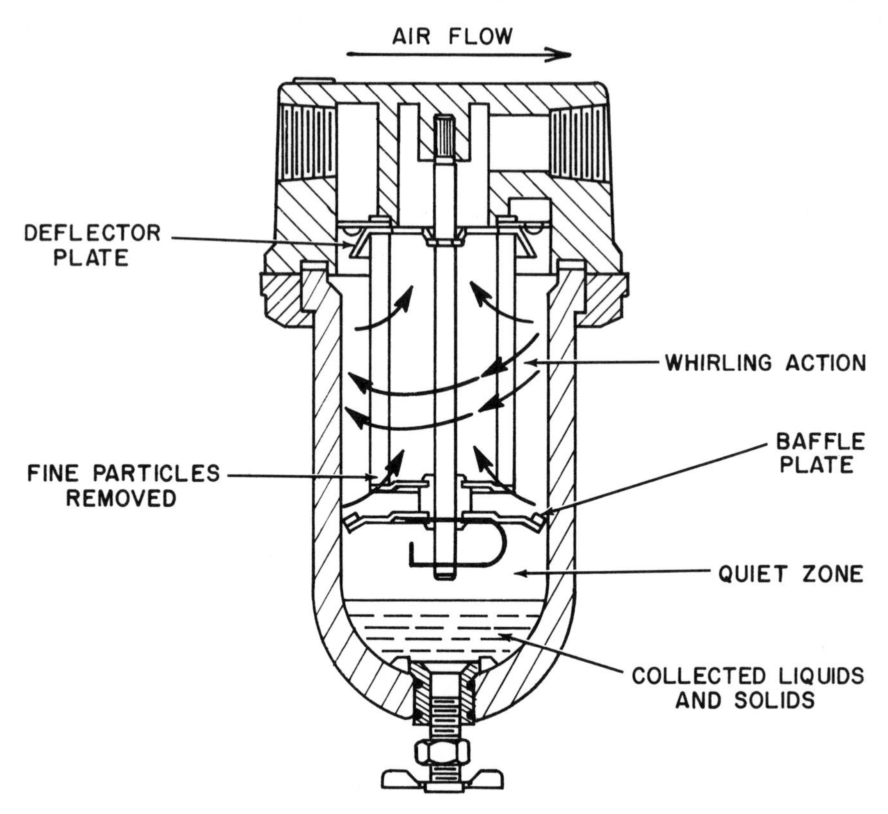

Fig. 6.5d Typical instrument air filter

makes servicing more convenient. A manually operated stopcock at the bottom of the bowl is for draining collected material. Units are also available with automatic drains triggered by a float. As the liquid in the bowl rises, the float lifts the drain valve, which in turn opens the drain. As the liquid is discharged, the float descends to close the drain. Because the liquid is not allowed to drain completely, no loss of compressed air occurs during the procedure.

Air filters should be sized according to air flow demand of the instrument. Oversize filters may not impart enough swirling action to separate all the droplets. Undersize filters require excessive operating pressure drops and may create too much turbulence in the air and prevent the liquids from settling.

Filters should be installed upstream and as close as possible to the instrument they serve. The filter bowl should be vertical and located where it can be easily observed for periodic inspection and draining. Filters should not be disassembled while under pressure and on installations where minimum maintenance is the goal; instead, they should have automatic drains.

6.6 AIR LUBRICATORS

Connection Sizes:	$\frac{1}{4}''$, $\frac{3}{8}''$, $\frac{1}{2}''$, $\frac{3}{4}''$ and 1″ NPT
Design Pressure:	To 200 PSIG at 175°F with polycarbonate bowl; to 250 PSIG with metal bowl.
Materials of Construction:	*Housing*—cast steel or brass. *Wetted parts*—brass, stainless steel or Monel. *Bowl*—transparent plastic
Cost:	$\frac{1}{4}''$ size in brass—about $15.
Partial List of Suppliers:	Generant Company, Inc., Parker Hannifin, Wilkerson Corp.

The most effective way to lubricate air cylinders is to inject oil directly into the operating air supply, which will transport it to the cylinder. The oil delivery rate is adjustable at the lubricator, but oil is only added when the air is in motion, i.e., air is being used.

Inline Lubricators

Air lubricators can either be installed directly in the air supply line or mounted remotely. For the inline lubricator, the flowing air lifts the oil from the reservoir or bowl and injects it into the air stream in the form of an oil fog or mist. The small oil particles collide as they travel with the air stream and occasionally drop out and collect on the tube walls. By the sweeping action of the air and by gravity the oil is moved along the air line to the air cylinder.

For an air motor or air tool, the air flow is unidirectional (Figure 6.6a), and for a dual-acting air cylinder (air can be loaded on both sides of the piston), air flow reversals occasionally occur (Figure 6.6b). For both of these installations, an inline lubricator is recommended, but for the bidirectional application the lubricator should be installed upstream of the diverting valve.

In some designs, the aspirator effect of the flowing air through a restriction causes the oil to move from the reservoir to the lower pressure *vena contracta*

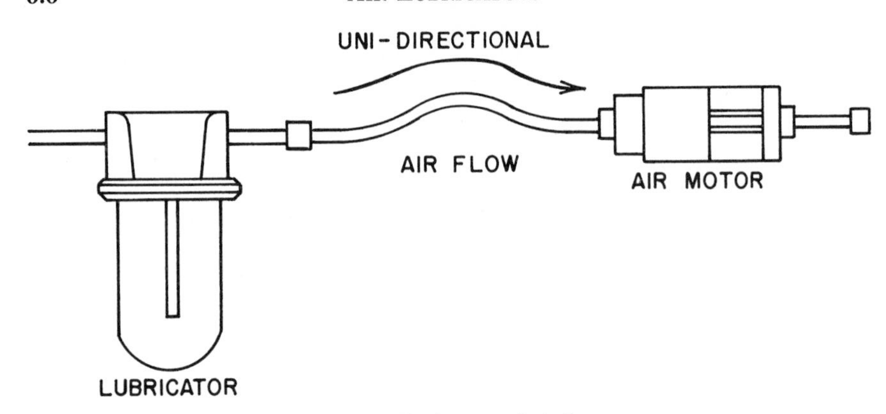

Fig. 6.6a Unidirectional air flow

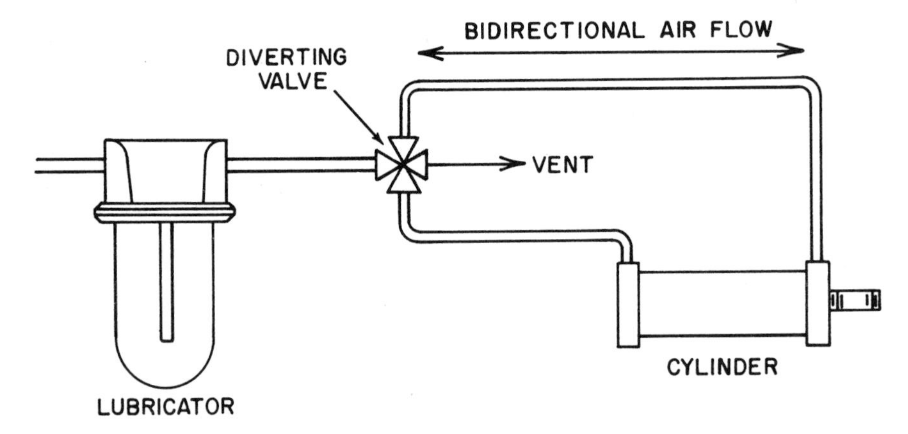

Fig. 6.6b Bidirectional air flow

point in the adjustable vane restrictor. Oil delivery rate can be measured by a sight ball in a graduated pickup tube or by a drip chamber where each drop of lubricant is observed as it enters the air stream.

If automatic oil refill is a special feature, several lubricators can be fed from a single pressurized oil source. The conventional manually filled inline lubricators can also be refilled without shutting down the air line.

Remote Lubricators

If the volume of the line between lubricator and air cylinder is not significantly smaller than the displacement of the cylinder itself or if there are air flow reversals and vertical upward runs, the inline lubricator cannot perform properly, and a remote lubricator should be used, in which oil is carried through small separate lines directly to the point of use with the oil feed rate being independent of the air flow.

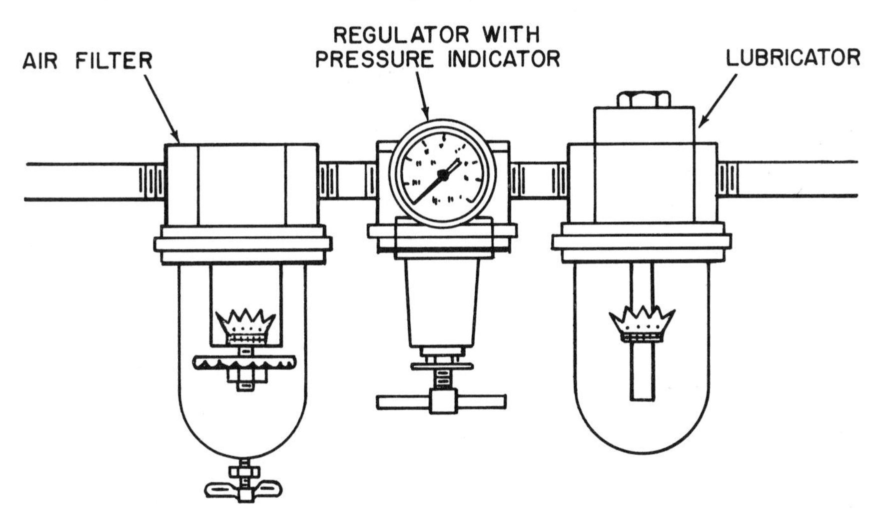

Fig. 6.6c Combination filter-lubricator-regulator unit

Lubricator Installation

If a filter, regulator and lubricator are required, the lubricator should be installed downstream from the other devices. Combination units (Figure 6.6c) are available. Lubricators with integral reservoirs should be mounted in the vertical position and located so that they can be visually inspected for refilling, without exposing them to mechanical damage. The lubricator bowl can be equipped with guard protectors perforated so as to allow visual inspection of the oil level.

Inline lubricators should be at the same height as or at a higher elevation than the air cylinder they serve. At startup, the lubricated air supply lines should be precoated with oil so as to minimize the lag time in having the oil reach its destination.

Oil delivery rates should be checked and readjusted as required to meet the variable air flow conditions. The exact amount of lubricant to each user is not as critical as is the need to have some oil in each air cylinder.

6.7 WIRES AND CABLES

Wire or Cable Types:

(a) Single conductor
(b) Multiple conductor
(c) Twisted pair (2 conductors)
(d) Multiple twisted pairs
(e) Two conductor, thermocouple
(f) Multiple pair, thermocouple
(g) Coaxial

Note: In the feature summary below, the letters (a) to (g) refer to the wire and cable types listed above.

Conductor Materials:

Bare, stranded or solid copper (a to g), tinned, stranded or solid copper (c,d,e), with smaller sizes of (a,b), silver coated, stranded or solid copper (g), iron-Constantan, chromel-alumel, copper-Constantan, solid only (e,f).

Conductor Sizes:

#18 AWG to #10 AWG (a) and from 2 conductor to 36 conductor (b); #24 AWG to #12 AWG (c); #24 AWG to #14 AWG and from 4 pair to 36 pair (d); #20 AWG to #16 AWG (e) and from 4 pair to 36 pair (f). Conductor size is included in cable designation for type (g) and need not be specified.

Conductor Insulation:

PVC—temperature ratings 60°, 75° and 105°C; voltage ratings 150, 300, 600 and 1,000 volts (a to f); Polyethylene—temperature ratings 60°, 75° and 90°C; voltage ratings 150, 300, 600 and 1,000 volts (a to g); Polypropylene—temperature ratings 60° and 75°C; voltage ratings 150,

300 and 600 volts (b,c,d,g), Teflon—temperature ratings 150°, 200° and 250°C; voltage ratings 300, 600 and 1,000 volts (a to g).

Shielding: Braided, bare or tinned copper wire provides 70 to 90% shield coverage (c,d,e,f,g); spiral wrapped bare copper or aluminum tape provides 80 to 95% shield coverage (c,d,e,f); aluminized mylar tape provides 100% shield coverage and includes copper drain wire for connection to shield (c to g); in addition to individual pair shields an over-all shield is also available for (d,f). Shielding is not generally available for (a,b).

Outer Jacket: PVC, polyethylene, neoprene and silicon rubber (a to g); Teflon, polypropylene, silicon impregnated braided fiberglass and nylon (a to g) but not in all sizes; armored cables (see next paragraph) available with an additional jacket over the armor for weather protection.

Protective Armor: Braided steel wire armor, square locked and interlocked galvanized steel armor, square locked and interlocked aluminum armor (b to f) and, to a limited extent, (g).

Support Method: Rigid steel, aluminum or PVC conduit, electric metallic tubing and enclosed metal wireways (a to g); solid bottom, ventilated bottom or basket type cable tray (b to g); ladder type cable tray (b,d,f); messenger wire, armored construction only (b,d,f); Direct burial underground (a to g), but cable construction must be specified by manufacturer as suitable for direct burial.

Approximate Costs: All costs are based on list prices *for ordering lengths of 1,000 feet* and do not include installation material or labor costs. Costs for ordering lengths less than 100 feet are generally $2\frac{1}{2}$ to 5 times higher per foot than when ordered in 1,000 foot lengths.

$35 (a) for single conductor #14 AWG solid bare copper, 600-volt polyethylene insulated, 75°C rated.

$25 (a) for single conductor #16 AWG bare stranded copper, 600-volt polyethylene insulated with PVC jacket, 90°C rated.

$700 (b) for 12 conductor #14 AWG bare stranded, copper, 600-volt polyethylene insulated with neoprene jacket, 90°C rated.

$1,400 (b) for 30 conductor #16 AWG tinned stranded copper, 600-volt polyethylene insulated with PVC jacket and aluminum interlocked armor, 90°C rated.

$85 (c) for 2 conductor, twisted, #16 AWG bare stranded copper, 300-volt PVC insulated with aluminized mylar tape shield and PVC jacket, 90°C rated.

$150 (c) for 2 conductor, twisted, #16 AWG tinned stranded copper, 300-volt PVC insulated with PVC jacket, braided steel wire armor and polyethylene outer jacket, 90°C rated.

$625 (d) for 20 twisted pairs #20 AWG bare stranded copper, 300-volt PVC insulated with over-all aluminized mylar tape shield and PVC jacket, 75°C rated.

$1,525 (d) for 16 twisted pairs #20 AWG bare stranded copper, 300-volt PVC insulated with aluminized mylar tape pair shields, over-all aluminized mylar tape shield, PVC jacket, galvanized steel interlocked armor and PVC outer jacket, 75°C rated.

$78 (e) for 2 conductor, twisted, #16

AWG solid iron-Constantan, PVC insulated with PVC jacket, 105°C rated.
$350 (e) for 2 conductor, twisted, #20 AWG solid chromel-alumel teflon insulated with braided bare copper wire shield and silicon impregnated braided fiberglass jacket, 200°C rated.
$900 (f) for 24 twisted pairs #20 AWG solid iron-Constantan, PVC insulated with over-all aluminized mylar tape shield and PVC jacket, 90°C rated.
$725 (f) for 4 twisted pairs #20 AWG solid copper-Constantan, PVC insulated with aluminized mylar tape pair shields, over-all aluminized mylar tape shield, PVC jacket, galvanized steel square locked armor and PVC outer jacket 105°C rated.
$120 (g) for type RG/58u, solid bare copper conductor, polyethylene insulated with braided tinned copper wire shield and PVC jacket, 75°C rated.
$600 (g) for type RG/9u, stranded silver coated copper conductor, polyethylene insulated with braided silver coated copper wire inner shield, braided bare copper wire outer shield and PVC jacket, 75°C rated.

Partial List of Suppliers:

Anaconda Wire and Cable Co., New York, N.Y. (a,b,c,d,g), Belden Manufacturing Co., Chicago, Ill. (a,b,c, d,g), Carol Cable Co., Pawtucket, R.I. (a,b,c,d,g), Cerro Wire and Cable Division, Cerro Corp., New Haven, Conn. (a,b,c,d,g), Chester Cable Operations, Cities Service Co., Chester, N.Y. (a,b,c,d,g), Consolidated Wire and Associated Corps., Chicago, Ill. (a,c,d,e,f,g), Dearborn Wire and Cable Co., Chicago, Ill.

(a,c,d,e,f,g), General Cable Corp., New York, N.Y. (a,b,c,d,g), General Electric Co., Wire and Cable Dept., Bridgeport, Conn. (a,b,c,d,g), Samuel Moore and Co., Mantua, Ohio (c,d,e,f), Standard Wire and Cable Co., Los Angeles, Cal. (a to g), Thermax Wire Corp., Flushing, N.Y. (a to g).

INDEX